Dragonflies and Damselflies of Texas and the South-Central United States

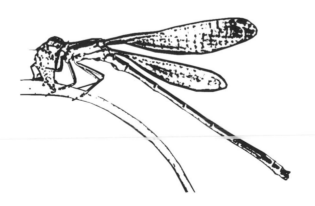

JOHN C. ABBOTT

Dragonflies and Damselflies of Texas and the South-Central United States

Texas, Louisiana, Arkansas, Oklahoma, and New Mexico

PRINCETON UNIVERSITY PRESS

PRINCETON AND OXFORD

Published by Princeton University Press, 41 William Street,
Princeton, New Jersey 08540

In the United Kingdom: Princeton University Press, 3 Market
Place, Woodstock, Oxfordshire OX20 1SY

Library of Congress Cataloging-in-Publication Data
Abbott, John C., 1972–
Dragonflies and damselflies of Texas and the South-Central
United States : Texas, Louisiana, Arkansas, Oklahoma,
and New Mexico / John C. Abbott.
p.   cm.
Includes bibliographical references.
ISBN 0-691-11363-7 (cl : alk. paper) — ISBN 0-691-11364-5
(pbk : alk. paper)
1. Odonata—Texas.   2. Odonata—Southwestern States.
I. Title.
QL520.2.U6A23   2005
595.7'33'09764—dc22   2004044531

British Library Cataloging-in-Publication Data is available

This book has been composed in ITC Stone

Printed on acid-free paper. ∞

nathist.princeton.edu

Printed in China

10   9   8   7   6   5   4   3   2   1

Contents

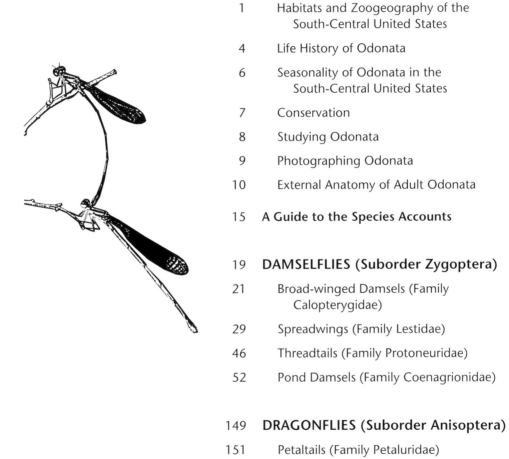

PREFACE

Thus the dragonfly enters upon a nobler life
than that it had hitherto led in the water
for in the latter it was obliged to live in misery,
creeping and swimming slowly,
but now it wings the air.
—Jan Swammerdam

 I became seriously interested in odonates (those insects belonging to the order Odonata, the damselflies and dragonflies) about 15 years ago. At that time there were still just a handful of scientists and a few amateur enthusiasts studying them. Recently, however, there has been an explosion of interest in these marvelous insects, fueled in part by the publication of Sidney Dunkle's *Dragonflies through Binoculars* in 2000, the first field guide for this group covering a significant geographic area. Many field guides, both regional and more cosmopolitan, are now available, and are beginning to fill what could only be described as a significant gap in our knowledge of North American natural history.

A major goal of this book is to facilitate the identification, both in the field and in the laboratory, of those odonates that occur in Texas and its surrounding states west of the Mississippi River (Louisiana, Arkansas, Oklahoma, and New Mexico). The region thus defined constitutes the south-central United States. All 263 species known to occur in this region (85 damselflies, 178 dragonflies), constituting over half of the 433 odonate species now known to occur in North America north of Mexico, are treated, and photographs, supplemental in many cases by line drawings, are provided for 262 of the 263 species.

This guide will fill a void in the resources currently available for identifying Odonata in the south-central United States. It addresses the needs and concerns of naturalists who are becoming interested in Odonata, and will serve as an up-to-date reference for odonate researchers interested in the fauna of this region. But in keeping with the ever-growing popularity of these insects, the book's primary intent is to aid laypersons who want to identify species seen in the field or in hand. Unlike Dunkle's popular *Dragonflies through Binoculars*, which treats all of North America, this book is specific to the south-central United States fauna, but it addresses damselflies as well as dragonflies, and offers larger photographs of each species. The only other relevant resource specific to the region is a 1972 publication by C. Johnson called *The Damselflies (Zygoptera) of Texas*. It is a technical volume that will not be readily available or accessible to most casual observers of Odonata, and it addresses only 53 of the 85 species that occur in the region. Recently, two technical manuals have been published for the identification of both dragonflies (by Needham, Westfall, and May) and damselflies (by Westfall and May) in North America. These volumes are wonderful compendiums, but contain few photographs and are written more with the professional odonatologist in mind.

Many odonates, particularly damselflies, are quite similar to one another and require careful examination in order to make accurate identifications. To aid with identifications, I have supplied keys, color photographs, detailed descriptions, and range maps for all species. In many cases, I have also supplied line drawings. Though my emphasis is on identification, I strive to include as much natural history and other biologically interesting information about each species as possible. In many cases, however, there is little, if any, such information to be passed along. But as odonates continue to gain a greater constituency of observers, so will our knowledge grow.

Many other odonates, particularly dragonflies, are unusual in the insect world in being readily identified in the field. A good pair of binoculars, keen eyes, and patience will allow you to begin learning many of the more common species fairly quickly. With time and experience you will be able to identify most of the dragonflies and many of the damselflies occurring in the south-central United States, in the field. Begin by trying to identify an odonate to family, then proceed to the likeliest group or genus within that family. Pay careful attention to head, eye, body, leg, and wing coloration; many of the species have distinctive, recognizable color patterns.

Some families, however, such as the clubtails and pond damsels, include many similar species and lack such notable characteristics. In all cases, though, I have attempted to point out features that are useful in the field, and have used these features in keys and tables wherever possible, but for many specimens examination under magnification (inverting your binoculars often works well) may be required. When trying to identify these difficult species, be sure to consider seasonality, habitat, and geographic range. Also note specific behaviors when observing them in the field. Do they perch on the ground or on vegetation? Do they perch horizontally or vertically? I strongly encourage the collection of specimens when done for a purpose and in moderation. Scientific collections are critical in the documentation of species' ranges and for the enhancement of knowledge of a particular group.

I have chosen to include both English (common) names and their Latin (scientific) counterparts, for families as well as genera and species. The Dragonfly Society of the Americas has formally adopted common names for the North American species. These names may seem more descriptive and easier to learn at first, but I encourage you to learn the scientific names as well. You will find these names to be every bit as descriptive, and their use will enable you to converse easily with scientists around the world about a particular species or group. For me, as a scientist, writing this book proved to be a useful exercise in learning the common names.

This book would not have been possible without the help and encouragement of people too numerous to mention. I do have to give special thanks to Sidney Dunkle for advice and encouragement, and for reviewing earlier drafts of the manuscript. I also thank David Flick, Rosser Garrison, John Matthews, and Dennis Paulson for reviewing parts or all of early manuscript drafts. Robert Behrstock generously allowed me to use many of his stunning photographs. I am also indebted to Robert Behrstock, Thomas Donnelly, Sidney Dunkle, Oliver Flint, and Dennis Paulson for supplying me with numerous geographic records and behavioral notes.

Much of the research leading to this book was done under the direction of Regents Professor Kenneth W. Stewart of the University of North Texas, Denton. Financial support came in part from the National Science Foundation, and from the Department of Biological Sciences and the Faculty Research Funds of the University of North Texas, and personal royalty funds came from Dr. Stewart. I am especially appreciative of the advice and friendship of Dan Petr, who assisted me in ways too numerous to mention, and the support and encouragement of my wife, DaLeesa Flick, and my parents, Bill and Ellie Abbott. Finally, I would like to thank Mark Bellis, Bill Carver, Dimitri Karetnikov, and Robert Kirk for all their help in seeing this book through its final editorial and production stages.

INTRODUCTION

Dragonflies and damselflies make up the insect order Odonata. They are found worldwide, except for Antarctica. Most treatments recognize three suborders within the Odonata. The names of these groups derive from the relative shapes of the fore- and hindwings. The dragonflies belong to the suborder Anisoptera, the hindwings of which are broader than the forewings. The damselflies, suborder Zygoptera, have fore- and hindwings of the same shape. The members of a third suborder, the Anisozygoptera, are often called "living fossils" and can be recognized by a mix of the characters seen in the Anisoptera and Zygoptera. This group contains only two extant species, both restricted to East Asia. In most parts of the world the term "dragonfly" refers to members of any of the three suborders, but in North America the term usually refers specifically to members of the Anisoptera. In this book I will refer to dragonflies and damselflies collectively as Odonata or odonates.

Because odonates are remarkably distinctive in appearance, and unique in many other aspects of their biology, they are seldom mistaken for other insects. Some 5,500 species are known worldwide, just 433 of them in North America north of Mexico. Odonata, especially in the tropics, is one of the larger aquatic insect orders, giving its members, wherever they occur, an ecologically important role in aquatic ecosystems. (Other insects that spend a substantial period of their life in the water include mayflies, order Ephemeroptera, and stoneflies, order Plecoptera.)

The Odonata represent one of the most primitive living insect groups. The earliest fossil odonates are some 250 million years old. Members of the order Protodonata, the probable ancestors of Odonata, lived more than 300 million years ago, and some had wingspans greater than 71 cm (2 feet)! The closest living relatives of the Odonata are the mayflies. Odonates exhibit many primitive features, including the inability to fold their wings flat and fanlike over the abdomen, a trait they share with the mayflies. The wings have dense venation, and each

wing is fully functional and independently movable. The robust thorax is strikingly skewed, forcing the legs forward and the wings backward.

Even the most casual observer must appreciate the phenomenal agility that odonates display in flight. Throughout human history they seem to have spawned not only interest but also fear. Their large size and fast, buzzing flight are the basis for such names as "devil's darning needles" and "horse stingers." Although some of the larger dragonflies can pinch a finger placed in the mouthparts, odonates do not normally bite humans. These names may have originated in part from occasional instances in which a female dragonfly mistakes the leg of a wader or river rafter for a plant when laying her eggs. Though I have never experienced this firsthand, those that have confirm that having a dragonfly attempt to lay eggs in you is painful!

Not only are odonates among the most beautiful of insects, they are also beneficial. Both dragonflies and damselflies are voracious predators, both as aquatic larvae and as adults, and one of their main prey items is mosquitoes.

Habitats and Zoogeography of the South-Central United States

The south-central United States, as defined herein, includes Texas and its four surrounding states. The area covered is approximately 1.2 million km², of which 695,000 km² are in Texas. The region encompasses ten biotic provinces (Fig. 1). The Mississippi River forms the eastern boundary of the region (which means that the few parishes of Louisiana that lie east of the Mississippi are not covered), and the Apachian and Navahonian biotic provinces bound the western edge.

A considerable amount of work has been done on the distribution of vegetation types in Texas (Bray 1901, 1905; Carter 1931; Tharp 1926, 1939), Louisiana (Viosca 1933; Holland 1944), Arkansas (Turner 1935; Stroud & Hanson 1981), and Oklahoma (Ortenburger 1928a,b; Bruner 1931). There is a tremendous variety in the environments available

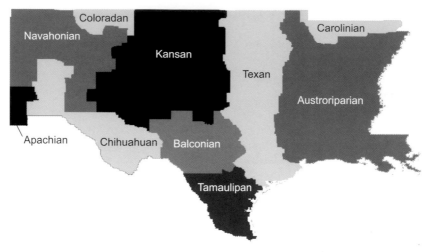

Fig. 1. The ten natural biotic provinces of the south-central United States (modified from Blair 1950, Dice 1943, Blair & Hubbell 1938).

for plant and animal communities, which are determined largely by climatic conditions and topography. A north–south line passing through central Oklahoma and Texas divides our region into areas of moisture sufficiency and moisture deficiency (Blair 1950), thus dictating plant and animal distributions in the region. Cope (1880) recognized three major biotas in Texas: a Sonoran fauna, an Austroriparian fauna, and a Neotropical fauna. I recognize a further division into ten distinct regional biotic provinces, as outlined by Blair and Hubbell (1938), Dice (1943), and Blair (1950), and as shown in Fig. 1. These provinces, which differ in topography, annual temperature range, vegetation, soil type, geology, and climate, have been given the names Apachian, Austroriparian, Balconian, Carolinian, Chihuahuan, Coloradan, Kansan, Navahonian, Tamaulipan, and Texan. The provinces are useful in detailing the distribution of Odonata, biologically vs. politically, within our region, as you will see in the individual species accounts.

Mean annual precipitation ranges from 147 cm/yr in the moist eastern parts of the region, such as New Orleans, to less than 25 cm/yr in the arid western areas, such as El Paso. Most of the precipitation falls during the months of March to May. Temperature is also an important factor in dictating the distribution of plant and animal communities, and ranges from an annual average of 22.8° C (73.4° F) in subtropical Brownsville, Texas, to 12° C (53° F) in the Texas pan-

handle, resulting in a shorter growing season in the latter. Major vegetation types include eastern pines and hardwoods, central prairies and grasslands, western semidesert areas, and western montane forests. Elevation ranges from sea level along the coastal areas to 4,011 m (Wheeler Peak, Taos County, New Mexico) in the Carson National Forest. The major watersheds in the region (Fig. 2) drain in an eastward or southeastward direction, and nearly all of them enter or approach the Austroriparian province. These stream systems provide important dispersal routes for the westward distribution of species of the Austroriparian province into the more arid, treeless environments (Blair 1950).

The Austroriparian province, as defined by Dice (1943), encompasses the Gulf coastal plain from extreme east Texas to the Atlantic Ocean. This biotic region's western boundary is demarcated by the availability of moisture. The typical vegetation types include longleaf pine (*Pinus palustris*) and loblolly pine (*P. taeda*) and hardwood forests variously consisting of sweetgum (*Liquidambar styraciflua*), post oak (*Quercus stellata*), and blackjack oak (*Q. marilandica*). The lowland hardwood forests of the southeastern portion of this province are typically characterized by magnolia (*Magnolia grandiflora*), tupelo (*Nyssa sylvatica*), and water oak (*Q. nigra*) in addition to those trees mentioned above. Other plants typical of this region include Spanish moss (*Tillandsia usneoides*) and palmetto (*Sabal minor*).

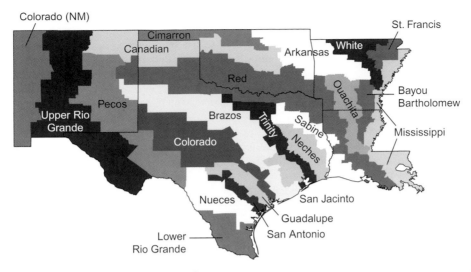

Fig. 2. The natural watersheds of the south-central United States.

The Texan biotic province constitutes a broad ecotone between the forests of the Austroriparian province in the eastern portion of this region and the western grasslands. The Balcones Escarpment forms an abrupt boundary to the west, otherwise delineated by a line based on soil type. This area was once characterized by tall-grass prairies supported by clay soils, but cultivation of much of the area has led to sandy soils characterized by combination oak-hickory forests, dominated usually by post and blackjack oaks and hickory (*Carya texana*). Thornthwaite (1948) classified this province as having a moist, humid climate, but receiving little water beyond that required for growth. The drainage pattern of the Texan province is an important biogeographical feature (Blair 1950). The Red and Trinity rivers, along with their tributaries, drain the northern part of this province. Both of these rivers enter the Austroriparian province before emptying into the Gulf. The southern portion of this province is drained largely by the Brazos, Colorado, San Marcos, and Guadalupe rivers.

One of the unique features in this province is the Arbuckle Mountains in south-central Oklahoma, just north of the Red River. This area is dominated by granite and travertine limestone geologic formations. One of the most prominent of these formations is in Turner Falls Park near Davis, Murray County, Oklahoma. Within the park, there is a 25 m waterfall on Honey Creek.

For the Kansan province, I follow Blair (1950), who delineated the province differently than did Dice (1943). Dice limited the province, excluding the Permian redbeds, while Blair included the areas north of the Edwards Plateau and south of the Red River. The province is characterized by a mixture of eastern forest species and western grassland species. Notable exceptions to the monotonous prairies of this province are Palo Duro Canyon State Park and Caprock Canyon State Park, which have been characterized as relict habitats. Moisture decreases from east to west in this province, and Thornthwaite (1948) considered the region moisture-deficient.

The Balconian biotic province is defined by the Edwards Plateau of Texas and derives its name from the Balcones fault zone that forms its southern and eastern boundaries (Blair 1950). It is characterized by scrub forests of juniper (*Juniperus* spp.) and oaks (*Quercus* spp.), including stunted live oak (*Q. virginiana*).

Farther south, the Tamaulipan province extends from southern Texas into eastern Mexico. This semiarid region is dominated by mesquite (*Prosopis glandulosa*), *Acacia* spp., *Mimosa* spp., and prickly pear cactus (*Opuntia* spp.). Thornthwaite (1948) noted a marked deficiency of moisture for plant growth, though some growth occurs year round. This province is drained in the north largely by the Nueces River and its tributaries, and is poorly drained in the southern portion by minor tributaries of the Rio Grande. In the Brownsville region, to the south, the province becomes subtropical.

The Chihuahuan province includes the Trans-Pecos area of Texas, excluding the Guadalupe Mountains. It extends southward into the Mexican states of Chihuahua and Coahuila and is drained largely by the Rio Grande. This biotic province is more diverse in physiographic features than all others in the region (Blair 1950). The climate in this area is arid and moisture-deficient (Thornthwaite 1948), and the vegetation is variable, but basin areas up to 1,500 m in elevation include grasses, desert shrubs, and creosote bush (*Larrea tridentata*). Streams in this area are usually small and intermittent; those that are permanent are usually spring-fed. The various mountains, including the Chisos and Davis ranges, show a vertical zonation of plant communities, with elevations above 1,500 m dominated by Emory oak (*Quercus emoryi*) and cedars (*Juniperus* spp.).

The Navahonian province includes most of New Mexico and barely enters the northern edge of western Texas (Culberson County) at the southern extension of the Guadalupe Mountains. A vertical zonation in elevation similar to that of the Chihuahuan province characterizes this area. Trees dominant at elevations above 2,500 m include various pines (*Pinus* spp.), oaks (*Quercus* spp.), and Douglas Fir (*Pseudotsuga menziesii*).

Small portions of the Carolinian (northern Arkansas), Coloradan (northwestern New Mexico), and Apachian (southwestern New Mexico) provinces are also found within the region covered by this book. These peripheral provinces are home to several odonates whose ranges barely extend into the south-central United States.

Life History of Odonata

As is true of any living organism, reproduction is a critical stage in the life cycle of odonates. It may take an individual adult anywhere from a day to several weeks, depending on the species, temperature, availability of food, and other environmental factors, to become sexually mature. Upon maturity, the male will, in most species, patrol a territory he has established over and around water. The male will search for females in this territory, and will defend it against other males of his species and occa-

sionally males of other species. At some point a receptive female looking for suitable egg-laying areas is seized by the male. This capture usually takes place in midair, the male flying over the female and grabbing her head and thorax with his legs. In damselflies, the male grasps the female's prothorax (the segment of the thorax behind the head) with the clasping structures at the tip of his abdomen (this is the ***tandem position***).

In dragonflies he grasps her head, and in aged females of some larger species (darners) the resulting damage to the eyes is often noticeable. The male will then curl his abdomen around so that sperm can be transferred from the genital pore on the ventral side of his segment 9 to the accessory genitalia located on the ventral side of his segment 2. This usually happens quickly, and the male then straightens his abdomen so that the receptive female can curl her abdomen around, such that her genital opening on the ventral side of segment 8 is in contact with his accessory genitalia on segment 2. It is not unusual to see odonates in this ***wheel position***. They may remain in this position, usually protected within vegetation, for minutes or hours, depending on the species. Still others copulate while in flight. (A number of the color photos in this book are of mating pairs.) Females will mate more than once, and in at least some species, males are known to remove the sperm from previous males before mating.

Egg-laying

Female damselflies, petaltails, and darners all lay their rod-shaped eggs using a specialized structure called the ***ovipositor***. They use the bladelike structures of their ovipositor to make slits in the soft tissues of plant stems, where they then lay their eggs. These eggs may be laid above, at, or below the water line. In other odonate families the eggs are generally more round in shape and often have a gelatinous covering. These eggs are laid in or near water or in an area that will fill with water. Female Spiketails employ a specialized ovipositor in sewing-machine fashion to deposit eggs in the substrate of shallow streams. The eggs will generally hatch in about a week. Eggs laid in temporary pools, however, will hatch more quickly, whereas others may be delayed for months. The male may continue to grasp the female (in the tandem position) while she lays eggs, or he may release her, but he will remain hovering or perched nearby to guard against intruding males. Females of some species of damselflies will actually submerge themselves underwater for an hour or more while ovipositing. Regardless of whether a female descends underwater or not, she, and often the male, are vulnerable to predation by fishes, frogs, birds, spiders, and other insects during this time.

Larvae

Dragonfly and damselfly larvae (also called nymphs or naiads), although most common in ponds, marshes, lakes, and streams, have exploited a wide range of permanent and temporary aquatic habitats, including brackish pools and estuarine habitats. Some larvae survive in moist substrates under rocks and in otherwise dry streambeds or ponds. When the larvae hatch from the eggs, they look very little like the adults they will eventually become (see Fig. 3). They are voracious predators, feeding on worms, small crustaceans, mosquito larvae, and other insects, and the larvae of larger odonate species may even take small fish and amphibians. They capture their prey by extending their lower lip (the **_labium_**), which is equipped with two movable toothed palps. The labium, which may reach one-third of the larva's body length, extends at lightning-fast speed. Because the larvae are generally shades of green and brown, they blend in well with their environment. Many are cylindrical in shape, but others are dorsoventrally flattened. The larvae of some species grow by molting their exoskeleton 7–18 times over a period of a few (typically 11–13) months to several years or more. Their wing pads will become more and more evident as they grow in size and approach adulthood (see Fig. 3).

Fig. 3. Odonata larvae: (A) Great Spreadwing (*Archilestes grandis*); (B) Common Green Darner (*Anax junius*); (C) Blue Dasher (*Pachydiplax longipennis*); (D) Skimmer (Libellulidae).

Damselfly larvae have three leaflike gills at the tip of the abdomen. Although the gills are used in respiration, they are also used like fins for swimming. Larvae have the ability to regenerate these gills, at least in part. Damselfly larvae also have the ability to absorb oxygen through the wall of their rectum. Dragonfly larvae lack the terminal gills and absorb oxygen through internal rectal gills instead.

Adults

An odonate larva will typically climb out of the water onto an emergent piece of vegetation, bridge pylon, or similar structure to begin its adult emergence. Some species are known to walk considerable horizontal distances (30 m) and heights (10 m) from the body of water in which they developed. Emergence generally occurs at night or in the early morning, under the cover of darkness. Clubtails and damselflies, however, often emerge during the day. A split develops along the dorsal side of the head and thorax as the larval skin dries. The adult will then start to pull itself free from the old larval exoskeleton, or **exuviae** (see photo 48f). This process may take some time, since the legs have to harden before it can pull itself completely free. Once it has done so, it will remain hanging from its perch until its wings have been inflated (via the veins) and the body has begun to harden. This newly emerged, or *teneral*, adult will not have the bright vibrant colors of its mature counterpart; the wings will appear cloudy or have a shimmer to them, and the tenerals are vulnerable during this time of limited mobility. Although they can fly, almost immediately, they usually do so only when disturbed. It may take several days or more than a week before sexual maturity is reached. During this time most odonates remain away from the water. Males usually mature more quickly than females and make their way back to suitable egg-laying habitats, where they set up territories and wait for females.

Odonates of a number of species are **crepuscular**: they feed actively at dusk. Occasionally, a dragonfly or damselfly will come to a light at night, but most of these are individuals roosting nearby and not actively flying. Odonates of some species, however, like the shadowdragons, may actually be nocturnal.

Dragonflies typically have no reason to fly great distances, but because they are accomplished gliders, they are quite capable of sustained flight, in some cases even across oceans, on the wings of prevailing wind currents.

Seasonality of Odonata in the South-Central United States

The flight season of most species of Odonata in this region extends from the spring through the summer months, and occasionally persists into the fall. The onset and duration of emergence, however, are both variable. Many species of temperate origin (e.g., Attenuated Bluet, Springtime Darner, Harlequin Darner, Banner Clubtail, Twin-spotted Spiketail, Stream Cruiser, Blue Corporal) have early (March–May) and explosive emergences in this area and then soon disappear. In the southern portions of the region, the year-round temperatures, averaging 23° C (73° F), and the subtropical climate allow several species to fly year round. Fifteen species (including Familiar Bluet, Fragile Forktail, Common Green Darner, Eastern Pondhawk) have been encountered as adults in every month. Other species (Fine-lined Emerald and Blue-faced Meadowhawk) are seen flying only later in the year. Because of the latitudinal gradient in temperature seen in the region, emergence occurs one to several weeks later in the more northern areas than in the subtropical southern areas.

Damselflies generally emerge as soon as temperatures permit in the spring, and continue to emerge throughout much of the summer. This results in a heterogeneous age structure, often allowing more than one generation per year. Many of the smaller pond damsels, for example forktails, have multiple (two or three) generations per year. The larger pond damsels, the spreadwings, and the broad-winged damsels generally require a full year for development.

Many dragonflies differ from damselflies in having an obligate larval diapause (a required period of arrested development), followed by a synchronous spring or early-summer emergence. This pattern results in a homogeneous age structure and a sudden disappearance later in the summer or fall. Most species require at least a year to develop, and some, for example spiketails, require longer (several years). A more or less typical adult lifespan, barring predation or other calamity, might be three months.

Development time is generally longer for those species restricted to running-water situations than it is for those found in ponds or lakes. A general seasonal progression, with the peak months in June and July, as seen for the entire south-central United States, was observed by Bick (1957) in Louisiana. Species present early in the year (January–February) were also generally present later in the year (November–December).

Conservation

Odonates play a major beneficial role as predators of mosquitoes and other biting insects, both as adults and as immatures. The larvae also form an important link in food chains for fish and other aquatic vertebrates. Historically, the Odonata have not been acknowledged as good indicators of water quality, but numerous authors (Castella 1987; Dolny & Asmera 1989; Bulankova 1997; Chovanec & Raab 1997) have recognized that capacity in this group recently. Schmidt (1985) presented a convenient working scheme for the rapid evaluation and characterization of aquatic habitats.

The most recent estimates (Dunkle 1995) are that about 15% of the dragonfly species and 6% of the damselfly species in North America have limited or restricted geographic ranges and may be at risk of extinction. Areas of particularly high endemism in the United States include the New England coast, Florida, the central Gulf of Mexico coast, and the Pacific Coast. A few species, the Oklahoma and Ozark Clubtails and the Texas Emerald, for example, are endemic to smaller regions within the south-central United States. Those species living in streams, rivers, and other flowing-water habitats are at greatest risk.

Sewage and other organic wastes run off into streams and promote bacterial growth that depletes the oxygen content of the water and in turn stresses or kills odonate larvae. Fertilizer runoff from agricultural fields leads to eutrophication, promoting algal growth that may lead to blooms removing oxygen from the system and preventing sunlight from penetrating the water. Pesticide runoff also kills larvae. Those species of Odonata living in ponds and lakes are generally at less risk. In fact, human activities resulting in the construction of new ponds, lakes, borrow pits, and even stock tanks may provide some Odonata with a new habitat, lacking competition, to colonize. Other human activities,

however, such as allowing livestock access to these areas, can severely impact these pond species.

The single most important factor in the conservation of Odonata is the protection of land and aquatic habitats. The efforts needed for the protection of these resources vary with the type of habitat. Removing the surrounding vegetation from streams by mowing will completely alter the composition of the water, effectively removing some species. A buffer zone of vegetation on either side of a stream (generally, at least 30 m is recommended) helps to prevent erosion. Construction of dams poses a real challenge to stream species. In all of Texas there is now only one natural lake (Caddo Lake), which indicates just how prevalent dams are in the state. The numerous manmade reservoirs may provide habitat for those few species that can breed in lakes, but they deprive other odonates of habitat. Rivers are seldom protected from human impact and disturbances, but groups in areas like the Big Thicket National Preserve in east Texas are working hard to secure these areas by purchasing riparian lands and creating corridors between their preserve units.

For some species, especially those living in more arid conditions, a small area of suitable habitat may be all that is needed for their continued existence. For most species, however, larger areas of suitable habitat are necessary. This is where a good network of local, state, and national parks is so important. Though such areas provide protected habitats for odonates, on the whole they protect relatively little land. Private groups like the Nature Conservancy are crucial for securing and protecting land. Time is of the essence, for whereas observations have shown some species expanding their ranges because of global warming, many others are left to compete for fewer habitats of poorer quality.

Collecting odonates on these protected lands requires special permission and often a scientific collecting permit. I have found that land stewards or agencies in charge of these protected areas generally have no problem with the collection of the voucher specimens needed to ensure proper documentation and identification. Often they encourage this practice in order to ensure that the fauna of the lands they manage and protect are properly documented. In Texas, like most other states, scientific collecting permits are required for any state park property or preserve, and are administered through a central agency (Texas Parks and Wildlife). Because National

Parks administer their own scientific collecting permits, each park must be contacted separately. In my experience the need for a scientific collecting permit in National Forests varies from state to state, despite the fact that they are all federally maintained. Some states require a fishing license if one is to collect insects on these or other lands. City and municipal parks may also be protected, and you should contact the appropriate administering agency in these areas to inquire about collecting. Before issuing a scientific collecting permit, the administering agency will require a proposal of the research being conducted, and will mandate the depositing of vouchers (specimens accessioned or numbered) in a legitimate and publicly accessible collection.

In addition to the preservation and management of natural habitats, the creation of new habitats, especially ponds, can play an important role in conservation. An artificial pond, especially one in an arid area, that has an assortment of aquatic vegetation and a lush riparian zone protected from livestock will provide productive breeding habitats for many species. Although many Odonata can coexist with fish in such ponds (as long as there is plenty of vegetation to hide in), the presence of fish will prevent some species from persisting in the habitat.

We still have a lot to learn about the specific microhabitat requirements of our individual Odonata species. Although many species are at risk, others (for example Great Spreadwing, Double-striped Bluet, Widow Skimmer, and Swift Setwing), seem to be expanding their ranges. It is unclear in most instances if the expansion of these species is at the expense of others. We need more refined methods of population estimation, ones that rely on exuvial and larval counts rather than adult counts, if we are to begin getting at these questions. Collection and photography of individuals will also play a critical roll in filling gaps in our knowledge and understanding of species distributions.

Both larvae and adults are recreationally important; fly fishermen have patterned tied flies after them, and the terrestrial adults are observed and studied by layman and scientists because of their colors, flying ability, and curious habits. Odonates have also served for centuries as favorite subjects of poets, naturalists, artists, and collectors. Particularly during this time of growing interest in the group, we should be vigilant in our attempts to conserve aquatic habitats. The British Dragonfly Society has published two pamphlets dealing with dragonflies and ponds (*Dig a Pond for Dragonflies* and *Managing Habitats for Dragonflies*). Both are available at their website, http://www.dragonflysoc.org.uk.

Studying Odonata

Because many of the more common species can be seen almost anywhere, even far from water, it is easy to begin studying odonates. And because they breed in water, any relatively nonpolluted body of water will provide a good opportunity to observe them. Many are large enough to be observed and identified readily with a good pair of close-focusing binoculars. But it is important to remember that not all species (particularly many damselflies) can be reliably identified without capture and closer examination. A small magnifying glass or jeweler's loupe can be useful for this kind of scrutiny in the field, and reversing a pair of binoculars will often work in a pinch.

Remember that, in addition to observing the physical appearance of an odonate, you should note the habitat, season, and observed behavior. (How and where does it perch? Does it spend most of its time perched or in flight?). Consideration of these characteristics jointly may yield a more confident identification. There are also certain physical features you will want to make careful notes or photographs of as well. The color of the eyes and front of the head, the pattern of thoracic stripes, the wing color or pattern, including the appearance of the pterostigma (colored cell at the tip of the wing), and the color pattern of the abdomen are useful identifying marks.

It may become necessary to capture an odonate, so as to examine it more closely in the hand. All you need if you are to capture one is an aerial net and a fair amount of hand-eye coordination. The best way to handle an odonate is to pinch its wings together above its back, so that they stand straight up. They are generally unharmed by the experience and can be released to carry on their activities after you have studied them.

At some point you may choose to expand your interests to the collection of odonates. Collecting small numbers of most odonates will have no harmful impact on their populations. Like other insects, they have a high reproductive potential. Most states do not require a collecting permit, so long as you are not collecting in state, federal, or, in some cases, municipal parks and preserves. Making a permanent a collection can be rewarding and will allow you to develop a greater appreciation for these insects. Unfortunately, unlike some other insects, such as beetles and butterflies, odonates do not hold their brilliant colors after death. The best way to preserve these colors is to immerse the odonate in acetone for 12–24 hours and then allow it to dry. This will fix the colors to some degree and will remove many of the fatty acids in the body that can discolor the specimen. Specimens for scientific study are generally placed in clear cellophane or polypropylene envelopes with a 3′ × 5′ card recording the locality and other data pertinent to the collection of the specimen.

With the increase in popularity of odonates, regional newsletters and societies are appearing, and in 1999 an active and informative international email discussion group was created. The International Odonata Research Institute (IORI) has a fairly complete assortment of field guides and books on Odonata for sale. There are currently two international dragonfly societies (the Worldwide Dragonfly Association and the International Odonatological Foundation, Societas Internationalis Odonatologica) that publish biannual and quarterly journals, respectively. Another society, the Dragonfly Society of the Americas, produces a quarterly newsletter and an occasional journal. Anyone interested in Odonata in North America will find the small yearly dues for this society well worth the price. A full list of these societies with their contact information is given at the back of the book in the Bibliography.

Photographing Odonata

It probably won't be long after you have started observing odonates that you will want to begin photographing them. Given their beauty, this is a natural extension of one's interest. Most of the photographs in this book were taken with a 35 mm SLR (single-lens reflex) camera and a 100 or 180 mm macro lens. Damselflies, not as easily disturbed, are generally easier to photograph, and if they do fly off, they generally don't fly far. Dragonflies, however, present a number of challenges. The longer the focal length of the lens, the greater the working distance you will have. A 180–200 mm macro lens or a close-focusing 300 mm telephoto lens equipped with a teleconverter will allow you to frame an odonate in your camera without having to get so close. Extension tubes can also be used with your macro lens to increase magnification and or working distance. There is a trade-off, however: these longer lenses generally require the use of a tripod to prevent a blurred image. A tripod can be useful or even critical in the field, but it is also another piece of equipment that can brush up against vegetation, perhaps scaring your subject away, and another thing to carry.

There are many films on the market and I find that photographers are generally committed to a favorite brand. Film speeds (ISO) are an indication of the film's sensitivity to light. The higher the ISO the more sensitive to light the film is, but also the grainier it is. I prefer Fuji films because of their saturated colors, which produce a vivid image. I use 50 or 100 ISO film, both of which work well on sunny days. With the advent of high-resolution SLR digital cameras, film may be less of an issue in the near future. A few of the photographs in this book were taken with a digital SLR, which I now use exclusively. The light sensitivity of the chip in the camera allows you to shoot at speeds as high as 400 and 800 ISO without the grain you would expect from film.

I often employ a macro ring flash and one or two slaved off-camera flashes. An electronic flash dedicated to your camera body with TTL (through-the-lens) metering can be useful in close-up photography, both as direct lighting and fill-flash. Using flash as direct lighting will allow you greater depth of field and eliminate blurring from camera shake, because you can shoot at a higher shutter speed, but some criticize the non-natural lighting that may result. It is also necessary to make sure that the background you are shooting against is not too distant, so that your resulting photo does not have a black background and appear as though it were taken at night.

When photographing odonates, it is important to make sure that your line of sight is completely perpendicular to the body surface you are photographing, so that the entire individual is in sharp focus. With large dragonflies, it is easy to have one end of the body—the head, say—in focus, but have the other end, the tip of the abdomen, a complete blur. Many cameras come with a depth-of-field preview feature, which is helpful in placing the complete insect in the focal plane. Another important criterion is the background. You may not always be able to choose this, but the better photographs have contrasting backgrounds (blue sky, green vegetation) that aren't too busy and distracting. Finally, be sure that in addition to your camera equipment, to take along a healthy dose of patience. Odonate photography can be rewarding, but it is not without frustrations!

External Anatomy of Adult Odonata

My purpose here is to introduce briefly the terminology I use in this field guide when describing adult odonates. Table 1 summarizes some of the major differences between adult dragonflies and adult damselflies; Figs. 4, 5, and 6 illustrate the particulars of odonate anatomy; and the Glossary furnishes definitions of most of the technical terms used in the figures and text.

Head
The large **compound eyes** are the most distinctive feature on an odonate's head. They make up the largest portion of the head, particularly in dragonflies. These eyes, composed of many small facets (ommatidia) fused together, secure odonates as dominant aerial predators and some of the most acrobatic and skillful fliers in the animal world.

The eye color is often one of the most distinctive features of an odonate. In many the dorsal color of the eyes will differ from their ventral color, the dorsal surface usually darker. Unless stated otherwise, when eye color is given in the species accounts, I am referring to the dorsal surface of the eye. Odonates see a wide spectrum of colors that includes not only the spectrum visible to us but also ultraviolet and polarized light. This explains why some female odonates try to oviposit on a shiny asphalt road or car hood. They are presumably mistaking the horizontally polarized reflection of these structures with that of water. In addition to their huge compound eyes, odonates have three **ocelli**, or simple eyes used for light detection, on top of the head. They also have a small pair of antennae.

Thorax
The thorax comprises three segments (pro-, meso- and meta-), each bearing a pair of legs. The first segment, the **prothorax**, is so reduced as to appear almost necklike, and articulates with the fused **mesothorax** and **metathorax**. The wings attach to the mesothorax and metathorax, which are much larger than the prothorax and sometimes jointly referred to as the **pterothorax**, because they are involved with wing function. Stripes are often visible on the pterothorax, and terms have been applied to them (starting anteriorly, the mid-dorsal, antehumeral, humeral, anterior lateral, and posterior lateral).

Wings
As just mentioned, two pairs of wings attach to the last two thoracic segments. The wings are made up of numerous **veins** enclosing **cells**. The arrangement of the veins and cells is often useful in identifying odonates. For that reason many of these, too, have been accorded specific terms. One need not become an expert on wing venation to identify most Odonata, but it is helpful to learn some of the general terminology used in this guide. I use the tra-

TABLE 1. DISTINGUISHING CHARACTERISTICS OF DAMSELFLIES AND DRAGONFLIES

Damselflies (Zygoptera)	Dragonflies (Anisoptera)
Of slighter build, and generally weaker fliers	Of robust build, and strong fliers
All wings similar in shape	Hindwings broader basally than forewings
Wings held closed or nearly so over the abdomen when at rest (except the spreadwings)	Wings held horizontally outward from the body when at rest
Eyes separated by at least their width	Eyes touching or at least not separated by their width
Males with two pairs of caudal appendages	Males with a pair of superior caudal appendages and a single inferior caudal appendage
Females with functional ovipositors	True ovipositor lacking or reduced in females (except the darners and petaltails)

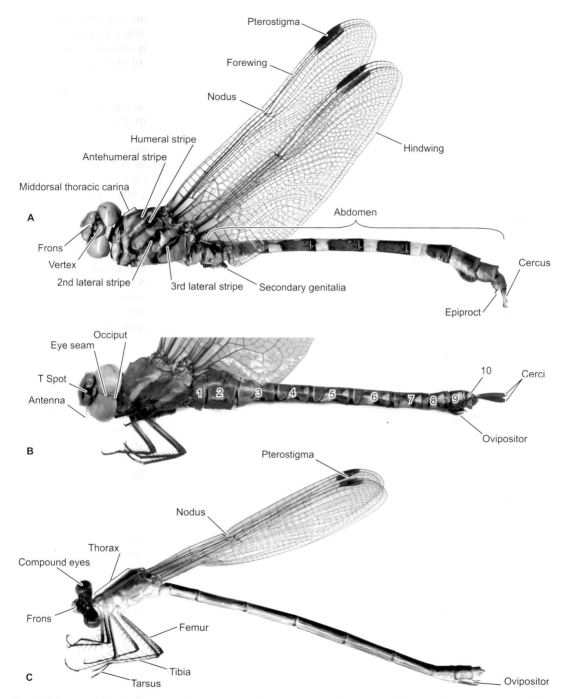

Fig. 4. Odonata adults, showing structures important in identification: (A) Four-striped Leaftail (*Phyllogomphoides stigmatus*), male; (B) Blue-eyed Darner (*Aeshna multicolor*), female; (C) Elegant Spreadwing (*Lestes inaequalis*), female.

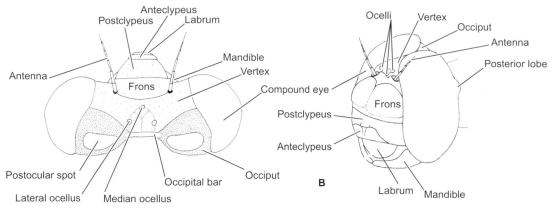

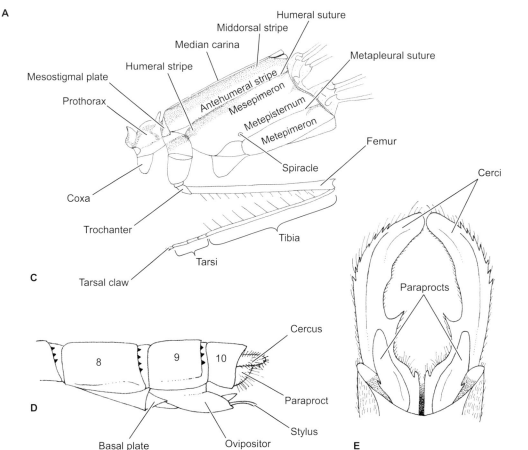

Fig. 5. Further structures important in odonate identification: (A) head of damselfly, dorsal view; (B) head of baskettail (*Epitheca*); (C) thorax of Variable Dancer (*Argia fumipennis*), lateral view; (D) posterior abdominal segments of female Swamp Spreadwing (*Lestes vigilax*), lateral view; (E) terminal abdominal segments of male Great Spreadwing (*Archilestes grandis*), ventral view.

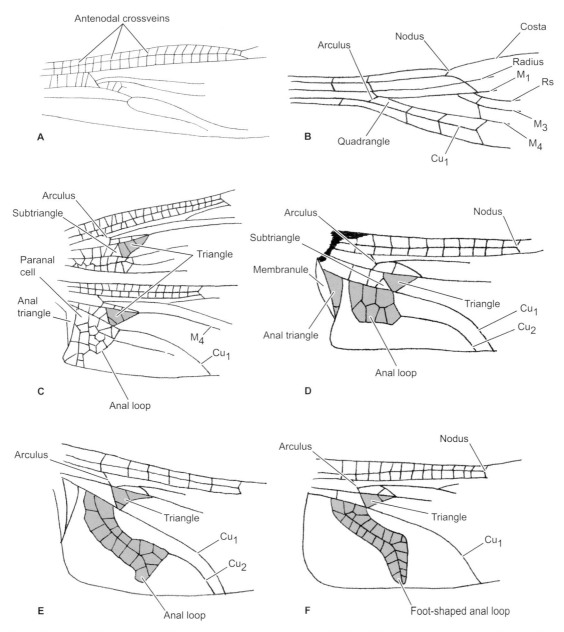

Fig. 6. Wing details in Odonata: (A) forewing of broad-winged damsels (Calopterygidae); (B) basal hindwing of Orange-striped Threadtail (*Protoneura cara*); (C) basal area of wings of Springtime Darner (*Basiaeschna janata*); (D) basal hindwing of river cruiser (*Macromia*); (E) basal hindwing of baskettail (*Epitheca*); (F) basal hindwing of skimmer (Libellulidae).

ditional Comstock-Needham system of naming veins here, but there are several alternative systems of nomenclature. A detailed comparison of these terms is given in Carle (1982). The strong anterior-most vein in each wing is called the **costa**. The costa is slightly notched toward the middle of its length at a region termed the **nodus**. The **pterostigma** (sometimes called the stigma) is a colored area at the tip of the wing along the costal margin. In dragonflies, there is a loop of veins near the base of the hindwing called the **anal loop**. These veins present different shapes in different families. There are also areas of cells in the fore- and hindwings that make up **triangles**, **supertriangles**, and **subtriangles**. In dragonflies the configuration of these cells is useful in distinguishing among the different families.

Abdomen

The abdomen, elongated in all odonates, comprises ten **segments**, the first and last of which are reduced and often hard to see. The color pattern on these segments is often useful in making field identifications, especially with many of the damselflies. (In some cases, especially the clubtail dragonflies, the terminal segments will be dilated into a clublike structure.) Females typically have slightly broader abdomens than males, and in damselflies, darners, and petaltails an **ovipositor** (a bladelike egg-laying structure) is evident ventrally on the female's segment 9. Those dragonflies lacking an obvious ovipositor have a **subgenital plate** (**vulvar lamina**) originating ventrally from segment 8 and extending to segment 9. It is often notched at its middle. A pair of **cerci** (sing. **cercus**), extending terminally, is usually visible, though in most groups these are reduced in the female.

Males have secondary, or accessory, genitalia, located ventrally on segments 2 and 3. In damselflies, the cerci are enlarged and form a dorsal pair of claspers. Below these is a pair of **paraprocts**. Together with the cerci, these structures form the caudal appendages, and are used for grasping the female during mating. In dragonflies, the paraprocts are reduced, and a single **epiproct** is located below the enlarged cerci.

Coloration

The beautiful and varied coloration of odonates is a significant factor in their appeal. In odonates, most of the colors seen result from pigment rather than from structural artifacts of the body. But the bright-blue color so characteristic of many damselflies is a result not of pigment, but rather of the scattering of light by tiny refractive granules located in epidermal cells; and the metallic coloration seen in many emerald dragonflies is also a result of the insects' surface structure. In many odonates a **pruinescence**, or waxy blue-white covering, will develop with age, and as noted above, the newly emerged, or teneral, adult will not have the vibrant colors of its mature counterpart.

In many cases, coloration is an important and useful tool for making field identification, but it is important to recognize some of the difficulties of using color. Odonates of many species are sexually dimorphic in this respect; the males and females differ in coloration. Moreover, many species of damselflies and darners are known to change colors with temperature; individuals will often become darker in color when exposed to cooler temperatures. This is important to remember when observing odonates on a cool early morning. Several species of pond damsels, such as Springwater Dancer, also become darker while in copulation.

A GUIDE TO THE SPECIES ACCOUNTS

The species accounts in this book treat 85 species of damselflies and 178 species of dragonflies, all occurring in the south-central United States. These 263 Odonata species are distributed among ten families and 69 genera. Two of the ten families, the pond damsels (Coenagrionidae, 66 spp.) and the skimmers (Libellulidae, 81 spp.), account for 56% of our total species and six of our 69 genera (*Lestes*, 10 spp.; *Ischnura*, 12 spp.; *Gomphus*, 12 spp.; *Libellula*, 17 spp.; *Argia*, 22 spp.; and *Enallagma*, 23 spp.) number ten

KEY TO THE FAMILIES OF SOUTH-CENTRAL ODONATA

1. Fore- and hindwings similar in size and shape, having quadrangles instead of triangles and subtriangles; eyes well separated on top of the head, by more than their own width — **Damselflies (Suborder Zygoptera, p. 19)** 2

1'. Fore- and hindwings not similar in size and shape, the hindwing considerably wider basally, each wing having a triangle and a subtriangle; eyes meeting middorsally on top of the head, or, if separated, then by less than their width — **Dragonflies (Suborder Anisoptera, p. 149)** 5

2(1). Numerous antenodal crossveins; postnodal crossveins not in line with veins posterior to them — **Broad-winged Damsels (Family Calopterygidae, p. 21)**

2'. Only 2 antenodal crossveins; postnodal crossveins in line with veins posterior to them — 3

3(2'). Veins M_3 and Rs arise nearer to the arculus than to the nodus — **Spreadwings (Family Lestidae, p. 29)**

3'. Veins M_3 and Rs (see Fig. 6) arise nearer to the nodus than to the arculus — 4

4(3'). Anal vein absent or greatly reduced; Cu_2 absent or at most only 1 cell long — **Threadtails (Family Protoneuridae, p. 46)**

4'. Anal vein and Cu_2 not absent or reduced — **Pond Damsels (Family Coenagrionidae, p. 52)**

5(1'). Eyes widely separated on top of the head — 6

5'. Eyes touching or only narrowly separated on top of the head — 7

6(5). Pterostigma at least 1/4 the distance from nodus to wing apex; body color gray and black; front margin of labium with a median cleft — **Petaltails (Family Petaluridae, p. 151)**

KEY TO THE FAMILIES OF SOUTH-CENTRAL ODONATA (*cont.*)

6'. Pterostigma not more than 1/6 the distance from nodus to wing apex; body color yellow or green with black (if gray, then total length less than 70 mm); front margin of labium without a median cleft — **Clubtails (Family Gomphidae, p. 176)**

7(5'). Eyes narrowly touching or barely separated on top of the head — **Spiketails (Family Cordulegastridae, p. 215)**

7'. Eyes broadly touching on top of the head — 8

8(7'). Triangles in fore- and hindwings similar in shape — **Darners (Family Aeshnidae, p. 153)**

8'. Triangles in fore- and hindwings dissimilar — 9

9(8'). Anal loop generally foot-shaped, with well-developed toe; no tubercle on rear margin of each compound eye — **Skimmers (Family Libellulidae, p. 240)**

9'. Anal loop either foot-shaped, but with little development of the toe, or circular and not foot-shaped; generally a tubercle on rear margin of each compound eye — **Emeralds & Cruisers (Family Corduliidae, p. 218)**

or more species each and collectively account for 38% of the total species.

Family and generic accounts introduce the respective groups and include keys to genera and species. Keep in mind that in many cases information given in these sections is restricted to what is applicable to the odonates of the south-central United States fauna. Species accounts include illustrations and photographs that will allow you to identify odonates in the region. I have intentionally cropped many of the photographs in an attempt to better show specific characters. This is especially true in the dragonflies, where in some areas I have cropped out one set of wings (especially when they were clear), to reveal more detail on the thorax and abdomen. Families are organized phylogenetically (along lines of their relationships), but for ease in accessing family members, the genera and species are arranged alphabetically by scientific name.

Individual species accounts consist of the following information: (1) common (English) name, taken from Paulson and Dunkle (1996) and updated when necessary at http://www.ups.edu/biology/museum/ NAdragons.html, (2) scientific name, as the species

is known throughout the world, (3) identifying numbers for figures, photos, and tables illustrating or characterizing the species, (4) measurements, including total length, length of the abdomen, and length of the hindwing, (5) lists of the regional biotic provinces and watersheds where the species is known to occur (see Figs. 1 and 2, in the Introduction), (6) the species' general distribution, (7) flight season within the region (state abbreviations are given in parentheses; early and late dates are reported from the literature and museum collections; in some cases, where specimens and data are scarce, only one date may be known for our region), (8) a description of the species, emphasizing tips on identification, (9) notes on similar species encountered in our region, with details on their differences, (10) a brief description of the species' preferred habitats, (11) a discussion of behavior and other matters, and (12) a selection of the relevant literature pertaining to the species (all sources cited in text are given in the Bibliography). Accompanying each species account is a map of the species' distribution within our region (updated through June 2002).

Figs. 7 through 32, falling at various points

throughout the text, present comparisons of certain body parts among related species, to assist in making identifications. Color photographs of living specimens of virtually all species treated, in many cases two or more photos per species, are gathered in a single section of the text, easily located; all 452 of them are referred to in the species accounts.

The preceding key will often be useful in initiating the task of identifying a specimen in hand. Similar keys are presented throughout the text.

Damselflies
(Suborder Zygoptera)

BROAD-WINGED DAMSELS
(Family Calopterygidae)

This family is represented in North America by two genera, the rubyspots (*Hetaerina*) and the jewelwings (*Calopteryx*), and both occur in our region. Jewelwings are confined largely to the Northern Hemisphere, whereas rubyspots are restricted to the New World and become most diverse in the tropics. The large size and strong iridescent coloration of these damsels has led to the characterization of this family as the "birds of paradise" among Odonata. The wings, broad and not stalked, as their common name implies, are colorful, and often displayed during courtship and territorial bouts. Adults typically exhibit strong sexual dimorphism, the males being brighter and more heavily pigmented.

Individuals of this family can be separated from all other North American damselflies by the numerous (five or more) antenodal crossveins found in all wings. In all other North American families, there are two antenodal crossveins. Further, the postnodal crossveins are not lined up with those posterior to them, as they are in our other damselfly families. The pterostigma is weakly formed or absent, especially in males.

Adults seldom stray far from the streams in which they have spent most of their lives, perching on the vegetation and surrounding rocks. They do not lay eggs in tandem, but, rather, the male will guard the female from a nearby perch, to fend off advances from other males. Elaborate courtship behaviors have been documented in jewelwings, and this family has been the subject of numerous other studies on territoriality and egg-laying that have provided great insight into damselfly behavior.

References. Alcock (1982, 1983, 1987a), Beatty and Beatty (1970), Grether and Grey (1996), Johnson (1962a,b), Waage (1974, 1979a, 1984).

KEY TO THE GENERA OF BROAD-WINGED DAMSELS (CALOPTERYGIDAE)

---

1. Median space, proximal to arculus, without crossveins; pterostigma absent in males, present and distinctly white in females   **Jewelwings (*Calopteryx*)**

1'. Median space with several crossveins; small pterostigma usually present in both sexes (absent in Canyon Rubyspot)   **Rubyspots (*Hetaerina*)**

Jewelwings
Genus *Calopteryx* Leach

The name *Calopteryx*, which means "beautiful wing," is appropriately descriptive of the group. These damselflies are large, and have brilliant iridescent green-and-blue bodies. They have long black legs armed with numerous spines. The wings of females are more uniformly colored and not as dark as those of males. The pterostigma, absent in males, is white and divided by crossveins in females.

This Holarctic genus is represented in North America by five species, two of which are found in our region. The wing patterns of these two species differ, and allow for easy field identification. Adults stay close to the streamside, moving from one bush or limb to another in a characteristic fluttering flight. Courtship has been well studied in this group; females of both of our species respond to male courtship with specific displays that signal differences in receptivity. Wing-spreading, for example, constitutes a rejection, whereas wing-flipping is an invitation.

Mate recognition in jewelwings is by color patterns and flight behavior. Diagnostic features for species recognition, therefore, consist of wing patterns, relative wing width, color of male ventral abdominal segments, and pterostigma size in females.

The distributions of Sparkling (*C. dimidiata*) and Ebony Jewelwing (*C. maculata*) widely overlap, and the species have similar courtship behaviors. Male territories center on egg-laying sites, where pairs initially encounter one another. Ebony Jewelwing males persist in courtship regardless of female response, whereas Sparkling Jewelwing males will stop court-

ing when presented with a rejection (wing spreading) or neutral response from the female. Evidence suggests that these differences are based on interspecific differences in egg-laying behavior.

One study has described the reproductive behavior of coexisting populations of Ebony and Sparkling Jewelwings in Massachusetts at the northern edge of the latter species' range. Ebony Jewelwings lay eggs at the water surface, which exposes females to disturbance by males attempting to mate. Females are therefore likely to re-mate to secure postcopulatory guarding when changing egg-laying sites, and males are expected to be persistent in courtship. Sparkling Jewelwings submerge to oviposit, which frees females from male disturbance and means that males have less control over female access to oviposition sites. Males therefore have less influence on mating by females, and would be expected not to persist in courtship of nonreceptive females.

References. Buchholtz (1955), Pajunen (1966), Waage (1984, 1988).

KEY TO THE SPECIES OF JEWELWINGS (*CALOPTERYX*)

---

1. Wings entirely dark brown or black; wings about 3 times as long as their greatest width   **Ebony (*maculata*)**

1'. Wings clear, with a dark band in the apical 1/3; wings more than 3.5 times as long as their greatest width   **Sparkling (*dimidiata*)**

Sparkling Jewelwing
Calopteryx dimidiata Burmeister
(photos 1a, 1b)

Size. Total length: 37–49 mm; abdomen: 29–40 mm; hindwing: 23–31 mm.

Regional Distribution. *Biotic Province:* Austroriparian. *Watersheds:* Brazos, Mississippi, Neches, Ouachita, Red, Sabine, Trinity.
General Distribution. Florida north to Massachusetts and westward to east Texas.
Flight Season. Mar. 1 (LA)–Sep. 10 (LA).
Identification. Essentially a costal-plain species, Sparkling Jewelwing is relatively uncommon in our region, and its westernmost records are limited to the Big Thicket Primitive Area of east Texas. The body and face are iridescent blue-green, the labium and antennae black. The wings are clear except for the apical 1/4 to 1/6. The pterostigma is lacking in males, and the wings are 3.5 to 4 times as long as wide. The apical black area may be faint or absent in females.

Similar Species. Ebony Jewelwing (*C. maculata*) has solid black wings. The wings of Smoky Rubyspot (*Hetaerina titia*) vary, but will at least have an extensive dark spot basally in each wing.

Habitat. Sandy-bottomed streams; occasionally rivers with little canopy cover.

Discussion. The variability, distribution, and taxonomy of Sparkling Jewelwing have been studied. Female wing-color patterns and pterostigma size show little or no seasonal variation, but rather occur in geographic clines. Data from six states suggests a general north–south trend toward larger size and an increase in numbers of andromorphic females (with apical bands) versus gynomorphic females (lacking apical bands) at southern latitudes.

Sparkling Jewelwing apparently has a narrower range of habitat requirements than Ebony Jewelwing and often exists in isolated colonies, as is seen in Texas and Louisiana populations. Sparkling Jewelwing does not disperse far from its breeding site.

References. Johnson (1972a, 1973a), Johnson and Westfall (1970).

Ebony Jewelwing
Calopteryx maculata (Beauvois)
(photo 1c)

Size. Total length: 37–57 mm; abdomen: 30–47 mm; hindwing: 25–37 mm.

Regional Distribution. *Biotic Provinces:* Austroriparian, Balconian, Carolinian, Kansan, Navahonian, Texan. *Watersheds:* Arkansas, Bayou Bartholomew, Brazos, Canadian, Cimarron, Colorado, Mississippi, Neches, Ouachita, Red, Sabine, San Jacinto, St. Francis, Trinity, White.

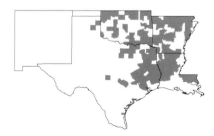

General Distribution. Eastern North America, including Canada; westward to Wisconsin and south to Texas.

Flight Season. Mar. 1 (LA)–Oct. 31 (LA).

Identification. This common, large, black damselfly is widely distributed in the eastern part of our region. The head is iridescent blue-green. The thorax is black, with strong iridescent blue-green coloration dorsally and on the sides. Males lack a pterostigma. Older, mature males have solid black wings; wings in teneral individuals are lighter and brown in color. Wings of females are usually paler, becoming progressively darker apically, with a conspicuous white pterostigma (enclosing numerous cells) that is distinctively widened at the middle. Wing length is about three times greatest width. The abdomen is iridescent blue-green dorsally, black ventrally, except for a white (males) or brown (females) area on the posterior of sterna 8 (ventral portion of segment) and segments 9 and 10.

Similar Species. Smoky Rubyspot (*Hetaerina titia*) is the only other damselfly that may have completely dark wings. It lacks the blue-green iridescence on the body, and the wings are only about 1/5 as wide as long. In Sparkling Jewelwing (*C. dimidiata*) only the apical 1/4 of the wings is black.

Habitat. Small, slow-moving, canopy-covered streams and occasionally exposed streams and rivulets.

Discussion. This species ranks among the most studied of damselflies in North America. Nymphs are local in occurrence and restricted to slow creeks and quiet areas of running streams. The primary factors affecting their distribution within streams are rate of flow, depth of water, and type of vegetation present, whereas adults occur along a wide variety of stream-riverine conditions and often disperse well away from water.

Males will compete vigorously among themselves for territories having submergent vegetation, the prime egg-laying habitat for females. Males attract females with a "cross display," where the male faces the female with his hindwings deflected downward at right angles to his body, and the forewings and abdomen are raised, revealing the ventral pale area of the abdomen. The majority of mating and egg-laying occurs in the early afternoon and a single male may guard multiple females, resulting sometimes in large congregations. Females will lay their eggs in submergent vegetation for 10 to 120 minutes and usually do not submerge themselves. The displays and behaviors of northern and southern populations may differ. For a summary of these behaviors, the reader is directed to Dunkle (1990). Female Ebony Jewelwing has been reported to use her ovipositor to steady herself on a leaf while feeding on a mayfly.

References. Alcock (1979, 1983, 1987a), Ballou (1984), Erickson (1989), Erickson and Reid (1989), Forsyth and Montgomerie (1987), Johnson (1962a,b), Johnson and Westfall (1970), Martin (1939), Mesterton-Gibbons et al. (1996), Pither and Taylor (1998), Tennessen (1998), Waage (1972, 1974, 1975, 1978, 1979a,b, 1980, 1983, 1984).

Rubyspots
Genus *Hetaerina* Hagen *in* Selys
(p. 27)

This is a primarily tropical genus, members of which are frequently encountered around streams and rivers. The group is represented in the United States by only three species, all of which occur in our region. The common name "Rubyspots" comes from the basal red spots on the wings of males. Members of this group can be separated from all other North American species by their relatively large size and the presence of this spot. Some specimens of the variable Smoky Rubyspot (*H. titia*), however, have entirely black wings.

Females usually have amber-colored wings and are more robust than males. The simple, small, white pterostigma is often absent in one or both sexes, but it is generally present in our species, with the exception of Canyon Rubyspot (*H. vulnerata*). The wings are narrower and the body is more slender than in jewelwings (*Calopteryx*). The thorax is usually iridescent red-brown and black in males and iridescent green and brown in females. In both sexes, the thorax is marked with pale lines, and the legs are long, slender, spiny, and black or brown in color. The arculus is strongly angled, there are several crossveins in the median space, and the quadrangle curves forward. The abdomen is iridescent dorsally and pale ventrally.

References. Garrison (1990), Johnson (1973b).

KEY TO THE SPECIES OF RUBYSPOTS (*HETAERINA*)

1. Basal spot of forewings usually bright red, that of hindwing much duller, often brown; sometimes with extensive dark-brown areas on one or both pairs of wings	**Smoky (*titia*)**
1'. Basal spots of both pairs of wings largely bright red (may be pink or orange-brown in tenerals); rest of wings clear or smoky, but not dark brown	2
2(1). Wings generally with pterostigma; male cerci with 1 or 2 toothlike medial lobes; abdomen green or dark with iridescent green highlights; common throughout area	**American (*americana*)**
2'. Wings without pterostigma; male cerci without distinct medial lobe; abdomen coppery brown, with at most an iridescent luster; south-central New Mexico south	**Canyon (*vulnerata*)**

American Rubyspot
Hetaerina americana (Fabricius)
(p. 27, photo 1d)

Size. Total length: 36–51 mm; abdomen: 29–40 mm; hindwing: 24–31 mm.

Regional Distribution. *Biotic Provinces:* Apachian, Austroriparian, Balconian, Carolinian, Chihuahuan, Coloradan, Kansan, Navahonian, Tamaulipan, Texan. *Watersheds:* Arkansas, Bayou Bartholomew, Brazos, Canadian, Cimarron, Colorado, Colorado (NM), Guadalupe, Lower Rio Grande, Mississippi, Neches, Nueces, Ouachita, Pecos, Red, Sabine, San Antonio, San Jacinto, St. Francis, Trinity, Upper Rio Grande, White.

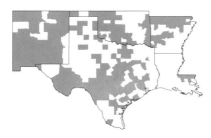

General Distribution. Throughout North America including southeastern Canada, south through Mexico, Guatemala, and Honduras.

Flight Season. Mar. 11 (TX)–Dec. 28 (TX).

Identification. Males are large, with an iridescent red head and thorax. The abdomen is iridescent green, getting darker with age. The caudal appendages are pale, and the wings have at least the basal 1/4 red, although the whole wing of teneral individuals may be amber. (A pterostigma is present, but see discussion below.) Females are largely iridescent green, but wing color is duller than in males. The abdomen is pale laterally, including the ovipositor. There is a pale, narrow middorsal line running the length of the abdomen.

Similar Species. This species is very similar to Canyon Rubyspot (*H. vulnerata*), which has a duller, coppery-brown abdomen, and females of which lack a pterostigma. Though the ranges of these two species overlap, Canyon Rubyspot is found only in the west. The head and thorax of the male Smoky Rubyspot (*H. titia*) are darker, the wing tips are brown, and the hindwing is variable, but brown (not red) basally. Female Smoky Rubyspots lack red in the basal area of the hindwing,

and the abdomen is brownish green, not distinctly green and tan.

Habitat. Wide, open streams and rivers.

Discussion. Males and females will perch horizontally on twigs and leaves of riparian vegetation, the females often higher. Sexes may also congregate near the water at night to roost. Numerous aspects of this species' distribution, behavior, and ecology have been well studied. Although this is primarily a stream species, it has been shown experimentally that larvae and teneral adults exposed to still-water returned to still water habitats after they had matured.

There is extensive variability in this species. There are populations that lack a pterostigma, but these seem to be most abundant west of the Sierra Nevada mountains in California, and all individuals seen by the author in our region have had pterostigmas. Apterostigmatous individuals have been collected as far east as Grant and Lincoln Counties, in southwestern New Mexico. An increase in the length of the basal red markings in the wings of males as the season progresses has been recorded in three different Texas populations. Basal red markings of males collected in April range from 20–35% of wing length. Males sampled at the same localities in September, however, had substantially larger red markings, ranging from 35–50% of wing length.

References. Bick and Sulzbach (1966), Calvert (1901–1908), Dunkle (1990), Garrison (1990), Grether (1995, 1996a,b), Grether and Grey (1996), Johnson (1961, 1962b, 1963, 1966a), Kellicott (1890), McCafferty (1979), Weichsel (1987).

Smoky Rubyspot
Hetaerina titia (Drury)
(p. 27, photo 1e)

Size. Total length: 39–53 mm; abdomen: 30–43 mm; hindwing: 25–31 mm.

Regional Distribution. *Biotic Provinces*: Austroriparian, Balconian, Carolinian, Chihuahuan, Kansan, Tamaulipan, Texan. *Watersheds:* Arkansas, Bayou Bartholomew, Brazos, Canadian, Colorado, Guadalupe, Lower Rio Grande, Mississippi, Neches, Nueces, Ouachita, Red, Sabine, San Antonio, San Jacinto, St. Francis, Trinity, Upper Rio Grande, White.

General Distribution. Southeastern United States north to Maryland; west to Wisconsin, Texas, New

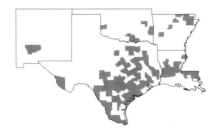

Mexico; extending southward through Mexico to Costa Rica.

Flight Season. Mar. 25 (TX)–Nov. 16 (TX).

Identification. This is a large damselfly widely dispersed east of the Kansan and Chihuahuan biotic provinces. The clypeus, labrum, anterior portion of the frons, and basal antennal segment in the male are light tan, giving way to darker brown or black on the head. The dorsum of the pterothorax ranges from deep iridescent red to black, with broad bronzy-green regions laterally. The wings are variable, but the basal fourth of both wings is red and often diffused with brown. Red veins may be present in this area, and the tips of both wings are usually brown. The remainder of the wings may be clear to smoky dark brown or entirely black, but the hind wing is usually much more extensively marked.

The female is similar to the male, but the head is lighter in front and variably iridescent green dorsally. The pterothorax is iridescent green, and the abdomen is iridescent green to brown. The wings are amber to brown. The dorsal carina of abdominal segment 10 usually ends in a prominent spine projecting well beyond the apical margin. Both sexes have a white pterostigma that darkens with age and surmounts one to three or more cells.

Similar Species. The body of Sparkling Jewelwing (*Calopteryx dimidiata*) is iridescent blue-green. American (*H. americana*) and Canyon Rubyspot (*H. vulnerata*) have bright-red basal spots in both wings and are never darkenend beyond the basal fourth of the wing. The range of Canyon Rubyspot does not overlap with with that of Smoky Rubyspot. The abdomen is uniformly dark brown with green reflections, not distinctly iridescent green dorsally and tan laterally as in American Rubyspot.

Habitat. Small to medium-sized streams and rivers with strong current.

Discussion. The variability of this species has led to confusion, and to the recognition of several races and forms. Johnson (1963) studied sympatric populations of American and Smoky Rubyspots on the Guadalupe River in Gruen and Comfort, Texas, and at the Llano River in Junction, Texas. A third population of Smoky Rubyspot was observed at Chinquapin Creek east of Lufkin, Texas. After observing differences in breeding and territorial behavior, Johnson found that the female body-color patterns were consistent, and that the females bred with their specific male type. He considered these forms to be possible valid species. He also found the flight seasons of the two forms to be different. Florida collections (FSCA) revealed occasional intergrades between the two female color patterns of the thorax, and showed no apparent difference in flight season, leading Johnson and Westfall (1970) to place these forms in a species complex. I follow Garrison (1990), who considers them forms of one species, based on the lack of detectable morphological differences between sexes of either form. The differences in degree of wing color in this species may be correlated with seasonality. Spring populations tend to have the dark markings in their wings restricted to the basal fourth of the wing, while midsummer pouplations show a considerable increase in the degree of brown in the wings, and fall generations have completely dark wings.

Like those of American Rubyspot, both sexes of Smoky Rubyspot perch horizontally on vegetation along the shore. They tend to prefer perches higher up, and are more wary than the former. Females will invite mating by hovering, and reject males with a display similar to that seen in Ebony Jewelwing (*Calopteryx maculata*), involving a simultaneous spreading of the wings and bending of the abdomen upward. Guarded above by males, females will spend up to 2 hours ovipositing underwater in wet wood.

References. Dunkle (1990), Harp (1986), Johnson (1961, 1963).

Canyon Rubyspot
Hetaerina vulnerata Hagen *in* Selys
(p. 27, photo 1f)

Size. Total length: 36–49 mm; abdomen: 28–41mm; hindwing: 25–32 mm.

Regional Distribution. *Biotic Provinces:* Apachian, Balconian, Chihuahuan, Navahonian. *Watersheds:* Colorado (NM), San Antonio, Upper Rio Grande.

Rubyspots (*Hetaerina*)

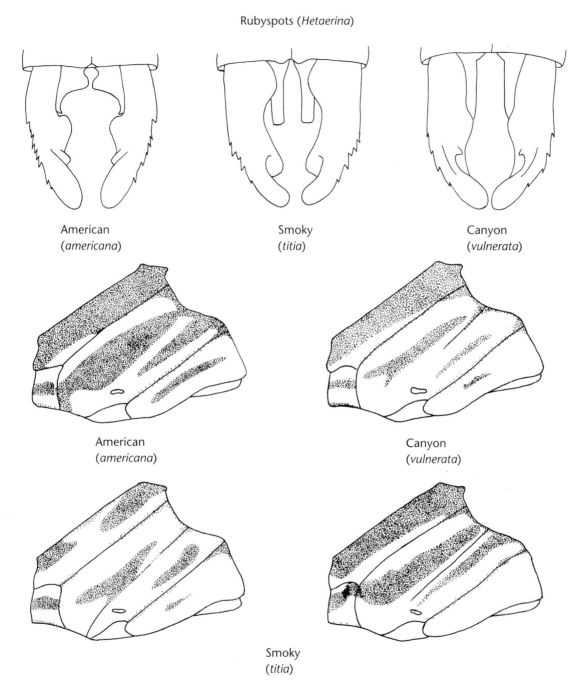

American
(*americana*)

Smoky
(*titia*)

Canyon
(*vulnerata*)

American
(*americana*)

Canyon
(*vulnerata*)

Smoky
(*titia*)

Fig. 7. Rubyspots (*Hetaerina*): male caudal appendages,
dorsal view; and female thoracic patterns, left lateral view
(pp. 24–28).

General Distribution. Southwestern United States; south through Mexico to Guatemala and Honduras.

Flight Season. May 18 (NM)–Oct. 19 (NM).

Identification. The male labrum is pale yellow with a central, dark brown spot. The clypeus and basal antennal segment are both tan in color. The top of the head is dark brown to black with iridescent bronze reflections. The pterothorax is deep iridescent red dorsally, with a black middorsal carina, and pale tan laterally. The red basal spots in both wings become more intense with age. The venation in the basal portion of the wing is white ventrally. The tips of the wings are often edged with brown and lack a pterostigma. The basal abdominal segments are dark brown-black dorsally, each with a narrow light-yellow band anteriorly. The caudal appendages are light tan to brown.

The female is similar to the male, but the lighter coloration is more extensive. The pterothorax is iridescent green dorsally with iridescent green lateral stripes, but these are less extensive than in males and may be absent. The wings lack the intense basal red spots of males, but are suffused with orange basally.

Similar Species. This southwestern species closely resembles American Rubyspot (*H. americana*), but it has a different abdominal pattern and lacks pterostigmas. (See the description of that species for futher distinctions.) Smoky Rubyspot (*H. titia*) has extensive brown in the wings, and its range does not overlap with that of Canyon Rubyspot.

Habitat. Streams and rivers in open-canopy woodlands.

Discussion. Canyon Rubyspot was thought to be completely apterostigmatous, but Garrison (1990) reported pterostigmatous specimens from Cañon Huasteca National Park in Nuevo León, Mexico. Though they show other differences, Garrison describes them as "easily referable to *H. vulnerata*." The ranges of American and Canyon Rubyspot overlap, and they are known to occur together. Males remain with the female after mating and will even adopt the unusual behavior of leaving their territory to accompany females in tandem on a search for egg-laying sites elsewhere. They will then perch and guard the female while she submerges underwater to oviposit. There are two records of male Canyon Rubyspots from Bexar Co., Texas. This Balconian locality is considerably east of the normal range for this species.

References. Alcock (1982), Johnson (1973b).

SPREADWINGS
(Family Lestidae)

The spreadwing family includes one of the largest zygopteran genera (*Lestes*) in the world. Spreadwings can generally be recognized by their unique resting posture. They perch with their wings held apart (45°), and not closed over their abdomens as generally seen in other damselflies (although pond damsels, Coenagrionidae, will occasionally perch with their wings apart). The eyes and face, especially in males, are blue.

Spreadwings are large to medium-sized. The wings are distinctly petiolate, with only two antenodal crossveins, and the postnodal crossveins are in a line with the veins directly below them. The pterostimas are relatively long, and veins M_3 and Rs originate closer to the arculus than to the nodus, distinguishing the spreadwings from all other North American families. The legs are armed with long spurs. The cerci are forceps-like. Females oviposit, usually in tandem, in submerged or floating vegetation.

KEY TO THE GENERA OF SPREADWINGS (LESTIDAE)

1. Hindwing 30mm or greater in length; vein M_2 arises about 1 cell beyond the nodus   **Stream (*Archilestes*)**

1'. Hindwing less than 30mm in length; vein M_2 arises several cells distal to the nodus   **Pond (*Lestes*)**

Stream Spreadwings
Genus *Archilestes* Selys

This is a small group of rather large damselflies with only two species in North America. California Spreadwing (*A. californica*) is found only in the southwest. The larger Great Spreadwing (*A. grandis*) occurs in the south-central United States. As its name implies, it may be readily distinguished from species of pond spreadwings by its large size and robust stature. Stream spreadwings are generally found perching on vegetation around pools or backwaters of slow-moving streams. The ventral margin of the ovipositor is much larger than in pond spreadwings, and has coarser teeth.

Great Spreadwing
Archilestes grandis (Rambur)
(pp. 5, 12, 30, 31; photos 2a, 2b)

Size. Total length: 50–62 mm; abdomen: 38–47 mm; hindwing: 31–39 mm.

Regional Distribution. *Biotic Provinces:* Apachian, Austroriparian, Balconian, Carolinian, Chihuahuan, Kansan, Navahonian, Tamaulipan, Texan. *Watersheds:* Arkansas, Brazos, Canadian, Cimarron, Colorado, Colorado (NM), Guadalupe, Neches, Nueces, Pecos, Red, San Antonio, St. Francis, Trinity, Upper Rio Grande, White.

TABLE 2. SPREADWINGS (LESTIDAE)

Species	Hindwing length (mm)	Antehumeral area (♂)	Marks on thorax[1]	Basal plate ovipositor[2]	Tibiae/tarsi	Rear of head (♂)	Abdomen color[3]	Distribution (biotic provinces)[4]
Great (*grandis*)	31–39	brown-green	–	truncate	dark/dark	yellow	metallic green	Ap,Au,B,Ca,Ch,K,N,Ta,Tx
Plateau (*alacer*)	19–25	blue/yellow	+	truncate	pale/dark	black	dark	Ap,B,Ca,Ch,K,N,Ta,Tx
Spotted (*congener*)	18–23	yellow	+	slight tooth	pale/dark	black	dark	Ca,Ch,Co,K,N
Common (*disjunctus*)	18–25	blue-green metallic	+/–	tooth	pale/dark	black	dark	Au,B,Ca,Ch,Co,K,N,Ta,Tx
Emerald (*dryas*)	19–25	green metallic	+	tooth	dark/dark	metallic green	metallic green	Co,N
Rainpool (*forficula*)	17–24	green metallic	+	tooth	pale/dark	yellow	dark	B,Ta,Tx
Elegant (*inaequalis*)	25–31	green	–	truncate	dark/dark	yellow	metallic green	Au,Ca
Slender (*rectangularis*)	20–25	pale blue	–	tooth	pale/pale	black	dark	Au,Ca,K,Tx
Chalky (*sigma*)	20–25	yellow/brown	+	tooth	pale/dark	yellow[5]	dark	A,B,Ta,Tx
Lyre-tipped (*unguiculatus*)	17–24	yellow-green	–	tooth	pale/dark	black	dark	Au,Ca,K,Tx
Swamp (*vigilax*)	23–27	rusty brown	–	tooth	dark/dark	black	dark	Au

[1] Paired dark markings posterior to legs on venter of thorax.
[2] Appearance of posterolateral margin of basal plate of ovipositor.
[3] Predominant color of abdomen dorsally.
[4] (Ap) Apachian, (Au) Austroriparian, (B) Balconian, (Ca) Carolinian, (Ch) Chihuahuan, (Co) Coloradan, (K) Kansan, (N) Navahonian, (Ta) Tamaulipan, (Tx) Texan.
[5] Black in older individuals.

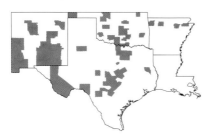

General Distribution. South Carolina north to Vermont, west to California, and south through Central America to Colombia and Venezuela.
Flight Season. Mar. 17 (TX)–Oct. 30 (OK).
Identification. Great Spreadwing is found commonly in the Texan biotic province westward. It is the largest damselfly in the United States and can be readily identified in the field on the basis of its large size, spreadwing perching behavior, and distinct bright-yellow thoracic stripes. The mesepisternum

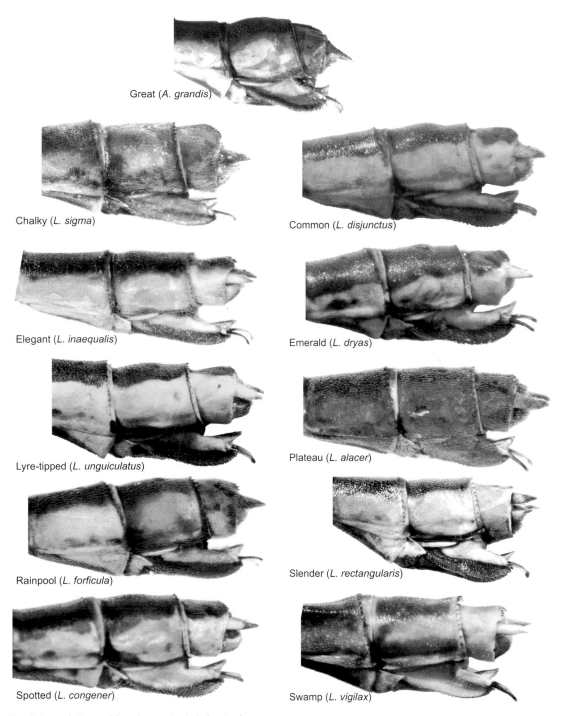

Great (*A. grandis*)

Chalky (*L. sigma*)

Common (*L. disjunctus*)

Elegant (*L. inaequalis*)

Emerald (*L. dryas*)

Lyre-tipped (*L. unguiculatus*)

Plateau (*L. alacer*)

Rainpool (*L. forficula*)

Slender (*L. rectangularis*)

Spotted (*L. congener*)

Swamp (*L. vigilax*)

Fig. 8. Lateral views of female terminal abdominal segments in various Pond Spreadwing species.

is a complete dark, metallic-green stripe no more than 1/2 the sclerite width. This stripe is generally not contiguous with the middorsal carina, though it is occasionally narrowed toward the middle. The mesepimeron is variable, with a metallic-green or black stripe. This stripe may extend the full length of the sclerite or slightly more than 1/2 its width. The remainder of the pterothorax is yellow, resulting in two broad, pale stripes. The wings are either clear or smoky, becoming darker at the tips. The first 1/3 to 1/2 of the abdomen is dark metallic green dorsally, becoming yellow or tan ventrolaterally. This pattern becomes obscured with age. On segment 9 there is a narrow, dark, middorsal stripe and black apical ventrolateral carinae. Segment 10 has a proximally directed dark median triangle on its dorsum. Mature males develop a white pruinosity laterally and basally on segments 1 and 2, all of segments 9 and 10, and on the sterna of segments 7 and 8.

Females are more robust than males. The general coloration is similar, but the head is paler. The dark mesepisternum and mesepimeron stripes are often narrower and sometimes nearly lacking. The proximal abdominal segments are pale bluish. Segment 10 is tan to nearly black. The basal plate of the ovipositor is truncated posterolaterally. The margins of the valves are strongly and coarsely toothed.

Similar Species. Pond spreadwings (*Lestes*) are smaller, the hindwing less than 30 mm long. No other damselfly in our region approaches the size of Great Spreadwing.

Habitat. Small permanent ponds or streams with slow or moderate flow.

Discussion. This species was known only from the southwest United States up until the 1920's, but it has since undergone a dramatic range expansion northward. It now occurs as far northeast as western New England. One study showed that neither males nor females exhibited any type of courtship behavior, and unreceptive females showed no refusal signs, but rather were simply not at the water, or escaped by rapid flight when unreceptive. No pairs were observed ovipositing from beginning to end, but the longest observed egg-laying time was 109 minutes.

References. Ahrens (1935), Bick and Bick (1970), Garman (1932), Gloyd (1980), Ingram (1976), Moskowitz and Bell (1998).

Pond Spreadwings
Genus *Lestes* Leach
(pp. 30, 31, 37, 42)

This globally distributed group includes medium- to large-sized damselflies. Ten species occur in our region, and all but two, Spotted (*L. congener*) and Emerald (*L. dryas*) Spreadwing, are found in Texas. The combination of metallic-green and bronze colors, elongated abdomen, and distinctive perching habit serve as good field-recognition characters. Males have a characteristic blue face and eyes, and often develop a distinct pruinose appearance toward the rear of the head, on the thorax, between the wings, and on posterior abdominal segments 9 and 10. The wings are clear and stalked basally. A long pterostigma, surmounting two or more cells, is characteristic. The legs are long, their tibial spurs longer than the intervening spaces, as in those of dancers. The paraprocts of all species in our region, except Spotted Spreadwing, are distinctly longer than 1/2 the length of the cerci, and the paraprocts of one species, Elegant Spreadwing (*L. inaequalis*), are longer than the cerci.

Females have brown eyes and are generally less pruinose and often pale in color. The thoracic color pattern is often characteristic and useful for making identifications in the field. The abdomen is generally uniformly dark and often distinctly ringed, particularly through the middle segments.

Pond spreadwings generally inhabit both permanent and ephemeral fishless bodies of standing water, including small sheltered lakes and ponds with an abundance of emergent vegetation. Oviposition usually occurs in tandem with the male and takes place in emergent vegetation. In the south-central United States, most species undergo egg diapause, and hatch in response to the warming temperatures of the encroaching spring. Elegant Spreadwing and a few other North American species are unique among the group in developing in slow-moving pools of streams. Slender Spreadwing (*L. rectangularis*) is one of several North American species capable of surviving in conditions of high salinity.

KEY TO THE SPECIES OF POND SPREADWINGS (*LESTES*)

MALES

1. Paraproct less than 1/2 as long as circus — **Spotted (*congener*)**

1'. Paraproct longer than 1/2 the length of circus — 2

2(1'). Paraproct distinctly longer than circus — **Elegant (*inaequalis*)**

2'. Paraproct at most 2/3 the length of circus — 3

3(2'). Paraproct slender in apical 1/2 and sigmoid in shape, its apex divergent — 4

3'. Paraproct may be slender, but never sigmoid in shape, and its apex not divergent — 5

4(3). In dorsal view, paraproct greater than 3/4 the length of cercus; basal tooth of paraproct narrow and acute; abdomen generally less than 31 mm long — **Lyre-tipped (*unguiculatus*)**

4'. In dorsal view, paraproct 3/4 the length of cercus; basal tooth of paraproct broad and blunt; abdomen usually greater than 31 mm long — **Chalky (*sigma*)**

5(3'). Paraproct wide distally, appearing boot-shaped dorsally — **Emerald (*dryas*)**

5'. Paraproct not expanded distally — 6

6(5'). Cercus with distinct basal tooth and similarly shaped distal tooth — 7

6'. Cercus with a distinct basal tooth only — 8

7(6). Hindwing less than 2/3 the length of abdomen; in lateral view, paraproct curved sharply downward in distal portion; abdomen generally greater than 35 mm long — **Slender (*rectangularis*)**

7'. Hindwing 2/3 the length of abdomen; in lateral view, paraproct straight; abdomen generally less than 35 mm long — **Common (*disjunctus*)**

8(6'). Inner margin of cercus sparsely serrate; in lateral view, paraproct slender — **Swamp (*vigilax*)**

8'. Inner margin of cercus strongly serrate; in lateral view, paraproct stout — 9

KEY TO THE SPECIES OF POND SPREADWINGS (*LESTES*) (*cont.*)

9(8'). Thin, metallic-green dorsal stripes on thorax not connected with middorsal carina; paraproct and circus equal in length — **Rainpool (*forficula*)**

9'. Broad, black dorsal stripe of thorax contiguous with middorsal carina; paraproct 3/4 or less the length of circus — **Plateau (*alacer*)**

FEMALES

1. Ovipositor distinctly longer than abdominal segment 7 — **Emerald (*dryas*)**

1'. Ovipositor shorter or no longer than segment 7 — 2

2(1'). Dorsum of thorax solid metallic green or with metallic-green stripes or spots — 3

2'. Dorsum of thorax lacking metallic-green areas, though dark areas may be bronze or coppery in appearance — 5

3(2). Basal plate on posterolateral margin of ovipositor rounded or broadly truncate — **Elegant (*inaequalis*)**

3'. Basal plate on posterolateral margin of ovipositor acutely angulate — 4

4(3'). A dark spot present above each metapleural carina; outer surface of tibiae generally pale; metallic-green stripe on mesepisterna no wider than 0.3 mm at narrowest point, widening only slightly posteriorly and less than 1/2 the width of the scderite; metallic-green stripe on mesepimeron, when present, parallel-sided — **Rainpool (*forficula*)**

4'. Dark spots absent from metapleural carina; tibiae uniformly brown in color (in young specimens tibiae may be pale); metallic-green stripe on mesepisterna parallel for most of its length and more than 1/2 the width of the sclerite — **Swamp (*vigilax*)** *in part*

5(2'). Dark stripes on mesepisterna only faintly distinguishable from otherwise pale thorax, in older specimens entire mesothorax may become heavily pruinose; abdomen dark, lacking strongly contrasting pale areas — **Chalky (*sigma*)**

5'. Mesepisterna and mesepimera both with complete dark stripes extending nearly full length, separated by a pale stripe at or just anterior to humeral suture, this pattern never entirely obscured even in older specimens; abdomen variable — 6

KEY TO THE SPECIES OF POND SPREADWINGS (*LESTES*) (*cont.*)

6(5'). A dark spot present on thorax above and/or below metapleural carina — 10

6'. No dark spot present on thorax above or below metapleural carina — 7

7(6'). Abdominal segment 7 1.5 times or more the length of ovipositor (excluding styli); abdomen greater than 35 mm long — 8

7'. Segment 7 less than 1.5 times the length of ovipositor (excluding styli); abdomen less than 35 mm long — 9

8(7). Tibiae and tarsi dark brown (tibiae may be pale in young specimens); antehumeral stripe reddish brown; postnodal crossveins generally 15 or more; abdomen 42-50 mm long — **Swamp (*vigilax*)** *in part*

8'. Tibiae and basal half of tarsi pale; antehumeral stripe blue-gray to yellow; postnodal crossveins generally no more than 12; abdomen 34-44 mm long — **Slender (*rectangularis*)**

9(7'). Posterior of head black except occiput; abdomen bronze-brown to black dorsally — **Common (*disjunctus*)**

9'. Posterior of head with distinct pale areas reaching compound eyes; abdomen generally metallic green dorsally — **Lyre-tipped (*unguiculatus*)**

10(6). Dark thoracic spot above and below metapleural carina — **Spotted (*congener*)**

10'. Dark thoracic spot only above metapleural carina — **Plateau (*alacer*)**

Plateau Spreadwing
***Lestes alacer* Hagen**
(pp. 30, 31, 37; photo 2c)

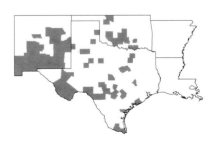

Size. Total length: 34–45 mm; abdomen: 28–36 mm; hindwing: 19–25 mm.

Regional Distribution. *Biotic Provinces:* Apachian, Balconian, Chihuahuan, Kansan, Navahonian, Tamaulipan, Texan. *Watersheds:* Arkansas, Brazos, Canadian, Cimarron, Colorado, Colorado (NM), Guadalupe, Lower Rio Grande, Nueces, Pecos, Red, San Antonio, Trinity, Upper Rio Grande.

General Distribution. Arizona, New Mexico, Oklahoma, and Texas as far east as the Texan biotic province, south through Mexico to Costa Rica.

Flight Season. Jan. 3 (TX)–Dec. 21 (TX).

Identification. The male has a blue face with a black head dorsally. The mesepisternum is black medially, pale blue or yellow laterally, and confluent with the dorsal 1/2 of the mesepimeron. The remaining ventral 1/2 of the mesepimeron is dark, and the remaining pterothorax is pale with a dark stripe of variable width running along the metapleural suture. There is a pair of dark marks ventrally on the thorax, posterior to the legs. The abdomen is slender, especially toward its middle. A ventrolateral spot is present on segments 3–5. Segments 6 and 7 are dark ventrolaterally. These areas, along with segments 1 and 2 and 8 and 10, become heavily pruinose with age.

Females and males are similarly colored. The markings above the female's mesopleural suture are similar to those of the male, but the pterothorax is pale ventrally except for a dark spot just above the anterior end of the metapleural carina and a posteromedial spot on the sternum. The coloration of abdominal segments 3–5 is similar to that of males. Segments 6–10 are similar, but the dark ventrolateral areas are less extensive. Segment 1 is pale in young individuals. The posterolateral margin of the ovipositor basal plate is acutely angulate.

Similar Species. Lyre-tipped Spreadwing (*L. unguiculatus*) lacks dark marks posterior to the legs, ventrally on the thorax. Spotted Spreadwing (*L. congener*) has dark spots on the metepimeron that are lacking in Plateau Spreadwing. Common Spreadwing (*L. disjunctus*) is larger, and the dark areas on the thorax are more extensive.

Habitat. Still or slow-moving waters.

Discussion. In our region this species emerges in early January and flies through December. These damselflies prefer laying eggs in rushes well above the waterline in irregular vertical rows.

References. Bird (1933).

Spotted Spreadwing
Lestes congener Hagen
(pp. 30, 31, 37; photos 2d, 2e)

Size. Total length: 32–42 mm; abdomen: 24–35 mm; hindwing: 18–23 mm.

Regional Distribution. *Biotic Provinces:* Carolinian, Chihuahuan, Coloradan, Kansan, Navahonian. *Watersheds:* Canadian, Colorado (NM), Pecos, Upper Rio Grande, White.

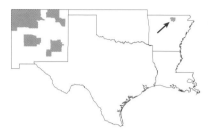

General Distribution. Transcontinental; southern Canada south to California and Arizona, east to Tennessee and Virginia.

Flight Season. Jul. 27 (NM)–Oct. (21).

Identification. The head of the male is dark and the back is black. The pterothorax is black dorsally with a thin, pale-yellow antehumeral stripe. The metepimeron is pale, and there is a prominent dark spot above and below the metapleural carina. The thorax is pruinose ventrolaterally (sometimes entirely in older males) and the abdomen is dark dorsally. There are pale-yellowish ventrolateral areas with ventrolateral black streaks, particularly on segments 5–10. Segments 1, 2, 9, and 10 and the ventrolateral areas of 6–8 are pruinose in older individuals.

The female is generally similar to the male, but overall, the pale areas are more extensive. The posterolateral portion of the ovipositor basal plate tapers to a tooth.

Similar Species. Common Spreadwing (*L. disjunctus*) has a blue-green antehumeral stripe, and Plateau Spreadwing (*L. alacer*) has a blue or yellow antehumeral stripe. Neither species has Spotted Spreadwing's distinctive dark spots on the metepimeron.

Habitat. Ponds, including saline waters.

Discussion. Within its range, this is one of the latest-flying damselflies. Spotted Spreadwing is found throughout northern North America. Females lay eggs in woody and herbaceous stems.

References. Sawchyn and Gillott (1974).

Common Spreadwing
Lestes disjunctus Selys
(pp. 30, 31, 37; photos 3a, 3b)

Size. Total length: 36–46 mm; abdomen: 28–36 mm; hindwing: 18–25 mm.

Pond Spreadwings (*Lestes*)

Plateau
(*alacer*)

Spotted
(*congener*)

Common
(*disjunctus*)

Emerald
(*dryas*)

Rainpool
(*forficula*)

Elegant
(*inaequalis*)

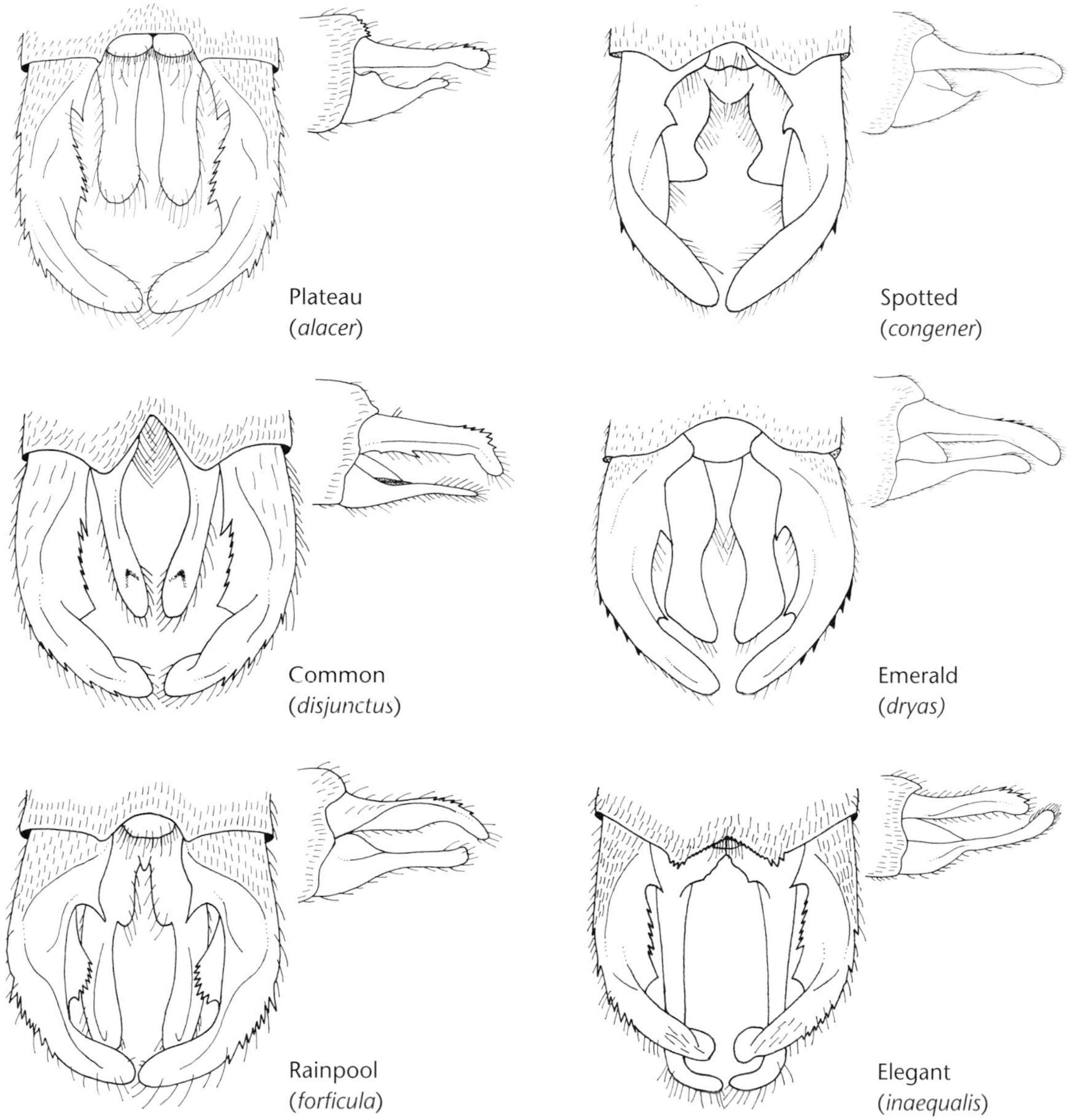

Fig. 9. Pond Spreadwings (*Lestes*): male caudal appendages; larger images are dorsal view, smaller are lateral view (pp. 35–40).

Regional Distribution. *Biotic Provinces:* Austroriparian, Balconian, Carolinian, Chihuahuan, Coloradan, Kansan, Navahonian, Tamaulipan, Texan. *Watersheds:* Arkansas, Bayou Bartholomew, Brazos, Canadian, Cimarron, Colorado, Guadalupe, Lower Rio Grande, Mississippi, Neches, Nueces, Ouachita, Pecos, Red, Sabine, San Antonio, San Jacinto, St. Francis, Trinity, Upper Rio Grande, White.

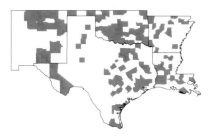

General Distribution. Throughout the United States and Canada.

Flight Season. Mar. 14 (TX)–Dec. 4 (LA).

Identification. This is the most widespread spreadwing in our region, occurring in all major watersheds and biotic provinces. The face and eyes of males are bright blue, with a small, pale spot lateral to each posterior ocellus. A pale blue-green stripe on the mesopleural suture extends the full length of the suture, and is confluent with the pale areas of the metepisternum at the border of the mesepimeron. The rest of the pterothorax varies from pale to black. Legs are dark brown. Older individuals become heavily pruinose ventrolaterally, the entire pterothorax occasionally becoming obscured (more commonly in northern populations). The abdomen is largely dark with a metallic-green luster dorsally. There are distinct dark ventrolateral spots on segments 6 and 7. The spots may also be present on segments 3–5, but they are generally less defined. Heavy pruinosity develops laterally on segments 1 and 2, ventrolaterally on segments 7 and 8, and completely on segments 9 and 10. The distal medial tooth of the cerci is acute, blunt, and distinctly smaller than the basal tooth. Paraprocts are nearly as long as cerci.

Females are generally similar to males; the pale stripe running along the mesopleural suture is complete and confluent with the metathoracic area. Females lack pruinosity on the abdomen, but the rear of the head becomes pruinose, along with the coxae and the ventrolateral margins of the pterothorax. Laterally, abdominal segments 7–10 are uniformly pale yellow, except for a ventrolateral rim of segment 9. The posterolateral corner of the basal plate of the ovipositor is produced to form a distinct acute tooth, longer than its basal width. The ovipositor reaches well beyond the margin of segment 10, but does not reach the tips of the paraprocts.

Similar Species. Common Spreadwing is smaller than most other spreadwings in the region. Females are similar to those of Slender Spreadwing (*L. rectangularis*) and may be easily mistaken for them, but for the larger size, more robust and longer abdomen, yellow tarsi, and tinted wingtips of Slender Spreadwing. The middorsum of the thorax is generally darker than in Plateau (*L. alacer*) and Rainpool (*L. forficula*) Spreadwings. The abdomen of Lyre-tipped Spreadwing (*L. unguiculatus*) is metallic green.

Habitat. Still, slow-moving waters, including permanent or ephemeral ponds, marshes, and lakes with moderate vegetation.

Discussion. This species has received considerable attention by workers, owing to both its abundance and its two distinct forms. The subspecies *Lestes disjunctus australis* Walker is found everywhere within the region except the extreme western edge, where it is replaced by the northern nominate form, *L. d. disjunctus* Selys. The nominate subspecies generally does not extend south of north-central New Mexico, but it can be separated from *L. d. australis* by a darker color pattern, smaller size (33–40 mm), and a more strongly developed distal tooth on the cerci. The fact that both of these forms are found together in some localities may imply that they are distinct species.

Common Spreadwing is colored like the more northern Sweetflag Spreadwing (*L. forcipatus*), and their caudal appendages closely resemble one another. Early records for both of these species must be viewed with caution as a result.

Common Spreadwing emerges early in the south-central United States, mid-March, and flies the rest of the year. This species does, however, seem to be most abundant in the fall. Males are not territorial, and individuals are often seen a considerable distance, several hundred meters, from any body of water. Mating activity in this species tends to peak in the late afternoon, ca. 5 P.M. Egg-laying occurs in tandem in the green stems of cattails and similar plants, above the waterline. Females will usually lay over a hundred eggs. The larvae can tolerate considerable salinity and may be a common inhabitant of saline lakes. A population study in eastern Ontario,

Canada, on sexual size dimorphism and sex-specific survival in adults, found that there was no difference in the mass of mated and unmated males, but that females were more than 50% heavier than males. It was also determined that males were eight times more abundant than females, but females were more active than males.

References. Anholt (1997), Bick (1957), Bick and Bick (1957, 1961), Bird (1932a), Cannings et al. (1980), Dunkle (1990), Ingram (1976), Sawchyn and Gillott (1974).

Emerald Spreadwing
Lestes dryas Kirby
(pp. 30, 31, 37; photo 2f)

Size. Total length: 32–40 mm; abdomen: 24–30 mm; hindwing: 19–25 mm.
Regional Distribution. *Biotic Provinces:* Coloradan, Navahonian. *Watersheds:* Colorado (NM), Pecos, Rio Grande.

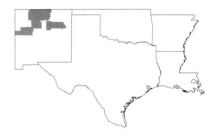

General Distribution. Throughout Canada and northern United States, extending south to California, eastward to New Mexico; also Europe and Asia.
Flight Season. Aug. (NM).
Identification. The head of the male, including the rear, is dark metallic green. The thorax is metallic green dorsally, occasionally with a narrow pale-yellow line along the mesopleural suture. The metepisternum is largely pale yellow. Dark spots may be present above and/or below the metapleural carina. The abdomen is metallic green dorsally. The posterior portion of the thorax, abdominal segment 1, lateral portions of 2, ventrolateral portions of 7 and 8, and all of 9 and 10 become pruinose in older individuals.

The female is generally similar to the male but paler, and the metallic-green areas are less extensive. The ovipositor is long, its valves extending to or beyond the posterior margin of the paraprocts.

The posterolateral corner of the ovipositor basal plate is acutely angulate, forming a distinctive tooth.
Similar Species. Emerald Spreadwing is smaller and stockier than most other spreadwing species. It is the only predominantly metallic-green species with dark marks posterior to the legs on the underside of the thorax. Both Elegant (*L. inaequalis*) and Swamp (*L. vigilax*) Spreadwings are substantially larger.
Habitat. Temporary ponds and pools and permanent waters with sufficient shade.
Discussion. Emerald Spreadwing is found in northern latitudes worldwide. This species is generally present early in the year, because it must complete its life cycle in the temporary pools in which it normally breeds before they dry up at the end of the summer. Eggs are laid in tandem in emergent plants, high above the waterline.

References. Sawchyn and Gillott (1974).

Rainpool Spreadwing
Lestes forficula Rambur
(pp. 30, 31, 37; photo 3c)

Size. Total length: 35–43 mm; abdomen: 28–36 mm; hindwing: 17–24 mm.
Regional Distribution. *Biotic Provinces:* Austroriparian, Balconian, Tamaulipan, Texan. *Watersheds:* Brazos, Lower Rio Grande, Nueces, San Antonio, San Jacinto, Trinity.

General Distribution. Texas, Mexico, and south through Central America and West Indies to Argentina and Brazil.
Flight Season. May 7 (TX)–Dec. 29 (TX).
Identification. This tropical species extends north into Texas. The middorsal carina is pale yellow, bordered on each side by a thin black line. The mesepisternum is pale blue to yellow with a thin metallic-green antehumeral stripe extending much less than

1/2 the width of the sclerite. The stripe is the same width for nearly its entire length, widening only slightly posteriorly; it lies closer to the middorsal carina but is widely separated from it and from the humeral suture. A second, thinner metallic-green stripe runs nearly the full length of the mesepimeron. Teneral individuals have dark spots above the anterior end of the metapleural carina on the otherwise pale-yellow thorax. Dark stripes develop along both the interpleural and metapleural sutures and expand with age. The ventral side of the pterothorax in mature individuals becomes heavily pruinose. Abdominal segment 1 is black, except for a small, pale, ventrolateral spot. Segments 2–6 are dark metallic green dorsally, expanding slightly subapically on segments 3–5. Segment 6 and beyond are considerably darker ventrolaterally, becoming even darker with age. Segments 1, 8–10, and the lateral part of 2 all become heavily pruinose. The paraprocts ar nearly as long as the cerci.

The general coloration of the head and the pterothorax in the female is similar to that of the male. The mesothorax is pale yellow, with a dark hairline and a posterior spot on the humeral suture. The thin metallic-green, antehumeral stripe is nearly absent in teneral individuals. There are dark hairline stripes along the interpleural and metapleural sutures, and a slight pruinosity develops with age on the coxae and metathorax. Abdominal segment 1 varies in color from blue to black. The middle segments, 2–6, are similar to those of the male, and segments 7–10 are dark brown dorsally, becoming paler laterally. The basal plate of the ovipositor is produced into a long, acuminate tooth.

Similar Species. Rainpool Spreadwing is the only spreadwing with a metallic-green antehumeral stripe and dark (not metallic-green) abdomen. The antehumeral stripes can become obscured in older males.

Habitat. Ponds, pools, other standing bodies of water; possibly slow reaches of streams, with heavy emergent vegetation.

Discussion. Females lay eggs both accompanied and unaccompanied by males, in sedges 20–25 cm above the water surface. This species probably flies year-round farther south in its range, and has been taken there in a variety of habitats, including rain pools, rivers, and sewage ponds.

References. Garcia-Diaz (1938), Paulson (1984).

Elegant Spreadwing
Lestes inaequalis Walsh
(pp. 11, 30, 31, 37; photos 3d, 3e)

Size. Total length: 45–60 mm; abdomen: 35–47 mm; hindwing: 25–31.

Regional Distribution. *Biotic Provinces:* Austroriparian, Carolinian. *Watersheds:* Arkansas, Canadian, Cimarron, Mississippi, Neches, Ouachita, Red, Sabine, St. Francis, White.

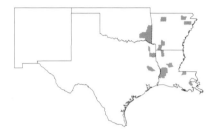

General Distribution. Florida north to Ontario; west to Minnesota and south to Texas.

Flight Season. Apr. 7 (LA)–Aug. 18 (LA).

Identification. This is a large eastern species that is confined to localized populations in the piney woods of the Austroriparian biotic province within our region. Males have a dark metallic-green head and generally pale spots lateral to each ocellus. The rear of the head is yellow, and the eyes are blue above and greenish yellow below. The pterothorax is metallic green, except for the black on the middorsal and antealar carinae and a thin line along the humeral suture. The mesepimeron is metallic green for its full width posteriorly, narrowing to 1/2 this width anteriorly. The pterothorax, yellow ventrolaterally, may become entirely pruinose with maturity, but the color pattern is never completely obscured. Abdominal segments 1–8 are metallic green dorsally, pale yellow laterally. The lateral areas on segments 8–10 vary from yellow to black, and the lateral parts of segment 1 and all of 9 and 10 become heavily pruinose with age. The males are unique among our species in having paraprocts longer than the cerci.

The coloration of the females is similar to that of males, but the metallic green on the pterothorax may be partly replaced by brown with bronze reflections. The middorsal carina and the narrow stripe along the humeral suture are always pale. The eyes are distinctly brown above, gradually becoming yellowish ventrally. The color pattern of the abdomen is like that of the male, with the following exceptions: segment 1 is metallic green dorsally on the apical 2/3 only; segments 7, 8, and 10 are pale laterally; and segment 9 may or may not have a basal black band and dark lateral markings. The posterolateral margin of the ovipositor basal plate is truncate and lacks a tooth.

Similar Species. Its large size and bright-green metallic color make Elegant Spreadwing easily recognizable in the field. It is similar only to Swamp Spreadwing (*L. vigilax*), which has rusty-brown antehumeral stripes, and the back of its head is black. Emerald Spreadwing (*L. dryas*) is substantially smaller.

Habitat. Canopy-covered permanent ponds, lakes, slow-moving streams, and marshes with plenty of emergent vegetation and heavily wooded shorelines.

Discussion. The reproductive behavior of these damselflies has never been reported. The diet of this large species includes smaller damselflies. They are easily disturbed and are generally found perching in shady areas during the heat of the day. They have the unique behavior of laying eggs in tandem in the upper surface of lily pads.

References. Dunkle (1990).

Slender Spreadwing
Lestes rectangularis Say
(pp. 30, 31, 42; photo 3f)

Size. Total length: 37–52 mm; abdomen: 30–44 mm; hindwing: 20–25 mm.

Regional Distribution. *Biotic Provinces:* Austroriparian, Carolinian, Kansan, Texan. *Watersheds:* Arkansas, Canadian, Cimarron, Mississippi, Red.

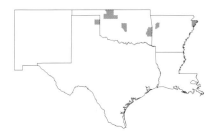

General Distribution. Florida north to Ontario, west to North Dakota, and south to Oklahoma and Texas.

Flight Season. May 24 (TX)–Aug. 1 (OK).

Identification. One of the most common damselflies in the northeast, Slender Spreadwing is found only in the northern parts of our region. It has an unusually long abdomen and distinctive pale-blue antehumeral stripes. The eyes of the male are blue. The rear of the head is black, becoming pruinose at maturity. The antehumeral area is black, except for a pale-blue middorsal and antealar carina. A pale-blue stripe on the lateral 1/4 of the mesepisterna extends onto the dorsal edge of the mesepimeron. The rest of the mesepimeron is black. The remaining lateral and ventral areas of the pterothorax are pale yellow, becoming slightly pruinose with age. Each femur has a black line running its length on the outside. The tibiae are nearly all yellow, with a black line apparent only anteriorly. Abdominal segments 1 and 2 are black dorsally and pale yellow or blue laterally. A dark dorsal line on segment 2 narrows at its middle to form two distinct spots, but this can be obscured in older individuals. A dorsal stripe on segments 3–7 is lighter, more tan or brown, and only slightly darker than the lateral pale areas. Segments 8–10 are black dorsally and pale yellow laterally. Segments 9 and 10 darken with age, so that only the pale-yellow apical and basal rings remain. This is the only spreadwing in our region with paraprocts distinctly and strongly curved downward.

The color pattern of the female is similar to that of the male, but the abdomen is uniformly black dorsally and tan laterally. The black of the antehumeral and metepisternal areas is less extensive than in that of males. Segments 9 and 10 are pale laterally, becoming darker ventrally. The posterolateral corner of the ovipositors' basal plate is acute, generally forming a distinctive tooth.

Similar Species. Males are easily recognized by a combination of an unusually long abdomen, blue antehumeral stripe, and paraprocts curving distinctly downward. Females are similar to females of Common Spreadwing (*L. disjunctus*), but have pale tarsi and a slightly longer abdomen.

Habitat. Lakes or ponds with regular shade and dense emergent vegetation; often found in bays and sand-bottomed lakes.

Discussion. Slender Spreadwing is relatively uncommon in the south-central United States. It was only recently discovered in the Texas Panhandle. In a western Pennsylvania population, more than 50%

Pond Spreadwings (*Lestes*)

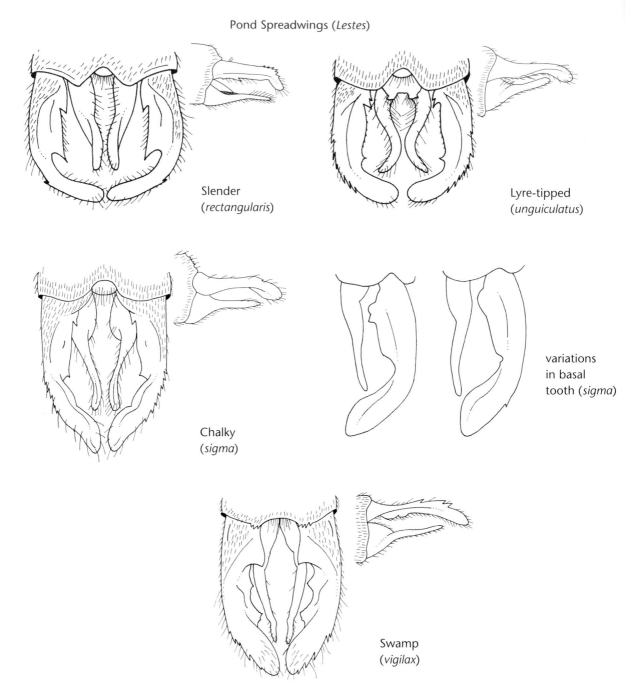

Slender
(*rectangularis*)

Lyre-tipped
(*unguiculatus*)

Chalky
(*sigma*)

variations
in basal
tooth (*sigma*)

Swamp
(*vigilax*)

Fig. 10. Pond Spreadwings (*Lestes*): male caudal
appendages; larger images are dorsal view, smaller are
lateral view (pp. 41–44).

of the individuals emerged during the first seven days of a 2–3-week emergence period. Individuals were most active in midafternoon, and females will lay eggs in tandem or alone, usually in cattail leaves. They are reluctant to fly over open water, and the female never submerges herself during egg-laying. Adults are most abundant in shade, and readily take shelter in thick vegetation during the heat of the day.

References. Abbott et al. (2003), Gower and Kormondy (1963), Walker (1953).

Chalky Spreadwing
Lestes sigma Calvert
(pp. 30, 31, 42; photos 4a, 4b)

Size. Total length: 39–47 mm; abdomen: 31–37 mm; hindwing: 20–25 mm.
Regional Distribution. *Biotic Provinces:* Balconian, Tamaulipan, Texan. *Watersheds:* Brazos, Colorado, Guadalupe, Lower Rio Grande, Nueces, Red, San Antonio.

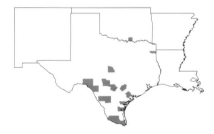

General Distribution. A southern species found in Oklahoma and Texas south through Mexico to Costa Rica.
Flight Season. May 9 (TX)–Sept. 7 (TX).
Identification. This Mexican species ranges as far north as the Red River. Males have extensive blue on the face. In some individuals, the spots lateral to the posterior ocelli may be expanded into triangles. The rear of the head is yellow in young individuals, quickly becoming pruinose and black with maturity. The eyes are blue dorsally, fading to paler white below. The pterothorax is gray to tan with extensive black markings, including an elongated antehumeral spot just above the humeral suture. The middorsal and antealar carinae are often black. There is a dark irregular stripe 1/2 the length of the mese-

pimeron just below the humeral suture. The dark stripe along the interpleural suture is often extensive, covering most of the metepisternum. There are two black spots, which may be connected, above the anterior and posterior ends of the metapleural carina, and another large spot below the carina; the black may become more extensive with age, eventually covering the entire pterothorax. The pruinescence may extend down to and include the legs. The pterostigma may be bicolored, dark brown, or black anteriorly and yellow posteriorly. The abdominal segments 3–7 are dark dorsally and confluent with ventrolateral markings. Segments 8–10 are completely black, except for lighter-brown posterodorsal and lateral spots on 10. Segments 1 and 2 and 8–10 become heavily pruinose, 10 remaining lighter than the others. The paraprocts are distinctly sigmoid in form.

The general color pattern of the female is like that of the male, but the black and the pruinescence are less extensive. Mature individuals become dark with age. The eyes are duller than in the male. The black on the dorsum of abdominal segment 1 encompasses only the apical 1/2 of the segment. The black on ventrolateral parts of segments 3–5 is generally not confluent with the dorsal stripe. Segments 1 and 2 and 8–10 become heavily pruinose with age. The basal plate of the ovipositor has a distinct, acuminate, posterolateral tooth.
Similar Species. Older males of Rainpool Spreadwing (*L. forficula*) are similar to Chalky Spreadwing, but the latter has distinctly sigmoid paraprocts and lacks a metallic-green antehumeral stripe. Lyre-tipped Spreadwing (*L. unguiculatus*) males also have sigmoid paraprocts, but are smaller in size and more northerly distributed. Lyre-tipped Spreadwing also lacks dark marks posterior to the legs on the underside of the thorax.
Habitat. Temporary pools and ponds.
Discussion. This species has been erroneously reported from New Mexico, but this species does not reach that far northwest. Marshall County, Oklahoma, is the farthest north this species has been documented, and the only record north of the Red River. Variation in the caudal appendages of this damselfly is equal to such differences across other spreadwing species. The basal tooth on the cerci can take the form of a distinct pointed tooth or a rounded lobe.

References. Bick (1978), Johnson (1975a).

Lyre-tipped Spreadwing
Lestes unguiculatus Hagen
(pp. 30, 31, 42; photos 4c, 4d)

Size. Total length: 31–44 mm; abdomen: 25–35 mm; hindwing: 17–24 mm.

Regional Distribution. *Biotic Provinces:* Austroriparian, Carolinian, Kansan, Texan. *Watersheds:* Arkansas, Mississippi, Red.

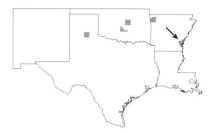

General Distribution. West Virginia north to Nova Scotia, west to Pacific Northwest and British Columbia, south to Oklahoma and Texas.

Flight Season. June (TX).

Identification. This chiefly northern species is found commonly throughout the United States and southern Canada. The head of the male is black, with a metallic-copper or -green luster that is generally obscured with age. The middorsal and antealar carinae are pale. There is a thin pale-yellow or green antehumeral stripe extending 2/3 to 4/5 the length of the humeral suture. The mesepimeron is mostly black. The yellow or pale-green metepisternum is divided into an anterior spot and a posterior stripe by a black diagonal stripe. The metepimeron is pale in young individuals, darkening with age. Distinct black lines are present on the outside of the femora and tibiae. The abdomen, relatively short among the spreadwings, is dark metallic green dorsally and pale yellow or blue ventrolaterally. The dark ventrolateral markings are faint on segments 5 and 6, becoming more pronounced posteriorly. The tip of the abdomen becomes pruinose with age.

The general color pattern of the female is similar to that of the male, but paler throughout. The pale-yellow antehumeral stripe extends the full length of the humeral suture. There are dark triangular markings on the posterodorsal corners of the metepisternum on the otherwise pale metathorax. The color pattern of the abdomen is similar to that of males, but the tip does not become pruinose. There is a distinct posterolateral tooth on the basal plate of the ovipositor.

Similar Species. Chalky Spreadwing (*L. sigma*) is more southerly distributed and has a pair of dark marks ventrally on the thorax, posterior to the hind legs. The abdomen of Common Spreadwing (*L. disjunctus*) is not metallic green.

Habitat. Open pools, ponds, sloughs, and slow reaches of streams.

Discussion. This species is scarce within the region, barely entering the northern limits. Tinkham (1934) reported three females from the Davis Mountains in west Texas, and Albright (1952) listed the species as a "record furnished by a letter from A.H. Ferguson" in an unpublished thesis on the Odonata surrounding San Antonio. The only verifiable records of Lyre-tipped Spreadwing in Texas are from Caprock Canyons State Park, Briscoe County.

In one study it was noted that unpaired males shift perch sites for no detectable reason about once every minute. Males infrequently wing-warned as they flew toward intruders, but a lack of aggressiveness resulted in loss of territory. Mating occurred in the early afternoon, between 1:30 and 3:00, and involved no courtship or display signals. Mating lasted an average of 25 minutes, but was never a continuous process, each pair momentarily breaking contact. Egg-laying generally occurs in tandem, but may occur alone, and lasts an average of 1.5 hours. Pairs generally oviposited in vegetation, 25–30 cm above the water surface, as is typical in pond spreadwings. Females, however, may submerge themselves underwater for short periods.

References. Bick and Hornuff (1965), Sawchyn and Gillott (1974).

Swamp Spreadwing
Lestes vigilax Hagen
(pp. 12, 30, 31, 42; photos 4e, 4f)

Size. Total length: 43–55 mm; abdomen: 36–45 mm; hindwing: 23–27 mm.

Regional Distribution. *Biotic Province:* Austroriparian. *Watersheds:* Brazos, Canadian, Cimarron, Mississippi, Neches, Ouachita, Red, Sabine, San Jacinto, Trinity.

General Distribution. Southeastern Canada south to Florida, west to Oklahoma and Texas, and north to Minnesota.

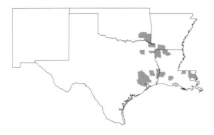

Flight Season. Mar. 31 (LA)–Oct. 31 (TX).

Identification. This large, metallic species is restricted to the eastern reaches of the region. The face of the male is pale blue in front, contrasting with green sides. The top of the head is dark metallic green or black, and the eyes are brilliant blue. The thorax and abdomen are dark metallic green or black, and there is an irregular pale-brown humeral stripe on the thorax. The middorsal and antealar carinae are both pale. The mesepimeron is dark, except for the pale anteroventral corner. The metepisternum and remainder of the pterothorax are pale yellow ventrally. The legs are dark externally. The abdomen is dark metallic green dorsally and pale yellow ventrally. In most individuals there is a fine pale line running dorsally down abdominal segments 2–5. Ventrolateral markings are absent, but the pattern expands subapically on segments 6–10.

In older individuals, pruinosity develops on segments 1 and 9 and the basal portion of 10. The paraprocts are thin and nearly as long as the cerci.

Females are similar to males, but the eyes are brown. The pale humeral stripe is wider and always runs the full length of the suture. The dark areas of the thorax are bronze or brown. The ventrolateral areas of the pterothorax are pale yellow. Abdominal segments 8–10 are pale laterally. There is a distinct posterolateral tooth on the basal plate of the ovipositor.

Similar Species. Elegant Spreadwing (*L. inaequalis*) is slightly larger than Swamp Spreadwing and has a brighter-green abdomen, and the sides of the thorax are more yellow. The rear of the head is black in Swamp Spreadwing and yellow in Elegant Spreadwing. Emerald Spreadwing (*L. dryas*) is substantially smaller.

Habitat. Generally found in shaded acidic waters such as bogs, lakes, swamps, oxbows, and slow streams. One study noted that this species prefers heavily shaded areas.

Discussion. Much of the biology and reproductive behavior of this species still remain unknown. Females lay eggs in tandem in emergent vegetation and never in submergent plants.

References. O'Briant (1972), Wright (1943b).

THREADTAILS
(Family Protoneuridae)

These slender damselflies are circumtropical in distribution. In the Northern Hemisphere, they are mainly restricted to Middle America. Only three species, in two genera, occur in North America north of Mexico, and all are limited to southern Texas. Adults and larvae generally resemble those of the closely related Pond Damsels. Threadtails may be readily recognized by a characteristically reduced venation and a thin, elongated abdomen. Males of our species are distinctively orange or red. Members of this family have a strongly rectangular and elongated quadrangle in each wing. The reduced venation results in the anal vein and Cu_2 being absent in some species. The area between Cu and the wing margin is devoid of crossveins. The pterostigma is rather short, generally subtending a single cell.

KEY TO THE GENERA OF THREADTAILS (PROTONEURIDAE)

1. Front of thorax mostly orange (males) or pale (females), but not black — **Robust Threadtails (*Neoneura*)**

1'. Front of thorax mostly black with a narrow red-orange or pale humeral stripe — **Orange-striped Threadtail (*Protoneura cara*)**

Robust Threadtails
Genus *Neoneura* Leach
(pp. 13, 48)

Robust threadtails are a large group of generally brightly colored damselflies. Two species occur in southern Texas. Because of their relatively thick abdomens, members of this genus resemble pond damsels more closely than do other threadtails. Larvae of this group are restricted to running waters, where they may be associated with leaf litter or cling to rocks. Adults prefer shaded areas of floating or emergent vegetation near the larval habitat.

References. Garrison (1999), Williamson (1917).

KEY TO THE SPECIES OF ROBUST THREADTAILS (*NEONEURA*)

1. Male abdominal segments 2 and 3 pale to bright orange; posterior margin of pronotum in female bears a symmetrical lobe on each side of the median lobe — **Amelia's (*amelia*)**

1'. Male abdominal segments 2 and 3 black dorsally; posterior margin of pronotum in female bears a median lobe only (lateral lobes, if present, are much reduced) — **Coral-fronted (*aaroni*)**

Coral-fronted Threadtail
Neoneura aaroni Calvert
(p. 48, photos 5a, 5b)

Size. Total length: 30–37 mm; abdomen: 23–30 mm; hindwing: 16–19 mm.
Regional Distribution. *Biotic Provinces:* Balconian, Tamaulipan. *Watersheds:* Colorado, Guadalupe, Lower Rio Grande, Nueces, San Antonio.

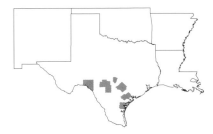

General Distribution. Texas; Nuevo León, Mexico.
Flight Season. May 17 (TX)–Sept. 17 (TX).
Identification. The face in mature males is bright orange. The rest of the head is orange and black. The middorsal thoracic carina and stripe are black, and the orange antehumeral stripe usually reaches to the anterior 1/2 of the humeral suture. The rest of the pterothorax is black with some less-defined paler areas. The legs are light brown, with dark stripes on the femora and tibiae. The thorax, including the legs, becomes pruinose in older males. The abdomen of teneral males is light brown, with dark apical bands on segments 3–6. Segments 9 and 10 are nearly all black dorsally. The abdomen is solid black in older males, with pale apical bands persisting in some individuals on segments 1 and 2 and 9 and 10. The abdomen becomes lightly pruinose with age. The paraprocts are notched apically, the ventral lobe distinctly hooked when viewed laterally. The cerci are longer and rounded apically.

The head and thorax of the female are pale brown. The middorsal carina is pale, with a fragmented dark middorsal stripe. The posterior tubercles and medial borders of the mesostigmal plates are both black. The hairline black humeral stripe is generally broken or lacking at its middle. The legs are pale. The abdomen is dark dorsally, with subapical bands and lateral stripes on the middle segments. Segments 8 and 9 and sometimes 10 are dark laterally.
Similar Species. Male and female Orange-striped Threadtails (*Protoneura cara*) have more black on

the thorax than orange. Male Amelia's Threadtail (*N. amelia*) generally has some orange dorsally on abdominal segments 1–3. The female of Amelia's Threadtail is nearly identical to the female Coral-fronted Threadtail, but Coral-fronted has darker markings on the thorax and abdomen, and Amelia's has a pair of dark spots on the posterior margin of the middle lobe of the pronotum. Orange Bluet (*Enallagma signatum*) is similar, but the orange on the face and thorax is not as bright or extensive.
Habitat. Protected areas of slow-moving rivers and streams with emergent or floating vegetation or detritus.
Discussion. In our region, Coral-fronted Threadtail has been collected from only a few widely dispersed Texas counties, spanning the Hill Country west to the Devils River and south to the Gulf Coast. It has been infrequently taken farther south.

References. Bick (1983), Calvert (1901–08).

Amelia's Threadtail
Neoneura amelia Calvert
(p. 48, photos 5c, 5d)

Size. Total length: 30–35 mm; abdomen: 24–29 mm; hindwing: 16–18 mm.
Regional Distribution. *Biotic Province:* Tamaulipan. *Watershed:* Lower Rio Grande.

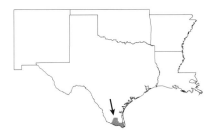

General Distribution. Texas; south through Mexico to Panama.
Flight Season. Apr. 25 (TX)–Nov. 5 (TX).
Identification. The eyes are brown above, fading to pale green below. The face and head are bright red, the area posterior to the ocelli black. The middorsal stripe and carina are black, and the antehumeral stripe is orange. A dark humeral stripe covers only the posterior 2/3 of the respective suture. The remaining pterothorax varies from orange to black laterally. A black stripe is present only on the anteri-

Orange-fronted Threadtails (*Neoneura*)

Coral-fronted (*aaroni*)   Amelia's (*amelia*)

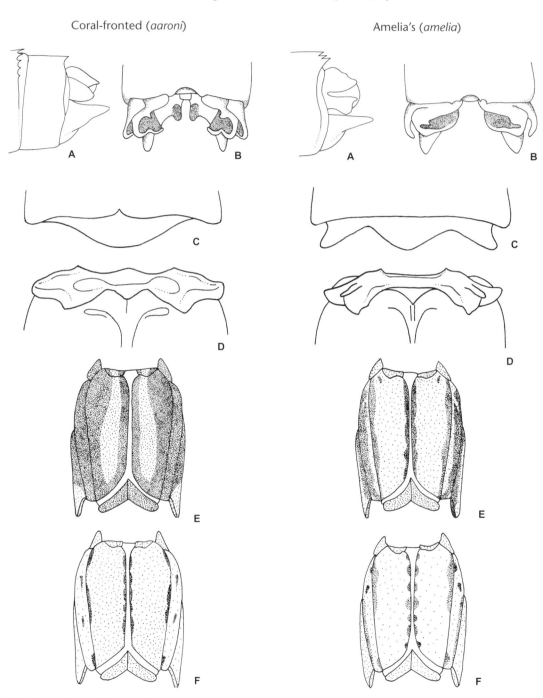

Fig. 11. Orange-fronted Threadtails (*Neoneura*): (A) lateral view of male caudal appendages; (B) dorsal view of male caudal appendages; (C) posterior pronotal margin of female: (D) mesostigmal plate of female; (E) dorsal thoracic pattern of male; and (F) dorsal thoracic pattern of female (pp. 46–49).

or mesepimeron. The metapleural suture and venter of the pterothorax vary from tan to black. The abdominal segments 1 and 2 are variable, but generally show orange dorsally. The rest of the abdomen is black, with narrow, pale apical bands on segments 3–9. Segments 1–6 are pale ventrolaterally. The cerci are deeply bifurcated.

The head and body of the female are tan in color, with only thin black markings. The middle lobe of pronotum bears a black spot on each side at the posterior margin. The abdomen is darker dorsally, with dark subapical bands and lateral stripes on the middle segments. Segments 8–10 are pale laterally.

Similar Species. Male and female Orange-striped Threadtail (*Protoneura cara*) have more black than orange on the thorax. Male Coral-fronted Threadtail (*N. aaroni*) lacks orange dorsally on abdominal segments 2 and 3. Female Coral-fronted Threadtail is nearly identical to Amelia's Threadtail, but Amelia's is generally paler on the thorax and abdomen, especially on the terminal segments, and Coral-fronted lacks the posterior markings on the middle lobe of the pronotum. Orange Bluet (*Enallagma signatum*) is similar, but the orange on the face and thorax is not as bright or extensive.

Habitat. Prefers protected, well-shaded areas of slow-moving rivers and streams with emergent or floating vegetation, detritus, or debris.

Discussion. Amelia's Threadtail is found throughout Middle America. It was only recently reported from the United States, at the southern border in Hidalgo County, Texas. It is found among floating debris in shady areas along the Rio Grande. Adults also perch on emergent vegetation of clear lakes in the immediate vicinity of stream outlets. Males are variable in the degree of black on the head, thorax, and first three abdominal segments, and I have seen individuals lacking any evidence of orange on the initial abdominal segments.

References. Westfall and May (1996), Williamson (1917).

Slender Threadtails
Genus *Protoneura* Selys

A single species of this genus ranges as far north as the Hill Country of central Texas. The males are readily distinguished in the field by an overall slender form and a long, thin, characteristically ringed abdomen. Most species, including those in our region, are brightly colored. The females are more robust and duller in color. The wings lack an anal vein beyond the anal crossing, a feature unique among damselflies in the south-central United States. The legs are armed with short tibial spurs.

Members of this genus inhabit small, slow streams, ditches, and seepages, as well as larger streams, and are often found on ponds or sheltered lakeshores with abundant litter or submerged vegetation.

References. Donnelly (1989), Williamson (1915).

Orange-striped Threadtail
Protoneura cara Calvert
(pp. 13, 50; photos 5e, 5f)

Size. Total length: 34–38 mm; abdomen: 27–32 mm; hindwing: 16–19 mm.

Regional Distribution. *Biotic Provinces:* Balconian, Tamaulipan. *Watersheds:* Colorado, Guadalupe, Lower Rio Grande, Nueces, Pecos, San Antonio.

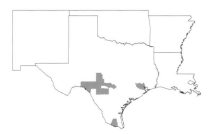

General Distribution. Texas; south through Mexico to Guatemala and Honduras.

Flight Season. Jun. 19 (TX)–Oct. 14 (TX).

Identification. The face of the male is bright orange, with a dark, black central spot. The top of the head is black, with orange spots lateral to each antenna. The antehumeral area is mostly black with a metallic luster. The middorsal carina and 2/3 of the antealar carina are orange. The mesepimeron is black with orange margins, and the mesopleural suture is

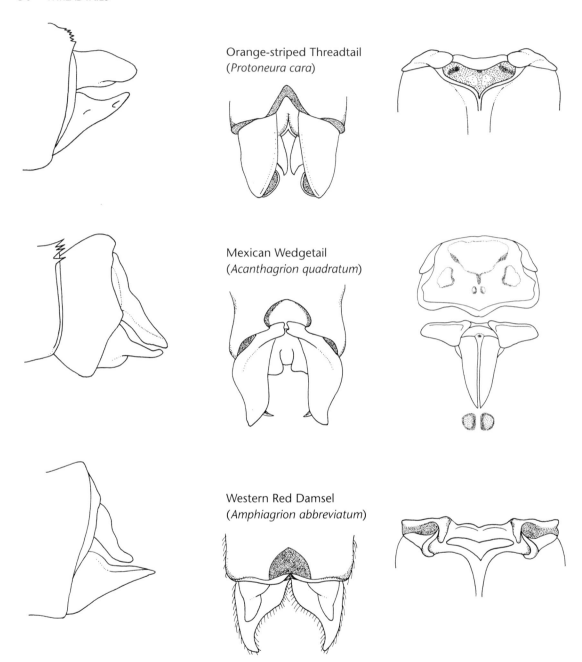

Orange-striped Threadtail
(*Protoneura cara*)

Mexican Wedgetail
(*Acanthagrion quadratum*)

Western Red Damsel
(*Amphiagrion abbreviatum*)

Fig. 12. Species in three genera, showing for each species, from left to right: lateral view of male caudal appendages; dorsal view of male caudal appendages; and mesostigmal plates of female (including prothorax for Mexican Wedgetail) (pp. 49–56).

traversed by a thin black line bordered by a wider orange stripe laterally. The legs are pale, darkening slightly at their apices. The first abdominal segment is black with a narrow, pale apical band. Segment 2 is pale orange, and segments 3–7 are dark dorsally, each with a pale basal ring. Segments 8 and 9 are black, each with pale ventrolateral margins and apical rings. Segment 10 is black, except for a pale band posterolaterally.

The female is stockier, but otherwise similar to the male. The abdomen differs only slightly in coloration. The dorsum of segments 1 and 2 and 10 are nearly all black, and the dark areas of the middle segments are expanded subapically.

Similar Species. Male Coral-fronted Threadtails (*N. aaroni*) and Amelia's Threadtails (*N. amelia*) have more orange on the face and top of the thorax. Orange Bluet (*Enallagma signatum*) is similar, but overall is paler, and the abdomen is not as long. Cherry Bluet (*E. concisum*) is deep red and does not overlap in distribution.

Habitat. Well-shaded, slow-moving streams with ample leaf litter and debris.

Discussion. This is a relatively uncommon species that has been documented only in a few counties within the Hill Country of central Texas and the lower Rio Grande Valley. A photograph of an ovipositing pair was taken at the Houston Arboretum in 1988. This pair was probably an accidental introduction with aquatic plants (R. Orr, pers. comm.). They have not been seen there since. Egg-laying typically occurs in tandem near the margin of slow-moving water, in floating and submerged vegetation and debris. The larva has not been discovered, but to judge from observations of related species, it may live in the leaf litter of small trickles and pools.

POND DAMSELS
(Family Coenagrionidae)

Pond damsels are found worldwide, and as the most diverse family of damselflies, they account for nearly 50% of the species belonging to the suborder. They are the dominant damselfly family both in North America and in the south-central United States. Pond damsels are generally not metallic, but males may be marked with bright blues, greens, yellows, oranges, or reds. Most are small and dainty and can be numerous in appropriate habitats. Many are pond inhabitants, but several, including most dancers (*Argia*), are found on streams. Many species represent a challenge to identification in the field because they are so similar to one another. Females are often dichromatic (having two color forms), one an andromorphic form (looking like the male), the other a gynomorphic form (differing from the male in both color and pattern). To identify many of these species accurately, careful examination in the hand may be necessary.

Pond damsels have clear petiolate wings that are held closed over the abdomen when at rest. There are only two antenodal crossveins, and the postnodal crossveins are generally in line with those below them. The veins M_3 and Rs originate much closer to the nodus than to the arculus. The pterostigma is short and generally surmounts only one or two cells. The legs are shorter than those of other damselfly families in North America, except perhaps the threadtails (Protoneuridae). The male paraprocts are shorter than abdominal segment 10 and are neither forceps-like nor pincers-like.

References. Bick (1972), Bick and Bick (1980).

KEY TO THE GENERA OF POND DAMSELS (COENAGRIONIDAE)

1. Spurs on 2nd and 3rd tibiae long, twice the length of the intervening spaces, at least proximally ... 2

1'. Spurs on 2nd and 3rd tibiae barely longer, at most, than the intervening spaces ... 3

2(1). Dorsum of thorax and abdomen metallic green to bronze, with some blue on abdominal segments 8–10 ... **Sprites (*Nehalennia*) (in part)**

2'. Dorsum of thorax and abdomen not metallic green ... **Dancers (*Argia*)**

3(1'). Prominent, moundlike ventral thoracic tubercle bearing numerous long, stiff setae; wings nearly equal in length to abdomen; robust red-and-black species ... **Western Red Damsel (*Amphiagrion abbreviatum*)**

3'. No prominent ventral thoracic tubercle; wings at most 3/4 the length of the abdomen; more slender species, usually not red and black ... 4

4(3'). Dorsum of thorax and abdomen metallic green to bronze with some blue on abdominal segments 8–10 ... **Sprites (*Nehalennia*) (in part)**

KEY TO THE GENERA OF POND DAMSELS (COENAGRIONIDAE) (*cont.*)

4'. Dorsum of thorax and abdomen with no metallic green ... 5

5(4'). Postocular area entirely pale or with pale spots ranging from narrow linear areas to large round spots ... 6

5'. Postocular area dark, without pale spots ... 15

6(5'). Vein M_2 arising proximal to or near 4th and 3rd postnodal crossveins in fore- and hindwings, respectively ... 7

6'. Vein M_2 arising near 5th and 4th postnodal crossveins or beyond in fore- and hindwings, respectively ... 9

7(6). Black humeral stripe divided along its entire length by a narrow, pale stripe ... **Bluets (*Enallagma*)** (in part)

7'. Black humeral stripe either entire along its length or lacking ... 8

8(7'). Thorax blue, face yellow (less obvious in females), and antehumeral stripe yellow-green ... **Caribbean Yellowface (*Neoerythromma cultellatum*)**

8'. Thorax not blue or, if so, then face and antehumeral stripe not yellow ... **Forktails (*Ischnura*)** (in part)

9(6'). Vein Cu_1 of forewing extending well beyond level of origin of M_{1a}, but Cu_1 of hindwing usually extending only to level of origin of M_{1a} ... **Painted Damsel (*Hesperagrion heterodoxum*)** (in part)

9'. Vein Cu_1 in fore- and hindwings similar, extending to or well beyond level of origin of M_{1a} ... 10

10(9'). Anal crossing links Cu to posterior border in fore- and hindwings ... 11

10'. Anal crossing links Cu to A, not reaching posterior border, in forewing and usually hindwing ... 12

11(10'). Abdominal segment 10 of male in lateral view 2.5 times higher than long, the dorsum projecting posterodorsally; female with a pit on each side of the middorsal carina; southwest Texas ... **Mexican Wedgetail (*Acanthagrion quadratum*)**

11'. Segment 10 of male in lateral view less than twice as high as long, the dorsum flat, directed straight rearward; female without pits near middorsal carina ... **Bluets (*Enallagma*)** (in part)

12(10'). Males ... 13

12'. Females	14
13(12). Abdominal segment 10 with a posterodorsally projecting bifid process at least 1/4 as high as the segment; pterostigma of forewing generally different in shape, color and/or size from pterostigma of hindwing	**Forktails (*Ischnura*) (in part)**
13'. Segment 10 with at most a low, widely bifid prominence less than 1/4 as high as the segment; pterostigma of fore- and hindwings similar in color, shape, and size	**Bluets (*Enallagma*) (in part)**
14(12'). Humeral suture usually pale, but *if* black stripe is present, *then* no apical spine on venter of abdominal segment 8	**Forktails (*Ischnura*) (in part)**
14'. Humeral suture usually with a black stripe, but *if* pale, *then* an apical spine on venter of segment 8	**Bluets (*Enallagma*) (in part)**
15(5'). Pale antehumeral stripe split into anterior and posterior spots	**Painted Damsel (*Hesperagrion heterodoxum*)** (in part)
15'. Antehumeral spot not divided into anterior and posterior spots	16
16(15'). Body predominantly blue and green, the metepimeron largely bright yellow	**Aurora Damsel (*Chromagrion conditum*)**
16'. Body predominantly red, without bright-yellow spot on metepimeron	**Firetails (*Telebasis*)**

Wedgetails
Genus *Acanthagrion* Selys

The wedgetails, a diverse tropical group, have been recognized as the ecological equivalent of the bluets in North America. A single species is found in southwest Texas. Most wedgetails are blue and black in color (some South American species are orange or yellow) and have distinctive postocular spots. The males may be readily distinguished by the characteristic downward sloping arrangement of the cerci and the elevated abdominal segment 10, providing the basis for this group's common name. Females may be readily separated from those of all other genera in the region by the cerci, which touch or nearly touch medially, and the distinctive mesepisternal pits, contiguous with the middorsal carina.

References. Westfall and May (1996).

Mexican Wedgetail
Acanthagrion quadratum Selys
(p. 50; photo 6a)

Size. Total length: 29–33 mm; abdomen: 23–27 mm; hindwing: 15–18 mm.
Regional Distribution. *Biotic Provinces:* Balconian, Tamaulipan. *Watershed:* Lower Rio Grande.

General Distribution. Texas and Mexico south to Nicaragua.

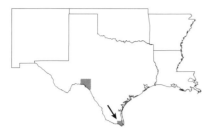

Flight Season. May 20 (TX)–Dec. 9 (TX).

Identification. The eyes of males are black above, fading to white below. The face is blue with varying degrees of black markings. They have large, distinctive, blue subcircular postocular spots outlined in black. There are generally visible pale spots lateral to each posterior ocellus. The middorsal carina varies from black to partly blue. The middorsal stripe is black and extends 2/5 to 1/2 the width of the mesepisternum. The antehumeral stripe is thin and blue and extends less than 1/2 the width of the middorsal stripe; the humeral stripe is wider and black. The remainder of the pterothorax is blue, fading to cream ventrally. The legs are blue basally and pale distally, with incomplete black stripes. Abdominal segments 1–3 and 7–9 are blue, segments 4–6 and 10 pale or cream in color. There is a black subquadrate spot dorsally on segment 1 and a black dorsal stripe extending the full length of segment 2. The entire dorsum on segments 3–6 is black, with the exception of a narrow basal ring on each segment. Most of segment 7 is black, except for blue ventrolateral margins and its apex. There is a black vertical streak on segments 8 and 9, and segment 10 is black for 1/2 its dorsal length and highly elevated. The cerci, distinctively black, slope downward at a 45°–60° angle. The paraprocts are pale with black apically.

The eyes of the female are brown dorsally, fading to cream ventrally. The color pattern of the head and thorax is similar to that of the male, but the pale color is more extensive. The mesostigmal plates are subtriangular, and a distinct sulcus outlines the posterior margin. The middorsal carina bifurcates at about 1/4 of its length from the anterior end; there are distinct pits just behind the bifurcation. The legs are similar to those of the male, but the black is more reduced. The color pattern of segments 1–6 is similar to that of the male. Segments 7 and 8 are black dorsally, except for pale basal and apical rings. Segment 8 is sometimes blue on the apical 1/4 of its length. Segment 9 is blue dorsally and black dorsolaterally. Segment 10 is blue dorsally and yellow ventrolaterally, occasionally with a narrow black apical ring.

Similar Species. Mexican Wedgetail is one of our few distinctive blue damselflies. Stream Bluet (*Enallagma exsulans*) seems similar to females, but its postocular spots are contiguous and its abdomen stockier.

Habitat. Weedy ponds and slow backwaters.

Discussion. This species has been reported in the state from only a few localities, including the Lower Rio Grande Valley and the Devils River. It is a common species throughout most of its range, and I anticipate that additional localities in the south-central United States will be revealed as more people continue to look at odonates.

References. Williamson (1916).

Red Damsels
Genus *Amphiagrion* Selys

Red damsels, a small group, are widely distributed throughout the United States, but only one species occurs in the south-central United States. Members of the group are instantly recognizable in the field by their stocky build and striking red-and-black coloration, the darker areas of which may become diffuse with age, and by their habit of flying low in vegetation. Red damsels also possess a distinctive large, heavily setose ventral tubercle posterior to the metathoracic legs. The costal margin of the pterostigma is much longer than the R_1 margin. The caudal appendages of the male are distinctive, the cerci sloping steeply downward.

There has been considerable confusion between the two currently recognized species in this group. For a long time the more robust Western Red Damsel (*A. abbreviatum*) was considered a variety of the slimmer Eastern Red Damsel (*A. saucium*). Leonora K. Gloyd was working on a revision of this group before her death, and at one time had planned on describing a third form from the midwestern states southwest into Arizona. It is still unclear what the relationships of these forms are. A specimen referable to the mid-American form was collected by G. Bick in Boiling Springs State Park, Woodward County, Oklahoma, and is now in the Florida State Collection of Arthropods.

Western Red Damsel
Amphiagrion abbreviatum (Selys)
(p. 50; photos 6b, 6c)

Size. Total length: 23–28 mm; abdomen: 17–21 mm; hindwing: 15–19 mm.
Regional Distribution. *Biotic Provinces:* Chihuahuan, Coloradan, Kansan, Navahonian, Texan. *Watersheds:* Canadian, Cimarron, Colorado (NM), Pecos, Red, Upper Rio Grande.

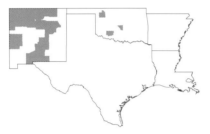

General Distribution. Western United States from Baja California through California north to British Columbia, east to Saskatchewan, and south to Oklahoma and New Mexico.
Flight Season. Apr. 4 (OK)–Aug. 7 (NM).
Identification. The head and thorax of the male are profusely hairy. The head is mostly dark red in younger individuals, changing to almost entirely black with age, though a small amount of red may persist around the antennal bases, ocelli, occipital ridge, and eyes. Postocular spots are generally lacking. The pterothorax is nearly all black, but the metathorax, except for the sutures, is sometimes red. There is a prominent ventral tubercle behind the metathoracic legs. The legs themselves are short and vary from red to yellow with black stripes, becoming increasingly darker with age, especially proximally. The wings are clear, with light-red venation and a dark pterostigma. The abdomen is noticeably short, not much longer than the wings, and red, with black apical rings on segments 3–6. Black dorsolateral spots, often visible, increase in size on segments 2–7. Segments 8–10 are black, with only a thin red middorsal line. The cerci and paraprocts are light red, and the cerci slant strongly downward.

The female is uniformly tan, often lacking dark pigmentation. There is generally a dark dorsolateral spot at the apices of the abdominal segments. Paired, dorsal, subapical spots are occasionally visible on segments 3, 4, or 5–9. There is a dark crescent on the frontal suture and darker spots between the ocelli. Occasionally, outlines of large postocular spots are visible. The caudal appendages are pale.
Similar Species. On Duckweed (*Telebasis byersi*) and Desert (*T. salva*) Firetails, a broad black arrow points posteriorly on the dorsum of the thorax. The tops of the head, legs, and thorax are pale in the red-form females of Painted Damsel (*Hesperagrion heterodoxum*).
Habitat. Sunlit, shallow, hard-bottomed marshy ponds or sloughs with vertical perches.
Discussion. Western Red Damsel is found only in the northern and western limits of our area. One study found Western Red Damsel to be conspicuously absent from forested areas, owing apparently to the absence of sunlit vertical perches. Western Red Damsel roosts parallel to these vertical perches at night, and presses its body closer to the perches in response to intruders during the early morning, before they are warm enough to fly.

References. Pritchard and Kortello (1997), Walker (1953).

Dancers
Genus *Argia* Rambur
(pp. 64, 68, 74, 76, 80, 84)

Dancers are a large group, confined entirely to the New World. Three characters can be used to separate members of this genus from all other pond damsels, though not all of these characters are readily visible in the field. First, the wings are stalked well before the level of the anal crossing (Ac), so that the distance between the ending of petiolation and the origination of Ac is greater than the length of Ac. Second, the tibial spurs in both sexes are longer than their intervening spaces, a character also seen in some species of sprite (*Nehalennia*). Third, males possess a pair of tori, specialized pad-like structures on the downward-sloping posterodorsal surface of abdominal segment 10. The cerci are short, the paraprocts are longer and often bifurcate, and females lack a vulvar spine on abdominal segment 8.

Adults of this group are often seen because of their habit of alighting on open sunny perches, such as along paths and streams. Most species are confined to streams or rivers, but some are associated with lakes and ponds. Females lay eggs in tandem or solitary. Individuals of some species, such as the common Powdered Dancer (*A. moesta*) of big rivers, may completely submerge themselves, remaining underwater for periods greater than a half hour.

Many species of dancer are similar in appearance and difficult to identify. This is especially true for the females. Making members of this group even more difficult to distinguish, it is common to find members of several species inhabiting the same stretch of stream or river. It will be necessary in many cases, especially when first starting out with this group, to capture specimens and critically examine them with a hand lens or microscope. Pay careful attention to the cerci and paraprocts of males, particularly any armature, such as teeth or lobes on the cercus. These appendages are often encrusted with dirt or excrement and may need to be brushed off to be clearly seen. Careful scrutiny of the mesostigmal plates is often required for the identification of females. With time, you will begin to recognize subtle color-pattern differences among species of this group as well. Unlike the keys in some other groups of Odonata, the keys below refer to many characters that can only be seen under a microscope or similar magnifying device.

References. Garrison (1994a), Gloyd (1958), Johnson (1972a).

KEY TO THE SPECIES OF DANCERS (*ARGIA*)

MALES

1. Middorsal area of thorax with separate lines on each side running parallel to, but not confluent with, the carina itself, these markings sometimes obscured by black in older individuals; abdomen 33–41 mm long — **Sooty (*lugens*)**

1'. Middorsal area of thorax without separate lines running parallel to one another, the thoracic markings seldom obscured by black or pruinosity; abdomen usually less than 35 mm long — 2

2(1'). Thoracic dorsum coppery, often showing metallic red reflections — **Coppery (*cuprea*)**

2'. Thoracic dorsum pale, not metallic — 3

3(2'). Abdominal segment 8 entirely black, dorsum of 9 and 10 blue — **Blue-tipped (*tibialis*)**

3'. Segment 8 pale, or, if not, segments 9 and 10 not blue — 4

4(3'). Dark humeral stripe divided longitudinally by pale stripe (may be obscured in older individuals); pale "M" pattern dorsally on abdominal segments 8 and 9 — **Dusky (*translata*)**

KEY TO THE SPECIES OF DANCERS (*ARGIA*) (*cont.*)

4′. Dark humeral stripe not divided longitudinally by pale stripe; segments 8 and 9 without pale "M" pattern dorsally — 5

5(4′). Dorsum of abdominal segments 3–5 with pale-dark-pale-dark pattern — **Kiowa (*immunda*)**

5′. Dorsum of segments 3–5 lacking above pattern — 6

6(5′). Dorsum of abdominal segments 3–5 largely pale — 7

6′. Dorsum of segments 3–5 largely dark — 15

7(6). Abdomen greater than 28 mm in length — 8

7′. Abdomen less than 28 mm in length — 10

8(7). Abdominal segments 4 and 5 with dorsolateral, subbasal black spot on each side — **Apache (*munda*)**

8′. Segments 4 and 5 without dorsolateral, subbasal black spots — 9

9(8′). Abdominal segments 3–6 with black lateral stripes, the stripes confluent with apical band — **Comanche (*barretti*)**

9′. Segments 3–6 without black lateral stripes — **Tonto (*tonto*)**

10(7′). Dark humeral stripe forked posteriorly (lower fork may be disconnected from stripe) — 11

10′. Dark humeral stripe not forked — 13

11(10). Ventrolateral dark stripe on abdominal segments 8–10 — **Variable (*fumipennis*)**

11′. No ventrolateral dark stripe on segments 8 and 9, and usually 10 — 12

12(11′). Pale areas blue; black on dorsum of abdominal segments 4–6 restricted to distal 1/5 of each segment — **Aztec (*nahuana*)**

12′. Pale areas violet; black on dorsum of abdominal segments 4–6 not restricted to distal 1/5 of each segment — **Lavender (*hinei*)**

13(10′). Pale color of abdomen reddish violet — **Amethyst (*pallens*)**

13′. Pale color of abdomen blue (but in cooler temperatures, abdomen may be darker) — 14

14(13′). Total length greater than 34 mm; distinct black stripe laterally on abdominal segment 2 — **Springwater (*plana*)**

14'. Total length less than 30 mm; black laterally on segment 2 restricted to posterior spot | **Seepage (*bipunctulata*)**

15(6'). Abdomen 28mm or greater in length | 22

15'. Abdomen less than 28 mm in length | 16

16(15'). Dark humeral stripe forked (evidence of fork may be restricted to small pale spot posteriorly) | 17

16'. Dark humeral stripe not forked | 21

17(16). Dorsomedial area of abdominal segment 2 black | **Blue-ringed (*sedula*)**

17'. Dorsomedial area of segment 2 blue | 18

18(17'). Wings amber | **Golden-winged (*rhoadsi*)** (in part)

18'. Wings clear | 19

19(18'). Pale postocular spots nearly connected by occipital bar | **Paiute (*alberta*)**

19'. Pale postocular spots widely separated from occipital bar | 20

20(19'). At its widest, dark humeral stripe as wide or wider than pale antehumeral stripe | **Vivid (*vivida*)**

20'. At its widest, dark humeral stripe narrower than pale antehumeral stripe | **Leonora's (*leonorae*)**

21(16'). Wings amber | **Golden-winged (*rhoadsi*)** (in part)

21'. Wings not amber | **Blue-fronted (*apicalis*)**

22(15).Abdomen dark violet | **Tezpi (*tezpi*)**

22'. Abdomen blue, brown, or white, never dark violet | **Powdered (*moesta*)**

FEMALES

1. Middorsal area of thorax showing a dark line along carina and a separate streak on each side, or these areas confluent only at their anterior and posterior ends | **Sooty (*lugens*)**

1'. Middorsal area of thorax showing a single dark line or stripe | 2

2(1'). Large species, abdomen generally longer than 32 mm; pterostigma surmounting more than 1 cell | **Powdered (*moesta*)**

KEY TO THE SPECIES OF DANCERS (*ARGIA*) (*cont.*)

2'. Smaller species, abdomen generally shorter than 32 mm; pterostigma surmounting no more than 1 cell — 3

3(2'). Middorsal and humeral stripes metallic red, with coppery reflections (metallic-red stripes become obscured in older individuals, but coppery reflections usually remain visible) — **Coppery (*cuprea*)**

3'. Middorsal and humeral stripes not metallic red, and lacking coppery reflections — 4

4(3'). Abdominal segments 8–10 pale dorsally — **Blue-ringed (*sedula*)**

4'. Segments 8, 9, and/or 10 with dark markings dorsally — 5

5(4'). Dorsum of abdominal segments 3–6 almost entirely black — 6

5'. Dorsum of segments 3–6 with pale areas more extensive — 12

6(5). Abdominal segments 8–10 with broad pale stripes medially, laterally, and ventrolaterally — 23

6'. Segments 8–10 without these pale stripes — 7

7(6'). Middorsal thoracic stripe less than 1/10 the width of thoracic dorsum — 8

7'. Middorsal thoracic stripe 1/4 or more the width of thoracic dorsum — 9

8(7). Wings distinctly amber; abdominal segments 9 and 10 largely pale dorsally — **Golden-winged (*rhoadsi*)**

8'. Wings clear; segments 9 and 10 black dorsally — **Blue-fronted (*apicalis*)**

9(7'). Postocular spots absent or reduced and scarcely larger than the ocelli; abdominal segment 8 mostly pale dorsally, segment 9 black; abdomen less than 24 mm long — **Seepage (*bipunctulata*)**

9'. Postocular spots present and at least twice the diameter of the ocelli; segment 8 with extensive dark areas dorsally *or*, if entirely pale, *then* segment 9 also pale; abdomen greater than 25 mm long — 10

10(9'). Forewing with 3 postquadrangular antenodal cells — **Leonora's (*leonorae*)**

KEY TO THE SPECIES OF DANCERS (*ARGIA*) (*cont.*)

10'. Forewing with 4 postquadrangular antenodal cells — 11

11(10'). Abdominal segments 3–7 with distinct distal black bands, complete and well-defined laterally; pale dorsal stripe on segment 9 generally not extending to base of segment, appearing as an apical spot — **Blue-tipped (*tibialis*)**

11'. Segments 3–7 with distal black bands either absent, divided laterally, or containing a pale lateral spot; pale dorsal stripe of segment 9 always extending to base of segment — **Blue-fronted (*apicalis*)** (in part)

12(5'). Mesepisternal pits prominent and deep, covered only slightly by the posterior lobes of the mesostigmal plates — **Apache (*munda*)**

12'. Mesepisternal pits not especially large, usually at least 1/2 covered by the lobes of the mesostigmal plates — 13

13(12'). Middorsal thoracic stripe less than 1/3 the width of the thoracic dorsum, often little more than a hairline — **Blue-fronted (*apicalis*)** (in part)

13'. Middorsal thoracic stripe greater than 1/3 the width of the thoracic dorsum — 14

14(13'). Abdominal segments 8 and 9 entirely pale dorsally, or with black spots on segment 9 — 15

14'. Segments 8 and 9 both with definite black areas — 16

15(14). Black markings on segment 2 consisting of an apical spot thinly connected to smaller basal spots — 21

15'. Black markings on segment 2 usually consisting of a stripe extending almost the length of the segment — **Kiowa (*immunda*)**

16(14'). Abdominal segments 3 and 4, and often 5, with subbasal black spots or streaks not confluent with the apical black band, and abdomen no longer than 28 mm — 17

16'. Segments 3 and 4 with black stripe usually more or less broadly confluent with the apical black band, or with the dorsum entirely black, but *if* these stripes are not confluent with apical band, *then* abdomen longer than 30 mm — 18

KEY TO THE SPECIES OF DANCERS (*ARGIA*) (*cont.*)

17(16). Middorsal thoracic carina bifurcating well behind the posterior border of the mesostigmal plates, and diverging widely	**Paiute (*alberta*)**
17'. Middorsal thoracic carina bifurcating only slightly behind the posterior border of the mesostigmal plates, and not diverging widely	**Aztec (*nahuana*)**
18(16'). Abdomen longer than 30 mm	22
18'. Abdomen 28 mm or less	19
19(18'). Posterior lobe of mesostigmal plate distinctly thickened distally; recurved mesal border of lobe forming a prominent tubercle, best viewed laterally	**Amethyst (*pallens*)**
19'. Posterior lobe of mesostigmal plate not distinctly thickened distally; tubercle formed by recurved mesal border of lobe not visible laterally	20
20(19'). Broad black band medially on top of head, with pale triangles encompassing the posterior ocelli	**Lavender (*hinei*)**
20'. Top of head mostly pale	**Variable (*fumipennis*)**
21(15). Black subapical spot and basal spot or complete stripe, dorsally on abdominal segment 2	**Springwater (*plana*)**
21'. Black subapical spot only, dorsally on abdominal segment 2	**Vivid (*vivida*)**
22(18). Dark middorsal thoracic stripe 1/4 the width of dorsum	**Comanche (*barretti*)**
22'. Dark middorsal thoracic stripe at least 1/3 the width of dorsum	**Tonto (*tonto*)**
23(6). Wings clear; dorsum of abdominal segment 10 with 2 black spots	**Dusky (*translata*)**
23'. Wings distinctly amber; dorsum of abdominal segment 10 pale	**Tezpi (*tezpi*)**

Paiute Dancer
Argia alberta Kennedy
(pp. 64, 68, 76; photo 6d)

Size. Total length: 27–32 mm; abdomen: 21–25 mm; hindwing: 16–20 mm.

Regional Distribution. *Biotic Provinces:* Austroriparian, Balconian, Carolinian, Chihuahuan, Kansan, Navahonian, Texan. *Watersheds:* Arkansas, Canadian, Pecos, Red, Upper Rio Grande.
General Distribution. Western United States east to Texas and Oklahoma.

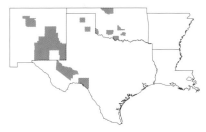

Flight Season. Apr. 26 (TX)–Nov. 13 (NM).

Identification. The labrum of the male is paler blue than the rest of the head. The eyes are dark blue dorsally, becoming paler ventrally. There is a wide, dark "T" spot anterior to the median ocellus. There are small blue postocular spots. The pterothorax is pale blue or violet, generally darker dorsally and becoming lighter laterally. There is a black middorsal stripe nearly twice the width of the pale antehumeral stripe. The black humeral stripe is forked at its upper 1/3 and is 1/2 the width of the pale antehumeral stripe. The legs are pale, but with blue basally on the femora and anteriorly on the inner surface of the tibiae. The tarsi are dark, often black. The abdominal segments 1 and 2 are blue; segment 1 has a black spot dorsobasally, and segment 2 has a nearly full-length black stripe laterally. Segments 3–7 are dark black dorsally with a basal blue ring on each segment. Segments 8–10 are blue with dark ventrolateral markings. The cerci are about 1/2 the length of segment 10 and are divergent dorsally, each with a prominent ventrally directed, internal hook at its apex. The paraprocts are distinctly bifid, the lower branch rounded and projecting only slightly posteriorly. The upper branch is more pointed and distinctly directed dorsally. There are three postquadrangular cells in each wing.

The head of the female is generally paler than that of the male. The postocular spots are much larger. The humeral stripe is narrower than that in males and symmetrically forked at its upper half. The legs are colored similarly to that of the male, but the black markings are less extensive. The medial posterior border of the mesostigmal plate is raised into a distinct rim. The middorsal thoracic carina bifurcates, diverging widely, well behind the posterior border of each mesostigmal plate. Mesepisternal tubercles are lacking.

Similar Species. Blue-ringed Dancer (*A. sedula*) is darker, its abdomen is predominantly black, including the dorsum of segment 3, and the humeral stripe is not as widely forked as in Paiute Dancer.

Blue-ringed Dancer also often has an amber tint to its wings.

Habitat. Small flowing streams or marshy springs.

Discussion. Paiute Dancer is primarily a Great Basin species whose distribution extends to western Oklahoma and Texas and to southern New Mexico in our region. It is most commonly seen at creeks, but northern specimens have been observed at hot springs, and it may be found associated with saline waters. Egg-laying occurs in tandem.

Blue-fronted Dancer
Argia apicalis (Say)
(pp. 64, 68, 76; photos 6e, 6f)

Size. Total length: 33–40 mm; abdomen: 26–32 mm; hindwing: 21–25 mm.

Regional Distribution. *Biotic Provinces:* Austroriparian, Balconian, Carolinian, Chihuahuan, Coloradan, Kansan, Navahonian, Tamaulipan, Texan. *Watersheds:* Arkansas, Bayou Bartholomew, Brazos, Canadian, Cimarron, Colorado, Guadalupe, Lower Rio Grande, Mississippi, Neches, Nueces, Ouachita, Pecos, Red, Sabine, San Antonio, San Jacinto, Saint Francis, Trinity, Upper Rio Grande, White.

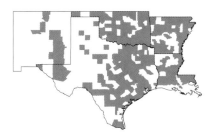

General Distribution. Generally east of the Rocky Mountains from Florida to Ontario, but also known from eastern Arizona.

Flight Season. Mar. 21 (TX)–Dec. 26 (TX).

Identification. The head of the male is pale blue anterior to the ocelli, and black behind the ocelli, with a stripe extending forward to the base of each antennae. The pale-blue occipital spots are sometimes connected. The lower lateral 1/3 and the ventral parts of the pterothorax are pale yellow, becoming pruinose and obscured with age. The black thoracic markings, however, are never obscured. The middorsal carina is narrowly outlined with black. The humeral suture is a hairline black stripe widening above and below, but more so anteriorly. The legs are pale, darkly marked

TABLE 3. DANCERS (*ARGIA*)

Species	Hindwing length (mm)	Blue dorsally (♂)[1]	Black laterally (♂)[2]	Abdomen color[3]	Humeral stripe forked[4]	Antehumeral color
Paiute (*alberta*)	16–20	8,9,10	8,9,10	blue	2	blue
Blue-fronted (*apicalis*)	21–25	8,9,10	8,9,10	blue/violet	0	blue
Comanche (*barretti*)	22–25	8,9,10	8,9,10	blue	0	blue
Seepage (*bipunctulata*)	13–18	—	—	blue	0	blue
Coppery (*cuprea*)	22–25	9,10	9,10	black	0	coppery
Variable (*fumipennis*)	18–23	8,9,10	8,9,10	violet	2	violet
Lavender (*hinei*)	17–25	8,9,10	8,9,10	violet	2	violet
Kiowa (*immunda*)	19–25	8,9,10	8,9,10	blue/violet	2	blue/violet
Leonora's (*leonorae*)	15–19	8,9,10	8,9,10	blue	1	blue
Sooty (*lugens*)	25–35	—	—	black	0	black
Powdered[8] (*moesta*)	22–29	—	—	black	0	pruinose
Apache (*munda*)	23–27	8,9,10	—	blue	0	blue
Aztec (*nahuana*)	18–23	8,9,10	8,9	blue	2	blue
Amethyst (*pallens*)	20–22	8,9,10	8	violet	0	violet
Springwater (*plana*)	22–25	8,9,10	—	blue/violet	0	blue/violet
Golden-winged (*rhoadsi*)	19–21	8,9,10	—	blue	1	blue
Blue-ringed (*sedula*)	17–21	8,9,10	8,9,10	black	1	blue
Tezpi (*tezpi*)	22–26	partial 9,10	—	dark violet	2	yellow

Species	Post-quadrangular cells[5]	Mesepisternal tubercles ($♀$)[6]	Distribution (biotic provinces)[7]
Paiute (*alberta*)	3/3	0	Ch,K,N,Tx
Blue-fronted (*apicalis*)	4/3	0	Au,B,Ca,Ch,Co,K,N,Ta,Tx
Comanche (*barretti*)	5/4	2	B,Ch,Ta
Seepage (*bipunctulata*)	3/3	0	Au,Ca,Tx
Coppery (*cuprea*)	5/4	2	B
Variable (*fumipennis*)	4/3	0	Ap,Au,B,Ca,Ch,Co,K,Ta,Tx
Lavender (*hinei*)	4/3	0	B,Ch,K
Kiowa (*immunda*)	3/3	0	Au,B,Ca,Ch,K,N,Ta,Tx
Leonora's (*leonorae*)	3/3	0	B,Ch,Ta,Tx
Sooty (*lugens*)	5–6/5	1–2	Ap,B,Ch,Co,K,N,Ta
Powdered[8] (*moesta*)	4–5/4	2	Ap,Au,B,Ca,Ch,Co,K,N,Ta,Tx
Apache (*munda*)	5/4	0	Ap,Ch
Aztec (*nahuana*)	4/3	0–1	Ap,Au,B,Ca,Ch,Co,K,N,Tx
Amethyst (*pallens*)	4/3	0	Ap,Ch
Springwater (*plana*)	4/3	2	Ap,Au,B,Ca,Ch,Co,K,N,Ta,Tx
Golden-winged (*rhoadsi*)	4/3	0–1	B,Ta
Blue-ringed (*sedula*)	4/3	0–1	Ap,Au,B,Ca,Ch,K,N,Ta,Tx
Tezpi (*tezpi*)	5/4	2	Ap

TABLE 3. DANCERS (*ARGIA*) (cont.)

Species	Hindwing length (mm)	Blue dorsally (♂)[1]	Black laterally (♂)[2]	Abdomen color[3]	Humeral stripe forked[4]	Antehumeral color
Blue-tipped (*tibialis*)	18–24	9,10	9,10	black	0♂, 1♀	yellow
Tonto (*tonto*)	25–29	8,9,10	8,9,10	pale violet	0	violet
Dusky (*translata*)	19–23	partial 8,9	8,9	black	0	yellow/black
Vivid (*vivida*)	18–25	8,9,10	8,9,10	blue	1♂, 2♀	blue

[1] Terminal (8–10) abdominal segments that are pale dorsally in male.

[2] Terminal (8–10) abdominal segments that have ventrolateral dark stripe.

[3] Predominant color on abdomen.

[4] Degree of fork in humeral stripe (0-none, 1-slight, 2-strong).

[5] Number of postquadrangular cells in the forewing/hindwing.

[6] Presence of mesepisternal tubercles (0-lacking, 1-reduced, 2-strong).

[7] (Ap) Apachian, (Au) Austroriparian, (B) Balconian, (Ca) Carolinian, (Ch) Chihuahuan, (Co) Coloradan, (K) Kansan, (N) Navahonian, (Ta) Tamaulipan, (Tx) Texan.

[8] Characters do not apply for teneral/young individuals.

with black on the outer surfaces of the femora and the inner surfaces of the tibiae and tarsi. The abdomen is black dorsally on segments 1–7, except for pale-blue basal rings on segments 3–7. Segments 8–10 are uniformly pale blue over their entire length; the ventrolateral areas of these segments are black. The cerci are short and blunt when viewed laterally, but when viewed posteriorly two decurved hooks are visible. The paraprocts extend out beyond the cerci, but both are shorter than, or subequal in length to, segment 10. The paraprocts are distinctly bifid.

The head of the female is colored similarly to that of the male. The dark humeral stripe is usually thin, as in the male, but occasionally it extends to the mesepimeron. The mesostigmal plates lack posterior lobes, but when viewed laterally the anterior carina is usually visible. Mesepisternal tubercles are small or lacking. The legs are colored similarly to those of the male. The abdomen is similar to that of the male, but often shows more extensive black markings. Segments 8–10 are black dorsally, and each bears a dark dorsolateral stripe.

Similar Species. Abdominal segment 8 of the Blue-tipped Dancer (*A. tibialis*) is black. Female Powdered Dancers (*A. moesta*) look very similar, but are larger. All other dancers have a broader middorsal thoracic stripe.

Habitat. Large rivers and occasionally streams, lakes, or ponds.

Discussion. A study of a population of this species, residing at a small pasture pond in Marshall County, Oklahoma, found that males and females reached reproductive age at an average of 8.4 and 7.0 days, respectively, with maximum activity occurring at noon. Males were at the water 44% of the days they lived, and females only 20%; males, however, mated on only 20% of these days, and females 89%. Repeat matings, in both sexes, were rare. The movement from roosting sites to the water was characterized by a random shifting, the males arriving at the water earlier than the females and spacing themselves at 2-meter intervals. Mating occurred a short distance from the main concentration of males and lasted 16 minutes. Females were gregarious laying eggs, either in tandem (75% of the time) or alone, in willow roots, boards, and sticks.

Further study revealed that males occurred in two color phases, with pale areas either bright blue or gray-black, and that color change could not be correlated positively with age or reproduction. In the same study, females occurred in three color phases: brown, turquoise, and gray-black. Further observations in the field have revealed that the gray-black phase is a temporary color change that occurs during mating and at cool temperatures.

Species	Post-quadrangular cells[5]	Mesepisternal tubercles (♀)[6]	Distribution (biotic provinces)[7]
Blue-tipped (*tibialis*)	4/3	0	Au,Ca,K,Tx
Tonto (*tonto*)	5(6–7),4	0	Ap
Dusky (*translata*)	5/4	2	Ap,Au,B,Ca,Ch,K,N,Ta,Tx
Vivid (*vivida*)	4(3,5)/3(4)	2	Ap,Co,N

Males live more than a month and exhibit intra- and interspecific competition, flicking their wings as a warning to intruders in their territory. On hot days they thermoregulate by raising their abdomen to a 60° angle, reducing the surface area exposed to the sun.

Variable thoracic stripe patterns are seen in individuals occuring in the southeast, one a broad-striped form having a wide, full-length humeral stripe, the other, more typical, form a reduced pattern, as seen in the south-central United States.

References. Bick (1963), Bick and Bick (1965a,b), Dunkle (1990), Johnson (1972b), Williamson (1906a).

Comanche Dancer
Argia barretti Calvert
(pp. 64, 68, 76; photo 7a)

Size. Total length: 38–43 mm; abdomen: 31–34 mm; hindwing: 22–25 mm.
Regional Distribution. *Biotic Provinces:* Balconian, Chihuahuan, Tamaulipan. *Watersheds:* Colorado, Guadalupe, Nueces, San Antonio, Upper Rio Grande.

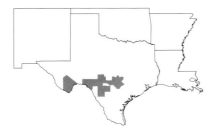

General Distribution. Texas south to San Luis Potosí and Tamaulipas, Mexico.
Flight Season. May 10 (TX)–Nov. 5 (TX).
Identification. The face of the male is bright blue, and there are large, pale postocular spots broadly confluent with the compound eyes. There is a distinct, thin, blue occipital bar between the postocular spots, broadly divided medially and appearing as two smaller spots. The pterothorax is blue, and there is a broad black middorsal stripe 1/2 as wide as the blue antehumeral stripe next to it. The humeral stripe is straight, but narrowing posteriorly. The legs are solid black exteriorly and pale medially. There are usually five and four postquadrangular cells in the fore- and hindwing, respectively. The abdomen is predominantly blue dorsally with black ventrolateral stripes widening posteriorly and becoming more pronounced on segments 6 and 7. The ventrolateral stripes form confluent rings around each segment posteriorly. Segment 7 is almost entirely black dorsally, and segments 8–10 are blue dorsally, with only a narrow black stripe laterally. The cerci are only 2/3 the length of the paraprocts, and distinctly bifid in dorsal view. The paraprocts are bifid, the upper branch directed dorsally and the lower branch directed posteriorly, the angle between the two approximately 90°.

The female color pattern is similar to that of the male, but paler, often with tan or light-brown colors replacing blues. The dark thoracic markings are less extensive. The humeral stripe is diffuse brown on the lower half of the suture and black on its upper half. The mesostigmal plates, seen in dorsal view, show large posterior lobes that are less than 1/2 the width of the plate itself. Mesepisternal tubercles are

Dancers (*Argia*)

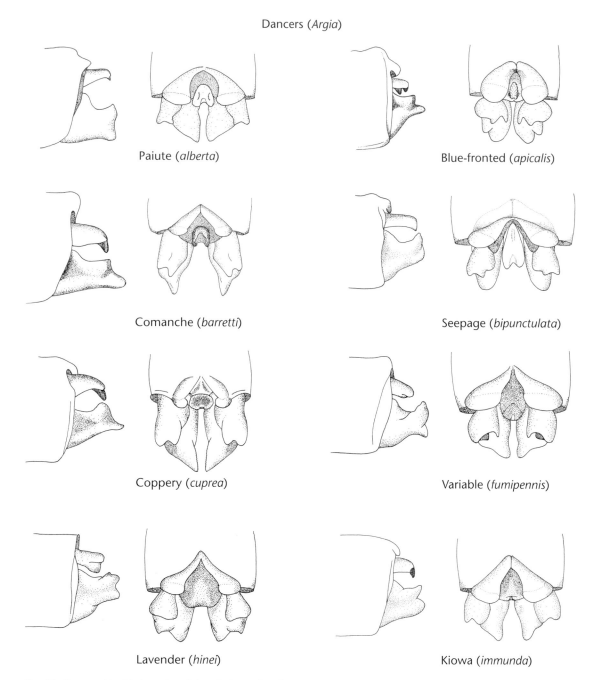

Paiute (*alberta*)

Blue-fronted (*apicalis*)

Comanche (*barretti*)

Seepage (*bipunctulata*)

Coppery (*cuprea*)

Variable (*fumipennis*)

Lavender (*hinei*)

Kiowa (*immunda*)

Fig. 13. Dancers (*Argia*): lateral and dorsal views of male caudal appendages (pp. 62–72).

present, but generally small and reduced in size. The legs are pale, with less extensive black markings than in the male.

Similar Species. Comanche Dancer is one of the largest and most easily recognized bright-blue damselflies in the region. Blue-fronted Dancer (*A. apicalis*) is smaller, and has a very narrow middorsal thoracic stripe. Paiute Dancer (*A. alberta*) is much smaller, and has a strongly forked humeral stripe. Big Bluet (*Enallagma durum*) is the same size, but of a paler blue, and has black dorsal markings, in the shape of spear points, on the abdomen.

Habitat. Rivers and streams.

Discussion. This species has a primarily Mexican distribution, occurring in the states of Nuevo Leon and Tamaulipas, but ranges northward to the Hill Country of Texas, where it can be a dominant species on wide, rocky streams. Females lay eggs in tandem on floating debris at river's edge. Details of its reproductive behavior are unknown.

References. Gloyd (1932).

Seepage Dancer
Argia bipunctulata (Hagen)
(pp. 64, 68, 76; photos 7b, 7c)

Size. Total length: 23–30 mm; abdomen: 18–24 mm; hindwing: 13–18 mm.

Regional Distribution. *Biotic Provinces:* Austroriparian, Carolinian, Kansan, Texan. *Watersheds:* Arkansas, Bayou Bartholomew, Canadian, Mississippi, Neches, Ouachita, Red, Sabine, Trinity.

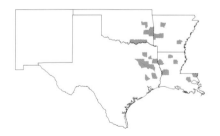

General Distribution. Eastern United States from Florida to New Hampshire and west to Kansas and Texas.

Flight Season. May 17 (TX)–Aug. 26 (OK).

Identification. The face of the male is light blue, and there is a single medial black spot at the base of the labrum and on either side of the postclypeus. The top of the head is black, and the postocular spots are lacking. The pale-blue occipital bar is broadly divided in the middle. The pterothorax is entirely black dorsally and pale blue laterally, with narrow, pale antehumeral stripes half the width of the black humeral stripe. The metapleural suture is narrowly outlined by black. The legs are pale, with black stripes on the outer surfaces of the femora and tibiae, and the tarsi are black. The wings are clear, and the pterostigma, light brown, rests on a single cell. There are three postquadrangular cells in both wings. Abdominal segments 1 and 2 are blue, becoming paler laterally. There is a single median black spot at the base of segment 1 and a pair of black spots at the posterior margin of segment 2; the latter spots may be confluent and often join with a black apical ring on segment 2. A dark band on the apical 1/4 of segment 3 surrounds a pale-blue spot laterally. Segments 4 and 5 have a black band for 1/3 the length of each. Segment 6 may have an apical band occupying half of its length. Segment 7 is entirely black except for a blue basal ring. Segments 8–10 are entirely blue. The cerci and paraprocts are of approximately the same length.

The female is similar to the male, but her pale colors are often yellow or tan. The small postocular spots may be fused with a stripe that is confluent with each compound eye. The mesostigmal plates have poorly developed lobes, and mesepisternal tubercles are lacking. The abdomen is mostly black dorsally, with only a light basal ring on segments 3–7. Segment 8 is pale dorsally and black laterally. Segments 9 and 10 are solid black.

Similar Species. Seepage Dancer is the smallest dancer in the south-central United States. Its size, its dark head and thorax, and its lack of postocular spots will separate it from any other dancer in the region. Double-striped Bluet (*Enallagma basidens*) is slightly smaller and has a thin pale stripe dividing each dark humeral stripe.

Habitat. Sunny sphagnum seepages, small lakes, ponds, and streams.

Discussion. This species often perches vertically on grass stems or other available perches in its habitat, only occasionally perching on the ground.

Coppery Dancer
Argia cuprea (Hagen)
(pp. 64, 68, 76; photos 7d, 7e)

Size. Total length: 39–42 mm; abdomen: 27–34 mm; hindwing: 22–25 mm.

Regional Distribution. *Biotic Province:* Balconian. *Watersheds:* Colorado, Nueces, San Antonio.

General Distribution. Texas south to Mexico, Venezuela, and Bolivia.

Flight Season. Apr. 12 (TX)–Aug. 16 (TX).

Identification. The male has brilliant ·cherry-red eyes. The front of the head, including the labrum, clypeus, and frons, is metallic coppery red. The pterothorax is a brilliant metallic red with coppery reflections, and the middorsal carina is thinly lined with black. The antehumeral stripe is lacking. The sides of the thorax are blue, fading to pale yellow ventrally. The legs are black, but with a pale brown stripe on the outer surface of the femora and tibiae. There are five and four postquadrangular cells in the fore- and hindwing, respectively. The first abdominal segment is blue, but has a black spot dorsally that may be confluent with lateral spots basally. Segments 2–7 are black with a shiny luster dorsally and basal blue rings. The black extends to the ventral side in the distal 1/5 of each segment. The rest of the abdomen is blue laterally, fading ventrally. Segment 8 is black dorsally with extensive blue basally and laterally; the blue is interrupted laterally by a black stripe extending to segment 10. Segments 9 and 10 are blue except for the above-mentioned black lateral stripe. The cerci are branched and about 3/4 the length of the paraprocts.

The female is generally similar to the male, but the eyes are not bright red and the front of the thorax lacks the extensive metallic coloration. The front of the head is pale yellowish. A pale antehumeral stripe extends approximately 1/2 the width of the middorsal stripe. The dark humeral stripe is twice the width of the antehumeral stripe and consists of an isolated spot at its upper end. The mesostigmal plates have strongly explanate lobes, much like those of the Dusky Dancer (*Argia translata*). Mesepisternal tubercles are present. Abdominal segments 2–8 are black dorsally, and there is a pale middorsal line extending the length of the segments. Segments 3–7 have a pale basal ring interrupted in the middle of segments 3 and 4. Segment 8 is black dorsally, often with two stripes isolated by a pale area extending the length of the segment, or two basal black spots that do not extend to the end of the segment. Segment 9 has two basal spots dorsally that extend 1/2 the length of the segment. Segments 8 and 9 also have a black lateral stripe. Segment 10 is nearly all blue.

Similar Species. Male Coppery Dancer is the only blue damselfly with red eyes occurring in the south-central United States. Females are similar to those of Dusky Dancer (*A. translata*), but in that species they lack the pale spot at the posterior end of the dark humeral stripe. Comanche Dancer (*A. barretti*) females are also similar, but have a much narrower humeral stripe, and again lack the pale spot at the posterior end.

Habitat. Rivers and streams.

Discussion. This species was not known from the United States until 1985. It has been taken from the Nueces and West Frio Rivers of the Texas Hill Country frequently since. Though it is the only species with red eyes and a metallic thorax, there is a similar species, the Fiery-eyed Dancer (*A. oenea*), in Arizona and Mexico.

Variable Dancer
Argia fumipennis (Burmeister)
(pp. 12, 64, 68, 76; photos 8a, 8b)

Size. Total length: 29–34 mm; abdomen: 23–28 mm; hindwing: 18–23 mm.

Regional Distribution. *Biotic Provinces:* Apachian, Austroriparian, Balconian, Carolinian, Chihuahuan, Coloradan, Kansan, Navahonian, Texan. *Watersheds:* Arkansas, Bayou Bartholomew, Brazos, Canadian, Cimarron, Colorado, Colorado (NM), Guadalupe, Mississippi, Neches, Nueces, Ouachita, Pecos, Red, Sabine, San Antonio, San Jacinto, St. Francis, Trinity, Upper Rio Grande, White.

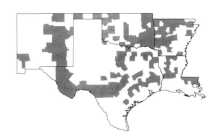

General Distribution. East of the Rocky Mountains throughout the eastern United States and southern Canada, south into Mexico and Guatemala.

Flight Season. Mar. 21 (TX)–Nov. 6 (TX).

Identification. The head of the male is violet, with large postocular spots contiguous with the eyes. The pterothorax is violet, fading to yellow ventrally and laterally. The antehumeral stripe is violet. The black humeral stripe is forked and 1/2 as wide as the antehumeral stripe, which extends 3/5 the length of the suture before forking (the smaller of the two forks is no more than a hairline and extends along the mesopleural suture). The legs are pale, but distinctly marked with black on the anterior surface of the femora, and to a lesser extent on the tibiae. There are four and three postquadrangular cells in the fore- and hindwings, respectively. The abdomen is distinctly violet in older individuals and segment 1 is marked with a transverse dark spot. Segment 2 has a large dark lateral spot on each side that may be confluent dorsoposteriorly. The middle segments, 3–6, each have a pair of dark triangular spots anterior to the dark posterior ring, with which they may be confluent. Segment 7 is entirely black, except for a posterior violet ring. Segments 8–10 are pale blue with a dark ventrolateral stripe. The cerci, when viewed dorsolaterally, have a ventrally projecting tooth originating from the subapical margin, and a more blunt lobe projecting ventrally from the medial margin. Laterally, the cerci slope distinctly downward; the paraprocts are longer than the cerci.

The female is much duller in color than the male. The dark markings on the head are lighter than in the male, and often there is more brown than black. The pterothorax is similar to that of the male, but with less extensive dark pigmentation. A posterior lobe on the mesostigmal plates is transversely flattened and slightly concave ventrally. Mesepisternal tubercles are reduced or entirely absent. The abdomen, paler than in the male and generally violet or brown, differs also in the following ways: segment 2 has a dark dorsolateral spot, the spot constricted or separated toward the middle of the segment; segments 3–6 each bear an elongate dorsolateral spot basally and a shorter, wider spot apically on each side; the dark basal and apical spots on segment 6 are confluent; on each side of segment 7 there is a full-length black lateral stripe; and segments 8–10 are light brown.

Similar Species. Variable Dancer may be easily confused with several other species in our region where older individuals of those species are generally more violet than blue. Lavender Dancer (*A. hinei*) is also violet, but the black on its head and the middorsal stripe of the thorax are more extensive in that species. Variable Dancer also has a dark lateral line extending the full length of segments 8 and 9, and these stripes are reduced and nearly lacking in Lavender Dancer. Leonora's Dancer is smaller, has arrow-shaped dark markings on the abdomen, and the fork in the humeral stripe is not as deep as in Variable Dancer. Aztec Dancer lacks full-length black stripes laterally on segments 8 and 9, and they are completely lacking on 10.

Habitat. Shallow streams with exposed rocks and small lakes.

Discussion. As its name implies, Variable Dancer exhibits significant geographic variation. Three different forms have been recognized, on the basis of wing color and body pattern. The only one occurring in the south-central United States is *A. fumipennis violacea*. It is the most widely distributed of the three subspecies and occurs throughout our region. Egg-laying usually takes place in pairs on submerged plants and debris, but females do not descend beneath the surface of the water. This species is most prevalent on exposed rocks in shallow streams where there is a gentle current. Small lakes along the course of dammed streams are also favorite haunts for this species. Adult males fly over the stream and settle on bare ground, while females are generally found a considerable distance from water.

References. Bick (1972), Bick and Bick (1982), Kellicott (1899).

Lavender Dancer
Argia hinei Kennedy
(pp. 64, 68, 76; photo 7f)

Size. Total length: 30–35 mm; abdomen: 24–28 mm; hindwing: 17–21 mm.

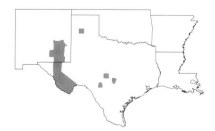

Regional Distribution. *Biotic Provinces:* Balconian, Chihuahuan, Kansan. *Watersheds:* Colorado, Guadalupe, Pecos, Red, Upper Rio Grande.

General Distribution. Southwestern United States; California east to Texas and south through Mexico.

Flight Season. Apr. 9 (TX)–Sep. 3 (TX).

Identification. The head of the male is violet, with a black line that runs through the ocelli and between the compound eyes; this line is sometimes interrupted at its middle. There are also large violet postocular spots. The middorsal stripe is black on a violet pterothorax, and the violet antehumeral stripe is nearly as wide as the middorsal stripe. The black humeral stripe is only 1/3 or less as wide as these stripes and is generally forked in the upper 1/2 to 1/3 of its length. The pterothorax is pale laterally and ventrally. The legs are pale, and there are black stripes on the outer surfaces of the femora, which are divided for their entire length by a pale stripe. The outer surfaces of the tibiae are entirely black. There are four and three postquadrangular cells in the fore- and hindwings, respectively. The abdomen is largely violet on segments 1–7, and the venter is pale yellow. There is a dark-black dorsal spot basally on segment 1. Segment 2 has a black lateral stripe that may be interrupted at its middle, forming two spots. There is a dark narrow ring apically on segments 2–7 and an apical black spot confluent with this ring on segments 3–6. Segment 7 is black, except for a pale basal ring. Segments 8–10 are blue, and there is a small ventrolateral black spot apically on 8. The paraprocts are forked and longer than the cerci, and the cerci are concave ventrally, with an outwardly directed tooth on the inner edge and a ventrally directed subapical black tubercle.

The head and thorax of the female are similar to those of the male, but the pale areas are largely brown instead of violet. The mesostigmal plates are distinctive, with a narrow, posteriorly directed lobe recurved medially. Mesepisternal tubercles are absent. The abdominal pattern is similar to that of the male, with segments 2–6 bearing a narrow, black apical ring and a dark lateral stripe that extends the full length of each segment, widening apically. The widened areas may become confluent dorsally on the posterior segments. Segments 2–7 also have a dark oblique stripe ventrolaterally. This stripe sometimes becomes confluent with the dorsolateral stripe. Segment 7 is black dorsally, with only a pale thin middorsal line and basal ring. Segments 8 and 9 bear a dorsolateral stripe that nearly reaches the apex of each.

Similar Species. Gloyd (1958) discussed the similarities and differences of Lavender Dancer and Variable Dancer (*A. fumipennis*), which may be difficult to distinguish from one another where they are found together. She described Lavender Dancer as a more slender species, with a more "delicate and orchid-like" purple color. Lavender Dancer lacks the complete ventrolateral dark stripe on segments 8 and 9 seen in Variable Dancer. Leonora's Dancer (*A. leonorae*) has distinctive black arrow- or spear-shaped markings dorsally on the abdomen. Aztec Dancer (*A. nahuana*) lacks dark markings anterolaterally on segments 4 and 5.

Habitat. Semidesert creeks, streams, and rivers.

Discussion. Lavender Dancer is found in the western desert areas of our region, where it is relatively uncommon and remains unknown biologically.

Kiowa Dancer
Argia immunda (Hagen)
(pp. 64, 68, 76; photo 8c)

Size. Total length: 33–38 mm; abdomen: 25–31 mm; hindwing: 19–25 mm.

Regional Distribution. *Biotic Provinces:* Austroriparian, Balconian, Carolinian, Chihuahuan, Kansan, Navahonian, Tamaulipan, Texan. *Watersheds:* Arkansas, Bayou Bartholomew, Brazos, Canadian, Cimarron, Colorado, Guadalupe, Lower Rio Grande, Neches, Nueces, Pecos, Red, Sabine, San Antonio, San Jacinto, Trinity, Upper Rio Grande.

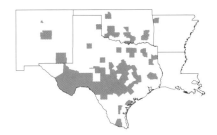

General Distribution. Southwestern United States from California eastward to South Dakota and Texas; also south through Mexico to Belize.

Flight Season. Apr. 2 (TX)–Dec. 28 (TX).

Identification. The male's head is largely pale violet. There is a distinct but irregular black bar between the eyes and the large violet postocular spots. The antehumeral stripe is pale blue or violet and of about the same width as the black middorsal stripe. The black humeral stripe is forked at its upper half, and the metapleural suture is outlined in black line. The bottom of the pterothorax generally has a mottled appearance. The femora are pale with dark stripes running lengthwise; the inner sides of the tibiae are pale, and the tarsi are black. There are generally only three postquadrangular cells in both the fore- and hindwing. The abdomen is pale violet or sometimes blue with black maculation. There is a basal black spot and small antero- and posterolateral spots on segment 1. Segment 2 has three spots, apical and basal superior spots and an inferior middle spot, all of which may be confluent. Segments 3–6 are marked with black postbasally and on the apical 1/3 of each segment laterally. On segments 4–6 these postbasal streaks are generally confluent with those of the opposite side, resulting in a distinct pale-dark-pale-dark pattern on the abdomen. Segment 7 is entirely black, with a thin, pale basal ring. Segments 8–10 are blue, with a black ventrolateral streak. The cerci are divided at their tip, and are sharply bifurcate, the lower lobe rounded and the superior lobe directed acutely dorsoposteriorly.

The female is similar to the male in color but paler. The forked humeral stripe is thin and scarcely covers the suture. The antehumeral stripe is pale and often marked with darker spots. The mesostigmal plates lack a posterior lobe, and their posterior margin is generally elevated medially. The middorsal thoracic carina bifurcates widely, well behind the posterior margin, and the mesepisternal tubercles are lacking. The abdomen is marked as in the male, but segment 7 is not as black and is similar to segment 6. Segments 8 and 9 are pale, with a dark ventrolateral stripe on each side. Segment 10 is entirely pale.

Similar Species. Variable Dancer (*A. fumipennis*) and Lavender Dancer (*A. hinei*) are similar, but lack the pale-dark-pale-dark abdominal pattern unique to Kiowa Dancer. Its size and this distinctive pattern should separate this species from all other damselflies in the region.

Habitat. Streams and rivers.

Discussion. Despite the fact that this species is widely distributed in our region, and commonly encountered, remarkably little is known of its biology. Though the alternating pale-dark-pale-dark pattern on the abdomen is always visible, it can become slightly obscured with age.

Leonora's Dancer
Argia leonorae Garrison
(pp. 64, 74, 76; photo 8d)

Size. Total length: 28–32 mm; abdomen: 21–26 mm; hindwing: 15–19 mm.

Regional Distribution. *Biotic Provinces:* Balconian, Chihuahuan, Tamaulipan, Texan. *Watersheds:* Brazos, Guadalupe, Lower Rio Grande, Nueces, San Antonio, Upper Rio Grande.

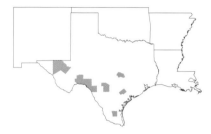

General Distribution. Texas and Nuevo León, Mexico.

Flight Season. May 26 (TX)–Sep. 15 (TX).

Identification. The male has a dark head and pale postocular spots, these variable in size. Sometimes, there is a blue line at the rear of the occiput, but when present the postocular spots never touch this line. Garrison (1994a) reported that on specimens from Reeves County, Texas, these postocular spots were considerably reduced and lacked the blue line at the rear of the occiput. The pterothorax is blue, gradually becoming paler laterally. There is a broad, black middorsal stripe and a dark humeral stripe forked at its upper third; the lower fork is half the width of the upper fork, and both of these may be broadly joined at their upper end. The legs are blue, becoming paler medially, and the inner and outer surfaces of the femora are black. There are postquadrangular cells in both fore- and hindwings. The abdomen is blue with black markings that become more pronounced posteriorly. Segment 1 is blue, with a single antehumeral black spot on each side, as well as a black spot dorsally on the basal 1/2 of the segment. Segment 2 bears an irregular dorsolateral stripe that is constricted medially and expanded in the posterior quarter of the segment, such that it becomes confluent with that of the other side. There is a small black spot laterally on half of segment 2. The apical 1/3 to 1/2 of segments 3–5 is black, narrowing to a point. Segments 6 and 7 are similar to 3–5, but with the black covering most of the segments. The appearance of these segments

Dancers (*Argia*)

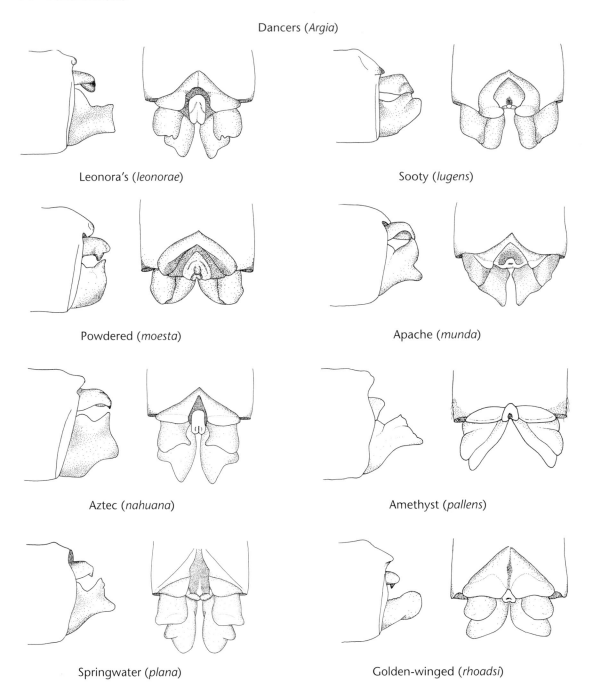

Leonora's (*leonorae*)

Sooty (*lugens*)

Powdered (*moesta*)

Apache (*munda*)

Aztec (*nahuana*)

Amethyst (*pallens*)

Springwater (*plana*)

Golden-winged (*rhoadsi*)

Fig. 14. Dancers (*Argia*): lateral and dorsal views of male caudal appendages (pp. 73–83).

when viewed dorsolaterally is of spear tips directed toward the head. Segments 8–10 are blue, with black on segment 8 extending ventrally to the anterior 1/2 of the segment. The cerci are dark, and the paraprocts are small and quadrad in shape. The cerci each bear a small lobe posteromedially, followed by a smaller, distal tooth. The paraprocts are forked and twice the length of the cerci, and their lower lobes are rounded.

The female, patterned similarly to the male, may be blue or tan. The pale postocular spots are larger, and the dark humeral stripe is forked at 1/2 its length. The mesostigmal plates are triangular and have a costate rim. There is a poorly developed mesostigmal lobe present. There is a small dorsobasal black spot on abdominal segment 1. On segment 2 there is a black dorsolateral stripe, as in the male, but it is reduced and separated dorsally and medially. Segments 3–5 are pale; the distal 1/3 of each segment is black and there are dorsolateral extensions to the basal part of segment. Segments 6 and 7 are like 3–5, but the black dorsolateral areas are confluent dorsally, obscuring the pale middorsal stripe. Segments 8 and 9 are pale, with a black dorsolateral stripe and a narrow pale middorsal line becoming wider on 9. Segment 10 is nearly all pale, its brown areas restricted to the dorsolateral areas. In some specimens in Texas, segments 3–7 are all black dorsally, except for the extreme bases.

Similar Species. Aztec Dancer (*A. nahuana*) is similar, but the dark rings on the middle abdominal segments are truncate and do not taper to points as they do in Leonora's Dancer. Double-striped Blue (*Enallagma basidens*) has a dark humeral stripe divided by a thin pale stripe. Female Variable Dancers (*A. fumipennis*) are essentially identical, and can be reliably separated from Leonora's Dancer only where they are found together, by examination of the mesostigmal plates.

Habitat. Small streams and seepages.

Discussion. This is a small, blue damselfly, uncommon but widely distributed in south-central and western Texas. Leonora's Dancer was first collected in Brooks County, Texas, in 1928. T.W. Donnelly and G.H. Beatty collected it in 1954 at Balmorhea State Park in Reeves County, Texas, and it was known as the "Balmorhea Damselfly" until Garrison formally described it in 1994.

Little is known about its biology. It frequents small streams and seepages, such as the "muddy banked rivulets" of Mustang Creek in Williamson County, Texas, and scattered sedge-ridden swales above the Rio Sabinal in Bandera County, Texas. I have found it associated with sedges in western and central Texas. Females are generally not found with males around water, unless they are mating or laying eggs, the latter of which they do accompanied by the male.

References. Garrison (1994a).

Sooty Dancer
Argia lugens (Hagen)
(pp. 64, 74, 76; photos 8e, 8f)

Size. Total length: 41–50 mm; abdomen: 32–41 mm; hindwing: 25–35 mm.

Regional Distribution. *Biotic Provinces:* Apachian, Balconian, Chihuahuan, Coloradan, Kansan, Navahonian. *Watersheds:* Brazos, Canadian, Colorado (NM), Pecos, Red, Upper Rio Grande.

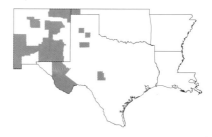

General Distribution. Southwestern United States and Mexico; Texas west to California and Oregon and south to Veracruz.

Flight Season. May 21 (TX)–Oct. 22 (TX).

Identification. The head pattern of the mature male is obscured by heavy black pruinescence. In younger specimens the head appears tan with black markings, and bears a pair of pale postocular spots separated by the occipital bar. The pterothorax is blue to tan, with a thin black middorsal line, and an additional lateral black line on each side contacts the middorsal carina above. The humeral stripe is black with a longitudinal stripe running down the middle of the mesepimeron. These stripes are often confluent at about 1/3 the length of the humeral stripe, isolating a pale spot anteriorly. In older individuals, the pterothorax darkens entirely, and the color pattern is obscured by pruinescence. The forewing has five or six postquadrangular cells. The abdomen is tan with black markings, becoming darker dorsally

Dancers (*Argia*)

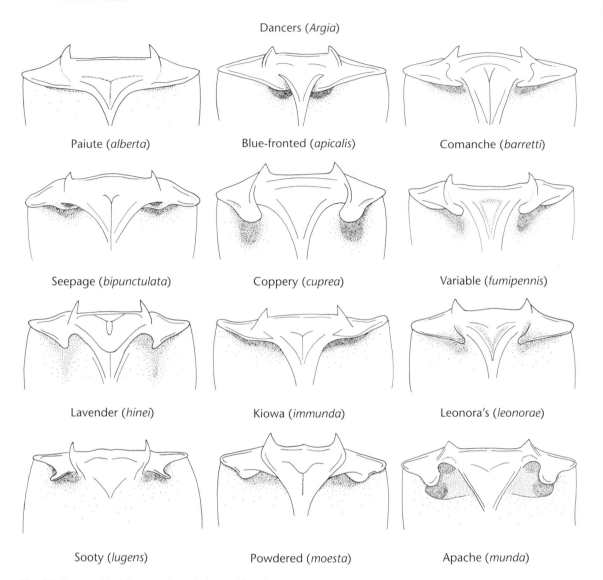

Paiute (*alberta*)   Blue-fronted (*apicalis*)   Comanche (*barretti*)

Seepage (*bipunctulata*)   Coppery (*cuprea*)   Variable (*fumipennis*)

Lavender (*hinei*)   Kiowa (*immunda*)   Leonora's (*leonorae*)

Sooty (*lugens*)   Powdered (*moesta*)   Apache (*munda*)

Fig. 15. Dancers (*Argia*): mesostigmal plates of females (pp. 62–78).

in mature individuals. Segments 1 and 2 are marked with black stripes extending the full length of the segment dorsolaterally. The stripe widens apically on segment 2 to meet, or nearly meet, on the other side. Segments 3–7 are marked with a dark apical ring, which covers 1/5 to 1/4 the length of the segment, and a dark lateral stripe that is confluent with the apical ring. On more mature individuals, the stripes may be more extensive dorsally. Segments 8 and 9 are nearly all black in mature individuals, with only a few pale spots. Segment 10 becomes

dark in older individuals, but usually retains its pale color laterally. The cerci and paraprocts are only 1/2 the length of segment 10. The cerci are bifid, with a pointed tooth on both lobes. The paraprocts each have a small black, dorsally directed tooth on the apex of the superior lobe.

Female coloration is similar to that of younger males, through the dark markings are less extensive. The dark lateral stripes are often contiguous with the middorsal carina, as in males, but are separated from any other black. The dark humeral stripe and the stripe in the middle of the mesepimeron are sometimes as in the male, but are usually touching at their origin. The mesostigmal plates have distinctly diverging thumblike posterior lobes. The mesepisternal tubercles are reduced. The legs are generally pale, with some dark markings on the femora. The abdominal pattern is similar to that of the male, with dark stripes laterally on segments 3–7, widening apically to become confluent dorsally. The subapical spots on the lower sides are often contiguous with the narrow apical ring of each segment. There is a dorsolateral stripe on segments 8 and 9.

Similar Species. Powdered Dancer (*A. moesta*) is somewhat similar, and mature males are almost white, rather than black. Females lack the additional dark stripes on either side of the middorsal stripe. Tonto Dancer (*Argia tonto*) is bluer and lacks the additional dark stripes on the middorsum of the thorax.

Habitat. Rocky desert rivers and streams.

Discussion. This species may be the most abundant damselfly at certain desert streams where it perches on emergent and marginal rocks. Nevertheless, its behavior has never been studied.

Powdered Dancer
Argia moesta (Hagen)
(pp. 64, 74, 76; photos 9a, 9b, 9c)

Size. Total length: 37–42 mm; abdomen: 28–37 mm; hindwing: 22–29 mm.

Regional Distribution. *Biotic Provinces:* Apachian, Austroriparian, Balconian, Carolinian, Chihuahuan, Coloradan, Kansan, Navahonian, Tamaulipan, Texan. *Watersheds:* Arkansas, Bayou Bartholomew, Brazos, Canadian, Cimarron, Colorado, Colorado (NM), Guadalupe, Lower Rio Grande, Mississippi, Neches, Nueces, Ouachita, Pecos, Red, Sabine, San Antonio, San Jacinto, St. Francis, Trinity, Upper Rio Grande, White.

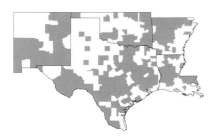

General Distribution. Throughout southern Canada and the United States, except for the Pacific Northwest; also south into Mexico.

Flight Season. Year-round in south Texas.

Identification. The head of young males is dark brown, with pale labium, labrum, clypeus, genae, and frons. Older individuals have a blue-white pruinosity dorsally on the head. There is a dark-brown middorsal thoracic stripe and a pale antehumeral stripe approximately 1/2 its width. A dark, narrow humeral stripe extends nearly to the antealar carina and is confluent with the mesepimeral stripe anteriorly. The legs are pale, with dark stripes on the outer surfaces of the femora, and the tarsi are black. The pterothorax, coxae, and femora are all pruinose in older individuals. There are four and five postquadrangular cells in the fore- and hindwings, respectively. The abdomen is abdominal segments 9 and 10 are brown basally and distally. The middle segments are much darker, almost black, and segments 1–6 are tan laterally. The abdomen in older individuals becomes black, except for pale areas laterally on segment 1 and the basal rings on segments 3–7. The cerci, when viewed laterally, are blunt and rounded, and each bears a dorsally directed tooth near its apex.

Females are patterned similarly to males, and can be tan or blue. The narrow middorsal carina is outlined in black, and a dark humeral stripe is barely visible. The metapleural suture is usually unmarked, but often bears a thin dark line and a brown spot in close proximity to the antealar carina. Each mesostigmal plates has a distinct posterior lobe bearing a low ridge that curves over the middle 1/2 of the plate, when viewed dorsally. Mesepisternal tubercles are present. The legs are pale, but with dark stripes on the outer surfaces of the femora and the inner surfaces of the tibiae; the tarsi and accompanying spurs are dark. The abdomen is pale blue or brown basally, becoming darker posteriorly, and a pale stripe runs its full length middorsally, but is constricted apically at each segment. A wide, dark-brown longitudinal stripe parallels the middorsal stripe, except for pale basal rings. Ventrolaterally, a brown spot is visible at the apex of segments 2–7. Segments 8–10 are pale with a dorsolateral dark stripe on segments 8 and 9.

Similar Species. Sooty Dancer (*A. lugens*) is larger and darker, with a dark stripe on each side of the middorsal thoracic stripe. Older males are also nearly black, lacking the white pruinescence seen in Powdered Dancer. The top of the middle abdominal segments in Tonto Dancer (*A. tonto*) are blue or violet. Female Blue-fronted Dancers (*A. apicalis*) are very similar, but smaller, and the last few abdominal segments are generally darker.

Habitat. Swift currents of rivers and lakes with emergent stones and rocky shores.

Discussion. Females in the southeastern United States may have more extensive black markings on the thorax and abdomen, such that they appear more similar to males. They have a wide black mid-dorsal stripe that is sometimes divided on each side by a thin pale stripe. The humeral stripe may also be wide and forked medially.

A study in an Ohio stream revealed that individuals will move as much as 185 m away from the stream. Females spend most of their time 75–150 m from the water. Males become pruinose, starting middorsally on the thorax, and become sexually mature when this pruinescence reaches beyond the black midfrontal stripe. Females are not receptive to males until they become blue. This color change occurs two days after maturation.

Next to Dusky Dancer (*A. translata*), Powdered Dancer is probably the most widely distributed species in the genus, occurring from about 45° N latitude in Canada to about 20° N in Mexico. Powdered Dancer is adaptable when invading new areas, by laying eggs in the previously unutilized surface of *Salix* roots. Mating and egg-laying average 22 and 47 minutes, respectively, and females turn dark while in tandem. Tandem pairs will aggregate in large numbers to lay eggs in roots, stems, debris, and algae, often submerging themselves more than a meter for periods up to an hour. Two additional studies have shown the first symphoretic association of Powdered Dancer with a nonbiting midge (*Nanocladius branchicolus*) and the effect on swimming speed of individuals losing their caudal lamellae. If two of these lamellae are missing, larger individuals swim faster than smaller ones, but statistically slower than individuals that retain two or three lamellae.

References. Bick and Bick (1972), Borror (1934), Dosdall and Parker (1998), Garrison (1994a), Johnson (1973c), Robinson et al. (1991), Walker (1913), Williamson (1906a).

Apache Dancer
Argia munda Calvert
(pp. 64, 74, 76; photo 9d)

Size. Total length: 36–40 mm; abdomen: 29–32 mm; hindwing: 23–27 mm.

Regional Distribution. *Biotic Provinces:* Apachian, Chihuahuan. *Watersheds:* Colorado (NM), Upper Rio Grande.

General Distribution. Southwestern United States (Arizona, New Mexico, and Texas) south into Mexico.

Flight Season. May 21 (TX)–Oct. 21 (TX).

Identification. The face of the male is bright blue, with a pale labrum. The postocular spots are large, blue, and connected by an occipital bar. The mid-dorsal stripe is black and about 1/4 as wide as the blue antehumeral stripe. The black humeral stripe starts narrowing posteriorly at 1/2 its length. The legs are blue, with black maculation on the outer femoral surfaces and inner tibial surfaces, and the tarsi are dark. There are five and four postquadrangular cells in the fore- and hindwings, respectively. The abdomen is blue, with segment 1 bearing a black spot at its extreme base. There is a black anteapical spot on each side of segment 2, and segment 3 has an apical black spot on each side. Segments 4–6 each have subbasal and subapical black spots that increase in size posteriorly. Segment 7 is nearly all black, except for a basal blue ring. Segments 3–6 each bear a distal black ring. Segments 8–10 are blue. The cerci are dome-shaped.

The coloration of the female is similar to that of the male, but the blues are replaced by paler tan and violet colors. The middorsal carina is pale and outlined on either side by a black stripe. There is a well-developed posterior lobe on each mesostigmal plate, but most notably a large, deep mesepisternal pit is visible below each plate. Mesepisternal tubercles are lacking. The legs are not marked with as much black as in the males. The abdomen is pale violet dorsally. Segment 1 is black basally, and segments 2–6 each have a black subbasal spot and a black subapical spot, and the two may be confluent. Segment 7 is completely black dorsally with only a blue basal ring. Segments 8–10 are completely pale.

Similar Species. Springwater Dancer (*A. plana*) is similar, but smaller in size, and the top of the head is largely black, not blue. The middorsal thoracic stripe is also as broad as the pale antehumeral area in Springwater Dancer. Comanche Dancer (*A. barretti*) has a black lateral stripe on abdominal segments 8–10, a stripe not present in Apache Dancer.

Habitat. Primarily found at desert streams.

Discussion. This is an average-sized southwestern species originally described from Arizona and Mexico as a variety of Vivid Dancer (*A. vivida*). It was first reported from Limpia Creek in Jeff Davis County, Texas, and has been infrequently collected from this stream, from Oak Creek in Big Bend National Park, and within the Guadalupe Mountains. It is also known from a couple of localities in New Mexico, but it seems never to be common where it is found.

References. Gloyd (1958).

Aztec Dancer
Argia nahuana Calvert
(pp. 64, 74, 84; photo 9e)

Size. Total length: 28–35 mm; abdomen: 23–28 mm; hindwing: 18–23 mm.

Regional Distribution. *Biotic Provinces:* Apachian, Austroriparian, Balconian, Chihuahuan, Coloradan, Kansan, Navahonian, Texan. *Watersheds:* Arkansas, Brazos, Canadian, Cimarron, Colorado, Colorado (NM), Guadalupe, Neches, Nueces, Pecos, Red, Sabine, San Antonio, Trinity, Upper Rio Grande.

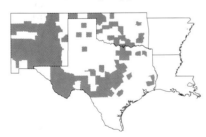

General Distribution. Western United States from Oregon south to California and east to Texas and Oklahoma; also south into Mexico.

Flight Season. Mar. 30 (TX)–Nov. 18 (NM).

Identification. The male's head is dark blue, with large postocular spots broadly connected. The pale-blue antehumeral stripe is nearly as wide as the black middorsal stripe, and the dark humeral stripe is forked posteriorly at about 1/2 its length. There are four and three postquadrangular cells in the fore- and hindwings, respectively. The legs are pale blue, with black stripes on the outer femoral and inner tibial surfaces. The tarsi, pale blue, are armed with black spurs. The abdomen is blue, with a large black anteapical spot and a smaller dark spot below on segment 2. The apical fourth of segments 3–6 is black. Segment 7 is generally black dorsally, with a blue middorsal streak of varying width and a blue basal ring. Segments 8–10 are blue, with black ventrolateral spots on segments 8 and 9 that may be confluent with one another. The cerci are distinct, in dorsal view, with a prominent medially directed lobe. Laterally, these appendages are no more than 2/3 the length of the paraprocts, and an apical, ventrally directed tooth is often visible.

The coloration of the female is similar to that of the male, but is pale brown instead of blue. The middorsal carina is usually pale brown, and the pale antehumeral stripe is nearly as wide as the dark middorsal stripe. The mesostigmal plates are recognizable because of the broad transverse expanse of each posterior lobe. Abdominal segments 3–6 bear a basal apical spot and an apical black spot that generally are not confluent with one another. Segment 7 is similar to that of the male but with a smaller apical spot below. Segments 8 and 9 both have a black spot dorsolaterally that may extend the full length of each segment. There is often an additional apical spot laterally on each of these segments. Segment 10 is pale.

Similar Species. The lateral black markings on the abdomen of Leonora's Dancer (*A. leonorae*) each taper to a point (apearing like spear tips dorsally), unlike those of Aztec Dancer. Variable Dancer (*A. fumipennis*) and Lavender (*A. hinei*) Dancer both have a complete, dark, ventrolateral stripe on segments 8–10 (Aztec Dancer lacks any ventrolateral markings on 10) and are largely violet. Springwater Bluet (*A. plana*) has dark markings anterolaterally on segments 5 and 6, and the postocular spots are not broadly joined. Double-striped Bluet (*Enallagma basidens*) has a pale stripe dividing the dark humeral stripe.

Habitat. Small, shallow, clear-water streams, fully exposed to sunlight with only moderate marginal vegetation.

Discussion. Bick and Bick (1958) studied the Odonata at Cowan Creek, Marshall County, in southern Oklahoma. Aztec Dancer was by far the dominant species at this creek, where they observed and documented its egg-laying behavior: "Male and female perched in full sunlight on a blade of grass six inch-

Dancers (*Argia*)

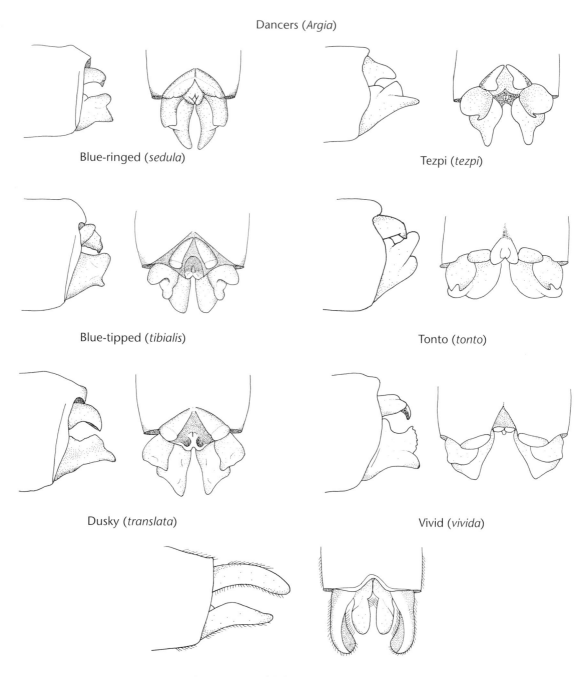

Blue-ringed (*sedula*)

Tezpi (*tezpi*)

Blue-tipped (*tibialis*)

Tonto (*tonto*)

Dusky (*translata*)

Vivid (*vivida*)

Aurora Damsel (*Chromagrion conditum*)

Fig. 16. Dancers (*Argia*): and Aurora Damsel
(*Chromagrion conditum*): lateral and dorsal views of male
caudal appendages (pp. 83–89).

es from the margin of the creek where the water was one inch deep. . . . The abdomen of the female was bent at a sharp angle, and its tip touched the plant one half inch below the water surface, where eggs were apparently deposited. She probed for a few seconds with the tip of her abdomen, remained motionless for two and one half minutes, probed briefly and remained motionless for five minutes. The pair visited three more blades of grass where the female alternately probed and remained motionless but for only 30 seconds at each blade."

Amethyst Dancer
Argia pallens Calvert
(pp. 64, 74, 84; photo 9f)

Size. Total length: 32–35 mm; abdomen: 25–27 mm; hindwing: 20–22 mm.

Regional Distribution. *Biotic Provinces:* Apachian, Chihuahuan. *Watersheds:* Colorado (NM), Upper Rio Grande.

General Distribution. Southwestern United States (Arizona, New Mexico, and Texas); also south through Mexico to Guatemala.

Flight Season. May 27 (TX).

Identification. The frons and top of the head in males is a reddish-violet color, largely unmarked except for a dark postocular bar extending medially from the margin of each eye. The back of the head is pale and the thorax is reddish violet. The pterothorax bears a dark, thin middorsal stripe about 1/4 as wide as the pale antehumeral stripe. The dark humeral stripe exists as a hairline and is never forked, but is usually widened at each end, occasionally with the middle of the stripe completely lacking. The legs are pale with brown stripes on the outer surfaces. There are four and three postquadrangular cells in the fore- and hindwing, respectively. The abdomen is reddish violet with a black spot laterally on the apical 1/4 of segments 2–6. Spots on segments 4–6 are often confluent with a black apical

ring. The lateral spot on segment 7 is drawn out anteriorly to form a stripe extending nearly the length of the segment. Segments 8–10 are bluish, with a dark lateral stripe that may or may not be present, especially on 10. The caudal appendages are pale, the cerci about 1/5 as long as the paraprocts.

The female is patterned similarly to the male, but the pale ground color is brown instead of violet. Segments 9 and 10 lack a ventrolateral stripe, though one may be faint on segment 8. The mesostigmal plates are bilobed posteriorly, and mesepisternal tubercles are absent.

Similar Species. Variable Dancer (*A. fumipennis*) is a deeper violet color and has a black ventrolateral stripe on segments 8–10, not just on 8 as in Amethyst Dancer. Segment 7 is largely black, and the humeral stripe is forked in both Variable Dancer and Lavender Dancer (*A. hinei*). The postocular spots in Lavender Dancer are also more broadly outlined with black.

Habitat. Small desert streams.

Discussion. This species was only recently found in west Texas and New Mexico, where it is no doubt more widely distributed than our current knowledge reveals. At present, details of its biology are unknown.

References. Abbott (2001).

Springwater Dancer
Argia plana Calvert
(pp. 64, 74, 84; photos 10a, 10b)

Size. Total length: 34–40 mm; abdomen: 26–33 mm; hindwing: 22–25 mm.

Regional Distribution. *Biotic Provinces:* Apachian, Austroriparian, Balconian, Carolinian, Chihuahuan, Coloradan, Kansan, Navahonian, Tamaulipan, Texan. *Watersheds:* Arkansas, Brazos, Canadian, Cimarron, Colorado, Colorado (NM), Guadalupe, Lower Rio Grande, Mississippi, Nueces, Pecos, Red, Sabine, San Antonio, San Jacinto, St. Francis, Trinity, Upper Rio Grande, White.

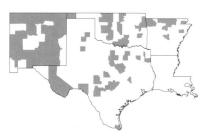

General Distribution. Central United States from Arkansas north to Wisconsin and west to Arizona; also south through Mexico to Guatemala.

Flight Season. Mar. 10 (TX)–Oct. 8 (OK).

Identification. The head of the male is blue, with a pair of postocular spots that are widely confluent with the compound eyes and separated by a pale occipital bar. The antehumeral stripe is pale blue, bordered by a dark middorsal stripe of about the same width. The black humeral stripe is unforked and about 1/2 the width of the humeral stripe at its widest; it narrows considerably at its middle and then widens again at the upper end. The pterothorax becomes paler laterally. The metapleural suture is narrowly outlined by a black line. The legs are blue, with heavy black markings on the outer femoral and inner tibial surfaces; the tarsi are black and armed with black spurs. There are generally four and three postquadrangular cells in the fore- and hindwings, respectively. The abdomen is bright blue, with a small black basal spot dorsally and a larger lateral spot apically on segment 1; on segment 2 the two spots may be connected to form a lateral stripe. The dark subbasal spot on segments 3–6 tapers apically on the posterior segments; the dark apical spots on segments 3–6 are confluent medially. Segment 7 is nearly all black, except for a pale apical ring and a middorsal line for half its length. Segments 8–10 are entirely pale blue. Laterally, there is a short, apical, ventrally directed black tooth on the cerci. The paraprocts are bifid, the superior lobe serrated and twice as long as the inferior lobe. Individuals from western New Mexico are violet, not blue.

The female is pale brown, the head and thorax pattern similar to that of the male. The mesostigmal plates are broadly flattened and unnotched, but slightly angulate posteriorly, and mesepisternal tubercles are present. The legs are much paler than in the male, with only limited black markings on the femora and tibiae, and the tarsi are brown. The abdominal color pattern closely approximates that of the male, but the black markings are more extensive and the general pattern is more variable. There is a basal black spot dorsolaterally on segment 9.

Similar Species. Aztec Dancer (*A. nahuana*) and Variable Dancer (*A. fumipennis*) both have a forked humeral stripe. Lavender Dancer (*A. hinei*) and Variable Dancer both have a black ventrolateral stripe on abdominal segments 8–10, this stripe absent in Springwater Dancer. Apache Dancer (*A. munda*) is larger, its legs are paler, and the middorsal thoracic stripe is of about the same width as the humeral stripe. Vivid Dancer (*A. vivida*) is very similar, and where these two species overlap, in the western part of our region, individuals should be checked carefully. There are no reliable field marks to separate these, but their ranges do not overlap extensively.

Habitat. Small, shallow, canopied spring seepages with clay substrate.

Discussion. Violet forms of Springwater Dancer appear in western Texas; I have collected them in Brewster and Jeff Davis counties. A study of egg-laying behavior of a population of this species at Cowen Creek in Marshall County, Oklahoma, revealed pairs in which the female clasped a small dead twig and the male was supported by only the female. The female curved her abdomen slightly and deposited eggs on or in the clay of this spring at a depth of no more than half an inch. The pairs remained motionless for 15 minutes with little probing. A female was also seen laying eggs in clay outside the spring itself.

Another study of a population in a southern Oklahoma stream found that males seized females predominantly at the water's edge as females approached. Mating occurred quickly at an average distance of 1.5 m from the water's edge. Pairs did not change perch location, but were seen shifting positions on the perch during mating, 2–7 times. Mating and egg-laying lasted an average of 27 and 47 minutes, respectively. Females were observed laying eggs almost exclusively in *Nasturtium* (watercress) and debris.

References. Bick and Bick (1958, 1965c, 1972), Gloyd (1958), Hornuff (1968).

Golden-winged Dancer
Argia rhoadsi Calvert
(pp. 64, 74, 84; photo 10c)

Size. Total length: 34–35 mm; abdomen: 27–28 mm; hindwing: 19–21 mm.

Regional Distribution. *Biotic Provinces:* Balconian, Tamaulipan. *Watershed:* Lower Rio Grande.

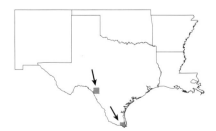

General Distribution. Texas south into Mexico.
Flight Season. Jul. 17 (TX)–Oct. 5 (TX).
Identification. The face of the male is blue, and there are two large postocular spots separated by a pale occipital bar. The antehumeral stripe is pale blue and equal in width to the black middorsal thoracic stripe. The black humeral stripe is approximately 1/3 the width of the antehumeral stripe and may be forked at its upper end. The lower fork is often faint and irregular in shape. There is a short black stripe at the upper end of the interpleural suture, contiguous with the black antealar carina. The legs are pale blue, with dark stripes on their outer femoral surfaces and inner tibial surfaces; the tarsi, brown, are armed with black spurs. The wings are flavescent or amber with a nearly black pterostigma. There are four and three postquadrangular cells in the fore- and hindwings, respectively. Abdominal segment 1 is entirely blue. Segment 2 is blue with a dark dorsolateral stripe widening posteriorly. Segments 3–7 are largely black, with metallic reflections dorsally. From above, each segment has a blue basal ring and a pale middorsal stripe that may not reach the apex. Segments 8–10 are entirely blue. The cerci and paraprocts are both entire and not bifid; the cerci are no more than 2/3 the length of the paraprocts and possess a ventrally directed anteapical tooth.

The female is patterned similarly to the male, but is tan, and the black markings are less extensive. The narrow black middorsal stripe is approximately 1/7 the width of the mesepisternum. There are short dark stripes posteriorly on the interpleural and metapleural sutures. The mesostigmal plates are mostly dark with pale lateral edges, and dorsally there is a posteromedially directed lobe projecting over a shallow dark pit. Mesepisternal tubercles are vestigial or entirely absent. The legs, tan with much less extensive black markings, are armed with black spurs. The wings are amber, with a paler pterostigma than in the male. The abdomen is tan in color with an ill-defined pattern. There is a dark stripe laterally, widening toward the apex of segment 2. Segments 3–7 each have a narrow blue basal ring and a pale middorsal stripe, running their entire length. Segments 8–10 are entirely light tan and unmarked.
Similar Species. The smaller Blue-ringed Dancer (*A. sedula*) may have tinted wings, but they are not as dark as in Golden-winged Dancer, and the thorax is darker. Male Blue-ringed Dancers also have black markings ventrolaterally on segments 8–10 that are not present in Golden-winged Dancer.

Habitat. Lagoons and pools formed at edges of streams and rivers.
Discussion. There is not much known about the biology of this Mexican species. The larva was described from exuviae and terminal instars collected at Laguna de Atezca, in the Mexican state of Hidalgo, where they were found clinging to roots of water hyacinth, *Eichhornia* sp., at the edge of the lagoon. Laboratory emergences occur in the late morning.

The species is presently known from only two localities in the United States. It has been found at Fort Clark Springs in Kinney Co., Texas, where a well-established poulation exists, and is historically known from the Lower Rio Grande Valley.

References. Novelo-G. (1992).

Blue-ringed Dancer
Argia sedula (Hagen)
(pp. 64, 80, 84; photo 10d)

Size. Total length: 29–34 mm; abdomen: 22–28 mm; hindwing: 17–21 mm.
Regional Distribution. *Biotic Provinces:* Apachian, Austroriparian, Balconian, Carolinian, Chihuahuan, Kansan, Navahonian, Tamaulipan, Texan. *Watersheds:* Arkansas, Bayou Bartholomew, Brazos, Canadian, Cimarron, Colorado, Colorado (NM), Guadalupe, Lower Rio Grande, Mississippi, Neches, Nueces, Ouachita, Pecos, Red, Sabine, San Antonio, San Jacinto, St. Francis, Trinity, Upper Rio Grande, White.

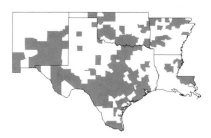

General Distribution. From Florida north to Pennsylvania, west to California, and south into Mexico.
Flight Season. Mar. 19 (TX)–Nov. 14 (OK).
Identification. The head of the males is dark, and the blue postocular spots are often obscured. The middorsal stripe is black. The broad humeral stripe is black and nearly twice as wide as the pale antehumeral stripe. The black stripe on the metapleural suture widens. The rest of the pterothorax is blue, becoming paler and almost yellow ventrally. The legs

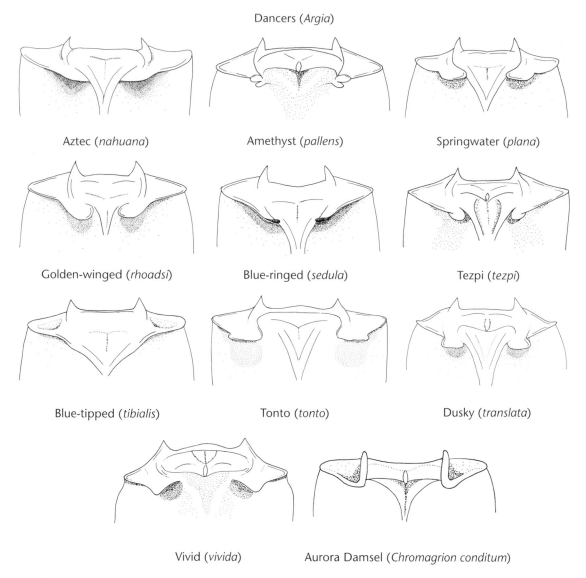

Fig. 17. Dancers (*Argia*) and Aurora Damsel (*Chromagrion conditum*): mesostigmal plates of females (pp. 79–89).

are largely black, except for the medial surfaces of the femora and the lateral surfaces of the tibiae and tarsi. There are four and three postquadrangular cells in the fore- and hindwings, respectively. Abdominal segment 1 is blue, with dark brown dorsobasally and laterally. Abdominal segment 2 is black dorsally, with a small pale-blue spot basally or on each side of the midline. Segments 3–7 are black dorsally, with a blue basal ring. Ventrolateral areas of the segments are paler and confluent. Segments

8–10 are blue, with a black ventrolateral stripe extending to the apical portion of segment 10. The cerci are straight in lateral view, with a ventrally projecting tooth. The paraprocts are bifid, with a superior lobe rounded and directed dorsally. The inferior lobe is strongly serrated and projects posteroventrally.

The head of the female is largely a combination of pale-brown and olivaceous markings. The pterothorax is brown dorsally, becoming paler ventrally.

The middorsal and humeral stripes are reduced to thin black hairlines, and the latter is often entirely absent. The legs, pale brown with dark stripes on the outer surfaces, are armed with black spurs. The mesostigmal plates are long and strongly erect, appearing almost perpendicular to the mesepisternum, when viewed laterally, and mesepisternal tubercles are small or entirely absent. The abdomen is pale brown dorsally on segments 2–7. The ventrolateral stripes and basal rings are often ill-defined, but touches of blue or green are evident. Segments 8–10 are pale and lack dark markings.

Similar Species. Golden-winged Dancer (*A. rhoadsi*) has deeper amber-colored wings and a bluer abdomen. Golden-winged Dancer also lacks dark ventrolateral stripes on segments 8–10. Dorsally, the dark middorsal stripe on abdominal segments 3–6 is divided by a pale area in Kiowa Dancer (*A. immunda*).

Habitat. Lakes, ditches, streams, and rivers with gentle current and dense vegetation.

Discussion. Blue-ringed Dancers are more prone to perching on vegetation, often in the shade, than most other dancers. Pairs require 10–15 minutes to mate, and egg-laying occurs in tandem, often in large numbers. A mark-recapture study on adult Blue-ringed Dancers at a small creek on the campus of the University of Texas at Arlington revealed that males had an average daily probability of survivorship of 0.79, and that activity was closely correlated with bright sun.

References. Dunkle (1990), Robinson et al. (1983).

Tezpi Dancer
Argia tezpi Calvert
(pp. 64, 80, 84; photo 10e)

Size. Total length: 35–41 mm; abdomen: 28–35 mm; hindwing: 22–26 mm.
Regional Distribution. *Biotic Province:* Apachian. *Watershed:* Colorado (NM).

General Distribution. Arizona and New Mexico south through Mexico to Costa Rica.
Flight Season. Jun. (NM).
Identification. This is a large, dark-black Mexican species found in our region only in Southwestern New Mexico. The face and head are dark, except for the small pale postocular spots and occipital bar. The antehumeral stripe is pale and narrow, only 1/10 or less as wide as the dark-violet or metallic middorsal stripe. The antehumeral stripe in females is broader, extending 1/4 to 1/2 the width of the middorsal stripe, and the humeral stripe is usually forked. In males the dark humeral stripe is wide, covering most of the mesepimeron. The legs are dark brown or black, and the wings are clear or often slightly amber. Abdominal segments 1–7 are black dorsally, with metallic reflections. Segments 3–7 each have a pale, narrow basal ring that may be interrupted dorsally. Segment 8 is entirely black, and segments 9 and 10 are largely black, usually covering more than 2/3 of the segments. The male cerci are 1/2 the length of the paraprocts, when viewed dorsally. The female mesostigmal plates each bear a small medially directed posterior lobe, and the mesepisternal tubercles are well developed.

Similar Species. Dusky Dancer (*A. translata*) is an equally large and dark species that may be found flying with Tezpi Dancer. It can be distinguished by its clear wings (usually amber in Tezpi Dancer) and divided humeral stripe.

Habitat. Streams and rivers of the arid Southwest.

Discussion. This is a Mexican species that barely ranges into the United States, where it is known from just Arizona and New Mexico. It may be locally common, but it is not widely distributed in our area. It is often found alongside Dusky Dancer throughout much of its range.

Blue-tipped Dancer
Argia tibialis (Rambur)
(pp. 66, 80, 84; photo 10f)

Size. Total length: 30–38 mm; abdomen: 23–30 mm; hindwing: 18–24 mm.
Regional Distribution. *Biotic Provinces:* Austroriparian, Balconian, Carolinian, Kansan, Texan. *Watersheds:* Arkansas, Bayou Bartholomew, Brazos, Canadian, Cimarron, Colorado, Mississippi, Neches, Ouachita, Red, Sabine, San Jacinto, St. Francis, Trinity, White.

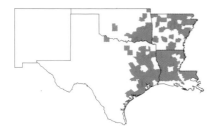

General Distribution. Eastern United States from Florida to Pennsylvania, west to Minnesota, and south to Texas.

Flight Season. Mar. 3 (LA)–Sep. 8 (TX).

Identification. The head of the male is largely black, and a pair of small postocular spots are separated by an occipital bar, visible in younger individuals. The antehumeral stripe is violet and nearly twice as wide as the black middorsal thoracic stripe. The broad black humeral stripe ranges from 1/2 to the entire width of the antehumeral stripe. The metapleuron and ventral side of the thorax are pale yellow, often becoming darker and covered with pruinescence in older males. There is a distinct black stripe along the metapleural suture. The legs are largely black, with a thin pale stripe on the outer surface of the femora and a pale broader stripe on the tibiae; the tarsi are black and armed with short spurs. There are four and three postquadrangular cells in the fore- and hindwings, respectively. The abdomen is largely black dorsally, with only a thin pale middorsal line on segments 1–7. Segments 3–7 each have a narrow pale basal ring, and their ventrolateral margins are pale for 3/4 of their length. Segment 8 is entirely black, and segments 9 and 10 are blue dorsally, with a wide black ventrolateral stripe. The cerci are short, only 1/3 the length of segment 10, and distinctly bifid dorsally. The paraprocts are also bifid, the superior lobe larger and generally extending beyond the inferior lobe.

The female color is generally light tan or pale blue. The head is colored much like that of a male, but the pale areas are generally more extensive. The postocular spots are larger and sometimes partially confluent with the occipital bar. There is a pair of pale dorsal spots on the prothorax in addition to the larger lateral spots. The pterothorax is like that of the male, but the wide humeral stripe is forked at its upper end. There are no posterior lobes on the mesostigmal plates, and the mesepisternal tubercles are lacking entirely. The abdominal color pattern is the same as that of the male on segments 1–7. Seg-

ments 8 and 9 are black. On segment 9 there is generally a pale middorsal spot that may be confluent with pale lateral areas apically. Segment 10 is pale, and dark basal spots are sometimes visible.

Similar Species. Blue-tipped Dancer is the only dancer in the south-central United States whose males have abdominal segment 8 black and segments 9 and 10 blue. Females are probably most easily distinguished from other dancers by the contrasting pale bottom and sides of the thorax and the generally black abdomen.

Habitat. Streams and rivers of various flows, also sloughs.

Discussion. Bick (1957) wrote that in Louisiana, the specimens he collected were ". . . from a wider range of habitats than any *Argia* in Louisiana." In the pinelands, Blue-tipped Dancer occurred along swift creeks (73%), sluggish streams (9%), and near sloughs, swamps, or ponds (5%). This species usually perches on the ground, but does perch on vegetation and in shade more than most other dancers. Egg-laying occurs in tandem, and eggs are generally deposited in floating eelgrass or debris, but sometimes in wet wood above the waterline. Members of this species often occur in large aggregations.

Tonto Dancer
Argia tonto Calvert
(pp. 66, 80, 84; photos 11a, 11b)

Size. Total length: 38–44 mm; abdomen: 30–35 mm; hindwing: 25–29 mm.

Regional Distribution. *Biotic Provinces:* Apachian, Navahonian. *Watershed:* Colorado (NM).

General Distribution. Southwestern New Mexico and Arizona south into Mexico.

Flight Season. Jul. (NM) Aug. (NM).

Identification. The is a large violet damselfly of the Southwest. The face of the male is blue, and the top of the head is violet with the typical black markings.

The pterothorax is violet and marked with a black middorsal stripe that is less than 1/4 as width as the pale antehumeral stripe. The humeral stripe is widest anteriorly and narrows posteriorly to form a thin line. The legs are largely black, and the wings are clear or occasionally amber. There are five and four postquadrangular cells in the fore- and hindwing, respectively. The abdomen is violet basally, becoming blue distally. Segments 4–6 each have an apical black spot confluent with an apical ring. Segment 7 is variable, the dorsum either all black or sometimes bearing a triangular spot. Segments 8–10 are blue, with a ventrolateral black stripe.

The female is similar, but the middorsal thoracic stripe is wider. Abdominal segments 3–5 are variable, with a dorsolateral stripe on each side. The posterior lobe of the prothorax is trilobed in the female. The mesostigmal plates bear posterior lobes that are 1/2 the width of the plates themselves, and mesepisternal tubercles are absent. Older males and females become pruinescent with age.

Similar Species. Powdered Dancer (*A. moesta*) is similar, but its middle abdominal segments are mostly dark or white, never violet. Both male and female Sooty Dancers (*A. lugens*) have a dark stripe paralleling the middorsal thoracic stripe on each side. The violet form of Springwater Dancer (*A. plana*) is smaller, and its middorsal thoracic stripe is broader, as wide as the pale antehumeral area.

Habitat. Large streams and rivers of the arid Southwest.

Discussion. Tonto Dancer is a beautiful large violet Mexican species whose northern range just enters southwestern New Mexico and Arizona. Males are often seen perched on exposed rocks in the middle of the streams or rivers they patrol along.

Dusky Dancer
Argia translata Hagen *in* Selys
(pp. 66, 80, 84; photo 11c)

Size. Total length: 32–38 mm; abdomen: 25–33 mm; hindwing: 19–23 mm.

Regional Distribution. *Biotic Provinces:* Apachian, Austroriparian, Balconian, Carolinian, Chihuahuan, Kansan, Navahonian, Tamaulipan, Texan. *Watersheds:* Arkansas, Brazos, Canadian, Cimarron, Colorado, Colorado (NM), Guadalupe, Lower Rio Grande, Mississippi, Nueces, Ouachita, Pecos, Red, San Antonio, San Jacinto, St. Francis, Trinity, Upper Rio Grande, White.

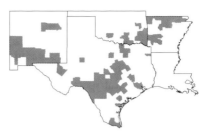

General Distribution. Georgia north to Ontario, Canada; east to Oklahoma and Texas and south through Central America to Argentina.

Flight Season. Mar. 17 (TX)–Dec. 26 (TX).

Identification. In males the front of the head is pale, becoming almost entirely black with age. The top of the head is largely black, with pale postoccipital spots and an occipital bar becoming obscured with age. The head is noticeably hairy. The black middorsal stripe is broad, and the pale-yellow antehumeral stripe is 1/3 the width of the middorsal stripe. In younger individuals the dark humeral stripe may be almost completely divided longitudinally by a pale stripe. These dark stripes become fused in older males. There is a black stripe on the metapleural suture. The rest of the pterothorax is pale yellow, becoming densely pruinose in older individuals. The legs are black, with a narrow pale stripe on their outer surface. There are five and four postquadrangular cells in the fore- and hindwings, respectively. Abdominal segment 1 is black, with a large pale lateral spot on each side and a laterally elongated apical spot. Segment 2 is also black, with a pale middorsal stripe often broken medially. Segments 3–7 are largely black, with a pale-yellow or blue basal ring and a pale ventrolateral stripe on each side. Segment 8 is black, with a strongly irregular blue ring basally whose sides are exaggerated and extend posteriorly. The top of segment 9 is blue for 1/4 or more of its length; the apical portion of the segment is black and strongly sinuate dorsally and laterally. Segment 10 is black, with a small pale spot laterally. The cerci are strongly decurved when viewed laterally. The paraprocts are dark and branched, the dorsally directed superior lobe rounded and the longer inferior lobe directed posteroventrally.

The female is similar to the male, with large pale postoccipital spots and an occipital bar. The posterior lobes of the mesostigmal plates are slightly constricted at their base when viewed dorsally; laterally, these lobes appear as a thin linear projection. Mesepisternal tubercles are well developed. The antehumeral stripe is pale and 2/5 as wide as the black middorsal stripe. The humeral stripe is most often divided, as in younger males. Abdominal segment 2 is as in younger males, the pale middorsal stripe often divided into a basal stripe and an apical spot. Segments 3–7 each have a pale basal ring that is contiguous with the pale ventrolateral stripe. A continuous pale middorsal stripe is widest on the anterior segments. Segments 8–10 are pale, with a lateral dark stripe.

Similar Species. Sooty Dancer (*A. lugens*) and Powdered Dancer (*A. moesta*) are both larger. In both sexes of Sooty Dancer, segments 8 and 9 are mostly black, and there is a dark line paralleling the middorsal thoracic stripe. The pale color pattern on segments 8 and 9 is different, and mature individuals of the Powdered Dancer become gray or whtie. Tonto Dancer (*A. tonto*) has a largely violet abdomen.

Habitat. Streams and rivers, generally with a lot of exposure to sun and only moderate vegetation.

Discussion. This species has the widest distribution of any dancer occurring in the United States, and is subject to a great amount of ontogenetic change as an early adult. This change is especially evident in the pterothoracic markings and on abdominal segments 8–10.

References. Donnelly (1961), Garrison (1994a), Walker (1941, 1953).

Vivid Dancer
Argia vivida Hagen *in* Selys
(pp. 66, 80, 84; photo 11d)

Size. Total length: 29–37 mm; abdomen: 23–32 mm; hindwing: 19–25 mm.

Regional Distribution. *Biotic Provinces:* Apachian, Chihuahuan, Coloradan, Kansan, Navahonian. *Watersheds:* Canadian, Colorado (NM), Pecos, Rio Grande.

General Distribution. Western Canada and United States east to New Mexico.

Flight Season. Mar. (NM)–Oct. (NM)

Identification. This is a medium-sized, but robust, blue species found within our region only in western New Mexico. The male's face is blue, the top of the head darker, almost violet. The postocular spots are large and broadly connected with the compound eyes. The pale-blue occipital bar is generally confluent with these spots. The pterothorax is blue, and the black middorsal stripe is broader than the pale-blue antehumeral stripe. The dark humeral stripe is widest at its ends, appearing as only a hairline in the middle. The legs are blue with black stripes exteriorly, and the tarsi are black. The wings are generally clear, but may show some amber. There are four (sometimes three or five) and three postquadrangular cells in the fore- and hindwings, respectively. The abdomen is blue, with a black spot dorsally on segment 1. Segment 2 generally has a full-length black stripe dorsolaterally. Segments 3–6 each have a black middorsal stripe extending anteriorly for most of each segment. Segment 7 is nearly all black, and segments 8–10 are nearly all blue with a black ventrolateral stripe.

The female is similar, but paler, and generally brown overall. The middorsal thoracic stripe is subequal in width to the antehumeral stripe. The femora are paler than in the male, and the tibiae are yellowish brown. The abdomen is generally like that of the male, but somewhat variable and usually darker. The hind margins of the mesostigmal plates develop into obtusely angular lobes projecting over the mesepisternal pits. Mesepisternal tubercles are present.

Similar Species. The blue form of Springwater Dancer (*A. plana*) is similar to Vivid Dancer, but their ranges do not overlap in most of our region. Where Vivid Dancer occurs, close examination of male caudal appendages and female mesostigmal plates will be necessary to separate these two species. Apache Dancer (*A. munda*) is another similar western dancer, but has more blue dorsally on the middle abdominal segments. (Look under the *Similar Species* of Springwater Dancer for additional differences from other species.)

Habitat. Streams and rivers of the arid southwest.

Discussion. Springwater and the less-common Apache Dancer were originally described as subspecies of Vivid Dancer. Many records of the latter in the early literature from the south-central United States are actually of Springwater Dancer. Within our region, Vivid Dancer is found only throughout northwestern New Mexico, where it behaves much like the closely related Springwater Dancer.

Aurora Damsel
Genus *Chromagrion* Needham

This distinctive genus of damselflies is represented by a single species endemic to the eastern United States and Canada. Though somewhat similar to bluets, their unique coloration, along with several distinct venational characteristics, separate the Aurora Damsel from all others. The unique venation characteristic of this genus includes the following: the anterior side of the triangle in both wings is two to four times longer than the proximal side; the anterior side of the hindwing quadrangle is three or more times longer than the proximal side; and the anal vein separates from the hind margin of both wings at the cubito-anal crossvein.

References. Needham (1903).

Aurora Damsel
Chromagrion conditum (Hagen *in* Selys)
(pp. 80, 84; photos 11e, 11f)

Size. Total length: 32–38 mm; abdomen: 26–32 mm; hindwing: 20–26 mm.
Regional Distribution. *Biotic Province:* Austroriparian. *Watersheds:* Mississippi, Ouachita.

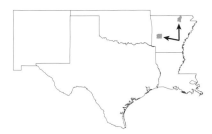

General Distribution. Eastern United States from Georgia north to Canada; west to Wisconsin and south to Arkansas.
Flight Season. May 18 (AR)–Jun. 7 (AR).
Identification. The head of the male is largely black dorsally, with a pale-blue face. The pterothorax is blue, with a black middorsal stripe widening posteriorly to the humeral suture. A black hairline humeral stripe extends the full length of the suture. The metepisternum is blue, becoming yellow anteriorly and extending into a bright-yellow metepimeron. The bottom of the thorax is white, becoming pruinose with age. There are three postquadrangular cells in both the fore- and hindwings. The legs are long and largely black, with pale yellow on the outer femoral and inner tibial surfaces. The abdomen is long, slender, blue, and spreadwing-like, with black maculation. The first abdominal segment is blue with a basal black spot. The black middorsal stripe on segment 2 extends the full length of the segment, widening posteriorly. Segments 3–7 are black dorsally, with only a pale basal ring. The lateral areas of the segments are blue, with black expanding apically from the top. Segments 8 and 9 are blue, with a thin black middorsal line widening apically. There is a pair of black subapical spots on segment 8 laterally. The ventrolateral region of segments 8 and 9 and all of 10 are black. The cerci are decurved, black, and distinctly forcipate when viewed from the top. The black paraprocts are long, blunt, and nearly straight.

The female is colored very much like the male, but with yellow-green rather than blue. The mesostigmal plates are broad and triangular, with a prominent ridge along the posterior and medial borders. The pale basal rings on abdominal segments 3–7 are generally less distinct. Segments 8–10 are entirely black, except for a pale ventrolateral stripe.
Similar Species. Aurora Damsel may initially appear as a bluet (*Enallagma*), but the bright yellow on the sides of the thorax is distinctive, and unique among pond damsels in the south-central United States. Caribbean Yellowface (*Neoerythromma cultellatum*) is a blue damselfly with yellow restricted to the face and the top of the thorax. Their ranges do not overlap. There are spreadwings with yellow on the sides of the thorax.
Habitat. Sheltered, slow-moving spring-fed streams, brooks, and occasionally pools and bogs.
Discussion. One report mentions males of this species patrolling low over ditches of a cranberry bog, "an environment that would seem not particularly well-suited to aquatic life." In the field, this species may be confused with spreadwings because of their general appearance and their tendency to spread their wings when perched. This characteristic is nearly unique among pond damsel genera.

One study found that males of a North Carolina population would often vacate their perches without defensive activity when apparently not searching for females. Mating averaged 36 minutes, with an additional 67 minutes spent in tandem, and almost 1/2 of this time was spent in exploratory activity. Eggs are laid with the male and female in tandem. Another study reported numerous pairs laying eggs, in aquatic plants just beneath the surface of the water. Individuals of neither sex ever completely submerges themselves.

References. Bick et al. (1976), Carpenter (1991), Walker (1953).

Bluets
Genus *Enallagma* Selys
(pp. 98, 101, 109, 112, 119, 121)

This large group of damselflies comprises some 80 species worldwide, occurring everywhere but Australia and the Orient. To the casual observer, they are among the most familiar damselflies, and *Enallagma* is the best-represented genus of damselflies in North America. Twenty-two species occur in the south-central United States. Most bluets are blue (as their common name implies), but others may be yellow, orange, or red in coloration. They all possess conspicuous postocular spots. The M_2 vein generally arises near the 5th and 4th postnodal crossveins in the fore- and hindwings, respectively, but at the 4th and 3rd in Double-striped Bluet (*E. basidens*). In the front row of tibial setae, the setae are less than twice the length of their intervening spaces. Females are generally present in two different color forms, and are thus dichromatic: The andromorphic forms are blue and black, and gynomorphic forms are brown and black; all have a vulvar spine.

Bluets are as diverse ecologically as they are morphologically. Some species, like Big Bluet (*E. durum*), live in brackish waters, and others, like Alkali Bluet, *E. clausum*), are found in desert alkaline ponds. Many species commonly fly close to the water's surface, often making themselves difficult to photograph or collect. Mature adults are commonly found around standing waters, and are most active during midday. Recently emerged teneral adults fly away from the water after 30 minutes, probably in an effort to avoid predation, and to contact mature adults during this vulnerable time. These recently emerged adults go through a maturation period of one to three weeks, during which they forage away from the water. Mature males will congregate around pools and ponds, while the females tend to remain away from the water, approaching only to mate and oviposit. Adults typically perch on standing sedges, cattails, or other riparian vegetation. Both sexes leave their perching sites near the water in late afternoon, probably to roost.

Courtship behavior has never been observed in this genus. Mating most often takes place during the active midday hours. Males may use their penes to remove sperm deposited by a previous male before depositing their own in the female spermatheca. Pairs of males and females may mate at or near the egg-laying site. Females, by themselves or in tandem, commonly lay eggs in living plant material, rotting wood, or algal mats. Females may submerge themselves for as much as an hour while laying their eggs. In most species, however, they lay their eggs at or near the surface of the water.

References. Chivers et al. (1996), Corbet (1963), Fincke (1994), Garrison (1984), Lombardo (1997), May (2002), McPeek (1989, 1990a,b, 1995, 1997, 1998), McPeek and Brown (2000), McPeek et al. (1996).

KEY TO THE SPECIES OF BLUETS (*ENALLAGMA*)

MALES

1. Pale colors on thorax and abdomen red, orange, or yellow   2

1′. Pale colors on thorax and abdomen blue or green   5

KEY TO THE SPECIES OF BLUETS (*ENALLAGMA*) (*cont.*)

2(1'). Thorax yellow, with dark humeral stripe reduced to a thin stripe or entirely absent — **Vesper (*vesperum*)**

2'. Thorax orange or red, with dark humeral stripe well developed — 3

3(2'). Dorsum of abdominal segment 9 pale orange — **Orange (*signatum*)**

3'. Dorsum of segment 9 black — 4

4(3'). Black humeral stripe at most about twice the width of the pale antehumeral stripe; larger, total length greater than 31 mm — **Cherry (*concisum*)**

4'. Black humeral stripe 4 times the width of the pale antehumeral stripe; smaller, total length less than 29 mm — **Burgundy (*dubium*)**

5(1'). Abdominal segment 8 violet or blue, with narrow black band extending distally 1/3 the length of the segment — **Neotropical (*novaehispaniae*)**

5'. Segment 8 pale or dark, but without narrow black band extending distally 1/3 the length of the segment — 6

6(5'). Black humeral stripe divided by a thin, pale longitudinal stripe — **Double-striped (*basidens*)**

6'. Black humeral stripe not divided by a thin, pale longitudinal stripe, and may be lacking — 7

7(6'). Top of head pale blue, the black around the ocelli limited to a thin outline of area — 8

7'. Top of head darker, the ocelli encompassed by black — 9

8(7). Abdomen unusually long, 28 mm or greater — **Attenuated (*daeckii*)**

8'. Abdomen shorter, less than 28 mm — **Slender (*traviatum*)**

9(7'). Abdominal segment 8 pale, with a broad black dorsal longitudinal stripe reaching distally 3/4 the length of the segment — 10

9'. Segment 8 entirely pale, without black longitudinal stripe dorsally — 11

10(9). Face orange — **Rainbow (*antennatum*)**

10'. Face blue — **Stream (*exsulans*)**

11(9'). Black area on dorsum of abdominal segment 2 occupying nearly 1/2 the length of the — 12

KEY TO THE SPECIES OF BLUETS (*ENALLAGMA*) (*cont.*)

segment or less, confined to the apical 1/2 or isolated near the center as a narrow, transverse band (ventrolateral arms may extend farther basally)

11'. Black area on dorsum of segment 2 occupying from 2/3 to the entire length of the segment   20

12(11). Paraprocts shorter than or subequal in length to cerci   13

12'. Paraprocts distinctly longer than cerci   17

13(12). Pale color of abdomen violet or purple   **Claw-tipped (*semicirculare*)**

13'. Pale color of abdomen bright blue   14

14(13'). Cerci, when viewed ventrolaterally, broadly emarginate and at least 2/3 the lateral length of abdominal segment 10   15

14'. Cerci, when viewed ventrolaterally, not broadly and deeply emarginate, and usually about 1/2 the lateral length of segment 10 (in Familiar Bluet 2/3 or more)   16

15(14). Dorsum of abdominal segment 7 blue on apical 1/3 to 1/2   **Azure (*aspersum*)**

15'. Dorsum of segment 7 black apically   23

16(14'). Upper arm of cerci more prominent than the lower arm, a pale tubercle not surpassing the tip of the dorsal arm; medial surface of cerci with a strong, pointed, recurved tooth at the base of the tubercle; black areas of abdominal segments 3–5 usually much less than 1/2 the length of their segments   **Familiar (*civile*)**

16'. Upper and lower arms of cerci nearly equal in prominence, a pale tubercle extending beyond the tips of the dorsal arm; medial surface of cerci lacking a sharp, recurved tooth; black areas of segments 3–5 variable   22

17(12'). Black marking on dorsum of abdominal segment 3 slightly greater than 1/2 the length of that segment; 4 or 5 postquadrangular cells   **Big (*durum*)**

17'. Black marking on dorsum of segment 3 generally much less than 1/2 the length of that segment; 3 or 4 postguadrangular cells   18

18(17'). Dorsally, cerci close together at base, almost touching   **Alkali (*clausum*)**

KEY TO THE SPECIES OF BLUETS (*ENALLAGMA*) (*cont.*)

18'. Dorsally, cerci widely separated	19
19(18'). Cerci, when viewed laterally, not upturned apically; the pale tubercle of the cerci located on the medial margin near the apex	**Boreal (*boreale*)**
19'. Cerci, when viewed laterally, upturned apically; the pale tubercle of the cerci located at the extreme tip (best seen dorsally)	**Northern (*cyathigerum*)**
20(11'). Ventrolateral black markings extending the entire length of abdominal segments 8 and 9, and nearly entire the length of segment 2	**Skimming (*geminatum*)**
20'. Segments 2, 8, and 9 lacking dark ventrolateral stripes	21
21(20'). Black humeral stripe, at midlength, 1/2 to 1/5 the width of the blue antehumeral stripe	**Arroyo (*praevarum*)** (in part)
21'. Black humeral stripe, at midlength, much more than 1/2 the width of the antehumeral stripe	**Turquoise (*geminatum*)**
22(16'). Black areas of abdominal segments 3 and 4 more than 1/2 the length of the segments; black area on dorsum of segment 7 not reaching extreme base of the segment (sometimes reaching as a hairline)	**Tule (*carunculatum*)**
22'. Black areas of segments 3 and 4 less than 1/2 the length of the segment; black on dorsum of segment 7 reaching extreme base as more than a hairline	**Atlantic (*doubledayi*)**
23(15'). Abdominal segments 4 and 5 with more black than blue dorsally	**Arroyo (*praevarum*)** (in part)
23'. Segments 4 and 5 with at least as much blue as black dorsally	**River (*anna*)**

FEMALES

1. Dark humeral stripe incomplete or divided longitudinally by light brown marking, or entirely light brown	2
1'. Dark humeral stripe entirely black and complete	8
2(1). A pale spot between the lateral ocelli, and middle prothoracic lobe without distinct dorsolateral pits; dorsum of abdominal segment 9 entirely pale or at most with a small basal black spot	3
2'. Area between lateral ocelli entirely black or, *if*	4

KEY TO THE SPECIES OF BLUETS (*ENALLAGMA*) (*cont.*)

small pale spot is present, *then* middle prothoracic lobe with distinct dorsolateral pit on each side; dorsum of segment 9 at least 1/3 black

3(2). Black extending entire length of the dorsum of abdominal segment 8; abdomen usually greater than 30 mm long

Attenuated (*daeckii*)

3'. Black extending at most 3/4 the distance from base to apex of dorsum of segment 8; abdomen generally less than 27 mm long

Slender (*traviatum*)

4(2'). Middle prothoracic lobe with a dorsolateral pit on each side; thoracic color usually predominantly yellow, green, or orange

Vesper (*vesperum*) (in part)

4'. Middle prothoracic lobe without dorsolateral pits; thoracic color predominantly grayish green, blue, or light brown

5

5(4'). Light-brown line dividing black humeral stripe distinct, full-length, and less than 1/2 the width of the entire stripe; area anterior to lateral ocelli with pale spots

Double-striped (*basidens*)

5'. Light-brown line dividing black humeral stripe without sharply defined edges, sometimes incomplete and varying in width from a hairline to nearly the full width of the stripe; area anterior to lateral ocelli black

6

6(5').Black humeral stripe usually divided by a brown spot only at its upper end; abdominal segment 9 in dorsal view black, but often with a median, pale, spindle-shaped spot

Rainbow (*antennatum*)

6'. Black humeral stripe usually divided by a brown stripe for most of its length; segment 9 variable but never with a median, pale, spindle-shaped spot

7

7(6'). Middorsal thoracic carina usually black; mesostigmal plates each with distinct posteromedial tubercle

Turquoise (*divagans*)

7'. Middorsal thoracic carina pale; mesostigmal plates without distinct posteromedial tubercles

Stream (*exsulans*)

8(1'). Middle lobe of prothorax with pair of dorsal or dorsolateral pits, the edges of which sharply defined

9

KEY TO THE SPECIES OF BLUETS (*ENALLAGMA*) (*cont.*)

8'. Middle lobe of prothorax without dorsal or dorsolateral pits, although shallow depressions may be present — 15

9(8). Abdominal segment 1 dorsally pale on the apical 1/3 to 1/2; dorsally, segment 8 with large pale areas — 10

9'. Segment 1 dorsally entirely black, except for the narrow, pale apical annulus; dorsally, segment 8 entirely black or nearly so — 12

10(9). Dorsally, abdominal segment 8 almost entirely pale; black humeral stripe at midlength less than 1/2 as wide as pale antehumeral stripe — **Alkali (*clausum*)**

10'. Dorsally, segment 8 with apical black spot extending anteriorly at least 1/2 the length of the segment; black humeral stripe usually more than 1/2 as wide as pale antehumeral stripe — 11

11(10'). Abdominal segment 7 in dorsal view largely pale in basal 2/3 or more, except along midline; abdominal segment 8 with a pair of large, blue, dorsolateral spots in the basal 1/2, the spots usually separate or nearly so from the pale ventrolateral areas — **Azure (*aspersum*)**

11'. Segment 7 in dorsal view mostly black basally; segment 8 with basal blue spots bordered laterally by a black stripe, the spots therefore not confluent with the pale ventrolateral areas — **Skimming (*geminatum*)**

12(9'). Dorsum of abdominal segment 10 almost entirely pale — 13

12'. Dorsum of segment 10 almost entirely black — 14

13(12). Black humeral stripe much less than 1/2 as wide as the pale antehumeral stripe — **Vesper (*vesperum*)** (in part)

13'. Black humeral stripe more than 1/2 as wide as the pale antehumeral stripe — **Orange (*signatum*)**

14(12'). Black humeral stripe at least 4 times as wide as the pale antehumeral stripe; prothoracic pits located near anterior margin of the middle lobe, and not bordered by a pale spot — **Burgundy (*dubium*)**

14'. Black humeral stripe at most twice as wide as the pale antehumeral stripe; prothoracic pits — **Cherry (*concisum*)**

KEY TO THE SPECIES OF BLUETS (*ENALLAGMA*) (*cont.*)

located nearer the midsection of the middle lobe, and usually bordered by a pale spot

15(8′). Large, stocky abdomen, at least 29 mm long | **Big (*durum*)**

15′. Smaller, slender abdomen, generally no longer than 28 mm | 16

16(15′). Abdominal segment 3 with spindle-shaped dorsal black stripe on its basal 2/3 and a transverse black stripe on its apical 1/6 to 1/4 | **Claw-tipped (*semicirculare*)**

16′. Segment 3 with dorsal black stripe continuous and wide from base to apex | 17

17(16′). Black humeral stripe with an anterior projection extending toward and sometimes reaching posterior margin of mesostigmal plates; abdominal segment 8 nearly always with a ventrolateral black stripe, typically confluent with apical transverse stripe | **Neotropical (*novaehispaniae*)**

17′. Black humeral stripe without a black projection extending toward mesostigmal plates; segment 8 without a ventrolateral black stripe | 18

18(17′). Face usually orange; abdominal segment 9 in dorsal view black, but often with a median, pale, spindle-shaped spot | **Rainbow (*antennatum*)**

18′. Face green, grayish blue, or tan; segment 9 in dorsal view *either* mostly pale *or* black without a pale median spot | 19

19(18′). Abdominal segment 10 entirely pale or with a small, black dorsal spot apically; segment 9 generally with a small, basal, bilobed black spot, only rarely extending the entire length of the segment; hind lobe of pronotum with a pale, median tubercle | **Turquoise (*divagans*)**

19′. Segment 10 usually with more extensive black markings; segment 9 usually with black from base to apex; hind lobe of pronotum lacking a median tubercle; | 20

20(19′). Mesostigmal plates with ridges forming narrow and sharply defined medial margins, and a large, smooth, oval depression that occupies the entire medial 1/2 of each plate | **Atlantic (*doubledayi*)**

20′. Mesostigmal plates with medial ridges different in shape, usually broader and less sharply de- | 21 fined, not delimiting such large, medial depressions

21(20′). Posterior border of each mesostigmal plate well defined by a narrow sulcus extending its entire length 22

21′. Posterior border of each mesostigmal plate indistinct over part of its length, the sulcus lacking 24

22(21). Mesostigmal plates with medial borders usually at least slightly convergent forward, each with a distinct, elongate depression confined to the anteromedial corner **Northern (*cyathigerum*)**

22′. Mesostigmal plates with medial borders parallel or slightly divergent forward, lacking such depressions 23

23(22′). Black stripe on dorsum of abdominal segment 1 generally almost reaching the apex of the segment and of uniform width; black stripe on dorsum of segment 8 usually not markedly constricted basally **Familiar (*civile*)**

23′. Black stripe on dorsum of segment 1 usually not nearly reaching the apex of the segment or, if so, constricted apically; black stripe on dorsum of segment 8 usually constricted basally **Arroyo (*praevarum*)**

24(21′). Dorsally, abdominal segment 8 with blue area more extensive than on segment 7 **Boreal (*boreale*)**

24′. Dorsally, segment 8 with blue area about as extensive as on segment 7 25

25(24′). Abdominal segment 1 with 2 black spots dorsally **River (*anna*)**

25′. Segment 1 with a single black spot dorsally **Tule (*carunculatum*)**

River Bluet
Enallagma anna Williamson
(pp. 98, 101, 112; photo 12a)

Size. Total length: 30–36 mm; abdomen: 23–28 mm; hindwing: 18–22 mm.
Regional Distribution. *Biotic Provinces:* Coloradan, Navahonian. *Watersheds:* Canadian, Colorado (NM).
General Distribution. Southern Canada and western United States west of the Rocky Mountains.

Flight Season. June. 23 (NM)–Sep. 14 (NM).

TABLE 4. BLUETS (*ENALLAGMA*)

Species	Hindwing length (mm)	Pale color	Middorsal thoracic carina	Tarsi	Pits on pronotum[1]	Vein M_2[2]	Distribution (biotic provinces)[3]
River (*anna*)	18–22	blue	blue	dark or pale	–	5/4	Co,N
Rainbow (*antennatum*)	15–21	orange-green	orange	pale	–	5/4	K,Tx
Azure (*aspersum*)	15–20	blue	black	dark	+	5/5	Au,Ca,Tx
Double-striped (*basidens*)	10–15	blue	blue	pale	–	4/3	Ap,Au,B,Ca, Ch,K,N,Ta,Tx
Boreal (*boreale*)	17–22	blue	blue	dark or pale	–	6/5	Ch,Co,K,N
Tule (*carunculatum*)	15–22	blue	blue	pale	–	6/5	Ap,Ch,Co,K,N
Familiar (*civile*)	16–21	blue	black	pale	–	5/5	Ap,Au,B,Ca,Ch, Co,K,N,Ta,Tx
Alkali (*clausum*)	16–23	blue	blue	pale	+	5/4	Co,K,N
Cherry (*concisum*)	13–17	red-orange	black	pale	+	5/4	Au
Northern (*cyathigerum*)	17–24	blue	blue	dark or pale	–	6/5	Ap,Ch,Co,K,N
Attenuated (*daeckii*)	19–25	blue	blue	pale	–	5/5	Au,Ca,K
Turquoise (*divagans*)	17–22	blue-violet	black	pale	–	5/4	Au,Ca,Tx
Atlantic (*doubledayi*)	16–21	blue	black	pale	–	5/4–5	Au,Tx
Burgundy (*dubium*)	12–17	red-violet	black	pale	+	4/4	Au
Big (*durum*)	17–25	blue	blue	pale	–	6/4	Au,Ta,Tx
Stream (*exsulans*)	17–21	blue	black	pale	–	5/4	Au,B,Ca,Ch,K, N,Ta,Tx
Skimming (*geminatum*)	12–17	blue	black	dark or pale	+	5/4	Au,Ca,Tx
Neotropical (*novaehispaniae*)	17–19	blue-violet	black	dark	–	5/4	B,Ch,Ta

TABLE 4. BLUETS (*ENALLAGMA*) (*cont.*)

Species	Hindwing length (mm)	Pale color	Middorsal thoracic carina	Tarsi	Pits on pronotum[1]	Vein M$_2$[2]	Distribution (biotic provinces)[3]
Arroyo (*praevarum*)	15–21	blue	blue	pale	–	5/4	Ap,B,Ch,Co,K, N,Ta
Claw-tipped (*semicirculare*)	15–24	blue-violet	black	dark	–	5/4	Ch
Orange (*signatum*)	15–21	orange	black	pale	+	6/5	Au,B,Ca,K,N, Ta,Tx
Slender (*traviatum*)	15–19	blue	black	pale	–	5/4	Au,Ca,Tx
Vesper (*vesperum*)	15–21	yellow-orange	black	pale	+	5/4	Au,B,Tx

[1] Presence of pits on middle lobe of pronotum in female.

[2] Postnodal crossveins where M$_2$ vein arises closest in the forewing and hindwing.

[3] (Ap) Apachian, (Au) Austroriparian, (B) Balconian, (Ca) Carolinian, (Ch) Chihuahuan, (Co) Coloradan, (K) Kansan, (N) Navahonian, (Ta) Tamaulipan, (Tx) Texan.

Identification. This is a large, blue western species. The face and thorax of the male are blue. The middorsal thoracic stripe is black and 1/2 the width of the mesepisterna. The pale-blue antehumeral stripe is not quite 1/2 as wide as the middorsal stripe, and the black humeral stripe is distinctly narrower than the antehumeral stripe. The legs are blue, with a black stripe, and the tarsi are pale or dark. The abdomen is bright blue dorsally, becoming paler laterally and ventrally. Segment 1 is black on the basal half dorsally. Segment 2 has a black subapical spot, and there is a hastate stripe on the apical 1/2 to 3/4 of segments 3–6. Segment 7 is black dorsally. Segments 8 and 9 are blue, with ventrolateral streaks or spots. Segment 10 is black dorsally. The cerci are black and protrude distinctly beyond the abdomen. Females are blue or tan with triangular-shaped mesostigmal plates, and the middle lobe of the pronotum lacks distinct pits. Segment 8 is variably marked with black, ranging from a hairline stripe on the basal 1/2 to a broad stripe extending the full length of the segment. Segment 9 and 10 are nearly all black dorsally.

Similar Species. Male Arroyo Bluets (*E. praevarum*) are shorter, not as robust, and overall darker. The black dorsal markings on abdominal segments 3–5 cover at least 1/2 the length of each segment in Arroyo Bluet, and these areas are restricted to 1/4 to 1/2 of each segment in River Bluet. Boreal Bluet (*E. boreale*) is stockier, and the black rings on the middle abdominal segments are truncate. The postocular spots in Northern Bluet (*E. cyathigerum*) are larger, and segment 10 is entirely black, not just dorsally. The cerci in Familiar Bluet (*E. civile*), Alkali Bluet (*E. clausum*), and Tule Bluet (*E. carunculatum*) do not protrude significantly (greater than the length of segment 10, beyond the abdomen. Female River Bluets are very difficult to distinguish from females of other species. Size, distribution, and association with the males will help, but they can only reliably be distinguished by the details of their mesostigmal plates.

Habitat. Slow streams and rivers, often associated with warm springs.

Discussion. Within our area this species is found only in far northwestern New Mexico, at the southern limits of its range. Females lay eggs in tandem and may submerge below water to lay eggs. In Utah, this species is common at elevations between 1,200 and 2,100 m.

References. Provonsha (1975), Roemhild (1975).

Rainbow Bluet
Enallagma antennatum (Say)
(pp. 98, 101, 112; photos 12b, 12c)

Size. Total length: 27–33 mm; abdomen: 21–27 mm; hindwing: 15–21 mm.
Regional Distribution. *Biotic Provinces:* Navahonian, Texan. *Watersheds:* Arkansas, Canadian, Cimarron.

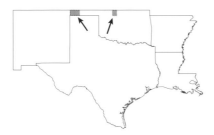

General Distribution. New York, Ontario, and Quebec, west to Montana and south to Oklahoma.
Flight Season. Jun. 15 (OK).
Identification. These are beautiful creatures. The males have a bright-orange face, a prominent blue-green occipital bar, and a pair of narrowed postocular spots that may be confluent with the bar. The eyes are brown on top, fading to green or yellow below. The middorsal stripe is black, and 1/2 the width of the mesepisterna. The antehumeral stripe is orange to greenish yellow, and the rest of the thorax is largely pale blue green fading extensively to yellow green ventrally. The legs are yellowish orange with black markings. Abdominal segments 1–7 are largely black dorsally; segments 4–7 are green laterally; and segments 1–3 and 8–10 are blue laterally. Segment 8 is entirely black dorsally, except for a narrow blue apical band. Segment 9 is generally completely blue, occasionally with a narrow black basal band. Segment 10 is black dorsally, becoming blue or yellow laterally and ventrally; the deeply bifurcated cerci are black and not more than 2/3 the length of the segment. The paraprocts are nearly all pale, becoming dark apically and curving dorsoanteriorly.

The female is colored and patterned similarly to the male on the head and thorax, but the middorsal carina is generally paler. The mesostigmal plates are bordered posteriorly by a distinct sulcus, though a deep pit is lacking. The plates are more or less triangular, with a pale tubercle at the posteromedial corners. Mesepisternal tubercles are lacking, but there is a transverse swelling medially behind each meso-stigmal plate. The abdomen is marked as in the male on segments 1–6. Segments 7–9 are generally all black, except for a narrow apical band, and there is generally a pale median spot dorsally on segment 9. Segment 10 is black, narrowing apically, but the black often does not extend the full length of the segment.
Similar Species. Because it is so colorful, Rainbow Bluet is more likely to be mistaken for a forktail (*Ischnura*) in the field, rather than for another bluet. None of the forktails in the south-central United States, however, has the orange face and blue-green postocular spots of this distinctive bluet.
Habitat. Slow streams, lakes, gravel pits, and borrow pits.
Discussion. This northern species ranges southward to Oklahoma. It is considered one of the most primitive of the bluets. It appears to be most closely related to Stream Bluet (*E. exsulans*) and Turquoise Bluet (*E. divagans*). One author described its habitat as the quiet reaches of streams where current is slow and where dense vegetation is lacking. Females may submerge themselves when laying eggs.

References. Donnelly (1963), Garman (1917), Kennedy (1919), Walker (1953), Westfall and May (1996).

Azure Bluet
Enallagma aspersum (Hagen)
(pp. 98, 101, 112; photos 12d, 12e)

Size. Total length: 27–34 mm; abdomen: 21–27 mm; hindwing: 15–20 mm.
Regional Distribution. *Biotic Province(s):* Austroriparian, Carolinian, Texan. *Watershed(s):* Arkansas, Canadian, Red, St. Francis, White.

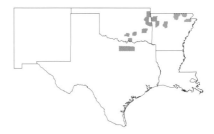

General Distribution. Eastern United States and southern Canada from Georgia north to Quebec; west to Wisconsin and south to Texas.
Flight Season. May 14 (TX)–Aug. 3 (TX).

Bluets (*Enallagma*)

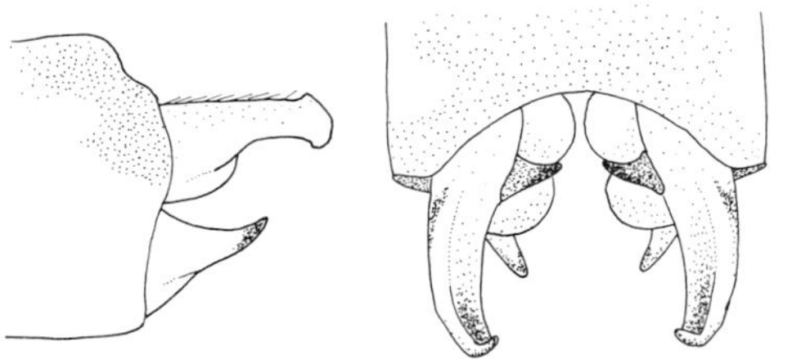

River (*anna*)

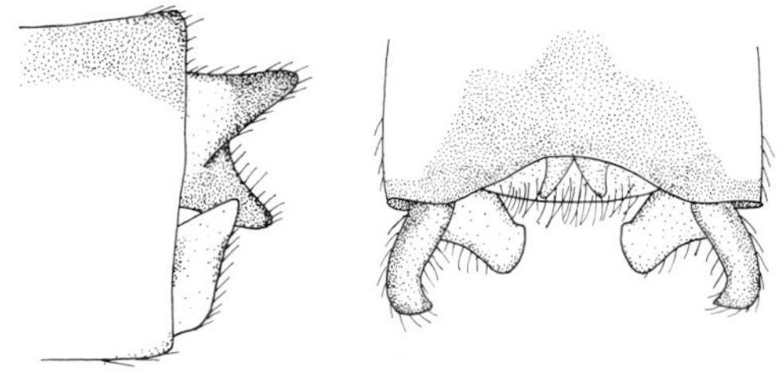

Rainbow (*antennatum*)

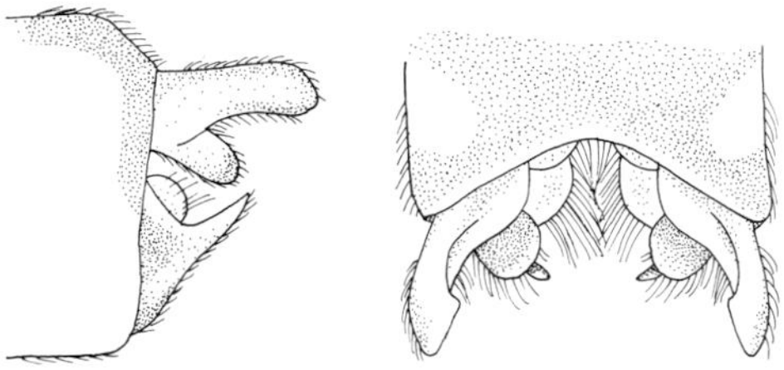

Azure (*aspersum*)

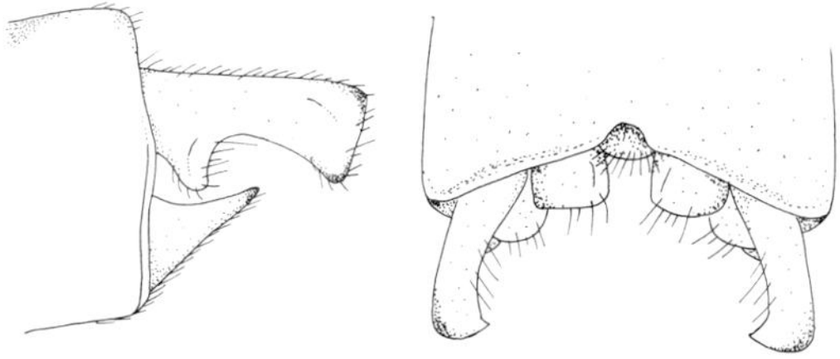

Double-striped (*basidens*)

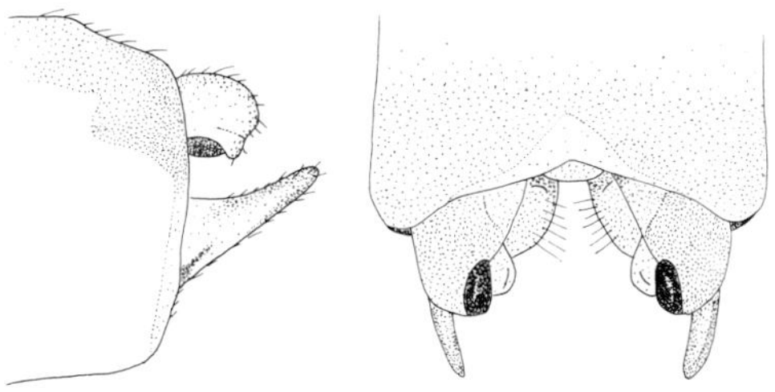

Boreal (*boreale*)

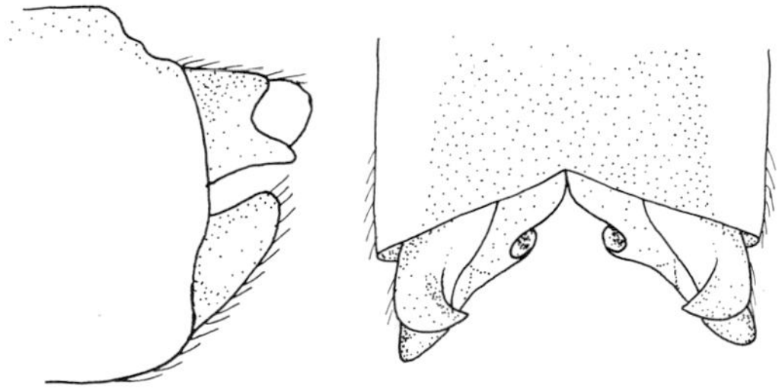

Tule (*carunculatum*)

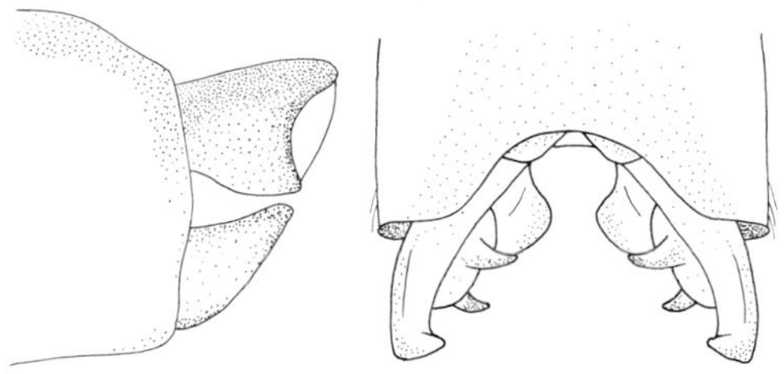

Familiar (*civile*)

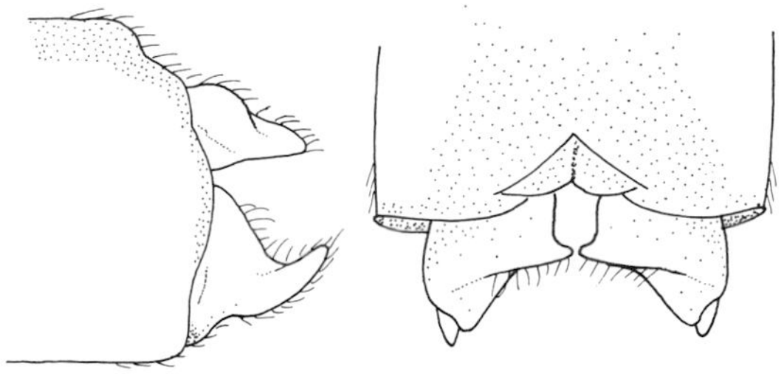

Alkali (*clausum*)

Fig. 18. Bluets (*Enallagma*): lateral and dorsal views of male caudal appendages (pp. 97–107).

Identification. The front of the head in the male is blue, with a distinct black line above the clypeus. The top of the head is black, except for a thin, pale occipital bar that is only narrowly separated from a pair of large blue, oval postocular spots. The pronotum is black dorsally, with a pair of medial blue spots. The middorsal carina and stripe of the pterothorax are black, the latter approximately 1/2 the width of the mesepisterna. The blue antehumeral stripe extends no more than 1/2 the width of the middorsal stripe. The black humeral stripe narrows posteriorly, often expanding anteriorly to the mesepimeron, but generally remaining narrower than the antehumeral stripe. The rest of the pterothorax is pale blue, fading ventrally. The legs are pale, with dark stripes laterally, and the tarsi are typically black with pale claws. The abdomen is blue above, fading laterally and ventrally. Segment 1 is black on the basal 1/2 dorsally, and a black spot on the apical 2/5 of segment 2 is confluent with the apical ring. There is a narrow, dorsal black stripe that starts basally on segment 3 and extends the full length of the segment, widening distally. The entire dorsum of segments 4–6 is black, except for a pale basal ring. Segment 7 is black on the basal 1/4 of the segment and blue apically and laterally. Segments 8 and 9 are entirely blue. Segment 10 bears a wide dorsal black stripe that narrows apically. The cerci are dark and distinctly bifurcated in lateral view. The upper arm of each cercus is much longer and more pronounced than the lower. The dark paraprocts curve dorsally to reach the lower arm of the appendages above.

The head and thorax of the female are similar in coloration to the male, but the pale-blue colors are generally replaced by green. The postocular spots are significantly smaller than in the male. There is a pair of kidney-shaped pits on the posterior 1/3 of the pronotum. The mesostigmal plates have a distinct posterior border. The abdomen is generally paler than in the male, and especially so ventrolaterally. Segments 1 and 3–6 are generally similar to those of the male. Segment 7 is nearly all black, with an apical blue ring. Segment 8 is black, with a pale narrow apical ring and a pair of pronounced blue spots on the basal 1/3 to 1/2 of the segment. Segments 9 and 10 are black dorsally, with only a pale-blue apical margin.

Similar Species. Familiar (*E. civile*), Alkali (*E. clausum*), Atlantic (*E. doubledayi*), and Boreal (*E. boreale*) Bluets are all a paler blue dorsally than Azure Bluet. This spe-

cies is restricted to the northeastern portion of our range. Skimming Bluet (*E. geminatum*) has a black ventrolateral line on segments 8–10, and 10 is black laterally.

Habitat. Fishless lakes and semipermanent ponds and bogs.

Discussion. In Texas, this species has been reported only in the north-central and northeastern portions of the state. It is generally restricted to fishless ponds and lakes, but has been reported occurring along shallow grassy or boggy shorelines.

Unpaired males seldom perch or maintain a territory prior to mating. One study found that females move away from the water between noon and 1:00 P.M. each day, only to reappear in numbers between 1:30 and 2:30 P.M. Males will often seize egg-laying females from the water. Sperm transfer generally occurs while in tandem and perching on vegetation, quickly followed by mating that lasts an average of 14 minutes. Females most often lay eggs completely submerged and unaccompanied.

Unlike most damselflies, the female Azure Bluet does not begin laying eggs above the water and then back down, but rather she determines an appropriate stem and immediately proceeds down it, head first. The male separates upon contact with the water and perches nearby. One study reported a female submerge as deep as 38 cm to lay eggs at the base of the plant, apparently as an adaptation to avoid having the eggs exposed during summer drought. Egg-laying generally lasts no more than 25 minutes. In Ontario, Canada, Azure Bluet was formerly restricted to bog-marginated lakes, but it seems to be expanding its habitat to include artificial ponds and calcareous and alkaline gravel pits.

References. Bick (1972), Bick and Hornuff (1965), Carpenter (1991), Catling and Pratt (1997), Ingram and Jenner (1976), Jacobs (1955), Morgan (1930).

Double-striped Bluet
Enallagma basidens Calvert
(pp. 98, 101, 112; photos 12f, 13a)

Size. Total length: 21–28 mm; abdomen: 17–22 mm; hindwing: 10–15 mm.

Regional Distribution. *Biotic Provinces:* Apachian, Austroriparian, Balconian, Carolinian, Chihuahuan, Kansan, Navahonian, Tamaulipan, Texan. *Watersheds:* Arkansas, Bayou Bartholomew, Brazos, Canadian, Cimarron, Colorado, Colorado (NM), Guadalupe, Lower Rio Grande, Mississippi, Neches, Nueces, Ouachita, Pecos, Red, Sabine, San Antonio, San Jacinto, St. Francis, Trinity, Upper Rio Grande, White.

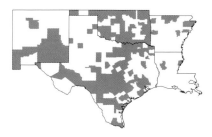

General Distribution. Florida to Ontario and west to Colorado and California; south to Texas and Mexico.
Flight Season. Feb. 18 (TX)–Nov. 6 (TX).
Identification. The front of the male's head is blue, with a broad black stripe. The top is largely black, with a pale occipital bar and small postoccipital spots. The middorsal thoracic carina is pale blue and bisects the broad black middorsal stripe, which is approximately 1/2 the width of the mesepisternum. The humeral stripe is black and subequal in width to the antehumeral stripe, but narrowed basally, and divided for nearly its entire length by a thin pale-blue line. The femora are pale blue or cream, with a dark stripe on their outer surface. The tibiae and tarsi are generally pale and lack dark markings. The abdomen is largely bright blue, becoming paler ventrally. The entire top of segment 1 is black, except for a narrow blue apical band. A black stripe runs the entire length of segments 2 and 3 dorsally and extends laterally to the last 1/4 of each segment. There is a black hastate stripe on the apical 1/2 to 2/3 of segments 4–6 dorsally. The entire dorsum of segment 7 is black, except for pale basal and apical rings. Segments 8 and 9 are generally all blue, and segment 10 has an irregular black stripe dorsally. The cerci are black and extend for approximately 1/2 the length of segment 10. They are sharply truncate when viewed laterally, and have a distinct basal, ventrally directed, lower lobe. The paraprocts are pale, becoming darker apically, curving upward and extending slightly more than 1/2 the length of the cerci.

The head and thorax of the female closely resemble those of the male, but the pale colors are more extensive. Females occur in three different color forms, their pale colors either blue, green, or brown. Generally, there are small pale spots anterior to each ocellus. The pale stripe dividing the humeral stripe is most often confluent anteriorly and posteriorly with the rest of the pterothorax. There is a distinct, high anterior ridge toward the middle of each depressed mesostigmal plate, and these plates have strong prominences at each posteromedial corner. The abdominal segments are similar to those of the male, but the dorsums of segments 3–6 are entirely black, except for a narrow, basal pale ring. Segment 8 is black dorsally, with only a narrow apical pale ring. Segment 9 is black dorsally, with a large blue spot in the basal 1/3 to 2/3 of each segment that narrows apically and is distinctly emarginate apically (this spot is sometimes divided into two separate triangles). The top of segment 10 is entirely blue.

Similar Species. Double-striped Bluet is the smallest of our bluets. The thin pale stripe that divides the dark humeral stripe longitudinally (and gives the species its name) is distinctive among all dancers (*Argia*) and bluets in our area.

Habitat. Various permanent and semipermanent ponds, lakes, and reservoirs, as well as slow reaches of streams and rivers.

Discussion. Double-striped Bluet, originally described from Texas, has expanded its range westward and northward, probably because of the suitable new breeding localities afforded by extensive irrigation. It has also expanded its range eastward this century, now reaching as far as Florida and the Carolinas, north to New York and Michigan. It was first reported from Canada in southwestern Ontario collections taken in 1985.

Females are often observed around water only while in tandem. Egg-laying occurs in tandem where floating masses of filamentous algae and other vegetation are preferred. Double-striped Bluet perches over water most often from 10:00 A.M. to 4:00 P.M.

References. Bird 1933, Cannings (1989), Dunkle (1990), Hornuff (1968), Montgomery (1942, 1966), Paulson and Garrison (1977).

Boreal Bluet
Enallagma boreale Selys
(pp. 98, 101, 112; photos 13b, 13c)

Size. Total length: 28–36 mm; abdomen: 22–29 mm; hindwing: 17–22 mm.

Regional Distribution. *Biotic Provinces:* Chihuahuan, Coloradan, Kansan, Navahonian. *Watersheds:* Canadian, Colorado (NM), Pecos, Rio Grande.

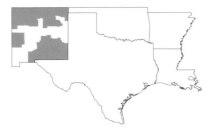

General Distribution. Northern Canada and United States south to California, New Mexico, Iowa, Pennsylvania, and Mexico.

Flight Season. June (NM).

Identification. The male's face is blue, with a black bar across the top of the frons. The top of the head is broadly black, with two pale, more or less oval, postocular spots that are not confluent with the compound eyes. The middorsal thoracic carina and stripe are black. The antehumeral stripe is pale blue and roughly 2/3 as wide as the middorsal stripe. The jagged black humeral stripe is subequal in width to the antehumeral stripe at its upper and lower ends, but is constricted toward the middle. The rest of the pterothorax is blue, with an abbreviated black stripe on the upper 1/4 of the interpleural suture. The legs are pale, with heavy dark stripes on the outer surface of the femora, and the tibiae and tarsi are also pale. The first abdominal segment is nearly all blue, with a small lateral black spot apically. Segment 2 is blue, with a distinctive black dorsal crescent on the apical 2/3 and a black apical ring. Segments 3–5 are blue, with a broad black apical ring for 1/4 of their length. Segments 6 and 7 each have a black band extending as much as 3/4 the segments' length. Segments 8 and 9 are entirely blue, and segment 10 is black dorsally. The cerci are dark and rounded apically, extending to approximately 1/3 the length of segment 10. A distinct medial tubercle is visible when viewed dorsally. The paraprocts are much longer, approximately 1.5 times the length of the cerci, and curved slightly upward.

The head and thorax of the female are colored similarly to those of the male, though the pale areas are more extensive. There are no distinct pits on the middle lobe of the pronotum. The hind margin of the mesostigmal plates is indistinct medially. The abdominal color pattern is similar to that of the male, but with a dark basal spot on the dorsum of segment 1. Segment 2 has a dark apical ring and a middorsal black stripe that extends the full length of the segment and is distinctly expanded apically. Segment 3 bears a basally tapering middorsal stripe that is confluent with a broad apical ring extending approximately 1/4 of the segment's length. Segments 4–7 have a broad, basally pointed dark stripe dorsally. Segment 8 bears a dark apical ring extending 1/3 of its length and a narrow middorsal stripe that is broadly confluent with the ring. Segments 9 and 10 are entirely black.

Similar Species. The pale-blue areas on the abdomen of Familiar Bluet (*E. civile*) are more extensive. In Tule Bluets (*E. carunculatum*) and Northern Bluets (*E. cyathigerum*) the black laterally on the abdomen is more extensive than that in Boreal Bluet, but this character is variable. The black spot dorsally on abdominal segment 2 in Alkali Bluet (*E. clausum*) is larger, extending to nearly 1/2 the segment's length. Critical examination with a hand lens or microscope may be necessary to distinguish these species, at least initially.

Habitat. Fishless ponds, lakes, slow-moving streams, and occasionally saline waters.

Discussion. This species is widespread, but at present is known only from New Mexico within the south-central United States. Boreal Bluet occurs in a variety of pond and lake habitats, but generally only when fish are lacking. Pairing begins shortly after emergence and continues throughout the summer. The female sometimes lays eggs unaccompanied by the male, but more often this is done in tandem. Emergent aquatic vegetation seems to be preferred, and egg-laying generally occurs just above the water level.

One Canadian study of larval Boreal Bluets found one larva occurring with the northern pike (*Esox lucius*), a predatory fish, and a second that did not. The population occurring with the pike adopted antipredator behavior in response to chemical stimuli from injured conspecifics and from chemical stimuli given off by the pike themselves. The study found that individuals that were previously unexposed and unresponsive to stimuli from the pike learned to recognize these stimuli after a single exposure.

References. Baker and Clifford (1982), Furtado (1973), Lebeuf and Pilon (1977), Lebuis and Pilon (1976), Logan

(1971), McPeek (1997), Paulson (1974), Provonsha (1975), Rivard et al. (1975), Robert (1963), Walker (1953), Wisenden et al. (1997).

Tule Bluet
Enallagma carunculatum Morse
(pp. 98, 101, 112; photo 13d)

Size. Total length: 26–37 mm; abdomen: 19–28 mm; hindwing: 15–22 mm.
Regional Distribution. *Biotic Provinces:* Apachian, Chihuahuan, Coloradan, Kansan, Navahonian. *Watersheds:* Brazos, Canadian, Colorado (NM), Pecos, Rio Grande.

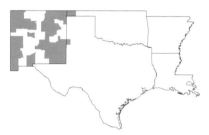

General Distribution. Transcontinental from southern Canada south to Baja California in the west, south to Pennsylvania in the east.
Flight Season. Apr. 17 (NM)–Aug. (NM).
Identification. The face of the male is blue, and the top of its head is black, with pale postocular spots and a widely separated occipital bar. The middorsal thoracic carina is pale blue and thinly visible in the center of the dark middorsal stripe, which is 1/2 the width of the mesepisterna. The antehumeral stripe is pale blue and nearly 1/2 the width of the middorsal stripe. The black humeral stripe is as wide as the antehumeral stripe at its widest. The rest of the pterothorax is blue, becoming paler below. The legs are blue or tan, with black stripes on the femora and tibiae, and the tarsi are pale. The abdomen is bright blue dorsally. The basal half of segment 1 is dark dorsally. Segment 2 has a black subapical spot or stripe extending 3/4 the length of the segment dorsally. Segments 3–6 each have a dark hastate stripe dorsally, extending at least 1/2 the length of the segment. Segment 7 is black dorsally, with a pale apical ring. Segments 8 and 9 are pale dorsally, and segment 10 is black dorsally. The cerci are half the length of segment 10, and a pale tubercle is visible laterally. The paraprocts are 2/3 the length of the cerci.

The female is pale blue or tan, the head and thorax similar in coloration to the male. The posterior borders of the mesostigmal plates are sinuate and indistinct. The middle lobe of the pronotum lacks distinct pits. Abdominal segment 1 is similar to that of the male, but with a larger dark spot dorsally. Segment 2 has a full-length black stripe dorsally, narrowing from the base. Segments 3–7 each have a hastate stripe. Segments 8–10 each have a full-length black stripe dorsally.
Similar Species. Familiar Bluet (*E. civile*) is very similar, but generally paler, and Tule Bluet often appears greenish compared to Familiar Bluet. Males can be reliably separated by careful in-hand examination of the cerci. In Familiar Bluet the tubercle does not extend beyond the cercus itself; it does in Tule Bluet. Northern Bluet (*E. cyathigerum*) has black laterally on segment 2. The black rings on the middle abdominal segments in male Boreal Bluets (*E. boreale*) are truncate, not pointed. Generally, no more than the apical 1/4 of segment 2 is black in Alkali Bluet (*E. clausum*), and Arroyo Bluet (*E. praevarum*) is thinner and darker overall. Familiar Bluet is the only other bluet in our region that has a tubercle on the male cerci.
Habitat. Slow reaches of rivers and occasionally lakes and ponds.
Discussion. Tule Bluet is known only from the extreme northwestern portions of our region. This species is known to tolerate waters with high salinity, though it is not found along the coast.

References. Logan (1971), Paulson (1974).

Familiar Bluet
Enallagma civile (Hagen)
(pp. 98, 101, 112; photos 13e, 13f)

Size. Total length: 29–39 mm; abdomen: 22–34 mm; hindwing: 16–21 mm.
Regional Distribution. *Biotic Provinces:* Apachian, Austroriparian, Balconian, Carolinian, Chihuahuan, Coloradan, Kansan, Navahonian, Tamaulipan, Texan. *Watersheds:* Arkansas, Bayou Bartholomew, Brazos, Canadian, Cimarron, Colorado, Colorado (NM), Guadalupe, Lower Rio Grande, Mississippi, Neches, Nueces, Ouachita, Pecos, Red, Sabine, San Antonio, San Jacinto, St. Francis, Trinity, Upper Rio Grande, White.

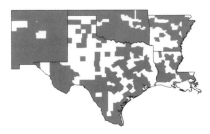

General Distribution. Throughout the United States and southern Canada except for the Pacific Northwest, and south through Mexico to Colombia and Venezuela.

Flight Season. Year-round (TX).

Identification. The front of the male's head is blue, striped with a broad black bar. The top of the head is largely black, except for two small pale postocular spots. The pronotum is largely black with patches of blue. The middorsal carina is nearly always black, bordered by a wide middorsal stripe 1/2 the width of the mesepisterna. The antehumeral stripe is blue, and no more than 1/2 the width of the middorsal stripe. The dark humeral stripe widens anteriorly and is usually only 2/3 as wide as the antehumeral stripe at that point. The rest of the thorax is pale to bright blue, fading ventrally. The legs are pale with broad black stripes on their outer surfaces, but the distal portions of the tibiae and all the tarsi lack black stripes. The abdomen is bright blue dorsally, marked with black and becoming pale ventrolaterally. Segment 1 is entirely blue, except for the basal 1/3 to 1/2, which is black dorsally. A large, irregular black spot occupies the apical 1/2 of segment 2 dorsally and is confluent with an apical ring. A similar spot on segments 3–5 extends dorsally for as much as 1/2 the length of each segment. A dorsal stripe extends 1/2 to 3/4 or more the length of segments 6 and 7. Segments 8 and 9 are entirely blue, and segment 10 is black dorsally. The cerci are uniformly black with a pale distal tubercle extending beyond the lower arm of the appendage, which is clearly encompassed by the upper arm when viewed laterally. The cerci are approximately 3/4 the length of segment 10. The paraprocts are pale with dark tips curved slightly upward.

Females may be either blue or tan. The head and thorax are marked similarly to the male. The middorsal carina may have a full-length hairline stripe. The legs are like those of the male. The middorsal lobe of the pronotum lacks any distinct pits. The mesostigmal plates are divergent anteriorly, the anterolateral corners elevated. The abdomen is generally marked with more black than in the male. Segment 1 is almost entirely black dorsally, with pale apical rings. There is a broad full-length black stripe dorsally on segment 2. Segments 3–6 each have a black hastate stripe running their entire length, or nearly so. Segment 7 is generally all black dorsally, with only a pale apical and sometimes basal ring. A full-length black stripe runs across segments 8–10 dorsally. This stripe is occasionally constricted basally on segment 8 and apically on segments 9 and 10.

Similar Species. Segment 7 in male Azure Bluets (*E. aspersum*) and Northern Bluets (*E. cyathigerum*) is all black, or nearly so,. Atlantic Bluet (*E. doubledayi*) is very similar, and careful examination of the male caudal appendages and female mesostigmal plates will be required for accurate identification. Male Familiar Bluets are one of two species in our region with a distal tubercle on the cerci. In the other, Tule Bluet (*E. carunculatum*) the tubercle protrudes beyond the end of the cerci. This is our most widely distributed bluet, but it is often found flying with bluets of other species, so look carefully at populations of little blue damselflies.

Habitat. Ephemeral or permanent ponds and lakes. Also slow-flowing streams, regardless of salinity and vegetation.

Discussion. This widely distributed southern species has extended its range dramatically, to extend now as far south as Colombia and Venezuela and as far north as southern Canada. In the contiguous United States it is absent only from the Pacific Northwest. It is only. It has entered California only in the last 80 years, and more recently has entered western Montana, British Columbia, Oregon, New York, and southern Ontario. It was accidentally introduced into Oahu in 1936, and now occurs commonly on all of the major Hawaiian Islands. Its success is probably due in part to its ability to colonize temporary and newly created aquatic habitats.

Prey of this species includes adult sweet potato whiteflies and other small flies and insects. One study included notes about a population observed at Sandpit Lake in Dallas County, Texas, where ". . . hundreds were in copula, ovipositing on small plants just below the surface of the water." Another study looked at oviposition site selection of Familiar Bluet and found that although aggregations reduced the risk of interference and may even lower predation risk, oviposition efficiency was also reduced. Peak activity occurs from midmorning into the afternoon. Males spend more days at the water (62%) than females (39%), but males mate on fewer of these days (14%) than females (79%) (Bick and Bick 1963). Mating, which involves no courtship, may last as long as 45 minutes, but usually is over in 20 minutes. Egg-laying occurs in tandem, with the male letting go to guard at a nearby perch before becoming completely submerged. Eggs are deposited in algae, roots, leaves, and upright stems at the surface of the water. Females usually remain submerged an average of just 12 minutes, but lay eggs for more than an hour below the surface, descending backwards.

A study of polymorphic head-color patterns found that the majority (73–86%) of mature adult males lacked a postoccipital bar, whereas it was present in most females. These females were split between having this postoccipital bar confluent (41–42%) or nonconfluent (50–52%) with the postoccipital spots.

One author found Familiar Bluet males hiding underwater in a laboratory aquarium after the room temperature had been accidentally left at cool temperatures overnight. This was apparently an attempt to escape the cold temperature, and may suggest how damselflies escape cold weather in nature.

References. Bick (1963, 1972), Bick and Bick (1963), Bird (1933), Cannings and Stuart (1977), Catling (1996, 1998), Dunkle (1990), Ferguson (1940), Garrison (1984), Hornuff (1968), Johnson (1964a), Johnson and Paulson (1998), Moss (1992), Polhemus and Asquith (1996), Roemhild (1975), Schaefer et al. (1996a), Scudder et al. (1976), Voshell and Simmons (1978), Wilson (1920), Zimmerman (1948).

Alkali Bluet
Enallagma clausum Morse
(pp. 98, 101, 112; photos 14a, 14b)

Size. Total length: 28–35 mm; abdomen: 22–28 mm; hindwing: 16–23 mm.
Regional Distribution. *Biotic Provinces:* Coloradan, Kansan, Navahonian. *Watersheds:* Canadian, Pecos.
General Distribution. Quebec to British Columbia; south through northern plains and Great Basin to New Mexico and Texas.
Flight Season. Jul. (TX).

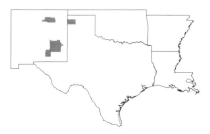

Identification. The male has a blue face. The top of the head is black, with a pale occipital bar and postocular spots. The middorsal thoracic carina is blue, surrounded by a black middorsal stripe 1/2 the width of the mesepisterna. The antehumeral stripe is blue and 1/2 or less the width of the middorsal stripe. The black humeral stripe is narrower than the antehumeral stripe for most of its length. The rest of the pterothorax is blue, becoming pale ventrally. The legs are blue or tan, with black stripes on the femora and tibiae. The abdomen is primarily blue dorsally. Segment 1 is black in the basal half. Segment 2 has a black spot on its apical third, and segments 3–5 have a spot on the apical 1/4 of each segment. Segment 7 is largely black dorsally. Segments 8 and 9 are blue, and segment 10 is black dorsally. The cerci are black, slightly upturned, and 1/3 the length of segment 10. The paraprocts are just longer than the cerci and upturned apically.

The female is tan or blue, with the head and thorax colored similarly to the male. The middorsal carina is generally pale. The humeral stripe is narrower than in the male and often appears only as a hairline in the middle 1/3. The middle lobe of the pronotum has a pair of distinctive pits. The mesostigmal plates are elongate, with a pinched middorsal carina at the base. Abdominal segment 1 is as in the male. Segment 2 may be entirely black dorsally or limited to a single spot on its apical half. Segments 3–6 each have a black stripe hastate extending 3/4 or more the length of the segment. Segment 8 is blue, and segments 9 and 10 are black dorsally.
Similar Species. Northern Bluets (*E. cyathigerum*) and Boreal Bluets (*E. boreale*) are very similar, and may be found flying with Alkali Bluet. Male Northern Bluets have the black on segment 2 expanded onto the sides. The black abdominal bands on the Boreal Bluet are truncate, not spear-shaped. Male Familiar Bluet (*E. civile*) and Tule Bluet (*E. carunculatum*) have a distal tubercle on the cerci. Careful examination of the female mesostigmal plates will often be required to accurately identify females.

Habitat. Ponds and lakes, especially those with saline or alkaline water.

Discussion. Alkali Bluet gained its name from the fact that it is often the only damselfly inhabiting saline or alkaline lakes. It is not restricted to these bodies of water, however, and may be found on freshwater lakes.

References. Kennedy (1915a).

Cherry Bluet
Enallagma concisum Williamson
(pp. 98, 109, 112; photo 14c)

Size. Total length: 27–32 mm; abdomen: 22–25 mm; hindwing: 13–17 mm.
Regional Distribution. *Biotic Province:* Austroriparian. *Watershed:* Mississippi.

General Distribution. Southeastern United States from Florida north to North Carolina and west to Louisiana.
Flight Season. Mar. 6 (LA)–Sep. 27 (LA).
Identification. The front of the male's head is bright red-orange, with a narrow black stripe. The top is black, with two transverse orange postocular spots confluent with the occipital bar to form a single narrow orange band. The middorsal carina and stripe are both black, the latter no more than 1/3 the width of the mesepisterna. The antehumeral stripe is red-orange, nearly 1/2 as wide as the middorsal stripe, and more than 1/2 as wide as the humeral stripe. There is an abbreviated black line on the interpleural suture and a slightly wider line on the metapleural suture. The remainder of the pterothorax is red-orange, quickly fading to yellow or cream color. The legs are entirely orange and armed with dark spurs. The abdomen is red-orange dorsally, becoming paler laterally, and marked with black. Except for the apical margin, the entire dorsum of segment 1 is black. A full-length black stripe extends dorsally on segment 2, the stripe thin basally, widening considerably to a large spot over the apical half of the segment. There is a dorsal black stripe on the apical 1/3 to 4/5 of segment 3, truncated basally. The entire dorsum of segments 4–7 is black, except for a broad, pale band basally. Segment 7 also bears a narrow pale apical ring dorsally. Segments 8–10 are black dorsally, although a pale narrow apical ring is sometimes present. The cerci are longer than segment 10 and orange or tan laterally, but distinctly white medially, with dark tips. The pale paraprocts are 1/2 to 2/3 the length of the cerci, and their upper surface is straight when viewed laterally.

Females are similar to males, but have darker heads. The pale areas are more yellow, and there are distinct small pits medially on the middle lobe of the prothorax. The rest of the thorax is patterned as in the male, but with a slightly wider humeral stripe and pale areas yellow or yellow-green. The mesostigmal plates are generally triangular, with a distinct pale tubercle on the posteromedial corners; the posterior and lateral borders are also pale. The legs are paler, more yellow or tan, and the femora have a dark stripe. The pale areas of the abdomen are more yellow-green or tan. Segments 1 and 2 are like those of the male, but the black stripe on the latter is not as narrowed basally. The dorsum of segments 3–10 is nearly all black, interrupted only by a pale ring on segments 7–10 and a basal ring on segments 3–7. There is a slight subapical expansion of the stripe on segments 3–6.
Similar Species. The red coloration of Cherry Bluet is unusual among our bluets, making it easier to identify. Burgundy Bluet (*E. dubium*) is smaller and has an orange, not red, antehumeral stripe. The metapleural suture is pale in Orange Bluet (*E. signatum*), whereas in Cherry Bluet there is a distinct dark stripe on this suture.
Habitat. Sand-bottomed lakes and ponds, generally with ample emergent vegetation and lily pads.
Discussion. This species is closely related to Orange Bluet, but little is known of its reproductive behavior and ecology. It is usually seen on emergent vegetation and lily pads, where females curl their abdomens to oviposit on the underside of the latter's leaves. It is relatively uncommon in our area, limited to the extreme southeastern parts of the region.

Bluets (*Enallagma*)

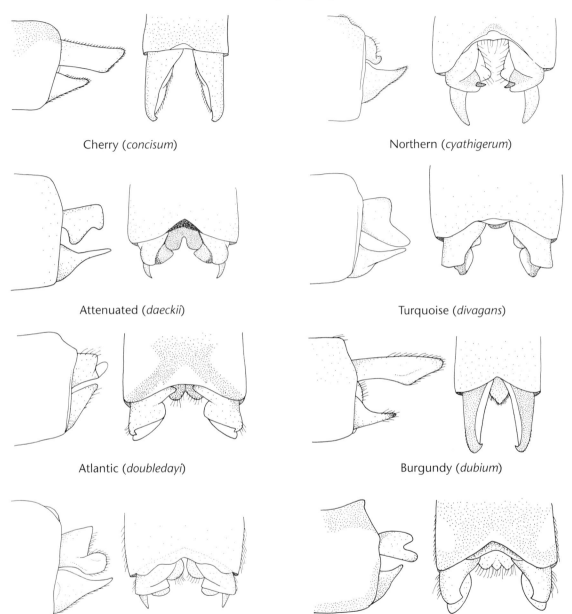

Cherry (*concisum*)

Northern (*cyathigerum*)

Attenuated (*daeckii*)

Turquoise (*divagans*)

Atlantic (*doubledayi*)

Burgundy (*dubium*)

Big (*durum*)

Stream (*exsulans*)

Fig. 19. Bluets (*Enallagma*): lateral and dorsal views of male caudal appendages (pp. 108–117).

Northern Bluet
Enallagma cyathigerum (Charpentier)
(pp. 98, 109, 112; photos 14d, 14e)

Size. Total length: 29–40 mm; abdomen: 23–32 mm; hindwing: 17–24 mm.
Regional Distribution. *Biotic Provinces:* Apachian, Chihuahuan, Coloradan, Kansan, Navahonian. *Watersheds:* Canadian, Colorado (NM), Pecos, Rio Grande.

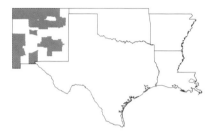

General Distribution. Northern United States and Canada south to New Mexico and Baja California; also Europe.
Flight Season. Jun. 26 (NM)–Aug. 7 (NM).
Identification. The male's face is blue, striped with black. The top of the head is largely black, with two large blue, teardrop-shaped postocular spots generally not confluent with the pale occipital bar. The middorsal carina of the pterothorax is generally blue, at least on the posterior half, surrounded by a black middorsal stripe approximately 1/2 as wide as the mesepisterna. The pale blue antehumeral stripe is slightly less than 1/2 the width of the middorsal and humeral stripe, narrowing posteriorly to 1/2 the width of the antehumeral stripe. The rest of the thorax is blue, fading ventrally to a pale white or cream color. The legs, either blue or tan, have a black stripe on their outer femoral and inner tibial surfaces; the tarsi may be pale or dark. The abdomen is mostly bright blue dorsally, fading laterally. The anterior 1/2 of segment 1 and the posterior 1/2 of segment 2 are black dorsally. There is an apical black spot on the dorsal 1/4 of segments 3–5; these spots often extend medially for 1/2 the length of the segments. A black dorsal stripe on segment 6 may extend from 1/2 to 3/4 the segment's length apically. The entire dorsum of segment 7 is black, with only a pale apical ring and a wider basal ring. Segments 8 and 9 are blue, and segment 10 is black dorsally. The cerci are black, and are no more than 1/2 the length of segment 10. The paraprocts are approximately twice the length of the cerci.

The female may be blue, green, or tan. The head is similar to that of the male, the only notable differences being smaller postocular spots. The humeral stripe is seldom wider than the antehumeral stripe. The middle prothoracic lobe lacks distinct pits, and the mesostigmal plates are generally subquadrate. The legs, especially the femora, are generally not as heavily marked with black as in the male. The abdomen is similar to that of the male, but the dorsal black stripe on segment 2 extends the full length of the segment, and is slightly expanded at about 3/4 its length apically. Segments 3–7 each have a broad black stripe that may either extend the full length of each segment or diminish anteriorly to only a hairline. Each stripe is widened considerably near the apex of the segment. Segment 8 may vary from entirely black to entirely blue. Generally, there is a dorsal black stripe, quadrate or triangulate in shape. Segments 9 and 10 each have a full-length black stripe dorsally.
Similar Species. The restricted distribution of Northern Bluet within our region helps to rule out many similar-looking species. Boreal Bluet (*E. boreale*) may be found alongside it, but the Boreal is stockier, and its black abdominal markings are generally more squared off. Males of both Familiar (*E. civile*) and Tule (*E. carunculatum*) have tubercles distally on the cerci. Careful examination of female mesostigmal plates may be necessary to identify congenerics.
Habitat. Commonly found in quiet waters, such as canals, marshes, ponds, lakes, and bogs, sometimes heavily vegetated.
Discussion. Northern Bluet is a variable species, and is truly Holarctic; it ranges throughout Europe and Asia to northern India and throughout the Northern Hemisphere. It is found from northern Mexico, Baja California, northward into Alaska, and occurs in every state west of the Rocky Mountains. It is absent only from the southeastern United States. Confirmed records are lacking from Oklahoma and Texas, but it is common westward.

Northern Bluet is generally not found around the typical acidic ponds and bogs frequented by Boreal Bluet, but one study found both species to be common at several fishless lakes in Michigan. Boreal Bluet may have a slightly later emergence, but there is considerable overlap. A mark-recapture study of a California population of Northern Bluet found their average longevity to be 4.68 days, but another study found it to be substantially longer, at 12.1 days.

References. Belle (1972), Doerksen (1979), Garrison (1978, 1984), Glas and Verdonk (1972), Juritza (1978), Macan (1974), McPeek (1989), Meulenbrock (1972), Parr (1976), Parr and Palmer (1971), Paulson (1974), Walker (1953).

Attenuated Bluet
Enallagma daeckii (Calvert)
(pp. 98, 109, 112; photos 14f, 15a)

Size. Total length: 35–47 mm; abdomen: 30–40 mm; hindwing: 19–25 mm.

Regional Distribution. *Biotic Provinces:* Austroriparian, Carolinian, Kansan. *Watersheds:* Canadian, Cimarron, Mississippi, Red, Sabine, Trinity, White.

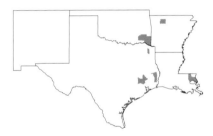

General Distribution. Southeastern United States from Florida north to Pennsylvania and west to Oklahoma and Texas.

Flight Season. Apr. 24 (LA)–Jun. (OK).

Identification. The entire head of the male, including the eyes in life, is pale blue, with an intricate pattern of thin black outlines. The postocular spots are large and narrowly outlined by jagged, often incomplete black lines. The thorax is nearly all pale blue. A narrow black middorsal stripe is reduced to thin lines on either side of the pale middorsal carina. The humeral stripe is a thin, dark line. The wings are unique among the bluets, in being petiolate out nearly as far as the anal crossing (Ac), and Cu_2 generally terminates much farther proximally than in other bluets. The legs are pale blue basally, becoming cream-colored distally, and a black stripe is present in varying degrees on the femora and tibiae; the tarsi are pale, and armed with black spurs. Abdominal Segments 1–6 are pale blue laterally and black dorsally. There is a dorsoapical blue ring on segment 1 and a full-length black stripe on segment 2, the latter expanded laterally at 3/4 its apical length. Segments 3–6 each bear a lateral, apical expansion of a dorsal stripe and a pale-blue basal ring.

Segment 7 is black dorsally, with blue on its apical 1/4, and segments 8–10 are entirely blue, with only a narrow black basal ring. The cerci are forked and short, approximately 1/3 the length of segment 10. Laterally, the cerci are brown, becoming darker apically, and each has a broad truncate upper arm and a more rounded lower arm. The paraprocts are upturned apically and only slightly longer than the cerci.

Females are pale blue, green, or tan, and the head and thorax are similar to those of male, but with overall less black. The middle lobe of the pronotum lacks prominent pits. The mesostigmal plates each bear a definite posterior margin and are concave medially, their anterolateral margins expanded and slightly raised. The abdominal pattern is similar to that of males on segments 1–6. Segments 7 and 8 are entirely black dorsally, except for a narrow apical pale-blue ring. Segments 9 and 10 are generally blue, but occasionally segment 9 has a small black spot, sometimes bilobed, extending as much as 1/2 the length of the segment.

Similar Species. Attenuated Bluet is easily recognizable in the field because of its long slender body. Its noticeably elongated abdomen (hence its name) makes it the longest pond damsel in all of North America. Its nearly all pale-blue head and thorax are also excellent field characters. The abdomen of Slender Bluet (*E. traviatum*), which also has a pale head, does not look unusually long, and there is generally more black on top of the head. The only other damselflies with unusually long abdomens in our area are the threadtails, which are all red or orange and found only in the western and southern portions of Texas.

Habitat. Swamp margins and shady, often heavily vegetated ponds, lakes, and stream backwaters.

Discussion. Populations of this infrequently seen species are restricted to the eastern Austroriparian province within the south-central United States. In Texas, populations are known only from Daingerfield State Park in Morris County, Sam Houston National Forest in Montgomery and San Jacinto counties, and Boykin Springs Recreation area in the Angelina National Forest, Jasper County. Much of the biology of this species remains unknown. Because it flutters in the shade like a "ghost" among the tangled vegetation of its habitat, it is difficult to spot.

References. Dunkle (1990).

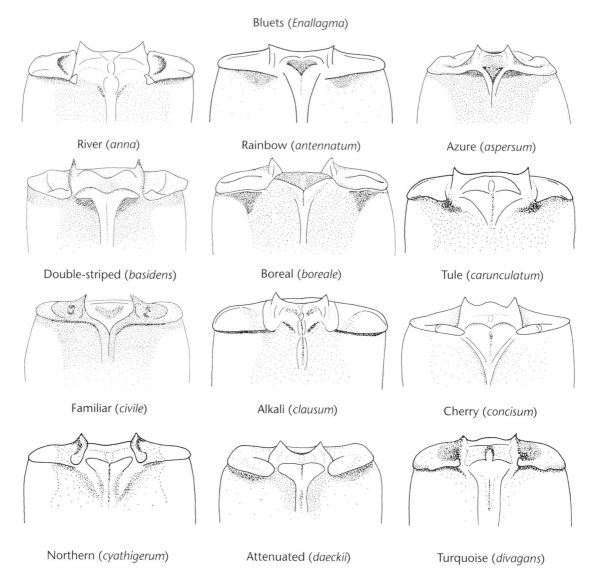

Bluets (*Enallagma*)

River (*anna*) · Rainbow (*antennatum*) · Azure (*aspersum*)

Double-striped (*basidens*) · Boreal (*boreale*) · Tule (*carunculatum*)

Familiar (*civile*) · Alkali (*clausum*) · Cherry (*concisum*)

Northern (*cyathigerum*) · Attenuated (*daeckii*) · Turquoise (*divagans*)

Fig. 20. Bluets (*Enallagma*): mesostigmal plates of females (pp. 97–113).

Turquoise Bluet
Enallagma divagans Selys
(pp. 98, 109, 112; photos 15b, 15c)

Size. Total length: 26–36 mm; abdomen: 22–30 mm; hindwing: 17–22 mm.

Regional Distribution. *Biotic Provinces:* Austroriparian, Carolinian, Texan. *Watersheds:* Arkansas, Bayou Bartholomew, Brazos, Canadian, Cimarron, Mississippi, Neches, Ouachita, Red, Sabine, San Jacinto, St. Francis, Trinity, White.

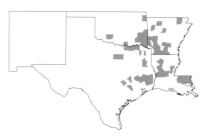

General Distribution. Eastern United States; Florida to Maine and west to Oklahoma and Texas.

Flight Season. Mar. 14 (LA)–Jun. 13 (OK).

Identification. The male's head is blue in front, crossed with a black stripe. The top is black, with a pair of large, elongated, blue postocular spots narrowly separated by the occipital bar. The middorsal stripe and carina of the pterothorax are black and vary from 1/3 to 1/2 the width of the mesepisterna. The antehumeral stripe is blue and no more than 1/2 the width of the middorsal stripe. The black humeral stripe is approximately equal in width to the antehumeral stripe and abruptly narrowed posteriorly. The rest of the pterothorax is turquoise-blue, becoming cream ventrally. There is an abbreviated stripe on the upper end of the interpleural suture and a dark spot on the metapleural fossa. The legs may be blue or tan, with black stripes on the femora and tibiae; the tarsi are pale. The abdomen is largely black dorsally and bright blue laterally, becoming paler ventrally. Segment 1 is black dorsally, with only a pale apical ring. A full-length black dorsal stripe on segments 2–7 expands subapically on each segment, and there is a narrow blue basal ring on segments 3–7. Generally, there is a small pale subapical spot on segment 7. Segments 8 and 9 are largely blue. Segment 8 has a wide basal triangle dorsally, and there is often an obscure lateral spot on segment 9. The dorsum of segment 10 is largely black. The caudal appendages are black, and each is approximately 1/2 the length of segment 10. The cerci are white medially, with a prominent dorsobasal lobe when viewed laterally. The paraprocts are widest basally and curve upward.

The female is generally paler than the male. The head and thorax differ from those of the male in the following ways: the postocular spots are larger, and the pale antehumeral stripe is generally wider; and the black humeral stripe is divided longitudinally by brown for its entire length, sometimes completely replacing the black. The middle lobe of the pronotum lacks distinct pits. The hind margin of the posterior prothoracic lobe bears a prominent, pale, median tubercle with long setae. The mesostigmal plates are more or less triangular, with a distinct posteromedial tubercle. Abdominal segments 1–7 are as in the male, the black dorsal stripes only slightly narrower. The black on the dorsum of segment 8 covers at least 2/3 of the basal length. Segment 9 is blue, with a basal black spot or stripe that is emarginate medially and extends to 1/2 the length of the segment. Segment 10 is entirely blue, with, at most, a black basal band.

Similar Species. Azure Bluet (*E. aspersum*) is blue dorsally on abdominal segment 7. Dorsally, abdominal segment 8 in Stream Bluet (*E. exsulans*) is black, and Skimming Bluet (*E. geminatum*) has a black stripe ventrolaterally on segments 8 and 9. Skimming Bluet also lacks a pale occipital bar.

Habitat. Shaded sluggish creeks and streams, sloughs, or lakes.

Discussion. This species is restricted to the eastern part of our region. Male and female Turquoise Bluets rarely stray far from water. Their flight is deliberate and slow. One study recorded that females submerged for up to 30 minutes while laying eggs. Another, which looked at the coexistence of Tennessee populations of Turquoise and Slender Bluet (*E. traviatum*), revealed that Turquoise Bluet larvae experienced significantly greater survival and biomass increase than their congener. Fecal-pellet analyses showed considerable dietary overlap and little evidence of resource partitioning between the two species.

References. Dunkle (1990), Johnson et al. (1984).

Atlantic Bluet
Enallagma doubledayi (Selys)
(pp. 98, 109, 121; photo 15d)

Size. Total length: 28–37 mm; abdomen: 22–30 mm; hindwing: 16–21 mm.

Regional Distribution. *Biotic Provinces:* Austroriparian, Texan. *Watersheds:* Neches, Red, Trinity.

General Distribution. Eastern United States from Florida to Pennsylvania and west to Texas; also Cuba and Jamaica.

Flight Season. May (TX).

Identification. The front of the head of the male is mostly blue, with an anteriorly rounded black stripe. The top is largely black, the pale blue postocular spots narrow and elongated. A pale occipital bar separates the spots, but is rarely confluent with them. The middorsal thoracic carina is black, with at most a hint of blue at its upper end. The black middorsal stripe is slightly less than 1/2 the width of the mesepisterna. The blue antehumeral stripe is 1/2 to 2/3 the width of the middorsal stripe, and the black humeral stripe is nearly equal in width to the antehumeral stripe. The rest of the pterothorax is blue, fading markedly toward the underside. The legs and coxae are largely pale blue to tan, with broad black stripes on the femora and tibiae; the tarsi are tan. The abdomen is blue, and segment 1 bears a basal black spot on the dorsum that extends posterolaterally. Segment 2 is black on the dorsoapical 1/2 of the segment. A black apical band extends for approximately 1/5 the length of segments 3–5. The black stripe on segment 6 is more extensive, tapering anteromedially to 3/4 the length of the segment. Segment 7 is black dorsally for its entire length, except for a narrow apical and sometimes basal pale ring. Segments 8 and 9 are blue, with a small apical and basal spot on the dorsum of 8. The dorsum of segment 10 is black and strongly emarginated laterally. The cerci are black and 1/2 or more the length of segment 10; each has a distinct pale apical tubercle, visible when viewed laterally. Dorsally, the upper arm of the cerci is distinctly wider than the tubercle. The paraprocts, broad basally, terminating rather bluntly, are pale with black tips and extend 2/3 to 3/4 the length of the cerci.

The female may be either pale brown, green, or blue. The general color pattern is similar to that of the male. The middle prothoracic lobe lacks distinct pits. The mesostigmal plates bear a distinct posterior border. The medial third of each plate is bordered by a distinct narrow ridge. The plates themselves are rather rectangular, with an oval depression medially. The pterothorax, wings, and legs are all generally like those of the male. The black stripes on the abdomen are more extensive. The stripe on segment 1 reaches apically for 3/4 the length of the segment. The remaining segments, 2–10, all bear a full-length black stripe, pale medial ring, and a medially interrupted basal ring.

Similar Species. Familiar Bluet (*E. civile*) is very similar, but the black markings on the abdomen are slightly more extensive on Atlantic Bluet. These two species have been taken together, though Familiar Bluet is far more abundant and widespread in the south-central United States, and there is a very good chance that Atlantic Bluet has been frequently overlooked in our region because of its similarity to the ubiquitous Familiar Bluet. Carefully examine the male caudal appendages and the female mesostigmal plates to distinguish these two species. Also, compare Atlantic with the *Similar Species* discussed under Familiar Bluet.

Habitat. Newly formed or ephemeral ponds and lakes, and occasionally sluggish streams.

Discussion. Atlantic Bluet is known west of the Mississippi only by a couple of collections from north-central Texas. This appears to be a dramatic range extension, to judge from previous collections, but Atlantic may occur in isolated populations eastward to the Mississippi River, and may simply have been overlooked because of its similarity to Familiar Bluet.

Atlantic Bluet shows some tolerance to saline waters along the coast, but is most typically found in new or ephemeral sandy-bottomed ponds. Males can generally be seen conspicuously perched on riparian vegetation at the water's edge. Pairs mate on emergent vegetation and females lay eggs in tandem. The female will remain underwater, laying eggs after pulling free from the male, for only a few minutes at a time. Males remain perched nearby, guarding the female. Leafhoppers are a common prey of Atlantic Bluet.

References. Dunkle (1990), Westfall and May (1996).

Burgundy Bluet
Enallagma dubium Root
(pp. 98, 109, 121; photo 15e)

Size. Total length: 25–30 mm; abdomen: 20–25 mm; hindwing: 12–17 mm.

Regional Distribution. *Biotic Province:* Austroriparian. *Watersheds:* Brazos, Canadian, Cimarron, Mississippi, Neches, Red, Sabine, San Jacinto, Trinity.

General Distribution. Eastern United States from Maryland south to Florida and west to Texas.

Flight Season. Apr. 9 (LA)–Sep. 23 (TX).

Identification. The male has deep-red eyes that become pale brown with age. The face is orange-red to violet, and the rest of the head is black with metallic reflections. The middorsal thoracic carina and stripe are black, with metallic purple or blue-green luster. The antehumeral stripe is orange-red and no more than 1/4 the width of the middorsal stripe. The broad humeral stripe is black with metallic reflections and nearly equal in width to the middorsal stripe. The upper end of the humeral stripe is confluent with the short black stripe of the interpleural stripe. Occasionally, there is a small isolated spot anterior to the stripe. The black stripe, at its widest, is equal in width to the antehumeral stripe. The area between the humeral and metapleural stripes is orange. The rest of the pterothorax is lighter, fading to yellow ventrally. The legs are orange-red, becoming paler distally. Occasionally, a black stripe is evident on the femora; the tarsi are pale, and armed with black spurs. The abdomen is orange-red and black. The entire dorsum is uniformly black, with only a narrow apical ring on segment 1 and basal rings on segments 3–7. The lateral pale areas are orange-red basally, fading to yellow on the distal segments. Occasionally, pale apical rings are evident on segments 7–9. The cerci, black laterally and pale medially, are generally as long as, if not slightly longer than, segment 10; there is a distinct ventrally projecting tooth at the midlength of each cercus. The pale paraprocts, no more than half the length of the cerci, are dark apically and project slightly posteroventrally.

The female is paler, with more orange-yellow or tan, and generally lacks any hint of metallic reflection in the black areas, as opposed to the male; otherwise, the female is similar in pattern to the male. The middle lobe of the pronotum bears large, distinctive pits near its anterior edge. The mesostigmal plates are noticeably triangular in shape, each having a prominent and deeply depressed posteromedial tubercle forming a distinct posterior border for most of the length of the plate laterally. There is a prominent tubercle immediately posterior to each mesostigmal plate on the mesepisternum. The abdomen is similar to that of the male, but the pale apical rings on segments 7–9 are often lacking.

Similar Species. Orange Bluet (*E. signatum*) is larger and lacks the dark stripe on the metapleural suture. Cherry Bluet (*E. concisum*) is also similar, but has a

pair of pale spots on the the prothorax, an area that is all black in Burgundy Bluet.

Habitat. Heavily vegetated black-water ponds, lakes, oxbows, sloughs, and slow reaches of streams, often associated with lily pads.

Discussion. Burgundy Bluets seem to be locally restricted, west of the Mississippi River, to Caddo Lake, southeastern piney woods, and the Big Thicket Region of eastern Texas and southeastern Oklahoma. It is infrequently encountered in these areas. Burgundy Bluet is generally found in areas where lily pads are abundant. It has been noted that these bluets can become amazingly inconspicuous, seemingly disappearing, as they enter shaded areas while patrolling. Mating pairs can be seen from midday into the afternoon on floating vegetation. Pairs prefer to lay eggs through holes in water lily leaves, where the female may submerge her abdomen to deposit eggs in semicircular rows on the underside of the leaf, a process that can take up to 30 minutes.

References. Dunkle (1990), Gloyd (1951), Westfall (1941).

Big Bluet
Enallagma durum (Hagen)
(pp. 98, 109, 121; photos 15f, 16a)

Size. Total length: 34–44 mm; abdomen: 28–35 mm; hindwing: 17–25 mm.

Regional Distribution. *Biotic Provinces:* Austroriparian, Tamaulipan, Texan. *Watersheds:* Brazos, Lower Rio Grande, Mississippi, Nueces, Sabine, Trinity.

General Distribution. Eastern coastal species from Maine to Florida, west to Texas and south into Tamaulipas, Mexico.

Flight Season. Mar. 6 (TX)–Sep. 27 (LA).

Identification. The male's face is largely blue, with a dark stripe across the front. The top of the head is a paler blue. The pale postocular spots are squared off and most often confluent with the occipital bar. The pterothorax is blue ventrolaterally. The middorsal carina is broadly pale blue, bisecting the entire length of the black middorsal stripe, which is 1/2 the width of the mesepisterna. The blue antehumeral stripe may be 1/2 to 3/4 the width of the middorsal stripe. The humeral stripe is black and narrows posteriorly, but is 1/3 to 1/2 the width of the antehumeral stripe at its widest. A black spot is present on the metapleural fossa. The rest of the pterothorax is blue, becoming much paler ventrally. The legs are pale, either tan or blue, with a black stripe on the outer femoral and inner tibial surfaces, and the tarsi are pale and armed with black spurs. The abdomen is largely bright blue dorsally, fading to cream ventrally. The basal 1/2 to 2/3 of segment 1 and the apical 2/5 to 1/2 of segment 2 are black. Segments 3–6 each have an apical black hastate spot that narrows medially, extending up to 1/2 the length of each segment. The entire dorsum of segment 7 is black, but the basal 1/5 is highly emarginated, so that the black is narrowly confluent with the anterior segment margin. Segments 8 and 9 are blue, and segment 10 is black dorsally. The cerci are tan, becoming dark dorsoapically, and are relatively short, extending no more than 1/3 the length of segment 10. Laterally, the cerci are truncated apically, with a rounded posteroventral lobe; dorsally, a pale internal tubercle is visible. The paraprocts are pale, with dark tips extending slightly beyond the cerci and appearing straight or slightly upturned.

The female is either blue or brown. The head is marked as in the male, but the black is less extensive. The middle lobe of the prothorax lacks distinct pits. The mesostigmal plates are quadrate and deeply concave medially, with a prominent posterior ridge. The pterothorax is similar to that of the male, but the humeral stripe is generally narrower than the pale antehumeral stripe. The legs are occasionally pale with black stripes; the apices of the tarsi and tarsal claws are dark. The abdominal pattern is generally similar to that of the male. The first segment is black dorsally for its entire length, with only a pale apical band. Segment 2 is black dorsally for its full length, with a slight lateral expansion apically at 3/4 of its length. Segments 3–7 are entirely black dorsally, but broadly emarginate basally. Segment 7 has a pale apical ring. Segments 8 and 9 are black dorsally, each with an apical pale ring. Segment 10 is generally blue dorsally, with a basal black spot extending rarely more than 1/3 the length of the segment.

Similar Species. Although generally larger, Big Bluet may be confused with Familiar (*E. civile*) and Atlantic (*E. doubledayi*) Bluets. The pale postocular spots and occipital bar are nearly confluent in Big Bluet, and generally separated by black in Familiar and Atlantic Bluets.

Habitat. Along the shores of lakes and rivers, often with brackish water and emergent vegetation.

Discussion. This species frequently inhabits brackish waters and seldom ventures far from the seacoast. It is considered to be one of the more primitive bluets because of its distinct venation and the details of its genitalia. Little is known about the reproductive behavior of Big Bluet. Females lay eggs head downward underwater while males guard from perching sites above. It has been suggested that the large size of Big Bluet may be an adaptation to the high winds on the open lakes and shorelines it inhabits.

References. Donnelly (1963), Dunkle (1990).

Stream Bluet
Enallagma exsulans (Hagen)
(pp. 98, 109, 121; photos 16b, 16c)

Size. Total length: 31–37 mm; abdomen: 24–30 mm; hindwing: 17–21 mm.

Regional Distribution. *Biotic Provinces:* Austroriparian, Balconian, Carolinian, Chihuahuan, Kansan, Navahonian, Tamaulipan, Texan. *Watersheds:* Arkansas, Bayou Bartholomew, Brazos, Canadian, Cimarron, Colorado, Guadalupe, Mississippi, Neches, Nueces, Ouachita, Red, Sabine, San Antonio, San Jacinto, St. Francis, Trinity, Upper Rio Grande, White.

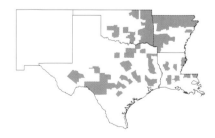

General Distribution. Eastern United States from Georgia north to Maine and southern Canada, west to Michigan and south to Texas and Mexico.

Flight Season. Apr. 12 (LA)–Sep. 15 (TX).

Identification. The face of the male is dark, and the top of the head is mostly black, only a narrow strip of blue visible. Two small teardrop-shaped pale postocular spots are narrowly separated or confluent with the pale occipital bar. The hind margin of the posterior pronotal lobe has a blue medial tubercle, bearing setae, and short stripes laterally. The lateral portions of the prothorax are pale blue. The middorsal carina and stripe of the pterothorax are generally black; the former sometimes tan posteriorly, the latter as much as 2/3 the width of the mesepisterna. The pale-blue antehumeral stripe is approximately 1/3 the width of the middorsal stripe. The broad black humeral stripe is equal to or as much as three times the width of the antehumeral stripe. The abbreviated black line on the interpleural suture is confluent with the humeral suture. The rest of the thorax, aside from a dark spot on the metapleural fossa, is blue, fading ventrally. The wings are clear or with only a slight smoky cast. The legs are pale, with a sometimes interrupted black stripe on the femora and tibiae; the tarsi are pale and armed with black spurs. The abdomen is blue, with a largely black dorsum. Segment 1 is entirely black dorsally, except for a pale apical ring. A black stripe extends the full length of segment 2, dorsally. Segments 3–6 each have a full-length black stripe narrowing basally to interrupt a pale-blue ring; the apical 1/5 of this stripe is expanded laterally on each segment. Segment 7 is either as the previous segments or with black extending over its entire dorsum. Segment 8 has a full-length black dorsal stripe, or sometimes only a basal triangle, extending for more than 1/2 the length of the segment. Segment 9 is entirely blue, with only a small basal black triangle. Segment 10 is entirely black dorsally and strongly emarginate laterally (sometimes the emargination envelops pale-blue spots laterally). The cerci are dark and distinctly forked, when viewed laterally; both lobes are directed posteriorly, but the lower lobe is noticeably longer than the upper one and approximately 1/2 the length of segment 10. The pale paraprocts are slightly shorter than the upper lobe of the cerci, and the dark apices are directed posterodorsally.

The female is generally more green than blue. The head pattern is like that of the male. The prothorax lacks any distinct pits on the middle lobe, and the hind margin of the posterior lobe has a median tubercle bearing numerous setae. The middorsal carina is distinctly tan and bordered by a narrower black middorsal stripe, the stripe generally not quite 1/2 as wide as the mesepisterna. The pale antehumeral stripe is 1/2 to 2/3 the width of the middorsal stripe. The black humeral stripe is generally less than twice the width of the antehumeral stripe and nearly always divided longitudinally (the brown sometimes replaces the black entirely). The metapleural stripe is not confluent with the humeral stripe. The rest of the pterothorax is as in the male. The mesostigmal plates are more or less triangular, with a distinct concavity just posterior to each plate. The abdominal pattern is similar to that of the male, but segment 8 bears a narrow pale apical ring. Segment 9 has two confluent basal, black spots or triangles, emarginated with blue medially. Segment 10 may be entirely blue, although occasionally a small black dorsal triangle is present.

Similar Species. Dorsally, abdominal segment 8 in Turquoise (*E. divagans*) and Skimming (*E. geminatum*) Bluets is pale blue, and the posterior portion of segment 7 is blue in Azure Bluet (*E. aspersum*). Skimming Bluet also lacks a pale occipital bar. Female Mexican Wedgetails (*Acanthagrion quadratum*) are similar to Stream Bluet, but the postocular spots are not contiguous in that species.

Habitat. Common along shores of slow-moving streams, rivers, and occasionally lakes.

Discussion. Stream Bluet is widespread throughout central Texas, but is more frequently encountered to the east, where it can be abundant. They are often sparse in the earlier parts of the day, but seem to become more numerous in the late afternoon as temperatures start to cool. In a study of the reproductive behavior of an Indiana population, mating lasted an average of 76 minutes, but was witnessed to last as long as 2 hours. Females submerge themselves to oviposit, sometimes while still in tandem with males, and other times with the male breaking away after contact with the water. Females remain submerged for 15–31 minutes, while the longest recorded time for a male was 9 minutes. Females have been observed probing backwards down stems of *Potamogeton* in a deliberate and repetitious manner, depositing eggs. One author noted capturing a female to whose thorax was attached the last abdominal segments of the male, a true testimony to the secure coupling mechanism of Stream Bluet!

References. Bick and Hornuff (1966), Eriksen (1960), Williamson (1906a).

Skimming Bluet
Enallagma geminatum Kellicott
(pp. 98, 119, 121; photos 16d, 16e)

Size. Total length: 19–29 mm; abdomen: 14–22 mm; hindwing: 12–17 mm.
Regional Distribution. *Biotic Provinces:* Austroriparian, Carolinian, Texan. *Watersheds:* Arkansas, Brazos, Canadian, Cimarron, Mississippi, Red, Sabine, San Jacinto, St. Francis, Trinity, White.

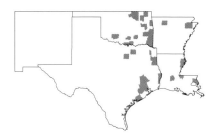

General Distribution. Eastern United States and southern Canada from Florida to Quebec; westward to Minnesota and south to Oklahoma and Texas.
Flight Season. Mar. 31 (LA)–Sep. 23 (TX).
Identification. The head and thorax and blue and black. The pale postocular spots are large and generally well separated by the absence of an occipital bar. The abdomen is more robust than in other bluets, and almost entirely black except for the basal and apical segments. In both sexes, there is a distinctive black diamond-shaped spot on segment 2 dorsally. In the male, segments 8 and 9 are pale blue, and in the female there is a pair of large blue basal spots; these are occasionally fused, but segment 9 is always black. The middle lobe of the female prothorax bears a pair of shallow posterolateral pits.
Similar Species. The larger Turquoise Bluet (*E. divagans*) lacks the diamond-shaped spot and ventrolateral stripe on abdominal segment 2. Azure Bluet (*E. aspersum*) has blue dorsally on segment 7, and Stream Bluet (*E. exsulans*) has black dorsally on segment 8. Lilypad Forktail (*Ischnura kellicotti*) is also similar, but it has larger postocular spots, segment 2 has a large dorsal basal blue spot and apical band, and segment 9 in females is either blue or red. Female Seepage Dancers (*Argia bipunctulata*) may also be mistaken for this species, but abdominal segment 8 is blue dorsally in Skimming Bluet. Females of the more western Neotropical Bluet (*E. novae-*

hispaniae) are similar, but there is some blue on segment 9.
Habitat. Prefers open, muddy, heavily vegetated ponds and lakes with fish, and more rarely slow-moving streams and swampy, smaller streams.
Discussion. One study found this species exclusively in lakes containing fish, and suggested that Skimming Bluet may be restricted to breeding in such situations to avoid predation by the abundant dragonfly larvae in fishless lakes. My observations agree with this, since large populations of the Skimming Bluet have been studied in a heavily vegetated southern Oklahoma pond with ample fish (including bass and *Lepomis* spp.) on a regular seasonal basis.

Skimming Bluet is most active in the morning, flying out over the water, perching on algae and other vegetation. Mating pairs aggregate on riparian branches and stems. The female lays eggs, unaccompanied by the male, in algae and floating debris. This species appears to be a poor disperser, the adults generally returning to their home waters.

References. McPeek (1989).

Neotropical Bluet
Enallagma novaehispaniae Calvert
(pp. 98, 119, 121; photos 16f, 17a, 17b)

Size. Total length: 29–35 mm; abdomen: 23–29 mm; hindwing: 17–19 mm.
Regional Distribution. *Biotic Provinces:* Balconian, Chihuahuan, Tamaulipan. *Watersheds:* Colorado, Guadalupe, Lower Rio Grande, Nueces, San Antonio, Upper Rio Grande.

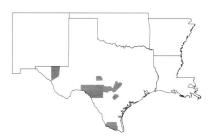

General Distribution. Texas south through Mexico to Peru and Argentina.
Flight Season. Mar. 21 (TX)–Dec. 28 (TX).
Identification. This species has a dark head, with large oval blue postocular spots that are usually not

Bluets (*Enallagma*)

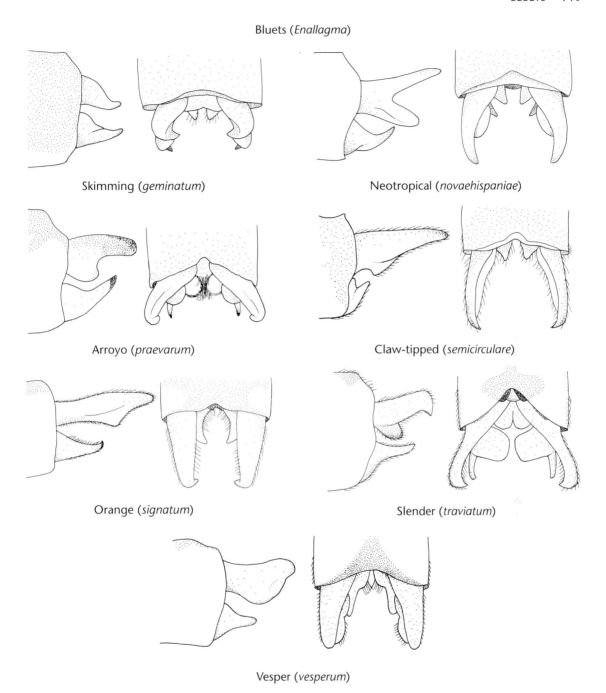

Skimming (*geminatum*)

Neotropical (*novaehispaniae*)

Arroyo (*praevarum*)

Claw-tipped (*semicirculare*)

Orange (*signatum*)

Slender (*traviatum*)

Vesper (*vesperum*)

Fig. 21. Bluets (*Enallagma*): lateral and dorsal views of male caudal appendages (pp. 118–124).

confluent with the occipital bar. The pterothorax is largely pale blue, with thin black middorsal and humeral stripes; both of these stripes are generally wider in the female. The abdomen is largely black dorsally, with blue or violet on the basal and apical segments. Segments 8 and 9 in the male are pale dorsally, with a black ventrolateral stripe. There is also a black apical band on segment 8. Segment 10 is black. The cerci are long and strongly forked, and clearly visible in the hand. Segment 8 in the female is pale, with an irregular black apical stripe. Segments 9 and 10 are black, each with a large pale apical lateral spot.

Similar Species. Female Skimming Bluet (*E. geminatum*) is similar, but abdominal segment 9 is all black in that species. Claw-tipped Bluet (*E. semicirculare*) is generally similar, but larger and overall more violet in color. The middorsal thoracic and humeral stripes in Dusky Dancer (*Argia translata*) are broader, much broader than the antehumeral area, and the distal abdominal segments in the female are longitudinally striped.

Habitat. Clear streams and rivers with a strong current.

Discussion. This is a tropical species, as its name implies, whose northern range includes the central and southern parts of Texas. Though it can be the most common damselfly at certain streams, the larva remains undescribed, and nothing has been reported on its biology.

Arroyo Bluet
Enallagma praevarum (Hagen)
(pp. 99, 119, 121; photo 17c)

Size. Total length: 26–35 mm; abdomen: 20–27 mm; hindwing: 15–21 mm.

Regional Distribution. *Biotic Provinces:* Apachian, Balconian, Chihuahuan, Coloradan, Kansan, Navahonian. *Watersheds:* Brazos, Canadian, Colorado (NM), Guadalupe, Nueces, Pecos, Upper Rio Grande.

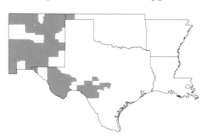

General Distribution. Western United States from North Dakota south to Texas and west to California; south through Mexico to Guatemala and Belize.

Flight Season. Apr. 10 (TX)–Nov. 18 (NM).

Identification. The male's face of this western species is blue. The top of the head is black, except for a pale occipital bar and teardrop-shaped postocular spots. The middorsal thoracic carina is blue, and the middorsal stripe is black and 1/2 the width of the mesepisterna. The antehumeral stripe is blue and usually slightly less than 1/2 the width of the middorsal stripe. The black humeral stripe is slightly narrower than the antehumeral stripe. The rest of the pterothorax is blue, becoming paler laterally and ventrally. The legs are blue or tan, with a black stripe on the femora and tibiae; the tarsi are pale. The abdomen is largely blue, marked with black in the following ways: segment 1 is black on the basal half; the dorsum of segment 2 may be 1/2 to entirely black; the black on segment 3 is limited to a small apical spot or a full-length stripe dorsally. Segments 4 and 5 have a dorsal spot on the apical forth of each segment, sometimes extending into a stripe. Segment 6 is black for at least 3/4 of its length. Segment 7 is nearly all black, and segment 10 is black dorsally. The cerci are 1/2 to 3/4 the length of segment 10. The paraprocts are 2/3 the length of the cerci and slightly upturned.

The female is very similar to the male, but may be either tan or blue. The postocular spots are generally larger and confluent with the occipital bar. The legs are generally paler than in the male. The mesostigmal plates are strongly divergent anteriorly. The middle lobe of the prothorax lacks distinct pits. Abdominal segments 3–7 each have a broad black stripe dorsally that narrows in the basal 1/5 of each segment. Segment 8 has a black triangle, and segments 9 and 10 each have a full-length black stripe, usually narrowing at its middle on 10.

Similar Species. Familiar Bluet (*E. civile*) is similar, but the black on the abdomen of Arroyo Bluet is generally more extensive, giving it an overall darker appearance. In the other similar western bluets, including Boreal (*E. boreale*), Northern (*E. cyathigerum*), and Alkali (*E. clausum*), the dark spot dorsally on segment 2 is more reduced and does not appear diamond-shaped.

Habitat. Common in ponds and slow reaches of streams.

Discussion. Arroyo Bluet prefers high, rather dry regions. It exhibits polymorphism in head-color pattern, individuals ranging from lacking an occipital

Bluets (*Enallagma*)

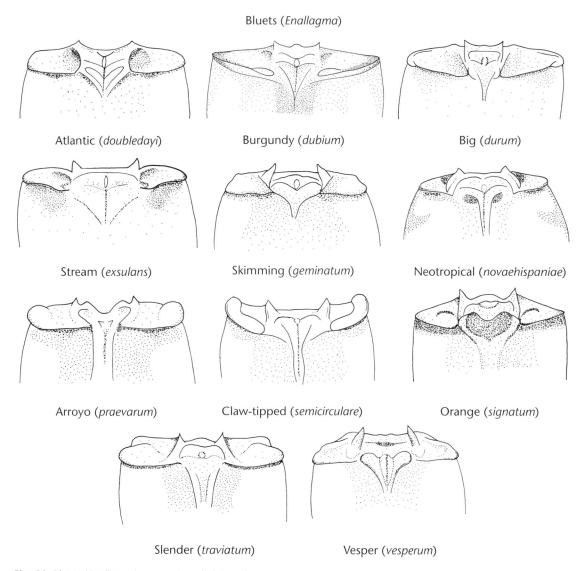

Atlantic (*doubledayi*)   Burgundy (*dubium*)   Big (*durum*)

Stream (*exsulans*)   Skimming (*geminatum*)   Neotropical (*novaehispaniae*)

Arroyo (*praevarum*)   Claw-tipped (*semicirculare*)   Orange (*signatum*)

Slender (*traviatum*)   Vesper (*vesperum*)

Fig. 22. Bluets (*Enallagma*): mesostigmal plates of females (pp. 113–124).

bar to having it confluent with postocular spots. One study showed that this variation, along with a dimorphic prothoracic color pattern, differs in frequency between the sexes. Another study in southeastern Arizona found that females were in greater abundance than males. The higher numbers of females is offset by a greater mating expectancy among males.

References. Donnelly (1968), Johnson (1964a,b).

Claw-tipped Bluet
Enallagma semicirculare Selys
(pp. 99, 119, 121; photo 18a)

Size. Total length: 29–33 mm; abdomen: 23–27 mm; hindwing: 15–24 mm.
Regional Distribution. *Biotic Province:* Chihuahuan. *Watershed:* Upper Rio Grande.
General Distribution. Arizona and New Mexico south through Mexico to Guatemala.

Flight Season. Jun. (NM).

Identification. The face of the male is violet or blue-violet. The top of the head is black, with a violet occipital bar and postoccipital spots that are usually confluent with one another. The middorsal thoracic carina is black. The black middorsal stripe is 1/3 as wide as the mesepisterna. The violet antehumeral stripe is 1/2 to 2/3 the width of the middorsal stripe. The black humeral stripe is 1/2 the width of the antehumeral stripe for most of its length. The rest of the pterothorax is pale. The legs are pale blue, with black stripes on the femora and tibiae; the tarsi are dark. The pale areas of the abdomen are bluish violet, often becoming pruinose on the basal segments. Segment 1 is black on the basal 1/3 to 3/4 of the segment dorsally. There is a variable dark spot dorsally on segment 2. Segments 3 and 4 each have a dark spot on the apical 1/5 of each segment dorsally. Segment 5 varies dorsally from a small dark spot to a full-length dorsal stripe. Segments 6 and 7 each have a nearly full-length dorsal stripe. Segments 8 and 9 are blue-violet, and segment 10 is black dorsally. The cerci are noticeably longer than segment 10. The paraprocts are short, 1/5 the length of the cerci.

The head and thorax are similar to those of the male. The mesostigmal plates are flat, with an oval swelling medially on the posterior margin. The middle lobe of the pronotum lacks distinct pits. Abdominal segment 1 is similar to that of the male. Segment 2 bears a full-length black stripe dorsally. Segment 3 has a black spot in the distal 1/5 of the segment and a narrowed stripe in its basal 2/3; these spots are generally connected by hairline. Segments 4 and 5 are generally as in the male. Segments 6 and 7 each have a full-length dorsal black stripe and a pale apical ring on 7. Segment 8 is pale dorsally. Segment 9 is black, with pale ventrolateral margins, and 10 has an irregular black dorsal stripe.

Similar Species. Neotropical Bluet (*E. novaehispaniae*) is similar, but is generally smaller and lacks black on the lateral portions of segments 9 and 10. Arroyo

Bluet (*E. praevarum*) has a diamond-shaped black spot dorsally on segment 2.

Habitat. Ponds.

Discussion. This Mexican species was only recently discovered in New Mexico, and is still known from only a single location in the region. M. Molles collected a male at Bosque del Apache at a salt cedar (*Tamarix*) flood site in 1994. The larva of this species remains undescribed and its general biology undocumented.

Orange Bluet
Enallagma signatum (Hagen)
(pp. 99, 119, 121; photos 17d, 17e)

Size. Total length: 28–37 mm; abdomen: 23–30 mm; hindwing: 15–21 mm.

Regional Distribution. *Biotic Provinces:* Austroriparian, Balconian, Carolinian, Kansan, Navahonian, Tamaulipan, Texan. *Watersheds:* Arkansas, Bayou Bartholomew, Brazos, Canadian, Cimarron, Colorado, Guadalupe, Lower Rio Grande, Mississippi, Neches, Nueces, Ouachita, Red, Sabine, San Antonio, San Jacinto, St. Francis, Trinity, White.

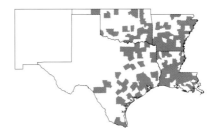

General Distribution. Eastern United States from Florida to Quebec; westward to South Dakota and south to Texas.

Flight Season. All year (TX).

Identification. The face of the male is orange, except for a narrow black stripe across the front. The top is black, with an orange occipital bar that is nearly confluent with two narrow postocular spots. The middorsal thoracic carina is black, and 1/3 as wide as the mesepisterna. The antehumeral stripe is orange and 1/2 the width of the middorsal stripe. The humeral stripe is black, and subequal in width to the antehumeral stripe. The remainder of the thorax is orange. The legs are orange, with the occasional black stripe on the femora and tibiae. The abdomen is orange dorsally and paler laterally. The dorsum of segments 1 and 2 are nearly all black.

Segments 3–8 are black, except for a pale basal ring that is generally lacking on 8. Segment 7 has a narrow apical ring. There is a pale basal ring on segment 9, and 10 is black dorsally. The cerci are subequal in length to segment 10. The pale paraprocts are 1/2 the length of the cerci. Teneral and young males are blue, rather than orange.

The female is paler in color, generally yellow-green. The head and thorax are similar to those of the male, but the pale areas are more extensive. The humeral stripe is generally narrower than the yellow antehumeral stripe. The pronotum has distinct pits on the anterior 1/3 of the middle lobe. The mesostigmal plates are nearly triangular, with a prominent tubercle at the posteromedial corner. The pale colors of the abdomen are yellowish green. Segments 1–8 are generally like those of the male. Segment 9 is black dorsally, and segment 10 is generally pale.

Similar Species. Cherry (*E. concisum*) and Burgundy (*E. dubium*) Bluets are both smaller and red, not orange. Teneral Vesper Bluets (*E. vesperum*) are similar, but the black humeral stripe is either lacking or narrowly reduced in that species. Our threadtail species (Family Protoneuridae) are orange and are found alongside Orange Bluet, but they have much longer and thinner abdomens (twice as long as the wings).

Habitat. Various ponds and lakes, as well as slow-moving streams and rivers.

Discussion. Found in a variety of habitats, Orange Bluet is unusual in that it is most active in the late afternoon. One study found that it never appeared before 2:30 P.M. In this respect, it is similar to Vesper Bluet. Females stay some distance from the water and are not often encountered, except in copula or tandem. Females are of one of three forms: the first remains blue throughout life, one becomes green, and the third becomes orange.

Males are often seen hovering low to the water, occasionally perching on water lilies or other emergent vegetation. A study on the reproductive behavior of Orange Bluet found that females will posture their unwillingness to mate. After mating, pairs will begin laying eggs in floating vegetation or debris, the male often accompanying his partner underwater. Females will remain underwater, ovipositing in the traditional manner, for up to 20 minutes at a time.

References. Dunkle (1990), Lutz and Pittman (1970), Tennessen (1975).

Slender Bluet
Enallagma traviatum Selys
(pp. 99, 119, 121; photo 18b)

Size. Total length: 29–32 mm; abdomen: 24–26 mm; hindwing: 15–19 mm.

Regional Distribution. *Biotic Provinces:* Austroriparian, Carolinian, Texan. *Watersheds:* Arkansas, Brazos, Canadian, Cimarron, Mississippi, Red, Sabine, Trinity, White.

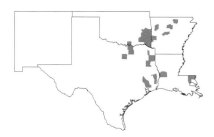

General Distribution. Eastern United States from Georgia to Ontario, westward to Wisconsin and south to Texas.

Flight Season. Apr. (LA)–Jul. 9 (OK).

Identification. The face and head of the male are mostly pale blue, with large distinctive pale postocular spots. These spots are thinly outlined by black and separated by an occipital bar. The pterothorax is pale blue with very little black. The middorsal stripe and carina are black, and the former is usually confined to the latter. There is often a yellow-brown discoloration lateral to the middorsal stripe, which may be incomplete and divided by discoloration. The black humeral stripe is narrow and often reduced to a hairline. The abdomen is largely black dorsally, and segments 8 and 9 are pale blue. On top, segment 8 usually has a small black triangular spot basally. The female is similar to the male, but segment 10 is blue.

Similar Species. Slender Bluet is distinctive because of its predominant pale-blue coloration and large postocular spots. Attenuated Bluet (*E. daeckii*) is similar, but is larger, with a longer abdomen, and the distal part of abdominal segment 7 is pale blue. This segment is all black in the Slender Bluet.

Habitat. Permanent ponds and lakes with sparse to abundant emergent vegetation.

Discussion. Slender Bluet is widespread throughout the eastern United States. The subspecies found in our region, *E. t. westfalli*, was described by Donnelly (1964) from a pond in east Texas near Cleveland, based on the relatively more robust posterodorsal arm of the male cerci. This subspecies and the nominate form are apparently geographically separated by the Appalachian Mountains. This species is uncommon in the eastern portion of our area, but I have seen it flying with the similar Attenuated Bluet.

References. Donnelly (1973).

Vesper Bluet
Enallagma vesperum Calvert
(pp. 99, 119, 121; photos 18c, 18d)

Size. Total length: 29–37 mm; abdomen: 24–30 mm; hindwing: 15–21 mm.
Regional Distribution. *Biotic Provinces:* Austroriparian, Balconian, Texan. *Watersheds:* Arkansas, Brazos, Cimarron, Colorado, Guadalupe, Mississippi, Neches, Ouachita, Red, Sabine, Trinity, White.

General Distribution. Eastern United States from Florida north to Maine, west to Wisconsin and south to Texas.
Flight Season. Mar. 31 (LA)–Oct. 21 (TX).
Identification. The face of the male is mostly yellow, with brown eyes that become paler laterally. The top of the head is black, with long postocular spots that are confluent with the occipital bar. The pronotum is largely pale. The middorsal thoracic carina is black. The black middorsal stripe is 1/4 to 1/3 the width of the mesepisterna. The antehumeral stripe is yellow-orange and 1 to 1.5 times the width of the middorsal stripe. The humeral stripe is black and largely reduced to a hairline. The rest of the pterothorax is orange or yellow. The legs are orange-yellow and occasionally bear a dark stripe on the femora. The abdomen is orange, becoming yellow or blue-green laterally. Segment 1 nearly all black. Segment 2 has a dark stripe dorsally. Segments 3–8 are nearly all black, except for a narrow pale basal ring on each segment. Segment 9 dorsally and 10 laterally are blue. The cerci are subequal in length to segment 10. The paraprocts are 1/2 the length of the cerci and nearly straight in profile.

The female is generally duller in color, more yellowish, than the male. The eyes are brown. The head and thorax are generally like those of the male, but without extensive black markings. There are distinct pits on the middle lobe of the pronotum. The mesostigmal plates are triangular, with a well-defined posterior margin, a strong medial ridge, and a pale posteromedial tubercle. The pale areas of the abdomen are yellowish, sometimes blue-green. Segments 1–8 are like those of the male. Segment 9 has a black triangular spot dorsally. Segment 10 is pale dorsally.
Similar Species. Vesper Bluet is the only largely yellow bluet in our region. Females of Orange Bluet (*E. signatum*) are similar, but the humeral stripe in Vesper Bluet is much narrower, and the prothoracic pits in Orange Bluet are positioned more anteriorly. No forktail in our region is yellow, with abdominal segments 9 and 10 blue.
Habitat. Found most commonly in heavily vegetated ponds and lakes, but occasionally in slow reaches of streams.
Discussion. Vesper Bluet is an unusual species, in the sense that it is most active in the late evening, as its name implies, and often does not appear over water until sunset. I have collected them at mercury vapor lights just after dusk. Their coloration and delicate shape allow them to take cover easily in vegetation during the day, but they may be seen sneaking about in the morning hours as well. Pairs may leave the water, for up to 20 minutes, to mate. Egg-laying occurs in tandem, occasionally after dark, in stems and other vegetation lying just below the surface.

References. Carpenter (1991), Dunkle (1990), Robert (1939).

Painted Damsel
Genus *Hesperagrion* Calvert

This is a monotypic genus that may be closely related to the forktails (*Ischnura*), but beyond superficial affinities, adults share few similarities with that group. A study of the penis morphology in this group associated the group with the Neotropical genera *Anisagrion*, *Apanisagrion*, and *Calvertagrion*. Male Painted Damsels lack the dorsoapical prominence of segment 10 typical of forktails, and the caudal appendages are sufficiently different. As its common name implies, it is colorful, rivaling all other damselflies in the region, and perhaps in North America, in beauty. Painted Damsels are found in streams of the arid southwestern United States and northern Mexico.

References. Calvert (1902), De Marmels (2002), Paulson and Cannings (1980).

Painted Damsel
Hesperagrion heterodoxum (Selys)
(pp. 133, 139; photos 18e, 18f)

Size. Total length: 28–35 mm; abdomen: 21–27 mm; hindwing: 16–21 mm.
Regional Distribution. *Biotic Provinces:* Apachian, Chihuahuan, Kansan, Navahonian. *Watersheds:* Colorado (NM), Pecos, Upper Rio Grande.

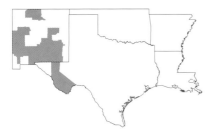

General Distribution. Southwestern United States and Mexico.
Flight Season. Apr. 9 (TX)–Nov. 11 (TX).
Identification. The head of the male is mostly black, except for two large bright-red postocular spots. The thorax is largely black, with a blue antehumeral stripe that is reduced to distinctive, elongated, anterior and posterior spots. The rest of the thorax is blue, fading below. The abdomen is largely black dorsally and blue-green laterally. The apex of segment 7 and most of the area on segments 8–10 are bright orange in the male. In females, the apical 3/4 of segment 7 is bright blue. Teneral individuals of both sexes are orange. Young individuals go through a number of intermediate color stages that may occur at different rates on the head, the thorax, and the abdomen. The male cerci are forked. The mesostigmal plates of the female are triangular, with a strong posterolateral flange and a distinct small knob bearing a setal tuft on each posteromedial corner.
Similar Species. Pacific (*I. cervula*) and Plains (*I. damula*) Forktails both have black thoraxes with a pale antehumeral stripe reduced to anterior and posterior spots. These spots are smaller than those in Painted Damsel, however, and neither of these species has bright-red postocular spots. Teneral females are orange, not red like firetails (*Telebasis*), and the dorsum of the thorax is pale in Painted Damsel. Females are also unique in having abdominal segment 7 pale.
Habitat. Permanent and ephemeral creeks and streams with moderate emergent vegetation.
Discussion. Painted Damsel is found commonly in the most western limits of our region. It often perches openly on emergent vegetation in rather large numbers. It has been reported as an inhabitant of permanent streams, but I have observed it in ephemeral streams of the Davis Mountains in West Texas. Despite its beauty, we don't know much about the behavior or egg-laying strategies of this species.

Forktails
Genus *Ischnura* Charpentier
(pp. 126, 133, 139)

These often brightly colored damselflies are among the smallest and most studied of the suborder, both within the region and in all of North America. Twelve of the 14 species found in the United States occur in the south-central region. The species in our region have distinctively eastern or western distributions, except for the northern Eastern Forktail (*I. verticalis*). The common name, forktails, refers to the forked posterodorsal projection on the apex of abdominal segment 10 in most males. This projection, lacking in Lilypad Forktail (*I. kellicotti*), gives the characteristic appearance of the genus. Forktails may be abundant around the dense vegetation of their typical habitat, generally including ponds, lakes, and marshes. Some species, however, prefer creeks and streams. Although many of the species

TABLE 5. FORKTAILS (*ISCHNURA*)

Species	Hindwing length (mm)	Antehumeral stripe (♂)	Antehumeral color	Dorsoapical projection (♂)[1]	Dorsoapical projection bifid (♂)[2]	Vulvar spine (♀)[3]	Projections pronotum (♀)[4]	Pale color of 8,9,10 (♂)[5]	Distribution (biotic provinces)[6]
Desert (*barberi*)	14–19	complete	yellow-green	1/4	–	+	–	8,9	Ap,B,Ch,Co,K,N,Tx
Pacific (*cervula*)	13–19	interrupted	blue	1/2	+	–	–	8,9	Ap,Ch,N
Plains (*damula*)	11–19	interrupted	blue	1/4	–	+	+	8,9	Ap,Ch,Co,K,N
Mexican (*demorsa*)	11–15	complete	blue	1/2	+	–	–	8,9	Ap,ChCo,K,N
Black-fronted (*denticollis*)	11–15	lacking	—	1/4	+	–	+	8,9	Ap,Ch,K,N,Tx
Citrine (*hastata*)	9–15	complete	yellow-green/ yellow	1	+	+/–	–	8,9,10	Au,B,Ca,Ch, Co,K,N,Ta,Tx
Lilypad (*kellicotti*)	12–18	complete	blue	0	–	+	–	8,9	Au,Tx
Western (*perparva*)	11–17	complete	blue/green	1/3–1/2	+	+/–	–	8,9	Ch,Co,K,N
Fragile (*posita*)	10–16	interrupted	yellow-green	1/4–1/3	+	–	–	—	Au,B,Ca,Ch, K,N,Ta,Tx
Furtive (*prognata*)	14–20	complete	blue-green	4/5	+	+/–	–	9	Au
Rambur's (*ramburii*)	15–19	complete	green/blue	<1/4	+	+	–	8	Au,B,Ca,Ch, K,Ta,Tx
Eastern (*verticalis*)	11–19	complete	yellow/ blue-green	1/4–1/3	+	+	–	8,9	Au,Ca,K,N,Tx

[1] Fraction of segment 10 length that projects dorsoapically from that segment.

[2] Dorsoapical projection on segment 10 clearly bifid (+) or entire (–).

[3] Vulvar spine on abdominal segment 8 present (+), always lacking (–), or minute and often absent (+/–).

[4] Middle lobe of pronotum with nipple or knoblike projections.

[5] Segments 8–10 that are predominantly (1/2 length of segment or more) pale.

[6] (Ap) Apachian, (Au) Austroriparian, (B) Balconian, (Ca) Carolinian, (Ch) Chihuahuan, (Co) Coloradan, (K) Kansan, (N) Navahonian, (Ta) Tamaulipan, (Tx) Texan.

are cosmopolitan in distribution, they are relatively weak fliers.

Forktails are fairly distinct, but resemble damselflies of other genera, like bluets (*Enallagma*) and Painted Damsels (*Hesperagrion*). The postocular spots are particularly well represented in some species, and an occipital bar may be present in some females. The thoracic color pattern of the group always involves a black middorsal carina and stripe, and usually a black humeral stripe. The antehumeral stripes of some males, when present, are pale, and may be reduced to anterior and posterior spots. The wings are clear, petiolate, and generally characterized by the separation of the anal vein from the wing margin before the anal crossing. The pterostigma in most males is of a different color and/or shape in the different wings. The abdomen of males is usually black dorsally, except in Citrine Forktail (*I. hastata*), and pale laterally with blue on segments 8 and 9, except in Fragile Forktail (*I. posita*).

A vulvar spine on abdominal segment 8 may or may not be present in females. Females are polymorphic (except for Fragile Forktail), often occurring in two or three color forms, which may in turn change with age. The various color forms generally include an andromorph (a form similar to the male) and one or two gynomorphs. Various explanations for the genetic basis of this balanced polymorphism in females have been proposed. In two of the western species, Mexican (*I. demorsa*) and Plains (*I. damula*) Forktails, the andromorphic state is recessive, less common than the gynomorphs and, because of its more conspicuous coloration, preyed upon more heavily. The less-colorful gynomorphs were also found to be at a disadvantage, however, because males frequently mated with the wrong species. Each color morph was determined to have an advantage in certain situations. Other studies, on Rambur's (*I. ramburii*) and Black-fronted (*I. denticollis*) Forktail, and on two European species, have shown that male interference of andromorphs during egg-laying is reduced at high population densities, allowing more time for egg-laying. Andromorphs, however, are at a disadvantage at low population densities because of the loss of mating opportunities when passed by males. At these low population densities the gynomorphs have an easier time laying eggs, because they are not harassed by males. All of the available data suggest that polymorphism in forktails is controlled by density-dependent factors. The mesostigmal plates of many of the females are similar. This, combined with the polymorphism of females, makes identifying them challenging.

Members of this group typically have longer flight seasons, as compared to other pond damsels, and multiple generations. The wide variation in size of many of the species is directly correlated to this pattern. Larger specimens are predominantly encountered in the spring, when larval durations are longer, and smaller individuals are more common in the summer and fall months. Most females of this genus lay eggs alone, not in tandem, in emergent plant stems. Exceptions in the region include the Black-fronted and Plains Forktails. Females typically mate in the early morning and often only once.

References. Aguilar (1992), Cordero (1990), Cordero and Andres (1996), Dunkle (1990), Fincke (1987), Hinnekint (1987), Johnson (1964c, 1966b, 1975b), Robertson (1985), Robinson and Allgeyer (1996), Robinson and Novak (1997).

KEY TO THE SPECIES OF FORKTAILS (*ISCHNURA*)

MALES

1. Dorsum of pterothorax solid black with metallic luster, but lacking pale antehumeral stripes — **Black-fronted (*denticollis*)**

1'. Dorsum of pterothorax with pale antehumeral stripes, sometimes represented by only a small anterior and posterior spot or stripe on each side — 2

2(1'). Forewing pterostigma separated from costa; abdomen mostly yellow dorsally; dorsoapical prominence on segment 10 spinelike and approximately 1/2 as long as segment 9 — **Citrine (*hastata*)**

2'. Forewing pterostigma not separated from costa; abdomen not yellow dorsally; dorsoapical prominence on segment 10 generally not spinelike (except in Furtive Forktail) and much shorter — 3

KEY TO THE SPECIES OF FORKTAILS (*ISCHNURA*) (*cont.*)

3(2'). Antehumeral stripe widely separated into equal anterior and posterior spots	4
3'. Antehumeral stripe complete or narrowly divided, appearing as an exclamation mark, with anterior spot much longer than the posterior	5
4(3). Posterior margin of mesostigmal plate black; dorsoapical prominence on segment 10 deeply bifurcated and 1/2 as high as the segment itself	**Pacific (*cervula*)**
4'. Posterior margin of mesostigmal plate pale; dorsoapical prominence on segment 10 not deeply bifurcated and extending 1/3 or less as high as the segment itself	**Plains (*damula*)**
5(3). Paraprocts deeply emarginate distally with ventral lobe curved medially	11
5'. Paraprocts not emarginated	6
6(5'). Cerci bifid, with a posteriorly directed lateral process and a ventrally directed medial process, the two subequal in length	7
6'. Cerci not bifid, but may be hooked downward	8
7(6). Upper and lower arms of cerci long and thin; dorsoapical prominence on abdominal segment 10 appears as an elongated spine; dorsum of abdominal segments 8 and 9 is black and blue, respectively	**Furtive (*prognata*)**
7'. Upper and lower arms of cerci short and thick; dorsoapical prominence on abdominal segment 10 is short and not spinelike; dorsum of segments 8 and 9 is largely blue and black, respectively	**Rambur's (*ramburii*)**
8(6'). Dorsoapical prominence on abdominal segment 10 low and not bifid; dorsum of abdominal segment 10 mostly blue	**Lilypad (*kellicotti*)**
8'. Dorsoapical prominence on segment 10 distinctly bifid; dorsum of segment 10 entirely black	9
9(8'). Abdominal segments 8 and 9 entirely blue, or with black only dorsobasally on 8, never laterally; cerci rounded apically	**Desert (*barberi*)**
9'. Segments 8 and 9 entirely black, or at least with extensive black areas laterally; cerci acute apically	10

10(9'). Abdominal segments 8 and 9 entirely black; paraprocts serrated, when viewed laterally; 2 postquadrangular cells in hindwing **Fragile (*posita*)**

10'. Segments 8 and 9 blue dorsally, with black stripes ventrolaterally; paraprocts not serrated, tapering to a blunt apical point; 3 postquadrangular cells in hindwing **Eastern (*verticalis*)**

11(5). Superior branch of paraprocts distinctly longer than inferior branch, neither branch with apical denticles or teeth; abdominal segment 9 is entirely blue, or bears only a small lateral black spot basally **Mexican (*demorsa*)**

11'. Superior and inferior branches of paraprocts subequal in length and one or both branches bear distinct denticles or teeth apically; segment 9 has black laterally for 1/2 its length **Western (*perparva*)**

FEMALES

1. Hind margin of prothorax with tongue-like median lobe bearing a tuft of long setae on either side **Pacific (*cervula*)**

1'. Hind margin of pronotum much broader than long, without tongue-like lobe and lacking long setae 2

2(1'). Middle prothoracic lobe with distinct nipple-like process on each side, generally pale in color 3

2'. Middle prothoracic lobe without distinct nipple-like process on each side, color variable 4

3(2). Prominent flange-like projection dorsally, along posterior margin of mesostigmal plate; hindwing generally greater than 14 mm long **Plains (*damula*)**

3'. No prominent flange-like projection dorsally, along posterior margin of mesostigmal plate, only low ridge; hindwing generally less than 14 mm long **Black-fronted (*denticollis*)**

4(2'). Abdomen usually more than 28 mm long; M_2 separates from M_{1-2} near the 5th postnodal crossvein in the forewing **Furtive (*prognata*)**

4'. Abdomen usually less than 27 mm long; M_2 separates from M_{1-2} closer to the 4th postnodal crossvein in the forewing 5

KEY TO THE SPECIES OF FORKTAILS (*ISCHNURA*) (*cont.*)

5(4'). Mesostigmal plates with a prominent ridge or flange extending above surface of pterothorax — 6

5'. Mesostigmal plates without a prominent ridge or flange extending above surface of pterothorax — 8

6(5). Prominent flange on mesostigmal plates restricted to lateral 2/3 of their width; a distinct tubercle in the posteromedial corner — **Mexican (*demorsa*)**

6'. Prominent flange or ridge on mesostigmal plates extending nearly their entire width — 7

7(6'). Hind margin of prothorax with prominent fringe of hairs across its entire width — **Eastern (*verticalis*)**

7'. Hind margin of prothorax without a prominent fringe of hairs, although a few lateral hairs may be present — 11

8(5'). Vulvar spine on abdominal segment 8 absent; antehumeral stripe generally divided with a longer posterior stripe and a smaller anterior spot (the two resembling an exclamation mark), although often obscured with age — **Fragile (*posita*)**

8'. Vulvar spine on segment 8 present, but may be small; dark humeral stripe absent or antehumeral stripe complete — 9

9(8'). Postocular spots large and conspicuous; dorsum of abdominal segment 2 black, with a blue or orange apical spot — **Lilypad (*kellicotti*)**

9'. Postocular spots small, often obscured in older individuals; dorsum of segment 2 entirely black or entirely orange — 10

10(9'). Hindwing generally no more than 15 mm long; middle prothoracic lobe with a pair of distinct pits; medial borders of mesostigmal plates nearly straight — **Citrine (*hastata*)**

10'. Hindwing generally 16 mm or longer; middle prothoracic lobe without distinct pits; medial borders of mesostigmal plates strongly concave — **Rambur's (*ramburii*)**

11(7'). Abdomen generally longer than 23 mm; black stripes on abdominal segments 3–5 constricted at no more than 3/4 their length — **Desert (*barberi*)**

11'. Abdomen less than 22 mm; black stripes on segments — **Western (*perparva*)**

Desert Forktail
Ischnura barberi Currie
(pp. 126, 133, 139; photos 19a, 19b)

Size. Total length: 28–35 mm; abdomen: 22–27 mm; hindwing: 14–19 mm.

Regional Distribution. *Biotic Provinces:* Apachian, Balconian, Chihuahuan, Coloradan, Kansan, Navahonian, Texan. *Watersheds:* Arkansas, Brazos, Canadian, Cimarron, Colorado (NM), Pecos, Red, Upper Rio Grande.

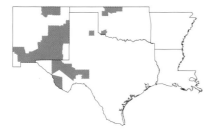

General Distribution. Western United States from Nebraska to California and south to Texas.

Flight Season. Jun. 5 (OK)–Nov. 7 (NM).

Identification. The face of the male is pale blue-green, heavily marked with black. Two small pale-blue postocular spots are generally confluent with a narrow occipital bar. The thorax is green, with a black middorsal stripe and a black humeral stripe; the latter is approximately 1/2 the width of the middorsal stripe. The abdomen is blue-green on the first two segments and part of segment 3. The remainder of the abdomen is yellow-orange dorsally marked with black. The posterolateral portion of segment 7 and all of segments 8 and 9 are blue. Segment 10 is blue, with a wide black dorsal stripe. The dorsoapical prominence on segment 10 is conspicuous, but does not extend posteriorly beyond segment 10. The cerci are not forked, but strongly directed downward. The paraprocts are gently upturned or straight, with only the apices upturned.

When viewed dorsally, the female mesostigmal plates each bear a flange extending posteromedially to the anterolateral corner. Andromorphic females are uncommon and nearly identical to males. Gynomorhpic females are orange or tan, often with a slight greenish cast to the abdomen. The abdomen is generally similar to that of the male, but with a black basal triangle and a subapical spot on the dorsum of an otherwise pale segment 8; these spots are occasionally narrowly confluent. There is a full-length black stripe on segment 9, and segment 10 bears a dorsal black triangle extending the entire length of the segment.

Similar Species. Male Rambur's Forktails (*I. ramburii*) are similar, but the top of abdominal segment 9 is black, and in Desert Forktail the entire segment is blue. The humeral stripe in Rambur's Forktail is also slightly wider. The postocular spots on female Desert Forktails are typically larger than those on Rambur's Forktail. The gynomorphic form of Rambur's Forktail is is typically brighter orange and has black dorsally on segment 8.

Habitat. Alkaline and saline desert springs, pools, irrigation ditches, and canals.

Discussion. Little is known of this species' ecology and behavior, but it is presumably similar to other forktails. It can be quite common around heavily vegetated alkaline and saline lakes.

Pacific Forktail
Ischnura cervula Selys
(pp. 126, 133, 139; photos 19c, 19d)

Size. Total length: 24–31 mm; abdomen: 19–26 mm; hindwing: 13–19 mm.

Regional Distribution. *Biotic Provinces:* Apachian, Chihuahuan, Navahonian. *Watersheds:* Colorado (NM), Pecos, Rio Grande.

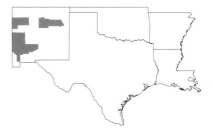

General Distribution. Alberta, Canada and Pacific Northwest south to New Mexico and Baja California.

Flight Season. Jul. 29–Aug. 6 (NM).

Identification. The male's face is yellow-green. The top of the head is largely black, except for two small blue postocular spots. The front of the thorax is black, with a pale-blue antehumeral stripe divided into small anterior and posterior spots. The sides of the thorax are pale blue. The pale areas on abdominal segments 1, 2, and 10 are blue-green, segments 3–7 are yellowish green, and segments 8 and 9 are blue. The anterior 3/4 of segment 1 is black dorsally. Segments 2–7 are black with pale basal ring on 4–7 and an additional apical ring on 7. Segment 8 is dark dorsally for as much as 1/5 of the segment's length. The lateral stripes on segment 8 extend 1/2 to 2/3 of the segment's length basally and 1/2 of its length on segments 9 and 10. The dorsoapical prominence on segment 10 is distinctly forked and extends 1/2 as high as the remainder of the segment.

Females have a distinct tongue-like flange on the posterior margin of the pronotum, and there is a distinct tuft of setae on each side of the median lobe. There is generally no vulvar spine. The andromorphic form is uncommon, but like the male, though the anterior mesepisternal spot is larger and sometimes elongated as in Fragile Forktail (*I. posita*). The postocular spots may be narrowly confluent. The abdomen is generally like that of the male, with segment 8 blue and bearing lateral stripes, but lacking a basal ring. Segments 9 and 10 are generally black dorsally, but 9 may be partly or entirely blue. Gynomorphic females are more common and are pale orange. A pale occipital bar is confluent with the large, subtriangular postocular spots. The pterothorax is black, with a middorsal stripe generally 1/3 as wide as the mesepisterna. The antehumeral stripe is orange-tan, and the humeral stripe is reduced and nearly absent. The abdomen is pale orange or tan. Segments 1–7 are generally patterned like those of the male. Segment 8 is blue, with a short lateral stripe. Segment 9 is entirely black dorsally, or with blue on the apical 2/3, and segment 10 is nearly all black.

Similar Species. On male Plains Forktail (*I. damula*) the pale antehumeral stripe is reduced to an anterior and a posterior spot like those on Pacific Forktail. The dorsoapical projection on abdominal segment 10 of Plains Forktail does not reach as high above the segment as it does in Pacific Forktail. The posterior margin of the the mesotigmal plate in males is black in Pacific Forktail and pale in Plains Forktail. Separating females where these two species overlap may require close in-hand examination. The entire front of the thorax is black in male Black-fronted Forktails (*I. denticollis*).

Habitat. Saline and alkaline ponds, as well as slow-moving streams.

Discussion. Within our region this species is restricted to western New Mexico.

References. Kennedy (1915b), Paulson (1974).

Plains Forktail
Ischnura damula Calvert
(pp. 126, 133, 139; photos 19e, 19f)

Size. Total length: 23–34 mm; abdomen: 18–27 mm; hindwing: 11–19 mm.

Regional Distribution. *Biotic Provinces:* Apachian, Chihuahuan, Coloradan, Kansan, Navahonian. *Watersheds:* Canadian, Cimarron, Colorado (NM), Pecos, Red, Upper Rio Grande.

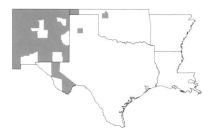

General Distribution. Eastern slope of Rockies and Great Plains from Texas to North Dakota and Wyoming, south to Arizona.

Flight Season. May (TX)–Sep. (TX).

Identification. Males have a dark thorax dorsally, with the antehumeral stripe reduced to a pair of small pale spots. The remaining areas of the thorax are blue. The abdomen is largely black, though segments 8 and 9 are blue, except for an abbreviated basal lateral stripe. The dorsoapical prominence on abdominal segment 10 is not forked and is about 1/4 the height of the rest of the segment. The caudal appendages are distinct, the cerci bearing a prominent posteroventral process and the paraprocts upcurved with bluntly pointed apices.

The female is one of only two species in the region with a prominent nipple-like process on each side of the pronotum. Andromorphic females are common and nearly identical to males, including the reduced antehumeral stripe. Gynomorphic females, by contrast, have a complete antehumeral stripe, and the pale abdominal colors are orange or tan with the occasional blue markings laterally on segments 1 and 2, the apex of 7, and all of 8–10.

Similar Species. On male Black-fronted Forktails (*I. denticollis*) the entire front of the thorax is black. Pacific Forktail (*I. cervula*) is similar, with the antehumeral stripe separated into distinct, pale posterior and anterior spots, but the dorsoapical projection on abdominal segment 10 does not reach as high above the segment in Plains Forktail as it does in

Forktails (*Ischnura*) and Painted Damsel (*Hesperagrion*)

Painted Damsel (*Hesperagrion heterodoxum*)

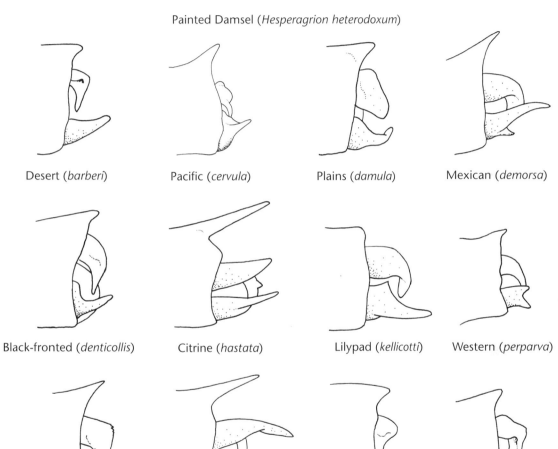

| Desert (*barberi*) | Pacific (*cervula*) | Plains (*damula*) | Mexican (*demorsa*) |

| Black-fronted (*denticollis*) | Citrine (*hastata*) | Lilypad (*kellicotti*) | Western (*perparva*) |

| Fragile (*posita*) | Furtive (*prognata*) | Rambur's (*ramburii*) | Eastern (*verticalis*) |

Fig. 23. Forktails (*Ischnura*) and Painted Damsel (*Hesperagrion heterodoxum*): lateral views of male caudal appendages (pp. 125–141).

that species. Close in-hand examination of the female mesostigmal plates will be necessary where these two species overlap.

Habitat. Ponds, springs, and slow-moving streams with heavy marginal vegetation.

Discussion. Egg-laying, which may occur unaccompanied by the male or in tandem, usually is done in emergent vegetation or algal mats. One study explored the genetic basis for the female polymorphism and its relationship to natural selection in this species. It found that the male-like andromorphic females were more vulnerable to predation than the cryptically colored and thus longer-lived gynomorphic forms. These gynomorphic forms, however, were found to engage in interspecific mating, which ultimately lowered their reproductive potential. These selective pressures explain the higher frequencies of andromorphic females in populations.

References. Johnson (1964c,d, 1965, 1966c, 1975b), Provonsha (1975).

Mexican Forktail
Ischnura demorsa (Hagen)
(pp. 126, 133, 139; photos 20a, 20b)

Size. Total length: 21–26 mm; abdomen: 17–21 mm; hindwing: 11–15 mm.

Regional Distribution. *Biotic Provinces:* Apachian, Chihuahuan, Coloradan, Kansan, Navahonian. *Watersheds:* Canadian, Cimarron, Colorado (NM), Pecos, Red, Upper Rio Grande.

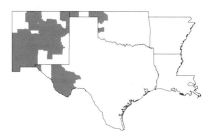

General Distribution. Utah and Kansas south to Texas, New Mexico, and Arizona, southward into Mexico.

Flight Season. Apr. 9 (TX)–Nov. 18 (NM).

Identification. The head and thorax of the male are blue-green and heavily marked with black. The antehumeral stripe is complete. Laterally, the abdomen is blue-green proximally, changing to lighter

yellow-green distally. Segments 8 and 9 are blue, each with a prominent lateral black stripe, the stripe extending more distally on segment 8. There is often a black dorsal stripe basally on 8, as well. Segment 10 is black dorsally and blue laterally, with a distinct, deeply forked, dorsoapical prominence, extending well above the segment and easily viewed in hand. The cerci are strongly curved downward. The upper arm of the deeply bifurcated paraprocts extends beyond both the lower arm and the cerci.

The andromorphic females are uncommon, and differ from males in having the entire dorsum of abdominal segment 10 blue. The more common gynomorphic females are pale orange or tan. The antehumeral stripe is sometimes reduced to a hairline. The abdomen is nearly all black dorsally, and segments 8–10 are sometimes like those of the andromorphic form. The mesostigmal plates are rather quadrate in shape, with only a moderate flange on the anterior border and the distal tubercle on the posteromedial corner. The females may become heavily pruinose after only a few days, and the pruinosity may entirely obscure the thoracic color pattern.

Similar Species. The postocular eyespots are much larger in Lilypad Forktail (*I. kellicotti*). In Mexican Forktail, the black ventrolateral stripe on segment 9 generally does not extend beyond 1/2 the length of that segment, and in Western Forktail (*I. perparva*) it often extends farther. Eastern Forktail (*I. verticalis*) is similar, but the dorsoapical projection in males is reduced considerably, such that, even in the hand, it appears wholly absent.

Habitat. Creeks, streams, springs, and slow reaches of rivers with moderate vegetation.

Discussion. A detailed study on the genetics of polymorphism in this species found the same process of natural selection operating as is discussed in Plains Forktail (*I. damula*), above. Gynomorphic forms are more cryptically colored and therefore less likely to suffer predation and live longer than the more colorful andromorphic forms.

References. Johnson (1965, 1966b,c, 1975b).

Black-fronted Forktail
Ischnura denticollis (Burmeister)
(pp. 126, 133, 139; photos 20c, 20d)

Size. Total length: 22–26 mm; abdomen: 17–21 mm; hindwing: 11–15 mm.

Regional Distribution. *Biotic Provinces:* Apachian, Chihuahuan, Kansan, Navahonian, Texan. *Watersheds:* Arkansas, Brazos, Canadian, Cimarron, Colorado (NM), Red, Upper Rio Grande.

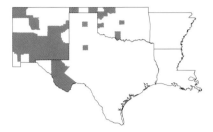

General Distribution. Western United States east to Oklahoma and Texas; south through Mexico to Guatemala.

Flight Season. Feb. 25 (NM)–Nov. 18 (NM).

Identification. The male has a solid metallic blue-black thorax, and the antehumeral stripe is entirely absent. The abdomen is dark dorsally, with a metallic blue-green luster. Segments 8 and 9 are blue dorsally, the former with a narrow dark basal ring. The dorsoapical prominence on segment 10 is distinct, but barely reaches above the height of segment 9, if at all. The tips of the cerci are directed anteroventrally. The upper arms of the forked paraprocts are slightly denticulate at their upper end. The lower arms project nearly straight posteriorly.

This is one of only two species in the south-central United States in which the females have distinct nipple-like processes immediately posterior to the deep pit on each side of the pronotum. The andromorphic form is rare, and differs from the male in the occasional presence of an antehumeral stripe and blue spot dorsally on segment 10. The gynomorhpic forms vary from pale blue to orange, the postocular spots larger than those in the male and separated by a pale occipital bar. The abdomen is patterned as in the andromorphic form.

Similar Species. Black-fronted Forktail (*I. denticollis*) is the only forktail in the region in which the front of the thorax is all black and the pale antehumeral stripe is completely lacking.

Habitat. Vegetated streams or ponds, often associated with springs, especially at northern latitudes.

Discussion. Black-fronted Forktail is more widely distributed than the other western forktail species. This species is in great abundance in Utah, in most aquatic habitats between 1,400 and 2,500 m where there is sufficient vegetation and a high-enough

minimum temperature to support damselflies. One author described Black-fronted Forktail as "undoubtedly the feeblest of all western Odonata . . ." In a Mexican population, survivorship rates for both sexes were among the lowest in the Odonata, and their ability to disperse is low.

Unlike most forktails, females will lay eggs in tandem, usually in emergent grasses or debris. The average mating time is 20 minutes, the shortest of any forktail reported. One study on seasonal variation and morphometric differentiation in sympatric populations of Black-fronted Forktail with the more western San Francisco Forktail, *I. gemina* (Kennedy), found that hybridization does occur, but that the evidence shows hybrid unfitness.

References. Aguilar (1992, 1993), Kennedy (1917), Leong and Hafernik (1992a,b), Provonsha (1975).

Citrine Forktail
Ischnura hastata (Say)
(pp. 126, 133, 139; photos 20e, 20f)

Size. Total length: 21–27 mm; abdomen: 16–22 mm; hindwing: 9–15 mm.

Regional Distribution. *Biotic Provinces:* Austroriparian, Balconian, Carolinian, Chihuahuan, Coloradan, Kansan, Navahonian, Tamaulipan, Texan. *Watersheds:* Arkansas, Bayou Bartholomew, Brazos, Canadian, Cimarron, Colorado, Colorado (NM), Guadalupe, Lower Rio Grande, Mississippi, Neches, Nueces, Ouachita, Pecos, Red, Sabine, San Antonio, San Jacinto, St. Francis, Trinity, Upper Rio Grande, White.

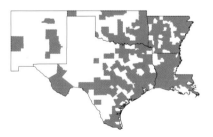

General Distribution. Florida to southern Ontario west to Colorado and California and south through Mexico to Central and South America.

Flight Season. Year-round (TX).

Identification. The male of this species is distinct because of a unique pterostigma, detached from the costa in the forewing. It is lighter in color and twice or more the size of its hindwing counterpart. No other damselfly in the world has this characteristic. The thorax of males is green, and the abdomen is bright yellow. The dorsoapical projection on segment 10 is strongly notched and prominent. The cerci, which project posteriorly, are rounded distally, with a ventrally directed medial projection off each. The paraprocts each have a short, rounded posteroventral lobe.

Females are red-orange with black stripes across the top of the head, middorsally on the pterothorax, and dorsally on abdominal segments 6–8. In older individuals a light pruinosity envelops the thorax and abdomen, but never completely obscures the pterothoracic pattern. Only gynomorphic forms are known in this species. A small vulvar spine may or may not be present on segment 8.

Similar Species. The female Citrine and Fragile Forktails (*I. posita*) are similar, but Fragile Forktails are generally darker, and even in pruinose females the distinctive exclamation-like antehumeral stripe is visible with the help of a little magnification. Fragile Forktail females always lack a vulvar spine on segment 8.

Habitat. Heavily vegetated ponds and lakes and other permanent or temporary bodies of water.

Discussion. Citrine Forktail is the smallest damselfly in North America, and as its wide distribution suggests, it is a cosmopolitan species with the ability to adapt readily to its environment. Citrine Forktail is found throughout the New World, but remarkably little has been written about its reproductive behavior or ecology. It is not unusual to find individuals far from water. They may be abundant in heavily vegetated areas with little or no water. Whether because of its small size or its secretive behavior, Citrine Forktail is seldom seen mating. Average mating time is 20 minutes, and females lay eggs unaccompanied by a male, in submerged vegetation just under the surface.

References. Carpenter (1991), Dunkle (1990), Walker (1913), Wilson (1911).

Lilypad Forktail
Ischnura kellicotti Williamson
(pp. 126, 133, 139; photos 21a, 21b)

Size. Total length: 25–31 mm; abdomen: 19–24 mm; hindwing: 12–18 mm.

Regional Distribution. *Biotic Provinces:* Austrroriparian, Texan. *Watersheds:* Arkansas, Bayou Bartholomew, Brazos, Canadian, Mississippi, Ouachita, Red, Sabine, San Jacinto, Trinity.

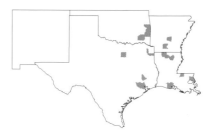

General Distribution. Eastern United States from Florida to Maine, west to Michigan, and south to Texas.

Flight Season. Apr. 1 (LA)–Oct. 1 (LA).

Identification. In males, the postocular spots are unusually large and the pterothorax is bright blue with a broad black middorsal stripe and pair of black humeral stripes. The abdomen is largely black, with blue on parts of segments 1 and 2 and 7–10. The forewing pterostigma is larger than its tan hindwing counterpart and becomes bright blue anteriorly with maturity. Males also lack a notable dorsoapical projection on segment 10. The cerci are distinctive, sloping ventrolaterally to form an acute apex. The paraprocts each have a lower appendage that projects posteroventrally.

Females exist in both a blue form and a red-orange form. Each is patterned like the male, with the male's pale colors replaced by red-orange. A small vulvar spine is usually visible on segment 8.

Similar Species. The bright blue or orange color of this damselfly may result in initial confusion with bluets like Skimming Bluet (*Enallagma gemnatum*). In Lilypad Forktail, however, the dorsum of abdominal segment 2 is largely blue, and the pale postocular spots are much larger than those in Skimming Bluet. No other forktails in our region have the large postocular spots seen in this species.

Habitat. Strongly associated with floating lily pads in lakes.

Discussion. This species is unique, among Nearctic odonates, in its obligatory relationship with water lilies (*Nuphar* and *Nymphaea*) in both the larval and adult stages. Williamson (1899a) was the first to report this little studied relationship. He stated that he ". . . never saw one at rest on any other location than a flat-floating leaf of the white water-lily. They were quarrelsome neighbors and frequently attacked [Skimming and Orange (*E. signatum*) Bluet], though apparently without serious injury." Larvae cling to the bottom of the lily pads and emerge by crawling on top. The adults are nearly always encountered perching or ovipositing on these plants. They sometimes exhibit a unique posture: while perching on a pad with the abdomen curled downward, they will tilt back on the abdomen with the front legs in the air, ready for an immediate getaway. Females, unaccompanied by males, take up to 20 minutes to deposit eggs.

One study on the color morphs in a north-central Texas population of this species found no evidence for dichromatic females, but concluded rather that the color change was ontogenetic. Young teneral females are orange, but with the onset of reproductive maturity become blue. The authors of the study found females of netted copulating pairs were nearly always of an intermediate color form. The young orange-color form, however, may be reproductively mature, as indicated by a photograph of a copulating pair in Westfall and May (1996). Interestingly, the same study found that although Lilypad Forktail is rarely harassed, females do not utilize an active mating-refusal display to thwart off nearby males, potentially explaining the photograph.

References. Dunkle (1990), Johnson and Westfall (1970), Robinson and Jordan (1996).

Western Forktail
Ischnura perparva McLachlan _in_ Selys
(pp. 126, 133, 139; photos 21c, 21d)

Size. Total length: 23–30 mm; abdomen: 18–24 mm; hindwing: 11–17 mm.
Regional Distribution. _Biotic Provinces:_ Chihuahuan, Coloradan, Kansan, Navahonian. _Watersheds:_ Canadian, Colorado (NM), Pecos, Rio Grande.

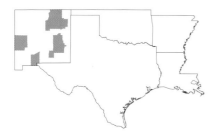

General Distribution. Northwestern United States and British Columbia south to Arizona, New Mexico, and Oklahoma.

Flight Season. July (NM).
Identification. The head and pterothorax of the male are blue-green, with considerable black markings. The black humeral stripe is as much as three times as wide as the antehumeral stripe. The abdomen is mostly yellow-green dorsally, with blue on segments 1 and 2, the base of 3, and all of 8–10. The dorsoapical prominence on segment 10 is forked and extends 1/3 to 1/2 again as high as the segment. The cerci curve gently downward, forming an acute process. The paraprocts are equally forked and project posteriorly.

Females are orange, becoming tan or olive with age, and the thoracic pattern is often obscured. The head and thorax are like those of the male, with the postoccipital spots becoming black dorsally on older individuals. Only gynomorhpic females are known.
Similar Species. Male Western Forktails most closely resemble Mexican Forktail (_I. demorsa_), but the dark ventrolateral stripe on segment 9 in that species is often reduced to less than 1/2 the length of the segment, and in Western Forktail it is generally greater than 1/2. The postocular spots are much larger in Lilypad Forktail (_I. kellicotti_). Eastern Forktail (_I. verticalis_) is similar, and the dorsoapical projection in males is so seriously reduced as to appear absent, even in the hand.
Habitat. Ponds, lakes, and slow-moving streams with heavy vegetation and muddy substrate; often found in alkaline or saline situations.
Discussion. The range of this species barely extends into the south-central United States, where it is known only from a few scattered counties in New Mexico and also from Oklahoma, with no additional information. It is common west of the Rocky Mountains. Females generally lay eggs in tandem, but may do so unaccompanied by the male.

References. Kennedy (1915b), Paulson (1974), Provonsha (1975).

Fragile Forktail
Ischnura posita (Hagen)
(pp. 126, 133, 139; photos 21e, 21f)

Size. Total length: 21–29 mm; abdomen: 16–22 mm; hindwing: 10–16 mm.
Regional Distribution. _Biotic Provinces:_ Austroriparian, Balconian, Carolinian, Chihuahuan, Kansan, Navahonian, Tamaulipan, Texan. _Watersheds:_ Arkansas, Bayou Bartholomew, Brazos, Canadian, Cimarron, Colorado, Guadalupe, Lower Rio Grande, Mis-

sissippi, Neches, Nueces, Ouachita, Red, Sabine, San Antonio, San Jacinto, St. Francis, Trinity, Upper Rio Grande, White.

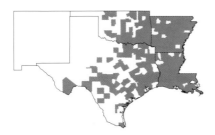

General Distribution. Eastern United States and Canada from Florida north to Newfoundland, west to North Dakota and south to Texas, and through Mexico to Belize and Guatemala; also Hawaii.

Flight Season. All year (TX).

Identification. Males are generally yellow-green, and mature females are blue. Both sexes are recognizable by a conspicuous division of the pale antehumeral stripe into an exclamation mark (stroke and spot) set against a nearly all-dark abdomen. The rest of the body is metallic black. This species is unique among forktails in having the pterostigma of all wings similar and having the fewest postquadrangular antenodal cells. The dorsoapical projection on segment 10 in males is forked, but short. The cerci and paraprocts are short and subequal in length. Only andromorphic females are known, but with age they become dark blue with heavy pruinosity.

Females lack the vulvar spine on segment 8 and the lateral corners of the subtriangular mesostigmal plates are raised.

Similar Species. The antehumeral stripe in Plains (*I. damula*) and Western (*I. perparva*) Forktails is divided into subequal anterior and posterior spots, and abdominal segments 8 and 9 of males are blue. Citrine Forktail (*I. hastata*) females are similar, but generally paler in color, and have a complete (though often obscured) antehumeral stripe, and may have a vulvar spine on segment 8.

Habitat. Heavily vegetated ponds, marshes, and slow-moving waters.

Discussion. This common, widespread species is found in every county in Arkansas and in all but eight parishes in Louisiana. It was introduced to Oahu in 1936 and is now found on all but one of the major Hawaiian islands, and as far north as Newfoundland and south into Mexico.

A study on the roosting behavior of a north-central Texas population found that unlike most odonates, both sexes of this species were regularly encountered at ponds during the day. Both sexes roosted at night on the same branches where they had perched earlier in the day, but significantly higher. At night, the body was found to be at a right angle to the stem, possibly allowing for a quicker escape from predation and more efficiency in warming.

References. Bick and Bick (1958), Mauffray (1997), Patrick and Lutz (1969), Polhemus and Asquith (1996), Robinson (1983), Robinson et al. (1985, 1991), Shaffer and Robinson (1993).

Furtive Forktail
Ischnura prognata (Hagen)
(pp. 126, 133, 139; photos 22a, 22b)

Size. Total length: 30–37 mm; abdomen: 24–31 mm; hindwing: 14–20 mm.

Regional Distribution. *Biotic Provinces:* Austroriparian. *Watersheds:* Mississippi, Red, Sabine, Trinity.

General Distribution. Eastern United States from Florida to New York and southwest to Texas.

Flight Season. Feb. 15 (LA)–Sep. 9 (LA).

Identification. Males are metallic black, with only a thin green antehumeral stripe and pale blue on abdominal segment 9 and laterally on 10. The pterostigma in the forewing is twice the size as that in the hindwing, and transparent in its outer 1/2. The dorsoapical projection on segment 10 is forked apically and readily visible in the hand. The cerci of the male are sharply forked, the upper branch projecting posteriorly and the lower ventrally. The paraprocts are stout and subequal to the upper branch of the cerci.

Only gynomorphic females are known. Young forms have an orange-red thorax, interrupted only by a black middorsal stripe. The abdomen is orange as far as segment 4, but becomes black apically. Older females become less vibrant and almost brown.

Forktails (*Ischnura*) and Painted Damsel (*Hesperagrion*)

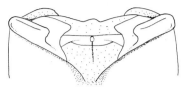

Painted Damsel (*Hesperagrion heterodoxum*)

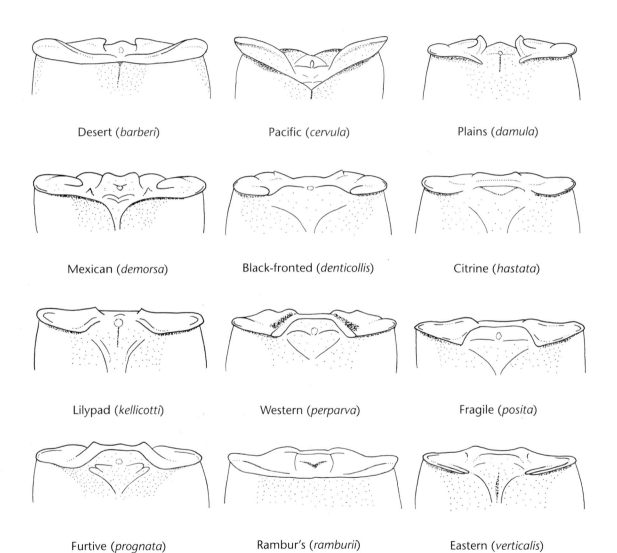

Desert (*barberi*) Pacific (*cervula*) Plains (*damula*)

Mexican (*demorsa*) Black-fronted (*denticollis*) Citrine (*hastata*)

Lilypad (*kellicotti*) Western (*perparva*) Fragile (*posita*)

Furtive (*prognata*) Rambur's (*ramburii*) Eastern (*verticalis*)

Fig. 24. Forktails (*Ischnura*) and Painted Damsel (*Hesperagrion heterodoxum*): mesostigmal plates of females (pp. 125–141).

There is no vulvar spine on segment 8. The posterior margin of the pronotum has a distinct V-shaped emargination at its middle. The mesostigmal plates are subtriangular and have small tufts of hair at their posteromedial corners.

Similar Species. Male and andromorphic female Rambur's Forktails (*I. ramburii*) both have the top of abdominal segment 8 blue, instead of 9. Only segments 1 and 2 are pale in the orange-red and olive-green forms of females. Segments 1–4 are pale in female Furtive Forktail (*I. prognata*).

Habitat. Heavily shaded ponds, swamps, and sloughs.

Discussion. This eastern species is uncommon throughout its range, and apparently finds its westernmost limit in the Sam Houston National Forest of east Texas. It has the longest abdomen of the forktails in the south-central United States. This species shares a behavioral similarity with many tropical damselflies, in that it will fly ghostlike from one stem to another in the shady forest undergrowth, foraging at a height of 2 m or more.

References. Dunkle (1990).

Rambur's Forktail
Ischnura ramburii (Selys)
(pp. 126, 133, 139; photos 22c, 22d, 22e, 22f)

Size. Total length: 27–36 mm; abdomen: 21–29 mm; hindwing: 15–19 mm.

Regional Distribution. *Biotic Provinces:* Austroriparian, Balconian, Carolinian, Chihuahuan, Kansan, Tamaulipan, Texan. *Watersheds:* Arkansas, Bayou Bartholomew, Brazos, Canadian, Cimarron, Colorado, Guadalupe, Lower Rio Grande, Mississippi, Neches, Nueces, Ouachita, Pecos, Red, Sabine, San Antonio, San Jacinto, St. Francis, Trinity, White.

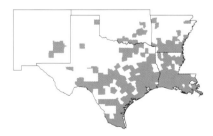

General Distribution. Florida to Maine, west to Illinois, south to Texas, and through Mexico to Chile; also Hawaii.

Flight Season. All year (TX).

Identification. The males have a dark head with small pale-green postocular spots. The thorax is green, with broad black middorsal and humeral stripes surrounding a thin pale antehumeral stripe. The abdomen is dark dorsally and pale laterally, with blue on all of abdominal segment 8 and anterolaterally on 9. The dorsoapical projection on segment 10 is slight. The cerci are short and blunt. The paraprocts are twice as long as the cerci, unforked, and projecting posteriorly.

Females are found in three color forms, including an andromorphic, male-like form. Gynomorphic forms include an orange-red form, where the thorax and abdominal segments 1 and 2 are entirely orange-red, except for a prominent black midbasal stripe on segment 1. The rest of the abdomen is dark. A second gynomorphic form is olive-green but patterned like the orange-red form. Because young individuals of the green form look like the red form, it is unclear whether a true genetically determined "red form" exists. There is a prominent vulvar spine on segment 8. The mesostigmal plates are subtriangular and a low, pale ridge forms a continuous posterior border.

Similar Species. Desert Forktail (*I. barberi*) can be distinguished from Rambur's Forktail by the continuous black stripe dorsally on the abdomen and the absence of blue dorsally on segment 9. Only abdominal segment 9 is blue in male Furtive Forktails (*I. prognata*), and only segments 1 and 2, never 3 and 4, are pale in young female Citrine (*I. hastata*) and Furtive Forktails.

Habitat. Heavily vegetated ponds, lakes, marshes, and slow reaches of streams exposed to sunlight, including brackish waters.

Discussion. Rambur's Forktail is the most widespread forktail of the New World, ranging as far north as Maine, southward to southern California, Mexico, and Central and South America. It occurs year-round in the southern parts of its range. It also inhabits the Hawaiian Islands, where it was introduced in 1973 (there may also have been subsequent introductions).

As widespread as this species is, surprisingly little has been written about its biology. Both sexes remain close to the water, and although males are not territorial, females are known to be highly predaceous and often cannibalistic. Males often do not release females from the wheel position for several hours, and sometimes as many as seven, so as to secure their genetic contribution. Red females will sometimes attack males, but more often curl their abdomen downward while fluttering their wings in a refusal display. Females often lay eggs late in the afternoon, unattended by males, on the underside of floating vegetation or debris, by curling the abdomens. There is an apparent lack of color preference by males, but there are selective advantages and disadvantages of various color forms in populations.

References. Dunkle (1990), Garcia-Diaz (1938), Hilton (1989), Polhemus and Asquith (1996), Robertson (1985), Wilson (1911).

Eastern Forktail
Ischnura verticalis (Say)
(pp. 126, 133, 139; photos 23a, 23b)

Size. Total length: 20–33 mm; abdomen: 15–26 mm; hindwing: 11–19 mm.

Regional Distribution. *Biotic Provinces:* Austroriparian, Carolinian, Kansan, Navahonian, Texan. *Watersheds:* Arkansas, Brazos, Canadian, Cimarron, Pecos, Red, St. Francis, White.

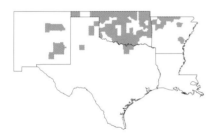

General Distribution. Eastern United States and Canada from Georgia to Newfoundland, and west to Montana and New Mexico.

Flight Season. Apr. 21 (AR)–Sept. 12 (TX).

Identification. Males are dark, with a narrow green antehumeral stripe and a yellow-green thorax laterally. The antehumeral stripe is often narrowed at about 2/3 its length. The abdomen is largely black, but segments 8 and 9 are bright blue dorsally. The dorsoapical projection on segment 10 is blunt and forked for 1/2 its length. It projects more dorsally than posteriorly, not extending beyond the hind margin of segment 10. The cerci slant downward in lateral view, terminating to an acute apex. The paraprocts have an abrupt dorsobasal process followed by a lower, posteriorly projecting process that curves upward.

The postocular spots of the less common andromorphic female are larger and confluent with the rear of the head. The blue on abdominal segments 8 and 9 is variable, and often restricted to the apical 1/2 of the segment. The more common gynomorphic females are orange, marked with black. The abdomen is largely black dorsally, especially on segments 4–10. Both color forms become dark and heavily pruinose with age. The mesostigmal plates are distinctly triangular, generally with a prominent posterior ridge or flange. There is a well-developed vulvar spine on segment 8.

Similar Species. Male Eastern Forktails most closely resemble Mexican (*I. demorsa*) and Western (*I. perparva*) Forktails. The dorsoapical projection on segment 10 in male Eastern Forktails, however, is reduced, whereas in Mexican and Western Forktails it is prominent and readily visible in the hand. Only Mexican Forktail overlaps slightly with Eastern Forktail's geographic range.

Habitat. Ponds, lakes, slow-moving streams, and marshes.

Discussion. One of the most common damselflies in the northeastern United States and throughout Oklahoma, this species is uncommon south of the Red River. Eastern Forktail has been variously reported from Louisiana, but the validity of these records remains questionable. There are no modern records in the state, but one study did not rule out its occurrence in the northern part of the state. In much of its range it is one of the first damselflies seen in the spring and last seen in the fall.

Various aspects of this ubiquitous species have been well studied. Among these, it has been suggested that strong winds may be the principal cause of its dispersal. Another study found that mating took place as early as four days after emergence, and that egg-laying began a few hours after mating. The well-documented behavior of Eastern Forktail females flexing the abdomen ventrally and rapidly beating their wings was determined to be a successful threat display, warding off intruders.

It has been shown that food intake is an important determinant of the number of eggs laid in Eastern Forktail, and that adult body size is relatively unimportant. Failure to find food on any one day has consequences not only for clutches laid the next day, but also for subsequent clutches. Unlike those of most damselflies, Eastern Forktail females tend to be monogamous, mating only once. A female may fertilize over a thousand eggs using the sperm from a single-male encounter without a drop in fertility.

References. Bick (1957, 1966), Calvert (1893, 1915), Fincke (1987), Foster (1914), Grieve (1937), Mauffray (1997), Mitchell (1962), Richardson and Baker (1997).

Sprites
Genus *Nehalennia* Selys
(p. 143)

This small genus of six species is largely confined to the New World, with only a single Palearctic species. The group consists of small, beautifully metallic-green or metallic-black damselflies characterized by a sharply angulate frons. All members lack postocular spots, but often have a visible occipital bar. The second antennal segment has a characteristic white ring. All of the above characters will readily separate this group from the closely related forktails (*Ischnura*). The legs are relatively short, with variable-sized tibial spurs. The wings are clear and characterized by 7–12 and 6–11 postnodal crossveins in the fore- and hind-wings, respectively. The abdomen is either metallic black or metallic green, becoming cream or tan ventrally. The male caudal appendages are distinct, but short and often difficult to see. The mesostigmal plates of the females are readily distinguished, one species to the next, by their heavy sculpturing.

This group occupies a specific ecological niche characterized by ponds, lakes, or slow-moving streams with variably dense, low, emergent vegetation where adults fly. The biology of most species is poorly known.

References. DeMarmels (1984), DeMarmels and Schies (1977).

KEY TO THE SPECIES OF SPRITES (*NEHALENNIA*)

1. Pale-blue antehumeral stripe present; thorax and abdomen metallic black — **Everglades (*pallidula*)**

1'. Antehumeral stripe absent; thorax and abdomen metallic green — **Southern (*integricollis*)**

Southern Sprite
Nehalennia integricollis Calvert
(pp. 143; photos 23c, 23d)

Size. Total length: 20–25 mm; abdomen: 16–20 mm; hindwing: 11–14 mm.
Regional Distribution. *Biotic Province:* Austroriparian. *Watersheds:* Mississippi, Red, Sabine, Trinity.

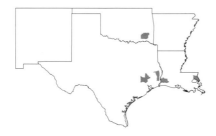

General Distribution. Eastern United States from New York to Texas.

Flight Season. Apr. 15 (TX)–Sep. 10 (LA).
Identification. This species is uncommon in the southwesternmost reaches of its range. It will not likely be confused with other damselflies in the region because of its small size, metallic-green coloration, and the presence of blue on abdominal segment 10. The cerci are short, 1/4 of the length of segment 10, and have a posteroventral apical tooth that may be visible only when viewed posteriorly. The serrated paraprocts are slightly longer than the cerci and have two or three acute teeth along the posterior margin.

Females appear like males in coloration. The mesostigmal plates are subtriangular, with a rounded posteromedial corner. Females lack a vulvar spine on abdominal segment 8.
Similar Species. Southern Sprite lacks pale antehumeral stripes and is dark metallic green, not black as in Everglades Sprite (*N. pallidula*). Female Citrine (*Ischnura hastata*) and Fragile (*I. posita*) Forktails have

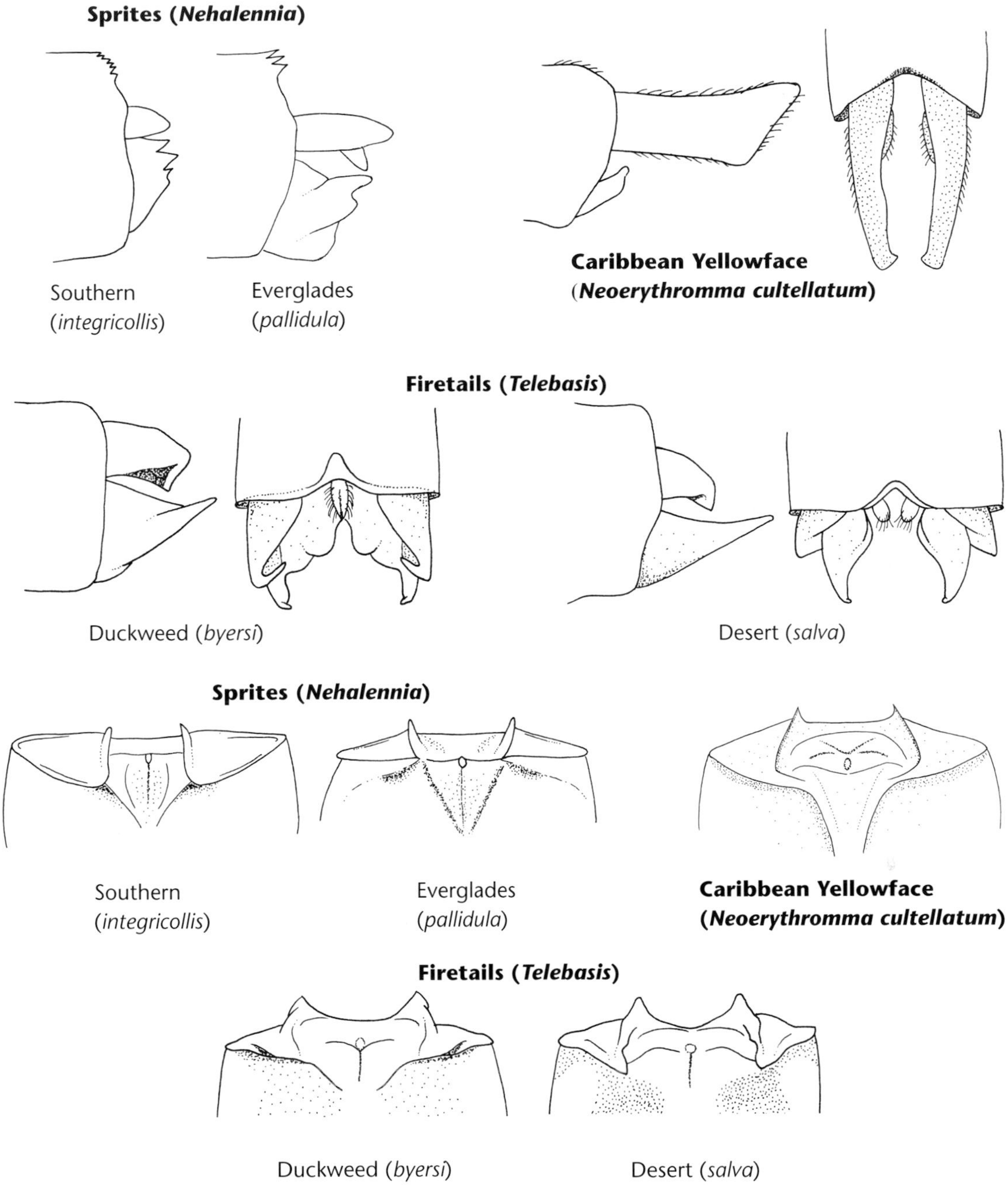

Fig. 25. Sprites (*Nehalennia*), Firetails (*Telebasis*), and Caribbean Yellowface (*Neoerythromma cultellatum*): male caudal appendages and female mesostigmal plates (pp. 142–147).

a pale antehumeral stripe, and the dorsum of abdominal segment 10 is black or pale orange.

Habitat. Ponds, lakes, bogs, and slow reaches of streams with moderately dense vegetation.

Discussion. This is one of the smallest and least-studied sprites occurring in North America. It has been reported from a diversity of habitats, including sandhill lakes in Florida and sphagnum bogs in New Jersey. The most recent records in our region have coincidentally been in Sam Houston State Park (Calcasieu Parish, Louisiana) and Sam Houston National Forest (San Jacinto County, Texas). It has been reported only from the southern extremities in Louisiana, where more looking will result in additional populations. Southern Sprite is generally found close to the ground, perching in thick clusters of sedges and grasses.

References. Mauffray (1997).

Everglades Sprite
Nehalennia pallidula Calvert
(pp. 143; photos 23e, 23f)

Size. Total length: 24–29 mm; abdomen: 19–23 mm; hindwing: 13–15 mm.

Regional Distribution. *Biotic Province:* Austroriparian. *Watershed:* Trinity.

General Distribution. Florida and historically Texas.

Flight Season. Oct. 13 (TX).

Identification. This small, dark damselfly appears nearly black from above. Much of the head, the top of thorax, and nearly all of the abdomen are metallic black dorsally. The posterior margin of the prothorax in the male in strongly elevated. Laterally, the pterothorax is blue. Abdominal segments 8–10 are dark dorsally and blue laterally. Segment 10 is mostly blue, with black apically.

The female is like the male but segment 10 is entirely blue.

Similar Species. Southern Sprite lacks the pale antehumeral stripes and is dark metallic green, not black. On similar female Forktails, Citrine (*Ischnura hastata*) and Fragile (*I. posita*), the dorsum of abdominal segment 10 is black or pale orange, but not blue.

Habitat. Primary habitat in Florida is the Everglades, but it is known from ponds and rock pits.

Discussion. This species has been reported as "The only damselfly endemic to Florida . . . geographically the most restricted and ecologically and behaviorally the least known of the genus." Recently, however, specimens were found in the Smithsonian that had been collected from Galveston in 1918 by Herbert Spencer Barber. The specimens appear to be properly labeled, though its presence, historically at least, in Texas represents an extension of its known range. It is unknown if a population still exists along the Texas coast today. In Florida it survives in moist refugia during the dry season and moves from sedges to the shade of trees if the day becomes too hot or windy.

References. Dunkle (1990), Flint (2000), Westfall and May (1996).

Yellowfaces
Genus *Neoerythromma* Kennedy

This small tropical genus comprises only two species, one of which ranges north to the lower Rio Grande Valley. The other species, *N. gladiolatum*, is found in western Mexico. Males of both species are recognizable by their bright-yellow faces. The placement of these two species has been uncertain because of their possible affinities with both bluets (*Enallagma*) and forktails (*Ischnura*). The two are dis-

tinct, however, and can be characterized by the origination of vein M_2 near the fourth and third postnodal crossveins in the fore- and hindwings, respectively. The pterostigma is relatively long. Females lack a vulvar spine on segment 8 and the ovipositor is short, not extending beyond the terminal abdominal segment. Both sexes, in both species, have characteristic yellow subapical bands on the femora. The biology and reproductive behavior of the two species is virtually unknown.

Caribbean Yellowface
Neoerythromma cultellatum (Selys)
(pp. 143; photos 24a, 24b)

Size. Total length: 27–31 mm; abdomen: 22–25 mm; hindwing: 13–16 mm.
Regional Distribution. *Biotic Province:* Tamaulipan. *Watershed:* Lower Rio Grande.

General Distribution. Florida and Texas, south through Mexico to Venezuela; also Cuba, Dominican Republic, Haiti, Jamaica, and Puerto Rico.
Flight Season. Apr. 28 (TX)–Oct. 18 (TX).
Identification. As its common name implies, males of this species have a distinctive bright-yellow face. This feature, combined with bright blue on the thorax and abdomen, will readily distinguish it from all other species in the region. Males, in addition to the yellow face, have a prominent bright-yellow antehumeral stripe and a pair of bright-blue postocular spots. The pterothorax is bright blue laterally, as are parts of abdominal segments 1 and 2 and all of segments 8 and 9. The cerci are long and black, with distinctly white dorsolateral-medial surfaces. The paraprocts are much shorter and strongly upturned to an apical process.

The face in females is blue-green, rather than yellow, and the abdomen is black with blue spots on segments 8–10. The mesostigmal plates are subtriangular in shape, but nearly flat and unsculptured. The absence of a vulvar spine on segment 8 will separate them from any potentially similar-looking bluets (*Enallagma*).
Similar Species. The face in Aurora Damsel (*Chromagrion conditum*) is blue, not yellow. No other damselfly in the south-central United States exhibits the unique combination of blue and yellow that Caribbean Yellowface presents.
Habitat. Ponds and slow reaches of streams or rivers with abundant floating debris.
Discussion. This species was only recently discovered in Texas. Within our region, it has been found only along the Rio Grande of extreme southern Texas. Virtually nothing is known about the life history of this tropical species. Tenerals mature some distance from water in forests. Adults are elusive, remaining some distance from the shoreline. They are often associated with floating debris or vegetation, where females may lay eggs, accompanied by males. I first saw these in Texas from a boat dock along the Rio Grande River where they were flying some distance from shore, no more than a few centimeters above the water, along with Amelia's Threadtail. Their elusiveness, and the difficulty in seeing them from shore without binoculars, has no doubt contributed to their having been discovered in Texas only recently.

References. Abbott and Stewart (1998), Dunkle (1990), Nikula (1998).

Firetails
Genus *Telebasis* Selys
(p. 143)

This is a species-rich group of 37 species extends from Argentina to the southern United States, becoming most diverse in the tropics. Males of this group are easily distinguished by their nearly unmarked bright-red abdomens hence the name. Most species, including the two in our region, lack postoccipital spots. The wings are clear, and females never have a vulvar spine on segment 8. Careful examination of the male caudal appendages and female mesostigmal plates is required for accurate discrimination between our two species where their ranges overlap. Members of this group are often found associated with emergent or floating vegetation, but life histories and behaviors of most species remain completely unknown.

References. Bick and Bick (1995).

KEY TO THE SPECIES OF FIRETAILS (*TELEBASIS*)

1. Cerci of male with 2 black subequal medial teeth; hind prothoracic lobe of female armed with prominent horns   **Desert (*salva*)**

1'. Cerci of male without distinct teeth, but with blunt apical medial projections; hind prothoracic lobe of female without prominent horns   **Duckweed (*byersi*)**

Duckweed Firetail
Telebasis byersi Westfall
(p. 143; photo 24c)

Size. Total length: 25–31 mm; abdomen: 20–24 mm; hindwing: 13–17 mm.

Regional Distribution. *Biotic Province:* Austroriparian. *Watersheds:* Bayou Bartholomew, Mississippi, Red, Sabine, Trinity.

General Distribution. Southeastern United States from Florida to Texas.

Flight Season. Jun. 14 (AR)–Jun. 22 (TX).

Identification. Males are bright red, including the eyes. There is a black middorsal stripe on each side of the carina that widens abruptly at the posterior end and then narrows again, so that, combined, the two stripes give the appearance of a posteriorly directed arrow. Most of the remaining pterothorax and all of the abdomen are bright red. The cerci are no more than 2/3 the length of segment 10, and are nearly uniform in height throughout their length. There are two black subapical medial teeth on each appendage, only the lower of which may be readily visible. The paraprocts are about four times as long as the cerci and have black apices.

The females are marked like males, but are tan in color. The abdominal segments 8 and 9 are black dorsally. The mesostigmal plates are subtriangular and only slightly sculptured.

Similar Species. The ranges of Duckweed and Desert Firetails (*T. salva*) do not overlap in the region, which will help to differentiate these closely related species.

Habitat. Swampy, partially shaded areas with abundant floating duckweed.

Discussion. Although locally common in the Southeast, this damselfly is rarely seen west of the Mississippi. The only known localities for this species in Texas are within the confines of Sam Houston National Forest, where it has been taken on at least two different occasions. It has been reported from eastern Louisiana on the basis of an early-instar larva and sight records. It is also known from southern Arkansas. Adults apparently mature in forests some distance from aquatic habitats. They may be surprisingly inconspicuous when perched on shady matted plants. They have a close association with duckweed, where the larvae live on the underside.

References. Dunkle (1990), Harp and Harp (1996), Lounibos et al. (1990), Mauffray (1997), Vidrine et al. (1992a).

Desert Firetail
Telebasis salva (Hagen)
(p. 143; photos 24d, 24e, 24f)

Size. Total length: 24–29 mm; abdomen: 19–22 mm; hindwing: 12–16 mm.

Regional Distribution. *Biotic Provinces:* Apachian, Austroriparian, Balconian, Carolinian, Chihuahuan, Kansan, Navahonian, Tamaulipan, Texan. *Watersheds:* Arkansas, Bayou Bartholomew, Brazos, Canadian, Cimarron, Colorado, Colorado (NM), Guadalupe, Lower Rio Grande, Nueces, Pecos, Red, San Antonio, San Jacinto, Trinity, Upper Rio Grande.

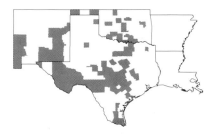

General Distribution. Southwestern United States from California to Texas; southward through Mexico to Venezuela.

Flight Season. Mar. 10 (TX)–Dec. 22 (TX).

Identification. Males and females are essentially identical in coloration to Duckweed Firetail (*T. byersi*). The male cerci are slightly longer than those in Duckweed Firetail, reaching 2/3 to 3/4 the length of segment 10. The dorsal surface of each appendage is straight for 1/4 of its length, when viewed laterally, then abruptly turns downward. Each appendage bears two separate, subapical, medial black teeth. The paraprocts are 1/2 again as long as the cerci and slightly more upturned than those in Duckweed Firetail.

The mesostigmal plates of the female are subtriangular with little sculpturing, closely resembling those of the Duckweed Firetail.

Similar Species. Desert Firetail is much more widely distributed than Duckweed Firetail. The species is found throughout the Texan biotic province, westward and southward through Central America to Venezuela. The two species apparently do not overlap in east Texas.

Habitat. Ponds, lakes, pools, springs, and slow reaches of streams with open sunlight and abundant emergent vegetation.

Discussion. This species is widespread throughout the southwestern United States. Several authors have noted the species' habit of flying low over the water, in and out of vegetation, almost literally taunting a prospective predator or collector. A study of a north-central Texas population revealed that mating lasted an average of 80 minutes. Egg-laying followed, the female accompanied by the male, and lasted 25 minutes on average. Preferred egg-laying substrates include stems, algal mats, and floating sticks. Females likely lay eggs at different localities on different days, improving the survival of their eggs. Interestingly, males abandoned their initial site if a female was not obtained on the first day.

References. Robinson and Frye (1986), Smith and Pritchard (1956).

Dragonflies
(Suborder Anisoptera)

PETALTAILS
(Family Petaluridae)

These large, prehistoric-looking dragonflies represent some of the oldest living members of this order. Fossil evidence indicates that this group flourished during the late Jurassic, one hundred fifty million years ago, and was once much more widespread than it is today. Today, the family is represented by only nine widely disjunct species, including four in Australia and New Zealand. The two species found in North America are uncommon, but may be locally abundant. Black Petaltail, *Tanypteryx hageni* (Selys), is endemic to the mountains of the Pacific Northwest and British Columbia, and the other, Gray Petaltail, *Tachopteryx thoreyi*, is found along the Atlantic seaboard west to Texas.

The family is distinctive, with a mosaic of characters seen in other families. The petaltails are clear-winged; the eyes are widely separated on top of the head, much like those of the clubtails; the occipital crest is rounded; and a median cleft divides the tip of the labium, as in the spiketails. The thorax and abdomen are compact and stout, giving the members of this family a robust, primitive, appearance. The last two abdominal segments lack a club and are subequal in length, but much shorter than those preceding them.

The accessory genitalia of segment 2 in the male are not prominent. The caudal appendages, however, are broad and strong, appearing petal-like in most species, hence the name. The ovipositor of the female, well-developed and strongly upcurved, resembles that of female darners, but is smaller. The venation in this family is variable in detail. There are usually two thickened antenodal crossveins, an extremely long, thin pterostigma surmounting five to nine cells, and a variously developed brace vein. The triangles in the fore- and hindwings are similar and equally distant from the arculus in the two North American species. Males generally have a strongly developed three-celled anal triangle. The larvae are found in bogs and spring seeps.

References. Carpenter (1992), Kennedy (1917), Rohdendorf (1991).

Gray Petaltail
Genus *Tachopteryx* Uhler *in* Selys

This genus is represented by a single large, gray-and-black eastern species, easily recognized by its large size and coloration. Kennedy (1917) stated that "*Tachopteryx* is, perhaps, the most primitive genus of living anisopterous dragonflies in North America."

Gray Petaltail
Tachopteryx thoreyi (Hagen *in* Selys)
(photo 25a)

Size. Total length: 71–82 mm; abdomen: 50–61 mm; hindwing: 45–56 mm.
Regional Distribution. *Biotic Provinces:* Austroriparian, Carolinian, Kansan, Texan. *Watersheds:* Arkansas, Cimarron, Mississippi, Neches, Ouachita, Red, Sabine, St. Francis, Trinity, White.

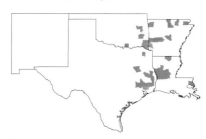

General Distribution. Southeastern United States west to eastern Texas.
Flight Season. Apr. 28 (TX)–Jun. 24 (TX).

Identification. The eyes are widely separated, and the face is pale with black crossbands. The pterothorax is grayish with black lateral stripes. The legs are entirely black. The venation is variable, but generally includes a well-developed brace vein under a noticeably long, black pterostigma. The anal triangle in the male usually has three cells, but may enclose as many as five. The abdomen is long and tapering. There are two distinct tufts of long white hairs dorsolaterally on segment 1. The female's ovipositor has blades resembling those of darners.

Similar Species. This species is the only gray-black dragonfly of its size in the region, and could not easily be confused with other species. The distinct behavior of perching vertically on tree trunks, discussed below, is a good field character.

Habitat. Permanent springs and seepages of hardwood forests.

Discussion. This species, although sometimes locally abundant, is uncommon across the eastern parts of the south-central United States. It is a rather bold dragonfly that does not shy away from people, even landing on them when they are motionless. It has a distinctive behavior of a lighting vertically on sunlit areas of tree trunks and cypress knees, where it can be cryptic. Although easily approached, these are strong fliers, and can be evasive. Individuals are occasionally seen lower to the ground or even on stones. They are predaceous on insects of various sizes, including large dragonflies. Males are often seen searching tree trunks for females, or waiting nearby, perched in sunlit areas. Mating occurs high in the forest canopy and females oviposit among roots in dense grasses and fallen leaves, or mud.

References. Barlow (1991), Dunkle (1981), Fisher (1940), Williamson (1900a).

DARNERS
(Family Aeshnidae)

These often brilliantly colored blue, green, and brown species are among the largest dragonflies flying today. They have large heads, their eyes making up the greatest portion and broadly meeting on top. The resulting anterior vertex and posterior occiput are reduced in size, to a small mound and a triangular space, respectively. Their long slender abdomens have led many to compare them with darning needles, hence the common name "darners." They are found worldwide and are among the strongest fliers of all dragonflies. The males are so strong that they often leave scars on the eyes of the females they have held during mating.

The legs are long, and used commonly for perching vertically on twigs or trunks of trees. The thorax is robust and the wings are clear, only occasionally becoming smoky in some species. Many similarities in wing venation nicely characterize the family. The fore- and hindwing triangles are similar in shape and are equally distant from the arculus. The pterostigma is of normal length and has a proximal brace vein. The subtriangles are only weakly developed. Vein M_2 is always strongly upwardly arched and does not parallel the radial sector (Rs), as it does in other families. Both radial and medial planates are strongly developed. There is generally a compact two-celled anal loop in the hindwing and an anal triangle in males consisting of two or more cells.

The long, slender abdomen is interrupted only by the swollen basal segments and a noticeably constricted segment 3. The long ovipositor has blades, resembling those of damselflies, that are used for ovipositing in vegetation. There is a variously armed plate projecting posteriorly from segment 10. Several of the genera in the region are represented by a single species.

References. Dunkle (1979, 1983).

KEY TO THE GENERA OF DARNERS (AESHNIDAE)

1. Thorax lacks lateral stripes	2
1'. Thorax with lateral stripes	3
2(1). Thorax brown with two pale-yellow spots	**Spotted (*Boyeria vinosa*)**
2'. Thorax uniformly green	**Green (*Anax*)**
3(1'). Small brown spots at base of wings extend to first antenodal crossvein	**Springtime (*Basiaeschna janata*)**
3'. Small basal brown spots absent in both wings	4
4(3'). Radial sector (Rs) not forked	9
4'. Radial sector forked	5
5(4'). Stalk of Rs originates under middle of pterostigma; fork of Rs encloses 2 rows of cells	**Pilot (*Coryphaeschna*)**

KEY TO THE GENERA OF DARNERS (AESHNIDAE) (*cont.*)

5'. Stalk of Rs originates proximal to pterostigma; fork of Rs usually encloses more than 2 rows of cells ... 6

6(5'). Two distinct conical projections between lateral ocelli ... 7

6'. Singular raised area, without 2 distinct projections, between lateral ocelli ... 8

7(6). Radial planate subtends a single row of cells; blue markings on top of abdomen ... **Cyrano (*Nasiaeschna pentacantha*)**

7'. Radial planate subtends more than a single row of cells; green markings on top of abdomen ... **Swamp (*Epiaeschna heros*)**

8(6'). Supertriangle not distinctly longer than midbasal space ... **Mosaic (*Aeshna*)**

8'. Supertriangle distinctly longer than midbasal space ... **Two-spined (*Gynacantha*)**

9(4). Hindwing less than 40 mm long; eastern species ... **Pygmy (*Gomphaeschna*)**

9'. Hindwing greater than 44 mm long; western species ... **Riffle (*Oplonaeschna armata*)**

TABLE 6. DARNERS (AESHNIDAE: TEN GENERA)

Species	Hindwing length (mm)	Fork of Rs[1]	Rpl rows[2]	Cells in ♂ triangle	Distribution (biotic provinces)[3]
Mosaic (*Aeshna*)	36–53	2–4	3–4	2–3	Ap,Au,B,Ch,Co,K,N,Ta,Tx
Green (*Anax*)	45–67	2	6–7	—	Ap,Au,B,Ca,Ch,Co,K,N,Ta,Tx
Springtime (*Basiaeschna*)	32–42	0	2–3	2	Au,Ba,Ca,K,Tx
Spotted (*Boyeria*)	39–46	2–3	1–2	3–5	Au,Ca,B,Tx
Pilot (*Coryphaeschna*)	54–60	2	4–6	2	Au,Ta,Tx
Swamp (*Epiaeschna*)	52–60	4	2	3	Au,B,K,Ta,Tx

TABLE 6. DARNERS (AESHNIDAE: TEN GENERA) *(cont.)*

Species	Hindwing length (mm)	Fork of Rs[1]	Rpl rows[2]	Cells in ♂ triangle	Distribution (biotic provinces)[3]
Pygmy (*Gomphaeschna*)	29–37	0	1	2	Au
Two-spined (*Gynacantha*)	46–56	3–4	4–6	3	Au,Ta
Cyrano (*Nasiaeschna*)	45–50	3–4	1	3	Au,B,C,K,Tx
Riffle (*Oplonaeschna*)	47–54	0	3	3	Ap,Ch,N

[1] Number of cell rows within the Rs fork.

[2] Rows of cells subtended by radial planate.

[3] (Ap) Apachian, (Au) Austroriparian, (B) Balconian, (Ca) Carolinian, (Ch) Chihuahuan, (Co) Coloradan, (K) Kansan, (N) Navahonian, (Ta) Tamaulipan, (Tx) Texan.

Mosaic Darners
Genus *Aeshna* Fabricius
(pp. 154, 157, 160)

Mosaic darners are found primarily in the Northern Hemisphere, and are the dominant North American group in this family, with 20 species. The nine species occurring in our region are all similar in appearance. The eyes adjoin on top of the head for a distance at least as long as the occiput, and generally longer. There is usually a distinct black "T" on the upper surface of the frons, and the thorax is usually brown with two pale-blue, green, or yellow middorsal stripes and two lateral stripes.

The wings are clear, the radial sector arching and in most species forking asymmetrically well before the pterostigma. The triangles usually comprise four or more cells, but almost always have two basal cells. The distinct color patterns and caudal appendages of males are the most useful characters for identification. The abdomen, usually spotted with blue or green in some females, is strongly constricted just beyond the basal segments.

Apparently different diurnal activity patterns in females and conspecific males are caused mainly by differences in body colors and color patterns. Abdominal heat gain was high in dull and dark-colored abdominal segments, owing to high light absorption, and low under Tyndall-blue spots, owing to high light reflection.

References. Sternberg (1996), Walker (1912).

KEY TO THE ADULT SPECIES OF MOSAIC DARNERS (*AESHNA*)

MALES

1. Anal triangle of hindwing with 3 or 4 cells   2

1'. Anal triangle of hindwing with 2 cells   8

2(1). Abdominal segment 1 with low ventral tubercle covered with small spines; segment 10 with a bare middorsal tubercle   3

KEY TO THE ADULT SPECIES OF MOSAIC DARNERS (*AESHNA*) (*cont.*)

2'. Segments 1 and 10 lacking tubercles	5
3(2). Neither dorsal nor ventral carinae of cerci sharply angulate or produced into spine	**Turquoise-tipped (*psilus*)**
3'. Dorsal carina of cerci sharply angulate, ventral carina angulate or produced into anteapical spine	4
4(3'). Cerci each with apex not strongly decurved, ventral carina not produced into a spine	**Arroyo (*dugesi*)**
4'. Cerci each with apex strongly decurved, ventral carina produced into a prominent spine	**Blue-eyed (*multicolor*)**
5(2'). Abdominal segments 4–6 with paired, ventral, pale spots; rear of head predominantly yellow or tan	**Shadow (*umbrosa*)**
5'. Segments 4–6 without paired, ventral pale spots; rear of head black	6
6(5'). Anterior and middle lateral spots absent or greatly reduced on abdominal segments 6–8	**Persephone's (*Persephone*)**
6'. Anterior and middle lateral spots well developed on segments 6–8	11
7(6'). Anteriormost pale lateral thoracic stripe with strongly sinuate margins	**Lance-tipped (*constricta*)**
7'. Anteriormost pale lateral thoracic stripe with nearly straight margins	**Paddle-tipped (*palmata*)**
8(1'). Lateral thoracic stripes narrow, especially at upper end, extending less than 1/4 the width of sclerite	**Variable (*interrupta*)**
8'. Lateral thoracic stripes wider, especially at upper end, extending 1/3 or more the width of sclerite	**Sedge (*juncea*)**

FEMALES

1. Abdominal segment 1 with low but distinct ventral tubercle; fork of vein Rs nearly symmetrical at base	2
1'. Segment 1 without ventral tubercle; fork of vein Rs distinctly asymmetrical at base	4
2(1). Cerci as long as abdominal segments 8–10; stem of "T" on frons nearly uniform in width	**Turquoise-tipped (*psilus*)**
2'. Cerci at most barely longer than segments 8 and 9; stem of "T" on frons distinctly widened from apex to base	3

KEY TO THE ADULT SPECIES OF MOSAIC DARNERS (*AESHNA*) (*cont.*)

3(2'). Anterior lateral pale stripe usually wider than 1.2 mm near lower end; genital valves each with more or less distinct lateral carina along most of its length — **Arroyo (*dugesi*)**

3'. Anterior lateral pale stripe usually no wider than 1.0 mm near lower end; genital valves each with lateral carina usually confined to apical 1/4 of its length — **Blue-eyed (*multicolor*)**

4(1'). Styli of ovipositor at least as long as top of abdominal segment 10; cerci widest at or basal to midlength — **Lance-tipped (*constricta*)**

4'. Styli of ovipositor distinctly shorter than tergum of segment 10; cerci widest distal to midlength — 5

5(4'). Posterior edge of basal plate of ovipositor straight or slightly rounded in ventral view — 6

5'. Posterior edge of basal plate of ovipositor distinctly bilobed in ventral view — **Sedge (*juncea*)**

6(5). Anterior margin of pale anteriormost lateral thoracic stripe not sinuate or divided and generally at least 1 mm wide — 7

6'. Anterior margin of pale anteriormost lateral thoracic stripe distinctly sinuate, strongly constricted, or divided into upper and lower spots — **Variable (*interrupta*)**

7(6). Paired, pale ventral spots on abdominal segments 4–6; rear of head yellow or tan — **Shadow (*umbrosa*)**

7'. No paired, pale ventral spots on segments 4–6; rear of head black — 8

8(7'). Anteriormost pale lateral thoracic stripe wide (at least 1.75 mm) — **Persephone's (*persephone*)**

8'. Anteriormost pale lateral thoracic stripe narrower (no wider than 1.5 mm) — **Paddle-tailed (*palmata*)**

TABLE 7. MOSAIC DARNERS (*AESHNA*)

Species	Hindwing length (mm)	Fronto-clypeal stripe[1]	Lateral thoracic stripes	Pale spot on s10[2]	Distribution (biotic provinces)[3]
Lance-tipped (*constricta*)	42–47	–	blue/yellow-green	–	Au,K
Arroyo (*dugesi*)	48–53	–	white-blue	+	Ap,Ch,K,N,Ta

TABLE 7. MOSAIC DARNERS (*AESHNA*) (*cont.*)

Species	Hindwing length (mm)	Fronto-clypeal stripe[1]	Lateral thoracic stripes	Pale spot on s10[2]	Distribution (biotic provinces)[3]
Variable (*interrupta*)	44–48	+	blue-green-yellow	+	Ca,Ch,N
Sedge (*juncea*)	39–45	+	blue-green-yellow/green-yellow	+	Co
Blue-eyed (*multicolor*)	42–47	–	white-blue	+	Au,B,Ch,K,N,Tx
Paddle-tailed (*palmata*)	42–46	+	blue-green/yellow	+	Ch,Co,N
Persephone's (*persephone*)	49–51	+	yellow	–	Ap,Na
Turquoise-tipped (*psilus*)	36–43	–	yellow-green	–	B,Ta
Shadow (*umbrosa*)	42–48	–	green-blue/green-yellow	–	Au,K,N,Tx

[1] Presence of dark stripe on fronto-clypeal suture (+ = present / – = absent or vestigial).

[2] Presence of pale spots dorsally on abdominal segment 10 (+ = present / – = absent or vestigial).

[3] (Ap) Apachian, (Au) Austroriparian, (B) Balconian, (Ca) Carolinian, (Ch) Chihuahuan, (Co) Coloradan, (K) Kansan, (N) Navahonian, (Ta) Tamaulipan, (Tx) Texan.

Lance-tipped Darner
Aeshna constricta Say
(pp. 157, 160; photos 25b, 25c, 25d)

Size. Total length: 65–72 mm; abdomen: 45–57 mm; hindwing: 42–47 mm.

Regional Distribution. *Biotic Provinces:* Austroriparian, Kansan. *Watersheds:* Arkansas, Pecos.

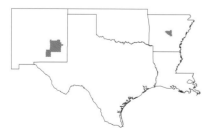

General Distribution. Northern North America, from southern Canada southward to northern Arkansas and New Mexico.

Flight Season. May (AR)–Sep. 19 (NM).

Identification. This is a robust, widely distributed, northern species. The thorax is reddish brown with yellowish stripes. The second lateral stripe is distinctly broader than the first. The wings are clear, and there are generally six paranal cells in the hindwing, including three within the anal loop. The abdomen is brown with bluish-green spots. The male cerci have a prominent posteroventrally projecting spine.

Similar Species. The male caudal appendages of Lance-tipped Darner are most similar to those of Shadow Darner (*A. umbrosa*), but the two can be distinguished in the hand by the color of the back of the head, which is pale in the Shadow Darner and black in Lance-tipped Darner.

Habitat. Open sunlit ponds, slow streams, and marshes with emergent vegetation.

Discussion. This species has been reported from Arkansas only once. It is a common northern species that is active during sunny days, but like most Mosaic Darners it is also active at dusk. Egg-laying requires a relatively longer period of time than in other species, and takes place in aquatic plants, up to 1 m above the waterline.

References. Harp and Rickett (1977), Walker (1958).

Arroyo Darner
Aeshna dugesi Calvert
(pp. 157, 160; photos 25e, 25f)

Size. Total length: 70–75 mm; abdomen: 49–55 mm; hindwing: 48–53 mm.

Regional Distribution. *Biotic Provinces:* Apachian, Chihuahuan, Kansan, Navahonian, Tamaulipan. *Watersheds:* Colorado (NM), Lower Rio Grande, Pecos, Upper Rio Grande.

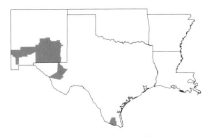

General Distribution. Southwestern United States from Texas to Arizona southward through Mexico to Oaxaca.

Flight Season. Jul. 4 (TX)–Sep. 14 (TX).

Identification. This large, western species is uncommon in southern New Mexico and western Texas, south to the lower Rio Grande Valley. It has blue eyes and a distinctly blue face. The pronounced black "T" on top of the frons has an exceptionally broad stem. The first blue lateral thoracic stripe widens downward while the second widens upward. The wings are clear, occasionally becoming tinged with brown, and the pterostigma is abbreviated, usually surmounting no more than two crossveins. The abdomen is dark brown with pale-blue spots and a distinct ventral tubercle on segment 1. Abdominal segment 10 is usually darker than the preceding segments and has a distinct middorsal tubercle, bordered laterally by a pair of large yellow spots. Females may have either blue markings (rare) or, most often, green-yellow thoracic stripes and muted yellowish-gray abdominal spots.

Similar Species. Arroyo Darner is very similar to the much more common Blue-eyed Darner (*A. multicolor*), but the former is generally more ro-

bust. Males lack the strongly forked cerci present in Blue-eyed Darner. The thoracic stripes are generally wider, and the anterior lateral thoracic stripe has a posterior extension at its upper end. It is easily distinguished from Variable Darner (*A. interrupta*), which has a black stripe on the fronto-clypeal groove and no tubercle on the venter of abdominal segment 1. Females are similar to Paddle-tailed Darner (*A. palmata*), but females of that species lack a ventral tubercle on segment 1. Blue-eyed Darner females are essentially identical and cannot be reliably separated.

Habitat. Pools of slow-flowing permanent mountain streams, rivulets, and arroyos. Often found higher up the watershed toward the headwaters than Blue-eyed Darner.

Discussion. This is a relatively uncommon species of the southwestern United States, but it can be a frequent visitor to arroyos and springs in open areas. It prefers pools edged with grass and trees rather than cattails (*Typha*).

References. Dunkle (2000).

Variable Darner
Aeshna interrupta Walker
(pp. 158, 160; photos 26a, 26b)

Size. Total length: 73–77 mm; abdomen: 50–58 mm; hindwing: 44–48 mm.

Regional Distribution. *Biotic Provinces:* Carolinian, Chihuahuan, Navahonian. *Watersheds:* Arkansas, Pecos, Upper Rio Grande, White.

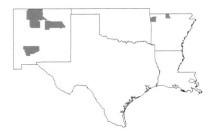

General Distribution. Northern North America southward to California, Arizona, and New Mexico.

Flight Season. Aug. 5 (NM)–Aug. 31 (NM).

Identification. This species is common throughout its range, but barely extends into the south-central United States. There is a black cross-stripe on the face, and the frontal thoracic stripes are reduced or

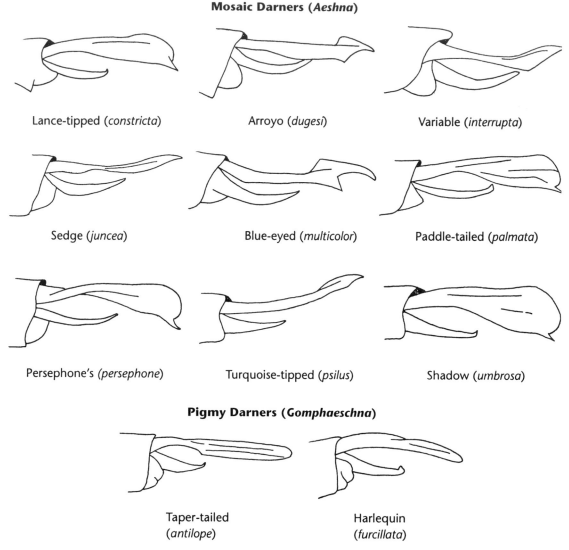

Mosaic Darners (*Aeshna*)

Lance-tipped (*constricta*)   Arroyo (*dugesi*)   Variable (*interrupta*)

Sedge (*juncea*)   Blue-eyed (*multicolor*)   Paddle-tailed (*palmata*)

Persephone's (*persephone*)   Turquoise-tipped (*psilus*)   Shadow (*umbrosa*)

Pigmy Darners (*Gomphaeschna*)

Taper-tailed
(*antilope*)

Harlequin
(*furcillata*)

Fig. 26. Mosaic Darners (*Aeshna*) and Pigmy Darners (*Gomphaeschna*): lateral views of male caudal appendages (pp. 155–172).

absent. Within our area, the lateral thoracic stripes are thin, straight, and bluish above and yellowish green below. Females may have blue, green, or yellow spots on the abdomen.

Similar Species. Males are distinctive because of the pattern of the lateral thoracic stripes and the reduced frontal thoracic stripes. Females cannot be readily separated from Paddle-tailed (*A. palmata*) and Blue-eyed (*A. multicolor*) Darners.

Habitat. Ponds, lakes, and slow streams, including those with high salinity and acidity.

Discussion. On the East Coast the lateral thoracic stripes of this species are divided into four spots, unusual among the mosaic darners. Though this species barely extends into the south-central United States it is one of the most common farther north, and it is often seen in large feeding aggregations in open areas and along roads.

Sedge Darner
Aeshna juncea (Linnaeus)
(pp. 158, 160; photos 26c, 26d)

Size. Total length: 65–74 mm; abdomen: 47–54 mm; hindwing: 39–45 mm.
Regional Distribution. *Biotic Province:* Coloradan. *Watershed:* Upper Rio Grande.

General Distribution. Throughout Canada and the Rocky Mountains of the United States.
Flight Season. Jul. (NM)–Aug. (NM).
Identification. This is a common species at higher elevations. It has a black cross-stripe on the face, and the frontal thoracic stripes are well developed. The lateral thoracic stripes are broad and straight. The pale anterior stripe is blue, narrowed above, and yellowish below. The pale lateral spots on abdominal segments 1 and 2 are yellowish green, the remaining pale spots blue. The male and the blue-form female are as above. Pale markings may also be green or yellow in females.
Similar Species. The bicolored appearance and the shape of the anterior lateral thoracic stripe is distinctive. Male Variable Darners (*A. interrupta*) have short frontal thoracic stripes and pale-blue spots laterally on segments 1 and 2.
Habitat. As its name implies, this dragonfly frequents marshes and ponds with sedges.
Discussion. Sedge Darner barely ranges into the south-central United States (Taos, northern New Mexico). They patrol all day until dark along forested edges.

Blue-eyed Darner
Aeshna multicolor Hagen
(pp. 11, 158, 160; photos 26e, 26f)

Size. Total length: 67–74 mm; abdomen: 45–52 mm; hindwing: 42–47 mm.
Regional Distribution. *Biotic Provinces:* Apachian, Austroriparian, Balconian, Carolinian, Chihuahuan, Coloradan, Kansan, Navahonian, Texan. *Watersheds:* Arkansas, Brazos, Canadian, Cimarron, Colorado, Colorado (NM), Pecos, Red, San Antonio, San Jacinto, Upper Rio Grande.

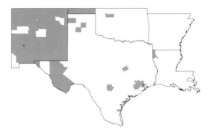

General Distribution. Central and western North America from southern Alberta and British Columbia to Texas and California; southward to Morelos, Mexico.
Flight Season. May 8 (NM)–Sep. 13 (TX).
Identification. This is a common, predominantly blue western species. The face, eyes, and pale spots are all brilliant blue. The black "T" on the top of the frons widens basally. Two pale-blue lateral stripes on the thorax are nearly the same width for their entire length. The wings are clear, and an abbreviated pterostigma surmounts two or three crossveins. The abdomen is long, and strongly constricted behind segment 3. There is a low ventral tubercle covered with small spines on segment 1 and a middorsal tubercle on segment 10. There are the usual pale-blue spots throughout the length of the abdomen. The male cerci are forked. Females may have blue or yellow-green thoracic stripes and abdominal spots.
Similar Species. Male Arroyo Darners (*A. dugesi*) appears similar but lack the forked cerci, and the anterior lateral thoracic stripe has a posterior extension at its upper end. Male Variable Darners (*A. interrupta*) are generally more muted in coloration and have a black line across the frons. Females of Paddle-tailed (*A. palmata*) and Variable Darner lack a tubercle on the venter of abdominal segment 1. Arroyo Darner females are probably not reliably separated in the field.
Habitat. Open sunlit areas of slow-flowing streams, sloughs, lakes, and ponds, including alkaline ones, with moderate vegetation.
Discussion. These are the most common darners found around still waters during the summer in the extreme western limits of our region. They tend to haunt almost any kind of standing water. I observed them to be so numerous around a Nebraska slough

that I counted 21 individuals perched on a single twig during the heat of the day. Kennedy (1917) once commented on their abundance near civilization, writing in Sacramento, California, "This species was observed catching insects on the market street of the city at twilight, they flew among the wagons and buggies, entirely indifferent to numerous passers-by. This habit of familiarity with man's haunts is very noticeable in *A. multicolor*. It is the most domestic of all the western Odonata."

Paddle-tailed Darner
Aeshna palmata Hagen
(pp. 158, 160; photos 27a, 27b)

Size. Total length: 65–73 mm; abdomen: 47–57 mm; hindwing: 42–46 mm.
Regional Distribution. *Biotic Provinces:* Chihuahuan, Coloradan, Navahonian. *Watersheds:* Colorado (NM), Pecos, Rio Grande.

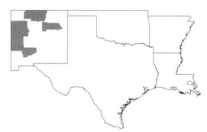

General Distribution. Western North America from Alaska south and westward to Nebraska and New Mexico.
Flight Season. Aug. (NM).
Identification. This is a common, widespread species in the west. There is usually a dark cross-stripe on the face. The broad, lateral thoracic stripes are straight throughout their length, and green or blue above, fading to yellow below. There is a pale vertical stripe on abdominal segment 1. Pale spots on the abdomen are blue in the male, abdominal segment 10 generally bearing two well-separated pale spots. Females usually have greenish-yellow thoracic and abdominal stripes and spots; more rarely, these are blue.
Similar Species. Shadow (*A. umbrosa*) and Lance-tipped (*A. constricta*) Darners are similar, but lack the dark stripe on the front of the face and a pale lateral stripe on abdominal segment 1. Persephone's Darner (*A. persephone*) has broad, yellow lateral tho-

racic stripes, and abdominal segment 10 is mostly black. Females are not easily separated from those of several other mosaic darners.
Habitat. Lakes, ponds, and slow streams with partial shade.
Discussion. Though this is often a locally abundant species, little has been written about its behavior. It may hunt low to the ground along wooded edges up until dusk, and females lay eggs in grass blades up to a meter above the water line.

References. Dunkle (2000).

Persephone's Darner
Aeshna persephone Donnelly
(pp. 158, 160; photo 28a)

Size. Total length: 73–78 mm; abdomen: 54–59 mm; hindwing: 49–51 mm.
Regional Distribution. *Biotic Provinces:* Apachian, Navahonian. *Watersheds:* Colorado (NM), Pecos.

General Distribution. Arizona, Colorado, New Mexico, and Utah south to Chihuahua, Mexico.
Flight Season. Aug. (NM).
Identification. This species is uncommon throughout its range. There is a black cross-stripe on the face and broad yellow thoracic stripes laterally, the latter straight throughout their length. The abdomen exhibits pale-blue spots in males and pale-greenish-yellow spots in females. The pale anterodorsal spots are well developed only on abdominal segments 4–6. Posterodorsal spots are well developed on segments 4–9. Midlateral spots are lacking.
Similar Species. Paddle-tipped Darner (*A. palmata*) is similar, but the lateral thoracic stripes are not as wide, and at least the upper halves are blue. Midlateral spots are also present on abdominal segments 6–8.
Habitat. Partially shaded desert mountain streams.

Discussion. Males fly erratically between shady canyons and sunny slopes, dropping close to the streams they patrol or flying high up the canyon walls.

Turquoise-tipped Darner
Aeshna psilus Calvert
(pp. 158, 160; photo 27d)

Size. Total length: 58–62 mm; abdomen: 41–51 mm; hindwing: 36–43 mm.

Regional Distribution. *Biotic Provinces:* Balconian, Tamaulipan. *Watersheds:* Guadalupe, Lower Rio Grande.

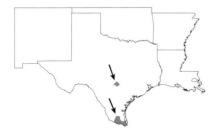

General Distribution. Southern Texas and southeastern Arizona southward through Mexico to Ecuador and Peru.

Flight Season. Mar. 10 (TX)–Oct. 19 (TX).

Identification. This Neotropical species is the smallest of the mosaic darners in our region. It reaches the lower Rio Grande Valley and southern Hill Country of Texas. It is slender, and has a blue face and a brown "T" spot on the top of the frons (the stem of the "T" is nearly parallel-sided). The eyes adjoin on top of the head for an unusually long distance. Both the middorsal and lateral thoracic stripes, green or blue, are present. The anterior lateral stripe is slightly zigzagged. The wings are clear, with six or seven paranal cells in the hindwing, including three encompassed by the anal loop. The pterostigma is short. The abdomen is brown, strongly constricted behind the proximal swollen segments, and has the typical blue spots down its length. This is the only mosaic darner with blue on the ventral side of abdominal segments 9 and 10. The cerci of the male are distinctly shorter than those in the female; they are nearly parallel throughout their length, but slope gently upward.

Similar Species. In the larger and more common Blue-eyed Darner (*A. multicolor*) and Arroyo Darner (*A. dugesi*) the lateral thoracic stripes are straight.

Habitat. Slow-flowing, open sunlit permanent and temporary streams and ponds.

Discussion. Turquoise-tipped Darner was only recently reported from United States for the first time, from two previously collected male specimens, both from Texas. One was from Landa Park in New Braunfels, the other from Brownsville, Texas. More recently, it has been found in southeastern Arizona and the lower Rio Grande Valley.

References. Abbott (1996).

Shadow Darner
Aeshna umbrosa Walker
(pp. 158, 160; photos 27c, 27e)

Size. Total length: 68–76 mm; abdomen: 49–59 mm; hindwing: 42–48 mm.

Regional Distribution. *Biotic Provinces:* Austroriparian, Coloradan, Kansan, Navahonian, Texan. *Watersheds:* Arkansas, Bayou Bartholomew, Canadian, Ouachita, Pecos, Upper Rio Grande, St. Francis.

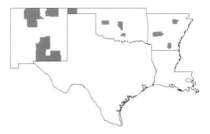

General Distribution. Throughout much of North America and Canada, extending southward to Georgia and westward through Oklahoma, Texas, New Mexico, and California.

Flight Season. Aug. 18 (AR).

Identification. This is a rather common, widespread midsummer species found in the northern parts of the region. It has a greenish-brown face, and the black "T" on the top of the frons is constricted slightly in the middle. The pterothorax is brown with parallel yellow-green thoracic stripes, both dorsally and laterally. The abdomen is strongly constricted beyond the proximal segments. The pale areas are blue (western form) or green (eastern form), and pale-blue spots are generally visible ventrally on segments 4–6 and sometimes 7. This pattern is distinctive to this species. Females are generally green, but some individuals are blue.

Similar Species. Within the eastern part of their range, Shadow Darners are the most similar to Lance-tipped Darners (*A. constricta*), but the back of the head is black, not pale, and Lance-tipped has the usual pale-blue abdominal spots. In the west, Paddle-tailed Darner (*A. palmata*) is similar, but it has a distinctive black stripe across the front of the face, and the lateral thoracic stripes are generally wider. Female Blue-eyed Darners (*A. multicolor*) have a ventral tubercle on abdominal segment 1. Lance-tipped females have blue spots on segment 10.

Habitat. Partly shaded, slow-flowing forest streams and ditches.

Discussion. Two forms of this species have been recognized, the typical "*umbrosa,*" found in our region and throughout the eastern United States, and a Pacific Northwestern form, "*occidentalis.*"

This species is unique in the group, tending to prefer shaded habitats. Females lay eggs in wet decaying wood and aquatic plants. One study found that prey consisted mainly of small flies.

References. Pritchard (1964).

Green Darners
Genus *Anax* Leach
(p. 154)

This cosmopolitan group of large dragonflies shows an unmarked green thorax. Males are unusual among the darners in lacking auricles on abdominal segment 2, and the basal margin of the hindwing is rounded. All four North American species occur in the south-central United States. The wings are clear, the pterostigma is long, and in males there is no anal triangle, owing to the rounded wing margins. There is a supplementary lateral carina on the posterior abdominal segments, and the epiproct of males is truncate.

KEY TO THE SPECIES OF GREEN DARNER (*ANAX*)

1. Large, greater than 88 mm long; abdomen long, 1.3 times the length of the wings — **Giant (*walsinghami*)**

1'. Not as large, less than 88 mm long; abdomen not noticeably long, about as long as the wings — 2

2(1'). No dark spots on top of the frons; occiput dark; hind femora unusually long, as long as the distance from the nodus to the end of the pterostigma in forewing; abdomen of males brick red — **Comet (*longipes*)**

2'. Dark spot on top of the frons; occiput pale yellow; hind femora not unusually long; abdomen of males not brick red — 3

3(2'). Frons with a round black spot encircled anteriorly by black or blue — **Common Green (*junius*)**

3'. Frons with a triangular dark spot with a dark triangle on either side, the two almost meeting anteriorly to it — **Amazon (*amazili*)**

Amazon Darner
Anax amazili (Burmeister)
(photo 28b)

Size. Total length: 70–74 mm; abdomen: 48–54 mm; hindwing: 48–52 mm.

Regional Distribution. *Biotic Provinces:* Austroriparian, Balconian, Chihuahuan, Tamaulipan, Texan. *Watersheds:* Brazos, Guadalupe, Lower Rio Grande, Mississippi, Nueces, Upper Rio Grande.

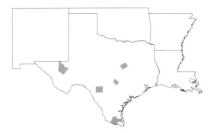

General Distribution. Florida, Louisiana, and Texas southward through Mexico and Puerto Rico to Argentina.

Flight Season. Jun. 8 (TX)–Aug. (TX).

Identification. The face is green, lightly marked with black, and there is a dark triangle on top of the frons. The pterostigma is brown, the costa green. The abdomen is mostly brown, with green on segments 1 and 2 and a distinctive white area on the postlateral rim of segment 1. Large green basal spots, blue in young individuals, are present on segments 3–6, giving the abdomen a ringed appearance.

Similar Species. This tropical species is similar to Common Green Darner (*A. junius*). The dark mark on top of the frons, however, is triangular rather than circular, and the abdomen looks distinctly ringed rather than striped. Comet Darner (*A. longipes*) has an all-red abdomen.

Habitat. Tropical ponds and lakes with weeds, including brackish pools.

Discussion. Breeding populations of this species in North America are apparently restricted to the Lower Rio Grande Valley of Texas. Curt Williams (pers. comm., August 1998) photographed a purple martin feeding a single female Amazon Darner to her young nestlings in his backyard in Marlin, Texas. This species feeds actively up until dark. Females lay eggs in vegetation that is either submerged or protruding a short distance above the waterline.

References. Dunkle (1989).

Common Green Darner
Anax junius (Drury)
(p. 5; photos 28c, 28d, 28e)

Size. Total length: 68–84 mm; abdomen: 46–60 mm; hindwing: 45–58 mm.

Regional Distribution. *Biotic Provinces:* Apachian, Austroriparian, Balconian, Carolinian, Chihuahuan, Kansan, Navahonian, Tamaulipan, Texan. *Watersheds:* Arkansas, Bayou Bartholomew, Brazos, Canadian, Cimarron, Colorado, Colorado (NM), Guadalupe, Lower Rio Grande, Mississippi, Neches, Nueces, Ouachita, Pecos, Red, Sabine, San Antonio, San Jacinto, St. Francis, Trinity, Upper Rio Grande, White.

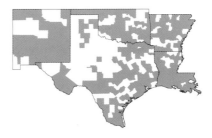

General Distribution. Throughout North America, including all 50 United States; also West Indies, Guatemala, and Belize south to Costa Rica; recently England.

Flight Season. Year-round (TX).

Identification. This widespread species is one of the most commonly seen in the region. The face is pale green, with a distinct black spot on the top of the frons, the spot bordered anteriorly by a blue semicircle conjuring the impression of a bull's eye. The thorax is green, with brown represented only lightly on the lateral sutures. The wings are clear, the costa yellow. The abdomen is mostly blue, with green on segment 1 in males, and greenish brown or reddish brown throughout in females. The brown superior caudal appendages in the male are long, about the length of segments 9 and 10 combined.

Similar Species. Amazon Darner (*A. amazili*) has a triangular spot on the top of the frons, and the abdomen appears ringed. Giant Darner (*A. walsinghami*) is much larger, and Great Pondhawk (*Erythemis vesiculosa*) is much smaller.

Habitat. Permanent and temporary ponds, lakes, bays, and slow-flowing streams with emergent vegetation.

Discussion. This is probably one of the most familiar dragonflies in all of North America. It is one of the few North American dragonflies that migrates, and is therefore most common in the spring and fall. It is a voracious predator commonly taking wasps, butterflies, mosquitoes, and other dragonflies on the wing (photo 28e). It has even been reported to attack hummingbirds, and can be cannibalistic. It is not uncommon to walk through an open field of tall grass in the early morning and have Common Green Darners flying up from their perches low to the ground, an unusual behavior among darners. Mating pairs may fall out of the air to the ground, or be seen hanging in bushes or trees. Females will lay eggs in tandem, an unusual behavior among the darners in the region. Individuals darken considerably in response to cold temperatures but regain their original color upon warming up. There is evidence that the migratory movements of this darner are strongly dictated by seasonal warm fronts.

References. Butler et al. (1975), Calvert (1929), Donnelly (1993), Dunkle (1989), Jordan and McCreary (1996), Kriegsman and Lutz (1965), May (1995a,b), Orr (1998), Trottier (1966, 1971), Young (1967).

Comet Darner
Anax longipes Hagen
(photo 28f)

Size. Total length: 75–87 mm; abdomen: 50–61 mm; hindwing: 46–56 mm.
Regional Distribution. *Biotic Provinces:* Austroriparian, Balconian, Carolinian, Tamaulipan, Texan. *Watersheds:* Arkansas, Bayou Bartholomew, Brazos, Colorado, Lower Rio Grande, Mississippi, Red, San Jacinto, Trinity, White.

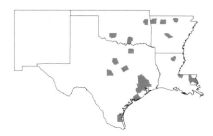

General Distribution. Ontario, Quebec, and throughout eastern United States, southward to Texas and Florida.

Flight Season. May 20 (TX)–Oct. 3 (TX).
Identification. This distinct darner has the typical green thorax of the genus, but the abdomen is brick red. The face is yellow-green, but the top of the frons is unmarked, unlike that of other green darners. The legs are unusually long, and the femora are red, except distally where they become black. The costa is green. The male abdomen is brick red beyond the basal green segments. Females are similar, but with greenish-brown spots, and the middle and posterior segments of young individuals are blue.
Similar Species. The brick-red abdomen, contrasted with the green thorax of older individuals, is a combination not seen in any other dragonfly species within the south-central United States. Amazon Darner (*A. amazili*) and female Common Green Darners (*A. junius*) both have markings on the frons.
Habitat. Primarily fishless temporary and semipermanent grassy ponds and pools.
Discussion. This is an uncommon species, but it is widely distributed throughout the eastern parts of the region. Its interactions with Common Green Darner are unclear. Comet Darner has been reported both outnumbering and being outcompeted by Common Green Darner in semipermanent artificial ponds. Individuals are not active as late in the day as Common Green Darners. Females lay eggs in submerged vegetation.

References. Beatty (1945), Johnson and Crowley (1980), Kielb and O'Brien (1996).

Giant Darner
Anax walsinghami McLachlan
(photo 29a)

Size. Total length: 88–116 mm; abdomen: 67–90 mm; hindwing: 56–67 mm.
Regional Distribution. *Biotic Provinces:* Chihuahuan, Kansan, Navahonian. *Watersheds:* Pecos, Upper Rio Grande.

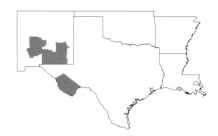

General Distribution. Southwestern United States, Mexico, Guatemala, and Honduras.

Flight Season. Jun. 26 (NM)–Oct. 4 (TX).

Identification. This is the largest dragonfly in North America. It is found only in the western extremes of our region and, due to its great size, is not likely to be confused with other species. The face is green, and the top of the frons is marked with a dark spot surrounded anteriorly by a semicircle of blue. The thorax is green, the legs dark brown becoming black at 2/3 the length of the femur. The costal vein is yellow. The abdomen is largely brown, heavily marked with blue in males and green in females. The prominent caudal appendages of the male are brown. The cerci are about twice as long as the epiproct.

Similar Species. The great size of this darner makes it nearly unmistakable. Common Green Darner is smaller and has a much quicker flight.

Habitat. Slow-flowing open, spring-fed streams, ponds, and pools.

Discussion. This species may be locally abundant and, despite its size, has a slower, more lumbering flight as compared to other green darners. Males consistently patrol low over the water, and fly with long wing beats over streams and ponds. Females are known to lay eggs in tandem, an unusual behavior among the darners in North America.

Springtime Darner
Genus *Basiaeschna* Selys
(p. 154)

This monotypic genus consists of a distinctive small, brown eastern species. The eyes are smaller than those in other darner genera, and there is a pair of pale-yellow stripes on each side of the pterothorax. The general color pattern is similar to that of the mosaic darners. The pterostigma is noticeably narrow, and both the triangles and the supertriangles usually have two crossveins. The caudal appendages in both males and females are no longer than abdominal segments 9 and 10 combined. This dragonfly frequents small forest streams.

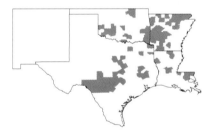

Springtime Darner
Basiaeschna janata (Say)
(p. 13; photo 29b)

Size. Total length: 50–67 mm; abdomen: 38–51 mm; hindwing: 32–42 mm.

Regional Distribution. *Biotic Province(s):* Austroriparian, Balconian, Carolinian, Kansan, Texan. *Watershed(s):* Arkansas, Bayou Bartholomew, Brazos, Canadian, Cimarron, Colorado, Guadalupe, Mississippi, Neches, Nueces, Ouachita, Red, Sabine, San Antonio, San Jacinto, St. Francis, Trinity, White.

General Distribution. Eastern United States and Canada westward to Oklahoma and Texas.

Flight Season. Mar. 19 (TX)–May 26 (AR).

Identification. This brown species is marked with blue abdominal spots and distinctive pale-yellow or cream-colored lateral thoracic stripes. The two pale middorsal thoracic stripes become obscured with age. The wings are clear, and a small, brown basal spot extends out to the first antenodal crossvein in each wing. The abdomen is brown with pale-blue spots, these becoming obscured in preserved specimens. Females have either blue or green spots.

Similar Species. Fawn Darner (*Boyeria vinosa*) has two pale-yellow spots on the side of the thorax and lacks blue spots on the abdomen. Harlequin (*Gomphaeschna furcillata*) and Taper-tailed (*G. antilope*) Darner both lack blue spots on the abdomen, and their lateral thoracic stripes are not straight.

Habitat. Small forest lakes and streams, and rivers with slow current.

Discussion. This common species is much like Fawn Darner in its habits. It is found earlier in the spring (hence the name) in many of the same locations that Fawn will visit later in the summer. Springtime is active during the day, sometimes well into the evening. It patrols streams and lakeshores at greater speed and at greater heights than Fawn Darner. Females lay eggs below the water surface in live plants and dead leaves or cattails.

References. Needham (1901), Needham and Westfall (1955), Walker (1958).

Spotted Darners
Genus *Boyeria* McLachlan
(p. 154)

This genus includes six distinctive species, four occurring in the Old World and two occurring in the eastern United States. Both North American species have two prominent pale spots on each side of the pterothorax. The thorax and abdomen are covered with fine hairs. In addition to the distinctive thoracic markings, no other North American dragonfly has crossveins in the midbasal space of the wings.

Members of this group are often abundant about forest edges and lakeshores, where they patrol with a characteristic butterfly-like flight pattern (with undulating up-and-down movements). They are inhabitants of woodland streams, much like Springtime Darner (*Basiaeschna janata*). Although they fly during the day in shaded areas, they are active at dusk and into the evening.

a dark basal spot in each wing. The wings themselves become strongly tinged with amber in older individuals. The pterothorax has two distinctive pale-yellow or cream-colored spots laterally, and there are rows of pale spots on the abdomen.

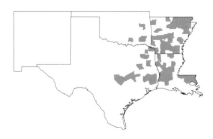

Habitat. Forest streams, rivers, and lakeshores with sufficient shade.

Discussion. This species could be considered crepuscular, its chief activity occurring well into the evening. One study reported that it often competes with bats for prey around lake edges. It is not uncommon, however, to see males patrolling along shaded areas of streams or lakeshores earlier in the day.

Fawn Darner has been called the most abundant and most widely distributed river dragonfly in the Great Plains. I have found them so abundant in some areas as to be overflowing the collection heads of a malaise trap (tent-style insect trap for collecting flying insects). Williamson commented on the discriminating nature of this dragonfly's feeding, occasionally rejecting some prey after seizing it.

References. Williamson (1932).

Fawn Darner
Boyeria vinosa (Say)
(photo 29c)

Size. Total length: 60–71 mm; abdomen: 45–56 mm; hindwing: 39–46 mm.

Regional Distribution. *Biotic Provinces:* Austroriparian, Balconian, Carolinian, Texan. *Watersheds:* Arkansas, Bayou Bartholomew, Brazos, Canadian, Colorado, Mississippi, Neches, Ouachita, Red, Sabine, San Jacinto, St. Francis, Trinity, White.

General Distribution. Eastern United States and Canada westward to Oklahoma and Texas.

Flight Season. May 3 (LA)–Nov. 15 (LA).

Identification. This common species is found throughout the eastern half of the region. Fawn Darner has a large head and green face. There is

1a. Sparkling Jewelwing (*Calopteryx dimidiata*), ♂. TX: San Jacinto Co., Sam Houston National Forest, posed.
1c. Ebony Jewelwing (*Calopteryx maculata*), ♂. TX: Anderson Co., Engeling Wildlife Management Area.
1e. Smoky Rubyspot (*Hetaerina titia*). Upper: ♂ with entirely dark wings, TX: Caldwell Co., San Marcos River; Lower: ♂ with partially dark wings, TX: Cameron Co., Arroyo Colorado.

1b. Sparkling Jewelwing (*Calopteryx dimidiata*), ♀. TX: San Jacinto Co., Sam Houston National Forest, posed.
1d. American Rubyspot (*Hetaerina americana*). Upper: ♂, AZ: Cochise Co., Muleshoe Ranch; Lower: ♀, TX: Terrell Co., Independence Creek.
1f. Canyon Rubyspot (*Hetaerina vulnerata*), ♂. AZ: Cochise Co., Garden Canyon.

2a. Great Spreadwing (*Archilestes grandis*), ♂. TX: Travis Co., Austin.

2b. Great Spreadwing (*Archilestes grandis*), ♀. AZ: Cochise Co., Muleshoe Ranch.

2c. Plateau Spreadwing (*Lestes alacer*), NM: Eddy Co., Rattlesnake Springs. Left: ♂; Right: ♀.

2d. Spotted Spreadwing (*Lestes congener*), ♂. CA: Butte Co., Oroville Wildlife Area.

2e. Spotted Spreadwing (*Lestes congener*), ♀. CA: Butte Co., Oroville Wildlife Area.

2f. Emerald Spreadwing (*Lestes dryas*), ♂. New Brunswick, Canada: Mt. Carleton Provincial Park.

a

b

c

d

e

f

3a. Common Spreadwing (*Lestes disjunctus*), ♂. TX: Bastrop Co., Bastrop State Park.

3c. Rainpool Spreadwing (*Lestes forficula*). Left: ♂, TX: Hidalgo Co., Bentsen Rio Grande Valley State Park; Right: ♀, TX: Hidalgo Co., McAllen, posed.

3e. Elegant Spreadwing (*Lestes inaequalis*), ♀. NJ: Morris Co., Lake Denmark.

3b. Common Spreadwing (*Lestes disjunctus*). ♀. TX: Harris Co., Houston.

3d. Elegant Spreadwing (*Lestes inaequalis*), ♂. NJ: Morris Co., Lake Denmark.

3f. Slender Spreadwing (*Lestes rectangularis*), KS: Stafford Co., Quivera National Wildlife Refuge. Left: ♂; Right: ♀.

4a. Chalky Spreadwing (*Lestes sigma*), ♂. TX: LaSalle Co., Chaparral Wildlife Management Area.

4c. Lyre-tipped Spreadwing (*Lestes unguiculatus*), ♂. NE: Cherry Co., Valentine.

4e. Swamp Spreadwing (*Lestes vigilax*), ♂. MA: Barnstable Co., Dennis.

4b. Chalky Spreadwing (*Lestes sigma*), ♀. TX: LaSalle Co., Chaparral Wildlife Management Area. Left: mature; Right: young.

4d. Lyre-tipped Spreadwing (*Lestes unguiculatus*), ♀. IL: Cook Co., Bluff Springs Fen.

4f. Swamp Spreadwing (*Lestes vigilax*), ♀. Upper: GA: Sumter Co., Flint River; Lower: NJ: Cape May Co., Beaver Swamp Wildlife Management Area.

5a. Coral-fronted Threadtail (*Neoneura aaroni*), ♂.
TX: Bexar Co., Medina River.
5c. Amelia's Threadtail (*Neoneura amelia*), ♂. TX:
Hidalgo Co., Rio Grande, posed.
5e. Orange-striped Threadtail (*Protoneura cara*), ♂.
TX: Val Verde Co., Devils River.

5b. Coral-fronted Threadtail (*Neoneura aaroni*), ♂ ♀
mating. TX: Bexar Co., Medina River.
5d. Amelia's Threadtail (*Neoneura amelia*), ♂ ♀ in
tandem. TX: Cameron Co., Resaca in Brownsville.
5f. Orange-striped Threadtail (*Protoneura cara*), ♂ ♀
laying eggs in tandem. TX: Val Verde Co., Devils River.

6a. Mexican Wedgetail (*Acanthagrion quadratum*), ♂. Honduras: Zamorano.

6c. Western Red Damsel (*Amphiagrion abbreviatum*), ♀. OR: Sherman Co., Rosebush Creek.

6e. Blue-fronted Dancer (*Argia apicalis*), ♂. TX: Bexar Co., Medina River.

6b. Western Red Damsel (*Amphiagrion abbreviatum*), ♂. OR: Sherman Co., Rosebush Creek.

6d. Paiute Dancer (*Argia alberta*). NM: Eddy Co. Upper: ♂; Lower: ♀.

6f. Blue-fronted Dancer (*Argia apicalis*), ♀. Upper: Blue form, TX: Harris Co., Jesse Jones Nature Center; Lower: Brown form, TX: Reeves Co., Balmorhea State Park.

7a. Comanche Dancer (*Argia barretti*). Upper: ♂, TX: Edwards/Real Co. line, Nueces River; Lower: ♀, TX: Uvalde Co., Nueces River.
7c. Seepage Dancer (*Argia bipunctulata*), ♀. FL: Clay Co., Gold Head Branch State Park.
7e. Coppery Dancer (*Argia cuprea*), ♂♀ in tandem. TX: Edwards/Real Co. line, Nueces River.

7b. Seepage Dancer (*Argia bipunctulata*), ♂. AR: Montgomery Co., Ouachita National Forest.
7d. Coppery Dancer (*Argia cuprea*), ♂. TX: Edwards/Real Co. line, Nueces River.
7f. Lavender Dancer (*Argia hinei*). Upper: ♂, AZ: Pima Co., Sabino Creek; Lower: ♀, TX: Presidio Co.

8a. Variable Dancer (*Argia fumipennis*), ♂. TX: Terrell Co., Independence Creek.

8c. Kiowa Dancer (*Argia immunda*). Upper: ♂, TX: Val Verde Co., Dolan Creek; Lower: ♀, TX: Terrell Co., Oasis Ranch.

8e. Sooty Dancer (*Argia lugens*), ♂. NM: Catron Co., Gila River.

8b. Variable Dancer (*Argia fumipennis*), ♀. NM: Eddy Co., Sitting Bull Falls.

8d. Leonora's Dancer (*Argia leonorae*), TX: Terrell Co., Independence Creek. Upper: ♂; Lower: ♀.

8f. Sooty Dancer (*Argia lugens*), ♀. NM: Catron Co., Gila River.

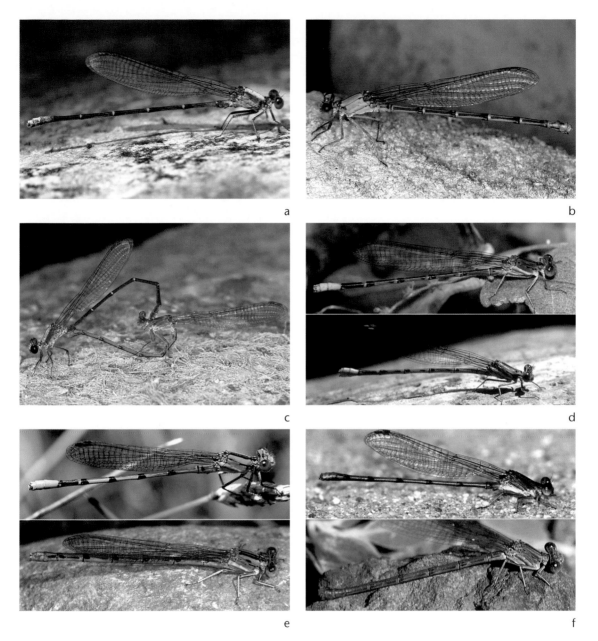

9a. Powdered Dancer (*Argia moesta*), ♂. TX: Blanco Co., Pedernales Falls State Park.

9c. Powdered Dancer (*Argia moesta*), ♂♀ mating. TX: Blanco Co., Pedernales Falls State Park.

9e. Aztec Dancer (*Argia nahuana*). Upper: ♂, TX: Val Verde Co., Devils River; Lower: ♀, TX: Presidio Co., Big Bend Ranch State Natural Area.

9b. Powdered Dancer (*Argia moesta*), ♀. TX: Blanco Co., Pedernales Falls State Park.

9d. Apache Dancer (*Argia munda*). Upper: ♂, AZ: Cochise Co., Coronado National Forest; Lower: ♀, AZ: Cochise Co., Chiricahua Mountains.

9f. Amethyst Dancer (*Argia pallens*). Upper: ♂, AZ: Pima Co., Sabino Creek; Lower: ♀, AZ: Yavapai Co., Sycamore Canyon.

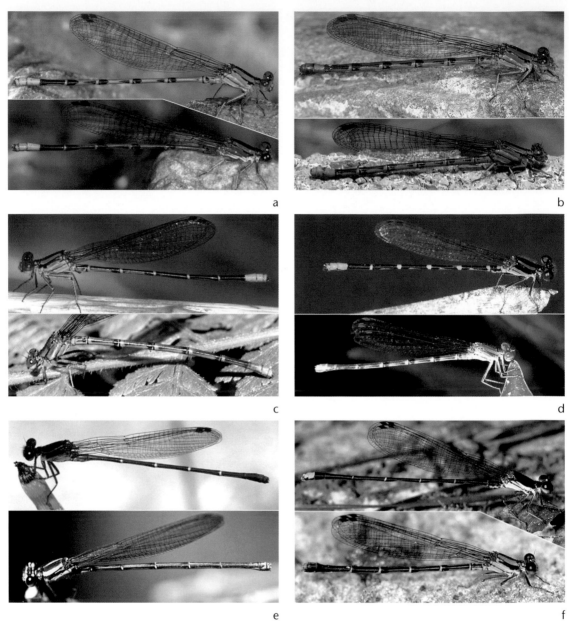

a

b

c

d

e

f

10a. Springwater Dancer (*Argia plana*), ♂. Upper: blue form, Mexico: Nuevo León, Monterrey; Lower: purple form, AZ: Cochise Co., Garden Canyon.

10c. Golden-winged Dancer (*Argia rhoadsi*). Upper: ♂, TX: Kinney Co., Ft. Clark Springs; Lower: ♀, Mexico: Tamaulipas.

10e. Tezpi Dancer (*Argia tezpi*). Upper: ♂, AZ: Cochise Co., Chiricahua Mountains; Lower: ♀, Honduras.

10b. Springwater Dancer (*Argia plana*), ♀. Upper: blue form, Mexico: Nuevo León, Monterrey; Lower: brown form, OK: Love Co., Thackerville.

10d. Blue-ringed Dancer (*Argia sedula*), TX: Val Verde Co., Devils River. Upper: ♂; Lower: ♀.

10f. Blue-tipped Dancer (*Argia tibialis*). Upper: ♂, TX: Polk Co., Horse Pen Creek; Lower: ♀, TX: San Jacinto Co., Sam Houston National Forest.

a

b

c

d

e

f

11a. Tonto Dancer (*Argia tonto*). Upper: ♂, NM: Grant Co., Meadow Creek; Lower: ♀, AZ: Cochise Co., Garden Canyon.

11c. Dusky Dancer (*Argia translata*), Upper: ♂, TX: Val Verde Co., Devils River; Lower: ♀, TX: Terrell Co., Oasis Ranch.

11e. Aurora Damsel (*Chromagrion conditum*), ♂. NJ: Morris Co., Lake Denmark.

11b. Tonto Dancer (*Argia tonto*), ♂♀ mating. AZ: Cochise Co., Garden Canyon.

11d. Vivid Dancer (*Argia vivida*). OR: Sherman Co., Rosebush Creek. Upper: ♂; Lower: ♀.

11f. Aurora Damsel (*Chromagrion conditum*), ♀. MA: Worcester Co., Petersham.

12a. River Bluet (*Enallagma anna*), ♂♀ mating. CA: Inyo Co., Big Pine.

12b. Rainbow Bluet (*Enallagma antennatum*), ♂. PA: Clinton Co., Kettle Creek State Park.

12c. Rainbow Bluet (*Enallagma antennatum*), ♀. NE: Cherry Co., Valentine, posed.

12d. Azure Bluet (*Enallagma aspersum*), ♂. NJ: Cape May Co., Dennisville.

12e. Azure Bluet (*Enallagma aspersum*), ♀. MA: Barnstable Co.

12f. Double-striped Bluet (*Enallagma basidens*), ♂. TX: Travis Co., Austin.

13a. Double-striped Bluet (*Enallagma basidens*), ♀.
OK: Love Co., Thackerville.

13c. Boreal Bluet (*Enallagma boreale*), ♀. New Mexico.

13e. Familiar Bluet (*Enallagma civile*), ♂♀ mating. TX:
Travis Co., Austin.

13b. Boreal Bluet (*Enallagma boreale*), ♂. New
Brunswick, Canada: Mt. Carleton Provincial Park.

13d. Tule Bluet (*Enallagma carunculatum*). Upper: ♂,
CA: Sacramento Co.; Lower: ♀, CA: Marin Co.

13f. Familiar Bluet (*Enallagma civile*). Upper: ♂, TX: Hidalgo Co.; Lower: ♀, Mexico: Nuevo León, Monterrey.

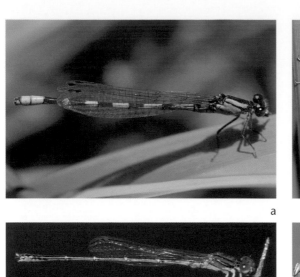
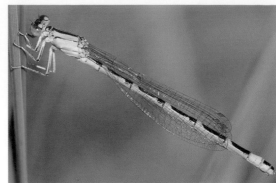

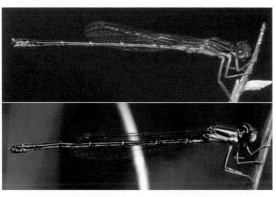
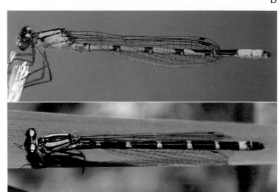

a

b

c

d

e

f

14a. Alkali Bluet (*Enallagma clausum*), ♂. NE: Cherry Co., Big Alkali Lake.

14c. Cherry Bluet (*Enallagma concisum*). Upper: ♂, FL: Leon Co., Little Dog Lake; Lower: ♀, FL: Clay Co., Gold Head Branch State Park.

14e. Northern Bluet (*Enallagma cyathigerum*), ♂♀ mating. British Columbia, Canada: Twin Lakes, Nakusp.

14b. Alkali Bluet (*Enallagma clausum*), ♀. NE: Cherry Co., Valentine.

14d. Northern Bluet (*Enallagma cyathigerum*). Upper: ♂, MA: Falmouth; Lower: ♀, WA: Stossel Creek.

14f. Attenuated Bluet (*Enallagma daeckii*), ♂. NJ: Cumberland Co., Cumberland Pond.

15a. Attenuated Bluet (*Enallagma daeckii*), ♀. NJ: Cumberland Co., Cumberland Pond.
15c. Turquoise Bluet (*Enallagma divagans*), ♂♀ mating. TX: San Jacinto Co., Big Creek Scenic Area.
15e. Burgundy Bluet (*Enallagma dubium*), ♂♀ laying eggs in tandem. FL: Alachua Co.

15b. Turquoise Bluet (*Enallagma divagans*), ♂. TX: San Jacinto Co., Sam Houston National Forest.
15d. Atlantic Bluet (*Enallagma doubledayi*). Upper: ♂, NJ: Cape May Co., Dennisville; Lower: ♀, MA: Barnstable Co., Dennis.
15f. Big Bluet (*Enallagma durum*), ♂. VA: Charles City Co., Chickahominy State Wildlife Management Area.

a

b

c

d

e

f

16a. Big Bluet (*Enallagma durum*), ♂♀ mating. FL: Alachua Co., Newnan's Lake, Gainesville.

16c. Stream Bluet (*Enallagma exsulans*), ♀. MA: Barnstable Co., Mashpee.

16e. Skimming Bluet (*Enallagma geminatum*), ♂♀ mating. OK: Love Co., Thackerville.

16b. Stream Bluet (*Enallagma exsulans*), ♂. TX: Val Verde Co., Devils River.

16d. Skimming Bluet (*Enallagma geminatum*), ♂. IL: McHenry Co., Moraine Hills State Park.

16f. Neotropical Bluet (*Enallagma novaehispaniae*), ♂ feeding on ♂. TX: Kinney Co., Ft. Clark Springs.

17a. Neotropical Bluet (*Enallagma novaehispaniae*), ♂.
TX: Kinney Co., Ft. Clark Springs.
17c. Arroyo Bluet (*Enallagma praevarum*). Upper: ♂,
TX: Jeff Davis Co., Phantom Cave Spring; Lower: ♀, AZ:
Cochise Co., San Pedro River.
17e. Orange Bluet (*Enallagma signatum*), IL: Cook Co.,
William Erickson Forest Preserve. Upper: mature ♂;
Lower: young ♂.

17b. Neotropical Bluet (*Enallagma novaehispaniae*), ♀.
TX: Kinney Co., Ft. Clark Springs.
17d. Orange Bluet (*Enallagma signatum*), ♂♀ in
tandem. TX: Fort Bend Co., Brazos Bend State Park.

18a. Claw-tipped Bluet (*Enallagma semicirculare*), ♂. AZ: Cochise Co., San Bernardino National Wildlife Refuge.

18c. Vesper Bluet (*Enallagma vesperum*). Upper: mature ♂, TX: Tyler Co., Martin Dies Jr. State Park; Lower: teneral ♂, VA: Gloucester Co., Beaverdam Park.

18e. Painted Damsel (*Hesperagrion heterodoxum*). Upper: mature ♂, NM: Eddy Co., Sitting Bull Falls, posed; Lower: teneral ♂, TX: Jeff Davis Co., Limpia Creek.

18b. Slender Bluet (*Enallagma traviatum*). Upper: ♂, MA: Middlesex Co., Holliston; Lower: ♀, MA: Barnstable Co., Brewster.

18d. Vesper Bluet (*Enallagma vesperum*), ♀. FL: Alachua Co., Watermelon Pond, near Gainesville.

18f. Painted Damsel (*Hesperagrion heterodoxum*). Upper: Male-form ♀, NM: Grant Co., Mangas Springs; Lower: Orange-form ♀, NM: Grant Co., Mangas Springs.

a

b

c

d

e

f

19a. Desert Forktail (*Ischnura barberi*), ♂. NM: Eddy Co., BLM overflow wetlands.

19b. Desert Forktail (*Ischnura barberi*), ♀. NM: Eddy Co., BLM overflow wetlands.

19c. Pacific Forktail (*Ischnura cervula*), ♂. CA: Humboldt Co., Aldergrove Marsh.

19d. Pacific Forktail (*Ischnura cervula*), ♀. CA: Humboldt Co., Aldergrove Marsh.

19e. Plains Forktail (*Ischnura damula*), ♂. TX: Jeff Davis Co., Limpia Creek.

19f. Plains Forktail (*Ischnura damula*). Upper: red-form ♀, AZ: Cochise Co., Wilcox; Lower: blue-form ♀, NM: Catron Co., Roswell.

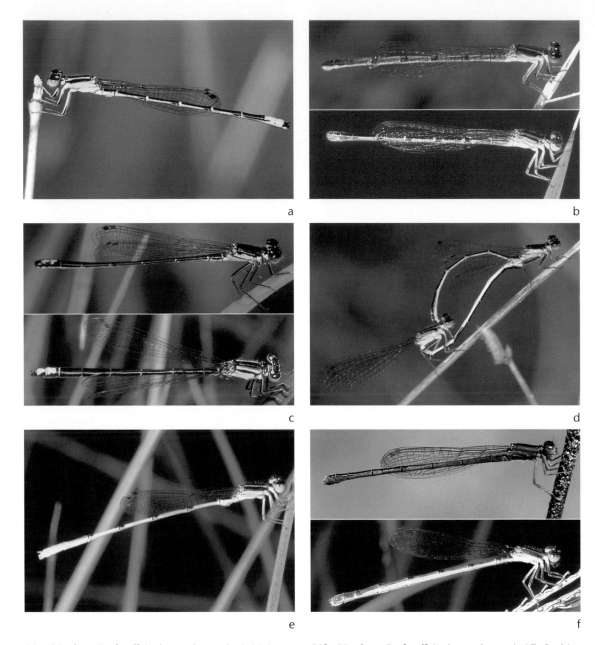

20a. Mexican Forktail (*Ischnura demorsa*), ♂. NM: Catron Co., Glenwood.

20c. Black-fronted Forktail (*Ischnura denticollis*). Upper: ♂, AZ: Cochise Co., San Bernardino National Wildlife Refuge; Lower: ♀, OR: Harney Co., Mickey Hot Springs.

20e. Citrine Forktail (*Ischnura hastata*), ♂. OK: Love Co., Thackerville.

20b. Mexican Forktail (*Ischnura demorsa*), AZ: Cochise Co., Muleshoe Ranch. Upper: blue-form ♀; Lower: red-form ♀.

20d. Black-fronted Forktail (*Ischnura denticollis*), ♂♀ mating. NM: Dona Ana Co., Las Cruces.

20f. Citrine Forktail (*Ischnura hastata*). Upper: Orange-form ♀, OK: Love Co., Thackerville; Lower: Blue-form ♂, TX: Anderson Co., Engeling Wildlife Management Area.

21a. Lilypad Forktail (*Ischnura kellicotti*), ♂. GA: Tift Co., Paradise Public Fishing Area.

21c. Western Forktail (*Ischnura perparva*), ♂. OR: Sherman Co., Rosebush Creek.

21e. Fragile Forktail (*Ischnura posita*), ♂. OK: Love Co., Thackerville.

21b. Lilypad Forktail (*Ischnura kellicotti*), ♀. NJ: Cape May Co., Tarkiln Pond.

21d. Western Forktail (*Ischnura perparva*). Upper: blue-form ♀, CA: Humboldt Co., Aldergrove marsh; Lower: orange-form ♀, CA: Marin Co., Port Reyes National Seashore.

21f. Fragile Forktail (*Ischnura posita*), ♀. GA: Sumter Co.

22a. Furtive Forktail (*Ischnura prognata*), ♂. FL: Leon Co.

22c. Rambur's Forktail (*Ischnura ramburii*), ♂. TX: Hidalgo Co., McAllen.

22e. Rambur's Forktail (*Ischnura ramburii*), orange-form ♀. TX: Hidalgo Co., McAllen.

22b. Furtive Forktail (*Ischnura prognata*), ♀. FL: Highlands Co., Highlands Hammock State Park. Upper: young; Lower: mature.

22d. Rambur's Forktail (*Ischnura ramburii*), ♀, male-like form. AZ: Cochise Co., San Bernardino National Wildlife Refuge.

22f. Rambur's Forktail (*Ischnura ramburii*), olive-form ♀. TX: San Jacinto Co., Double Lake Recreation Area.

23a. Eastern Forktail (*Ischnura verticallis*), ♂. NJ: Morris Co., Lake Denmark.

23c. Southern Sprite (*Nehalennia integricollis*), ♂. TX: Jasper Co., Boykin Springs Recreation Area.

23e. Everglades Sprite (*Nehalennia pallidula*), ♂. FL: Dade Co.

23b. Eastern Forktail (*Ischnura verticallis*). Upper: blue-form ♀, IL: Cook Co., Crabtree National Preserve; Lower: orange-form ♀, VA: Stafford Co., Aquia Creek.

23d. Southern Sprite (*Nehalennia integricollis*), ♀. TX: Jasper Co., Boykin Springs Recreation Area.

23f. Everglades Sprite (*Nehalennia pallidula*), ♀. FL: Dade Co.

24a. Caribbean Yellowface (*Neoerythromma cultellatum*), ♂. TX: Cameron Co., Resaca in Brownsville.
24c. Duckweed Firetail (*Telebasis byersi*), ♂. FL: Leon Co., Natural Bridge State Historic Site.
24e. Desert Firetail (*Telebasis salva*), ♀. TX: Travis Co., Austin.

24b. Caribbean Yellowface (*Neoerythromma cultellatum*), ♂♀ in tandem. TX: Cameron Co., Resaca in Brownsville.
24d. Desert Firetail (*Telebasis salva*), ♂. TX: Travis Co., Austin.
24f. Desert Firetail (*Telebasis salva*), ♂♀ mating. TX: Travis Co., Austin.

25a. Gray Petaltail (*Tachopteryx thoreyi*), ♀. TX: San Jacinto Co., Big Creek Scenic Area.
25c. Lance-tipped Darner (*Aeshna constricta*), blue-striped ♀, posed.
25e. Arroyo Darner (*Aeshna dugesi*), ♂. AZ: Pima Co., Tucson, Rose Canyon Lake.

25b. Lance-tipped Darner (*Aeshna constricta*), ♂, posed.
25d. Lance-tipped Darner (*Aeshna constricta*), yellow-striped ♀. MA: Middlesex Co., Acton, posed.
25f. Arroyo Darner (*Aeshna dugesi*), ♀. AZ, preserved specimen.

26a. Variable Darner (*Aeshna interrupta*), ♂. MA: Worcester, Ashburnham, posed.

26c. Sedge Darner (*Aeshna juncea*), ♂. Yukon Territory, Canada: Old Crow.

26e. Blue-eyed Darner (*Aeshna multicolor*), ♂. NE: Cherry Co., Duck Lake.

26b. Variable Darner (*Aeshna interrupta*), ♀. MA: Worcester, Ashburnham, posed.

26d. Sedge Darner (*Aeshna juncea*), ♀. Yukon Territory, Canada: Old Crow.

26f. Blue-eyed Darner (*Aeshna multicolor*), ♀. TX: Randall Co., Palo Duro Canyon State Park.

27a. Paddle-tailed Darner (*Aeshna palmata*), ♂.
British Columbia, Canada: Socco Lake.
27c. Shadow Darner (*Aeshna umbrosa*), ♂. MA:
Barnstable, Eastham.
27e. Shadow Darner (*Aeshna umbrosa*), ♀. WA: King
Co., Seattle.

27b. Paddle-tailed Darner (*Aeshna palmata*). ♀,
posed.
27d. Turquoise-tipped Darner (*Aeshna psilus*), ♂.
Nicaragua: Matagalapa Dept., Selva Negra.

28a. Persephone's Darner (*Aeshna persephone*), ♂. Arizona, posed.

28b. Amazon Darner (*Anax amazili*), ♀. Costa Rica, posed.

28c. Common Green Darner (*Anax junius*), ♂. TX: Travis Co., Hornsby Bend.

28d. Common Green Darner (*Anax junius*), ♀. AR: Montgomery Co., Ouachita National Forest.

28e. Common Green Darner (*Anax junius*) feeding on Roseate Skimmer (*Orthemis ferruginea*) while seemingly stuck in the mud. TX: Hidalgo Co., World Birding Center, Edinburg.

28f. Comet Darner (*Anax longipes*), ♀. TX: Travis Co., Hornsby Bend.

29a. Giant Darner (*Anax walsinghami*), ♂. NM: Eddy Co., Sitting Bull Falls.

29c. Fawn Darner (*Boyeria vinosa*), ♂. MA: Worcester, Ashburnham, posed.

29e. Regal Darner (*Coryphaeschna ingens*) ♂. FL: Monroe Co., Key West Botanical Garden.

29b. Springtime Darner (*Basiaeschna janata*), ♂. TX: Tarrant Co., Trinity River, posed.

29d. Blue-faced Darner (*Coryphaeschna adnexa*), ♂. Honduras, posed.

29f. Swamp Darner (*Epiaeschna heros*), ♂. OK: Love Co., near Thackerville.

30a. **Taper-tailed Darner** (*Gomphaeschna antilope*), ♂, posed.

30c. **Bar-sided Darner** (*Gynacantha mexicana*), ♂. TX: Hidalgo Co., Santa Ana National Wildlife Refuge.

30e. **Cyrano Darner** (*Nasiaeschna pentacantha*), ♂. AL: Escambia Co., Burnt Corn Creek, posed.

30b. **Harlequin Darner** (*Gomphaeschna furcillata*), ♂. AL: Hale Co., Talladega National Forest, posed; Inset: ♀, AL: Hale Co., Payne Lake, Talladega National Forest, posed.

30d. **Twilight Darner** (*Gynacantha nervosa*), ♂. Mexico: Colima, posed.

30f. **Riffle Darner** (*Oplonaeschna armata*), ♂. AZ: Pima Co., Marshall Gulch at Summerhaven, posed.

31a. Broad-striped Forceptail (*Aphylla angustifolia*),
♂. TX: Travis Co., Austin.
31c. Narrow-striped Forceptail (*Aphylla protracta*), ♂.
TX: Hidalgo Co., World Birding Center, Edinburg.
31e. Stillwater Clubtail (*Arigomphus lentulus*), ♂. TX:
Harris Co., Houston.

31b. Broad-striped Forceptail (*Aphylla angustifolia*),
♀. TX: Travis Co., Austin.
31d. Two-striped Forceptail (*Aphylla williamsoni*), ♂.
GA: Sumter Co., Lake Philema.
31f. Bayou Clubtail (*Arigomphus maxwelli*), ♂. TX:
Chambers Co., Turtle Bayou in White Memorial Park.

32a. Jade Clubtail (*Arigomphus submedianus*), ♂. TX: Bastrop Co., Buescher State Park.
32c. Southeastern Spinyleg (*Dromogomphus armatus*), ♀. FL: Okaloosa Co., Blackwater River.
32e. Black-shouldered Spinyleg (*Dromogomphus spinosus*), ♀. TX: San Jacinto Co., Big Creek Scenic Area.

32b. Unicorn Clubtail (*Arigomphus villosipes*), ♂. TN: Carroll Co., Maple Creek Lake.
32d. Black-shouldered Spinyleg (*Dromogomphus spinosus*), ♂. TX: Hardin Co., Roy E. Larsen Sandyland Sanctuary.
32f. Flag-tailed Spinyleg (*Dromogomphus spoliatus*), ♂. TX: Val Verde Co., Devils River.

33a. White-belted Ringtail (*Erpetogomphus compositus*), ♂. NM: Grant Co., Gila River.

33c. Eastern Ringtail (*Erpetogomphus designatus*), ♂. TX: Travis Co., Hornsby Bend.

33e. Blue-faced Ringtail (*Erpetogomphus eutainia*), ♂. TX: Gonzales Co., Guadalupe River, Independence Park.

33b. Yellow-legged Ringtail (*Erpetogomphus crotalinus*), ♂. NM: Eddy Co.

33d. Eastern Ringtail (*Erpetogomphus designatus*), ♀. TX: Travis Co., Hornsby Bend.

33f. Blue-faced Ringtail (*Erpetogomphus eutainia*), ♀. TX: Gonzales Co., Guadalupe River, Independence Park.

34a. Dashed Ringtail (*Erpetogomphus heterodon*), ♂. New Mexico.
34c. Banner Clubtail (*Gomphus apomyius*), ♂. NJ: Cumberland Co.
34e. Plains Clubtail (*Gomphus externus*), ♀. New Mexico, posed.

34b. Serpent Ringtail (*Erpetogomphus lampropeltis*), ♂. AZ: Cochise Co., San Pedro River.
34d. Plains Clubtail (*Gomphus externus*), ♂. TX: Montgomery Co., Jones State Forest.
34f. Tamaulipan Clubtail (*Gomphus gonzalezi*), ♂. TX: Starr Co., Falcon Dam.

a

b

c

d

e

f

35a. Pronghorn Clubtail (*Gomphus graslinellus*), ♂.
SD: Custer Co., Black Hills National Forest, posed.
35c. Cocoa Clubtail (*Gomphus hybridus*), ♂. NC: Lenoir
Co.
35e. Ashy Clubtail (*Gomphus lividus*), ♂. WV:
Pocahontas Co., Cranberry Glade.

35b. Pronghorn Clubtail (*Gomphus graslinellus*), ♀.
SD: Custer Co., Black Hills National Forest, posed.
35d. Cocoa Clubtail (*Gomphus hybridus*), ♂♀ mating.
SC: Darlington Co., Big Pee Dee River.
35f. Ashy Clubtail (*Gomphus lividus*), ♀. WV:
Pocahontas Co., Cranberry Glade.

36a. Sulphur-tipped Clubtail (*Gomphus militaris*), ♂.
TX: Fort Bend Co., Brazos Bend State Park.
36c. Gulf Coast Clubtail (*Gomphus modestus*), ♂. MS:
Benton Co., posed.
36e. Oklahoma Clubtail (*Gomphus oklahomensis*), ♀.
AR: Montgomery Co., Ouachita National Forest.

36b. Sulphur-tipped Clubtail (*Gomphus militaris*), ♀.
TX: Travis Co., McKinney Falls State Park.
36d. Oklahoma Clubtail (*Gomphus oklahomensis*), ♂.
AR: Montgomery Co., Ouachita National Forest.
36f. Ozark Clubtail (*Gomphus ozarkensis*), ♂. AR:
Montgomery Co., Ouachita National Forest.

37a. Ozark Clubtail (*Gomphus ozarkensis*), ♀. AR:
Montgomery Co., Ouachita National Forest.
37c. Rapids Clubtail (*Gomphus quadricolor*), ♀. VT:
Rutland Co., West Haven, posed.
37e. Cobra Clubtail (*Gomphus vastus*), ♀. TX: Tarrant
Co., Trinity River, posed.

37b. Rapids Clubtail (*Gomphus quadricolor*), ♂. NY:
Saratoga Co., Lake Luzerne, posed.
37d. Cobra Clubtail (*Gomphus vastus*), ♂. TX:
Gonzales Co., Palmetto State Park.
37f. Dragonhunter (*Hagenius brevistylus*), ♀. TX: Kerr
Co., Guadalupe River.

38a. Arizona Snaketail (*Ophiogomphus arizonicus*), ♂.
NM: Catron Co., San Francisco River.
38c. Pale Snaketail (*Ophiogomphus severus*), ♂. NE:
Cherry Co., Niobrara National Wildlife Refuge, posed.
38e. Westfall's Snaketail (*Ophiogomphus westfalli*), ♂.
MO: Phelps Co., Bourbeuse River.

38b. Arizona Snaketail (*Ophiogomphus arizonicus*), ♀.
Arizona, posed.
38d. Pale Snaketail (*Ophiogomphus severus*), ♀. NE:
Cherry Co., Valentine.
38f. Westfall's Snaketail (*Ophiogomphus westfalli*), ♀.
AR: Fulton Co., South Fork of Spring River.

39a. Five-striped Leaftail (*Phyllogomphoides albrighti*).
Upper: ♂, TX: Kerr Co., Guadalupe River; Lower: ♀, TX:
Travis Co., Hornsby Bend.
39c. Gray Sanddragon (*Progomphus borealis*), ♂. TX:
Presidio Co., Alamito Creek.
39e. Common Sanddragon (*Progomphus obscurus*), ♂.
TX: Travis Co., McKinney Falls State Park.

39b. Four-striped Leaftail (*Phyllogomphoides
stigmatus*), ♂. TX: Travis Co., Hornsby Bend.
39d. Gray Sanddragon (*Progomphus borealis*), ♀. CA:
Monterey Co., Salinas River.
39f. Common Sanddragon (*Progomphus obscurus*),
♂ ♀ mating. TX: Blanco Co., Pedernales Falls State Park.

40a. Least Clubtail (*Stylogomphus albistylus*), ♂. WV: Pocahontas Co., Elk River.

40c. Brimstone Clubtail (*Stylurus intricatus*), ♀. TX: Presidio Co., Rio Grande River at Big Bend Ranch State Natural Area.

40e. Laura's Clubtail (*Stylurus laurae*), ♀. TX: San Jacinto Co., Big Creek Scenic Area.

40b. Brimstone Clubtail (*Stylurus intricatus*), ♂. NE: Cherry Co., Niobrara River, Smith Falls State Park, posed.

40d. Laura's Clubtail (*Stylurus laurae*), ♂. FL: Gadsden Co., posed.

40f. Arrow Clubtail (*Stylurus spiniceps*), ♂. WI: posed.

41a. Russet-tipped Clubtail (*Stylurus plagiatus*), ♂.
TX: Gonzales Co., Guadalupe River, Independence Park.
41c. Pacific Spiketail (*Cordulegaster dorsalis*), ♀. CA:
Inyo Co., Antelope Springs.
41e. Arrowhead Spiketail (*Cordulegaster obliqua*), ♀.
NC: Chatham Co.

41b. Apache Spiketail (*Cordulegaster diadema*), ♂. AZ:
Coconino Co., posed.
41d. Twin-spotted Spiketail (*Cordulegaster maculata*),
♂. WV: Pocahontas Co., Cranberry Glade, posed.
41f. Stream Cruiser (*Didymops transversa*), ♂. TX:
Travis Co., Austin, posed.

42a. Allegheny River Cruiser (*Macromia alleghaniensis*), ♂. TN: Cumberland Co., posed.
42c. Illinois River Cruiser (*Macromia illinoiensis*), ♂. TX: Travis Co., Austin, posed.
42d. Gilded River Cruiser (*Macromia pacifica*), ♂. TX: McLennan Co., posed.

42b. Bronzed River Cruiser (*Macromia annulata*), ♂♀ mating. TX: Bastrop Co., Buescher State Park.
42e. Royal River Cruiser (*Macromia taeniolata*), ♂. FL, posed.

43a. Stripe-winged Baskettail (*Epitheca costalis*), ♂.
OK: Love Co., Thackerville, posed.
43c. Common Baskettail (*Epitheca cynosura*), ♂ with
basal wing markings. AL: Bibb Co., Talladega National
Forest, posed.
43e. Prince Baskettail (*Epitheca princeps*), ♂. TX:
Travis Co., Hornsby Bend.

43b. Common Baskettail (*Epitheca cynosura*), ♂ with
reduced wing markings. NJ: Morris Co., Denmark.
43d. Dot-winged Baskettail (*Epitheca petechialis*), ♂.
TX: McLennan Co., posed.
43f. Mantled Baskettail (*Epitheca semiaquea*), ♂. OK:
Love Co., Thackerville, posed.

44a. Robust Baskettail (*Epitheca spinosa*), ♂. NJ: Gloucester Co., Malaga.

44b. Florida Baskettail (*Epitheca stella*), ♂. FL: Alachua Co., posed.

44c. Selys' Sundragon (*Helocordulia selysii*), ♂. TX: San Jacinto Co., Big Creek Scenic Area.

44d. Uhler's Sundragon (*Helocordulia uhleri*), ♂. MA: Hampshire Co., Amherst, posed.

44e. Alabama Shadowdragon (*Neurocordulia alabamensis*), ♀. FL: Alachua Co., posed.

44f. Smoky Shadowdragon (*Neurocordulia molesta*), ♂. FL: Liberty Co., posed.

a

b

c

d

e

f

45a. Umber Shadowdragon (*Neurocordulia obsoleta*), ♂. NC: Chatham Co., Pittsboro.
45c. Orange Shadowdragon (*Neurocordulia xanthosoma*), ♂. TX: Val Verde Co., Devils River.
45e. Coppery Emerald (*Somatochlora georgiana*), ♂. MA: Middlesex Co., Holliston, posed.

45b. Cinnamon Shadowdragon (*Neurocordulia virginiensis*), ♀. FL: Alachua Co., posed.
45d. Fine-lined Emerald (*Somatochlora filosa*), ♂. FL: Alachua Co., posed.
45f. Mocha Emerald (*Somatochlora linearis*), ♂. MA: Middlesex Co., Holliston, posed.

46a. Texas Emerald (*Somatochlora margarita*), ♂. TX: San Jacinto Co., posed.
46c. Clamp-tipped Emerald (*Somatochlora tenebrosa*), ♂. AL: Bibb Co., Talladega National Forest, posed.
46e. Red-tailed Pennant (*Brachymesia furcata*), ♂. TX: Kleburg Co., King Ranch.

46b. Mountain Emerald (*Somatochlora semicircularis*), ♂. WA: Summit Lake.
46d. Clamp-tipped Emerald (*Somatochlora tenebrosa*), ♀. AL: Bibb Co., Talladega National Forest, posed.
46f. Red-tailed Pennant (*Brachymesia furcata*), ♀. TX: Hidalgo Co., Santa Ana National Wildlife Refuge.

47a. Four-spotted Pennant (*Brachymesia gravida*), ♂.
TX: Travis Co., McKinney Falls State Park, posed.
47c. Tawny Pennant (*Brachymesia herbida*), ♂.
Mexico: Veracruz.
47e. Gray-waisted Skimmer (*Cannaphila insularis*), ♂.
TX: Kinney Co., Ft. Clark Springs; Inset: ♀, Honduras,
posed.

47b. Four-spotted Pennant (*Brachymesia gravida*), ♀.
TX: Starr Co., Falcon Dam, posed.
47d. Pale-faced Clubskimmer (*Brechmorhoga
mendax*), ♂. TX: Travis Co., Hornsby Bend, posed.
47f. Amanda's Pennant (*Celithemis amanda*). Left: ♂,
FL: Washington Co.; Right: ♀, FL: Okaloosa Co.

a

b

c

d

e

f

48a. Red-veined Pennant (*Celithemis bertha*). Left: ♂, GA: Lee Co; Right: ♀, FL: Putnam Co., posed.
48c. Calico Pennant (*Celithemis elisa*), ♀. TX: Montgomery Co., Jones State Forest.
48e. Banded Pennant (*Celithemis fasciata*), ♂. TX: Jasper Co., Boykin Springs Recreation Area, posed.

48b. Calico Pennant (*Celithemis elisa*), ♂. AR: Montgomery Co., Ouachita National Forest.
48d. Halloween Pennant (*Celithemis eponina*), ♂. TX: Fort Bend Co., Sugarland.
48f. Banded Pennant (*Celithemis fasciata*), ♀ with exuviae. AL: Elmore Co.

49a. Faded Pennant (*Celithemis ornata*), ♂. TX: Harris Co.
49c. Double-ringed Pennant (*Celithemis verna*), ♂. NJ: Cumberland Co.
49e. Mayan Setwing (*Dythemis maya*), ♂. TX: Presidio Co., Big Bend State Natural Area.

49b. Faded Pennant (*Celithemis ornata*), ♀. TX: Montgomery Co., Jones State Forest.
49d. Double-ringed Pennant (*Celithemis verna*), ♀. SC: Marlboro Co., White's Creek.
49f. Mayan Setwing (*Dythemis maya*), ♀. Mexico: Nayarit.

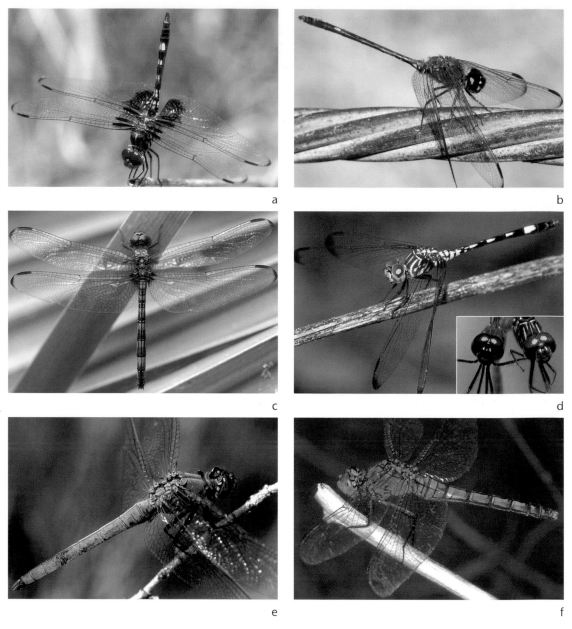

50a. Checkered Setwing (*Dythemis fugax*), ♂. TX: Val Verde Co., Devils River.
50c. Black Setwing (*Dythemis nigrescens*), ♀. TX: Hidalgo Co., Rio Grande, posed.
50e. Western Pondhawk (*Erythemis collocata*), ♂. NM: Eddy Co.

50b. Black Setwing (*Dythemis nigrescens*), mature ♂. TX: Hidalgo Co., Anzulduas County Park.
50d. Swift Setwing (*Dythemis velox*), ♀. TX: Travis Co., Austin; Inset: front view of Black (*D. nigrescens*) and Swift (*D. velox*) Setwings.
50f. Western Pondhawk (*Erythemis collocata*), ♀. NM: Eddy Co.

a

b

c

d

e

f

51a. Flame-tailed Pondhawk (*Erythemis peruviana*), ♂. Mexico: Campeche.
51c. Pin-tailed Pondhawk (*Erythemis plebeja*), young ♂. Mexico: Yucatan, Rio Lagartos.
51e. Eastern Pondhawk (*Erythemis simplicicollis*), mature ♂. TX: Travis Co., Austin.

51b. Pin-tailed Pondhawk (*Erythemis plebeja*), ♂. TX: Kleburg Co., King Ranch.
51d. Eastern Pondhawk (*Erythemis simplicicollis*). Upper: ♀, TX: Fort Bend Co., Sugarland; Lower: young ♂, TX: Harris Co.
51f. Great Pondhawk (*Erythemis vesiculosa*), ♂. TX: Fort Bend Co., Brazos Bend State Park.

a

b

c

d

e

f

52a. Plateau Dragonlet (*Erythrodiplax basifusca*), ♂.
TX: Presidio Co., Alamito Creek.
52c. Seaside Dragonlet (*Erythrodiplax berenice*), ♀.
NM: Chaves Co., Bitter Lake National Wildlife Refuge.
52e. Black-winged Dragonlet (*Erythrodiplax funerea*),
♂. Mexico, posed.

52b. Plateau Dragonlet (*Erythrodiplax basifusca*), ♀.
AZ: Pima Co., Sabino Canyon.
52d. Seaside Dragonlet (*Erythrodiplax berenice*),
mature ♂. TX: Chambers Co., Anahuac National Wildlife
Refuge.
52f. Red-faced Dragonlet (*Erythrodiplax fusca*). Left:
young ♂, TX: Edwards Co., Little Hickory Creek; Right:
mature ♂, Venezuela, Rio Verde.

53a. Little Blue Dragonlet (*Erythrodiplax minuscula*),
TX: Harris Co.; Left: mature ♂; Right: young ♂.
53c. Band-winged Dragonlet (*Erythrodiplax umbrata*),
♀. TX: Hidalgo Co., Santa Ana National Wildlife Refuge.
53e. Blue Corporal (*Ladona deplanata*), ♀. TX: Harris
Co.

53b. Band-winged Dragonlet (*Erythrodiplax umbrata*),
♀. TX: Harris Co.
53d. Blue Corporal (*Ladona deplanata*), ♂. OK: Love
Co., Thackerville, posed.
53f. Golden-winged Skimmer (*Libellula auripennis*),
♂. TX: posed.

54a. Bar-winged Skimmer (*Libellula axilena*), ♂. TX: Tyler Co., John H. Kirby State Forest.
54c. Comanche Skimmer (*Libellula comanche*), ♀. TX: Reeves Co., Balmorhea State Park.
54e. Bleached Skimmer (*Libellula composita*), ♀. NM: Chaves Co., Bitter Lake National Wildlife Refuge, posed.

54b. Comanche Skimmer (*Libellula comanche*), ♂. TX: Reeves Co., Balmorhea State Park; Inset: ♂ face, TX: Val Verde Co., Dolan Creek.
54d. Bleached Skimmer (*Libellula composita*), ♂. NM: Chaves Co., Bitter Lake National Wildlife Refuge, posed.
54f. Neon Skimmer (*Libellula croceipennis*), ♂. TX: Harris Co., Vellaire.

55a. Neon Skimmer (*Libellula croceipennis*), ♀. TX: Val Verde Co., Dolan Creek.
55c. Spangled Skimmer (*Libellula cyanea*), ♀. TX: Newton Co., Caney Creek; Inset: young ♂, NJ: Cumberland Co.
55e. Yellow-sided Skimmer (*Libellula flavida*), ♀. AR: Montgomery Co., Ouachita National Forest.

55b. Spangled Skimmer (*Libellula cyanea*), mature ♂. MD: Ann Arundel Co.
55d. Yellow-sided Skimmer (*Libellula flavida*), ♂. NJ: Ocean Co.
55f. Eight-spotted Skimmer (*Libellula forensis*), ♂. CA: Marin Co., Alpine Lake.

56a. Slaty Skimmer (*Libellula incesta*), mature ♂. TX: Gonzales Co., Palmetto State Park.
56c. Widow Skimmer (*Libellula luctuosa*), ♂. TX: Blanco Co., Pedernales Falls State Park.
56e. Needham's Skimmer (*Libellula needhami*), ♀. TX: Starr Co., Falcon Dam, posed.

56b. Slaty Skimmer (*Libellula incesta*), ♀. TX: Fort Bend Co., Brazos Bend State Park.
56d. Widow Skimmer (*Libellula luctuosa*), ♀. TX: Travis Co., Austin.
56f. Hoary Skimmer (*Libellula nodisticta*), ♂. CA: Sonora Co., Mayacamas Mountain National Audubon Sanctuary.

57a. Twelve-spotted Skimmer (*Libellula pulchella*), ♂. IL: Cook Co., Winnetka.
57c. Flame Skimmer (*Libellula saturata*), ♂. NM: Eddy Co., Sitting Bull Falls, posed.
57e. Great Blue Skimmer (*Libellula vibrans*), ♂. TX: Gonzales Co., Palmetto State Park.

57b. Four-spotted Skimmer (*Libellula quadrimaculata*), ♂. IL: Cook Co., Winnetka.
57d. Painted Skimmer (*Libellula semifasciata*), ♂. NC: Bladen Co.
57f. Great Blue Skimmer (*Libellula vibrans*), ♀. TX: Hardin Co., Turkey Creek; Inset: older ♂, TX: Montgomery Co.

a

b

c

d

e

f

58a. Marl Pennant (*Macrodiplax balteata*), ♂. NM: Chaves Co., Bitter Lake National Wildlife Refuge.

58c. Ivory-striped Sylph (*Macrothemis imitans*), ♂. Brazil: Rio de Janeiro.

58e. Jade-striped Sylph (*Macrothemis inequiunguis*), ♂. Costa Rica.

58b. Marl Pennant (*Macrodiplax balteata*), ♀. NM: Chaves Co., Bottomless Lake State Park.

58d. Straw-colored Sylph (*Macrothemis inacuta*), ♂. TX: Hidalgo Co., World Birding Center, Edinburg.

58f. Hyacinth Glider (*Miathyria marcella*), ♂. TX: Kleburg Co., King Ranch.

59a. Spot-tailed Dasher (*Microthyria aequalis*), Lower: ♂, TX: Hidalgo Co.; Upper: ♀, Mexico: Colima.
59c. Thornbush Dasher (*Microthyria hagenii*), ♂. TX: Travis Co., Austin.
59e. Roseate Skimmer (*Orthemis ferruginea*), ♂. TX: Travis Co., Austin.

59b. Three-striped Dasher (*Microthyria didyma*), ♂. Mexico: Veracruz, Catemaco.
59d. Orange-bellied Skimmer (*Orthemis discolor*), ♂. TX: Kleburg Co., King Ranch.
59f. Roseate Skimmer (*Orthemis ferruginea*), ♀. TX: Travis Co., Austin.

60a. Blue Dasher (*Pachydiplax longipennis*), ♂. TX: Travis Co., Austin; Inset: ♀, TX: Travis Co., Hornsby Bend.
60c. Wandering Glider (*Pantala flavescens*), ♂. TX: Dimmit Co., Chaparral Wildlife Management Area, posed.
60e. Slough Amberwing (*Perithemis domitia*). TX: Hidalgo Co., Left: ♀; Right: ♂.

60b. Red Rock Skimmer (*Paltothemis lineatipes*), ♂. AZ: Pima Co., Sabino Canyon.
60d. Spot-winged Glider (*Pantala hymenaea*), ♂. TX: Val Verde Co., Devils River, posed.
60f. Mexican Amberwing (*Perithemis intensa*). Upper: ♂, AZ: Santa Cruz Co., Patagonia Lake State Park; Lower: ♀, CA: Imperial Co., Salton Sea National Wildlife Refuge.

a

b

c

d

e

f

61a. Eastern Amberwing (*Perithemis tenera*). Upper: ♀, TX: Harris Co., Houston Arboretum; Lower: ♂, TX: Travis Co., Austin.
61c. Common Whitetail (*Plathemis lydia*), ♀. TX: Travis Co., Austin.
61e. Desert Whitetail (*Plathemis subornata*), ♀. NM: Grant Co.

61b. Common Whitetail (*Plathemis lydia*), ♂. TX: Travis. Co., Austin.
61d. Desert Whitetail (*Plathemis subornata*), mature ♂. NM: Grant Co.; Inset: young ♂, AZ: Cochise Co., San Bernardino National Wildlife Refuge.
61f. Filigree Skimmer (*Pseudoleon superbus*), ♂. TX: Val Verde Co., Dolan Creek.

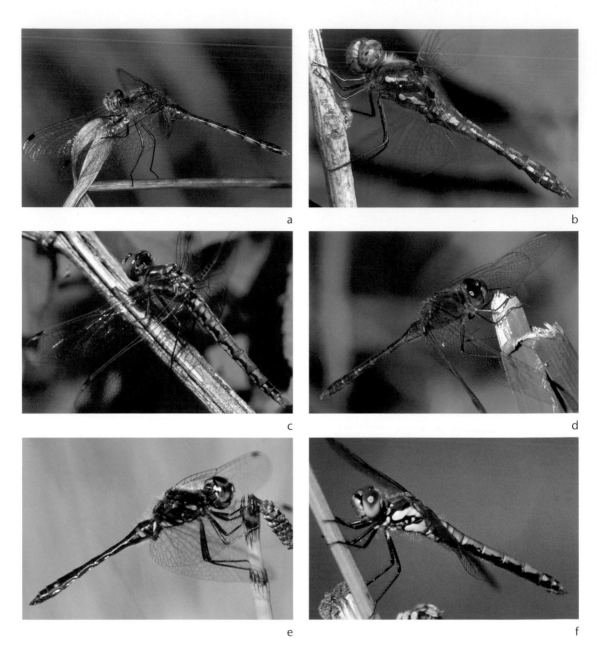

62a. Blue-faced Meadowhawk (*Sympetrum ambiguum*), ♂. TX: Harris Co.
62c. Variegated Meadowhawk (*Sympetrum corruptum*), ♀. TX: Cooke Co., posed.
62e. Black Meadowhawk (*Sympetrum danae*), ♂. WA: Swamp Lake.

62b. Variegated Meadowhawk (*Sympetrum corruptum*), ♂. TX: Randall Co., Palo Duro Canyon State Park, posed.
62d. Saffron-winged Meadowhawk (*Sympetrum costiferum*), ♂. WA: Jameson Lake.
62f. Black Meadowhawk (*Sympetrum danae*), ♀. WA: Kittitas Co., Ellensburg.

63a. Cardinal Meadowhawk (*Sympetrum illotum*), ♂.
Nicaragua: Matagalpa Dept., Selva Negra, posed.
63c. Striped Meadowhawk (*Sympetrum pallipes*), ♂.
CA: Butte Co., Oroville Wildlife Area.
63e. Yellow-legged Meadowhawk (*Sympetrum vicinum*), NC: Granville Co. Left: mature ♂; Right: young ♂.

63b. Cherry-faced Meadowhawk (*Sympetrum internum*), ♂♀ mating. NY: Broome Co.
63d. Band-winged Meadowhawk (*Sympetrum semicinctum*), ♀. NE: Cherry Co., Big Alkali Lake.
63f. Aztec Glider (*Tauriphila azteca*), ♂. Mexico, posed.

64a. Evening Skimmer (*Tholymis citrina*), ♂. Brazil, posed.

64c. Carolina Saddlebags (*Tramea carolina*), ♂. OK: Love Co., posed.

64e. Black Saddlebags (*Tramea lacerata*), ♂. TX: Newton Co.

64b. Striped Saddlebags (*Tramea calverti*), ♂. TX: Hidalgo Co., Santa Ana National Wildlife Refuge.

64d. Antillean Saddlebags (*Tramea insularis*), ♂. TX: Brewster Co., posed.

64f. Red Saddlebags (*Tramea onusta*), ♂. TX: Starr Co., Falcon Dam, posed.

Pilot Darners

Genus *Coryphaeschna* Williamson

(p. 154)

Pilot Darners are large, primarily Neotropical, dragonflies. Four species occur in North America, including three in the east and a single western species. Two species occur in our region. These dragonflies are high, swift fliers. The large eyes constitute a major part of the head and meet dorsally for a considerable distance, reducing the occiput to a small triangle. The thorax is largely green, with brown lateral thoracic stripes. The legs are shorter than in other members of the family. The wings are clear in males, but the wings of the female become tinted with age. The basal portion of the wings is tinted in younger individuals, and the distal portion is tinted in older ones. The triangles are long and the Rs vein forks below the pterostigma. The abdomen is brown, and long and narrow beyond the first few swollen segments. The caudal appendages are unusually long, and often broken off in older females. The two species in our region can be readily distinguished by size and the extent of the brown lateral thoracic stripes.

KEY TO THE SPECIES OF PILOT DARNERS (*CORYPHAESCHNA*)

1. Abdomen less than 55 mm in length; brown lateral thoracic stripes thin, much narrower than broader green areas — **Blue-faced (*adnexa*)**

1'. Abdomen greater than 65 mm in length; brown lateral thoracic stripes nearly as wide as green stripes — **Regal (*ingens*)**

Blue-faced Darner
Coryphaeschna adnexa (Hagen)
(photo 29d)

Size. Total length: 66–70 mm; abdomen: 48–52 mm; hindwing: 42–45 mm.
Regional Distribution. *Biotic Province:* Tamaulipan. *Watershed:* Lower Rio Grande.

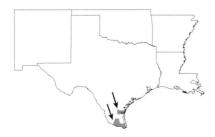

General Distribution. Florida, south Texas, and Mexico south to Brazil.
Flight Season. June (TX).
Identification. The face is bright sky blue, with a dark black "T" spot on the frons. The rear of the head is pale blue in both sexes. The thorax is green, with narrow brown stripes. The legs are black. The abdomen is dark brown marked with green. In females, the lateral abdominal spots on segments 5–10 are crimson.
Similar Species. This species is much smaller than Regal Darner (*C. ingens*), and the brown lateral thoracic stripes are distinctly narrower than the broad intervening green areas. Blue-eyed Darner (*Aeshna multicolor*) has blue lateral thoracic stripes and a blue abdomen.
Habitat. Heavily vegetated lakes, canals, and marshy areas.
Discussion. This species was only recently reported from Texas, where J.S. Rose discovered it at Santa Ana National Wildlife Refuge. A breeding population is known along creeks in the King Ranch (Kingsville) area of south Texas. Males patrol in short wing beats 1–2 m over marshy breeding areas.

References. Dunkle (1989).

Regal Darner
Coryphaeschna ingens (Rambur)
(photo 29e)

Size. Total length: 85–90 mm; abdomen: 64–78 mm; hindwing: 54–60 mm.
Regional Distribution. *Biotic Provinces:* Austroriparian, Tamaulipan, Texan. *Watersheds:* Bayou Bartholomew, Guadalupe, Mississippi, Neches, Nueces, Ouachita, Red, Sabine, San Jacinto, Trinity.

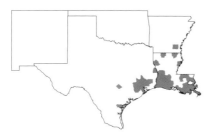

General Distribution. Southeastern United States, Cuba, and Bahamas.
Flight Season. Apr. 17 (TX)–Oct. 3 (LA).

Identification. The eyes are green in young individuals and males, but become deep blue in older females. The wings are clear, but in females they change with age (as described for the genus). The abdomen is brown, with narrow green markings. The basal segments are swollen, and there is no noticeable constriction following them.
Similar Species. This large coastal species may be confused with the equally large Swamp Darner (*Epiaeschna heros*). The latter, however, has a brown pterothorax with green stripes, while that of the Regal Darner is green with brown stripes.
Habitat. Lakes and slow-flowing streams with heavy vegetation.
Discussion. This species is unusual in that males do not defend or patrol territories, and it is not unusual to see large numbers of males in feeding swarms. They are strong fliers that rarely perch, but rather have a tendency to flush out prey by flying close to vegetation. Like several other darners, they may be rather active at dusk.

References. Dunkle (1989).

Swamp Darner
Genus *Epiaeschna* Hagen
(p. 154)

This is a monotypic genus, its lone member common throughout the Midwest and eastern United States. As its common name implies, it haunts swamps and bogs. The single large species is similar in coloration to Cyrano Darner (*Nasiaeschna pentacantha*). The large compound eyes occupy much of the head. The thorax and abdomen are robust. The wings are clear in young individuals, but become heavily tinted with age. The outer side of the triangle is sinuous, and the small anal loop generally comprises five cells in two rows. The caudal appendages are long and noticeably hairy in the male, and flattened in the female.

Texan. *Watersheds:* Arkansas, Bayou Bartholomew, Brazos, Canadian, Cimarron, Colorado, Guadalupe, Mississippi, Neches, Nueces, Ouachita, Red, Sabine, San Jacinto, St. Francis, Trinity, White.

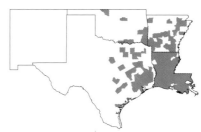

General Distribution. Eastern United States and Canada westward to central Oklahoma and Texas, straying south into Mexico.
Flight Season. Feb. 25 (LA)–Oct. 28 (LA).
Identification. This large, common species has brilliant-blue eyes and a brown body with green thoracic stripes and narrow green abdominal rings. The wings are often heavily tinged with amber. The abdomen is

Swamp Darner
Epiaeschna heros Fabricius
(photo 29f)

Size. Total length: 80–94 mm; abdomen: 63–72 mm; hindwing: 52–60 mm.
Regional Distribution. *Biotic Provinces:* Austroriparian, Balconian, Carolinian, Kansan, Tamaulipan,

brown, long, and nearly parallel-sided posteriorly. The caudal appendages are long in both sexes. The male appendages are complex and distinctly hairy. The female appendages are flattened, appearing petiolate.

Similar Species. This species is equally large as, but more widespread than, the similar Regal Darner (*Coryphaeschna ingens*), which has a largely green thorax with brown stripes and blue eyes only in mature females. Cyrano Darner (*Nasiaeschna pentacantha*) is similar, but it is smaller and has a striped abdomen.

Habitat. Heavily wooded ponds, streams, and oxbows, including ephemeral pools and ponds.

Discussion. This species is among the largest in North America. It is unusual in that, like the Regal Darner, males do not defend or patrol territories. They are, however, often seen swarming in large numbers, feeding on flying insects at dusk, both high in the air or lower to the ground, such as over culverts. This darner seems to enter open windows and buildings with some frequency, perhaps owing to the similarity of such places to its naturally shaded haunts. Incidental collections of Swamp Darners have been made in traps designed for arboreal beetles, in which the trap opening was slightly smaller than the wingspan of the victims. Females lay eggs in mud or vegetation, often some distance above the waterline, or in areas that will fill with water after heavy rains.

References. Ferguson (1940), Schaefer et al. (1996b), Walker (1958).

Pygmy Darners
Genus *Gomphaeschna* Selys
(pp. 155, 160)

Pygmy Darners make up a group of two small North American species that occur in the eastern United States. They are largely dark darners with a greenish-gray face and green eyes at maturity. The wings are clear, with relatively reduced venation. There is generally a single crossvein distal to the brace vein below the pterostigma. The two-celled anal loop is preceded by two or three paranal cells, and the triangle is usually only two-celled. The male caudal appendages are long, and the epiproct is distinctly bifurcated, a condition not found in other darners. The female ovipositor is short and truncate, but bears a pair of long, slender palps.

KEY TO THE SPECIES OF PYGMY DARNERS (*GOMPHAESCHNA*)

1. A single bridge crossvein in both wings; hindwing at level of nodus as wide as or wider than the distance in forewing from nodus to pterostigma .... **Taper-tailed (*antilope*)**

1'. More than 1 bridge crossvein in both wings; hindwing at level of nodus not wider than the distance in forewing from nodus to pterostigma .... **Harlequin (*furcillata*)**

Taper-tailed Darner
Gomphaeschna antilope (Hagen)
(p. 160; photo 30a)

Size. Total length: 53–60 mm; abdomen: 38–46 mm; hindwing: 30–37 mm.

Regional Distribution. *Biotic Province:* Austroriparian. *Watersheds:* Mississippi, Red.

General Distribution. Southeastern United States from New Hampshire south to Florida and west to Louisiana.

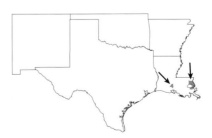

Flight Season. Apr. 11 (LA)–Jun. 2 (LA).

Identification. This small dark darner has green eyes

in life, and the costal margins of the wings are yellow. The first and third or fourth antenodal crossveins are usually thickened, and there is a single bridge crossvein. The middle 1/2 of the female forewing is generally tinted amber. The posterior abdominal segments in the male have green spots, and the middle abdominal segments in the female each have white spots laterally and brown-orange spots dorsally. The cerci in the male are flattened abruptly at about 1/3 of their length, forming a distinct inferior angle not present in Harlequin Darner (*G. furcillata*).

Similar Species. Harlequin Darner is similar, but not as dark or dull in overall color. The male abdomen is not tapered in Harlequin Darner. The lateral spots on the middle abdominal segments and the dorsal spots on segments 2 and 3 are rusty orange in Taper-tailed Darner.

Habitat. Shallow sphagnum bogs and swamps.

Discussion. Females lay eggs in wet wood just above the waterline. This species is unusual among darners in that it readily perches on tree trunks. Feeding swarms are not uncommon and are usually made up entirely of males.

References. Dunkle (1989, 2000), Gloyd (1940).

Harlequin Darner
Gomphaeschna furcillata (Say)
(p. 160; photo 30b)

Size. Total length: 52–60 mm; abdomen: 39–46 mm; hindwing: 29–36 mm.

Regional Distribution. *Biotic Province:* Austroriparian. *Watersheds:* Bayou Bartholomew, Mississippi, Ouachita, Red, St. Francis, Trinity.

General Distribution. Eastern United States from Ontario southward to Florida; westward to Texas and Wisconsin.

Flight Season. Feb. 3 (LA)–Apr. 4 (TX).

Identification. This is the only pygmy darner in Texas. The face and thorax are greenish-brown, the latter covered with numerous gray-white hairs. The abdomen is dark with pale marks, as in Taper-tailed Darner (*G. antilope*). The male cerci are smoothly curved to their tip, thus lacking the distinct inferior angle seen in Taper-tailed Darner.

Similar Species. Taper-tailed Darner is similar, but duller and darker in overall color. The male abdomen is not tapered in Harlequin Darner. The lateral spots on the middle abdominal segments and the dorsal spots on segments 2 and 3 are rusty orange in the Taper-tailed Darner.

Habitat. Shallow sphagnum bogs and swamps.

Discussion. Harlequin Darner emerges in the early spring, well before Taper-tailed Darner. Males are unusual in often patrolling over land, and are commonly seen flying on windy days. They can fly long distances and are often found feeding at forest edges. Males may patrol nonstop for an hour or more at midday, flying just a few centimeters above the water. Like Taper-tailed Darner, they will readily perch on tree trunks, but also occasionally on overhanging branches.

References. Dunkle (1989), Kennedy (1936), Williams (1979a).

Two-spined Darners
Genus *Gynacantha* Rambur
(p. 155)

Two-spined darners make up a large tropical genus represented by two species in the south-central United States. They are mostly greenish or brown, with clear or tinted wings. The triangles are long, and the radial sector is forked well before the pterostigma in the hindwing and usually in the forewing. The abdomen is slender and tapers posteriorly. The genus is named for the two spines on the ventral process of abdominal segment 10, which females use as a fulcrum while laying eggs. This group is most active just after sunrise and before sunset.

References. Williamson (1923a).

KEY TO THE SPECIES OF TWO-SPINED DARNERS (*GYNACANTHA*)

1. Thorax without lateral dark stripe or bar on lower posterior area; abdominal segments 1–3 without blue markings — **Twilight (*nervosa*)**

1'. Thorax with lateral dark stripe or bar on lower posterior area; abdominal segments 1–3 with blue markings — **Bar-sided (*mexicana*)**

Bar-sided Darner
Gynacantha mexicana Selys
(photo 30c)

Size. Total length: 70–76 mm; abdomen: 55–58 mm; hindwing: 46–50 mm.
Regional Distribution. *Biotic Province:* Tamaulipan. *Watershed:* Lower Rio Grande.

General Distribution. Lower Rio Grande Valley of Texas south through Mexico.
Flight Season. Jun. 25 (TX)–Oct. 25 (TX).
Identification. This is a largely brown dragonfly marked with pale green. The eyes are brown or green. There is a wide dark-brown stripe on the lower rear of the thorax. There is sometimes a brown stripe along the front edge of the wings. The abdomen is narrowly constricted at abdominal segment 3. There are pale blue markings on segments 1–3. The cerci of the females is longer than the combined length of segments 9 and 10. Male cerci are more than twice the length of the epiproct.
Similar Species. Other brown darners include Fawn Darner (*Boyeria vinosa*), which has two pale-yellow spots on each side of the thorax. Twilight Darner (*G. nervosa*) is similar but lacks blue on abdominal segments 1–3, and has a dark-brown thoracic stripe and only a slight constriction of abdominal segment 3. The epiproct is also longer in the male Twilight Darner, and the cerci in the family are longer than those in Bar-sided.

Habitat. Ephemeral ponds and pools.
Discussion. This species was only recently discovered in far south Texas, and it is still poorly known there, though the evidence seems to support breeding populations there. Its behavior is presumably similar to the more well-studied Twilight Darner.

Twilight Darner
Gynacantha nervosa Rambur
(photo 30d)

Size. Total length: 75–80 mm; abdomen: 50–57 mm; hindwing: 47–56 mm.
Regional Distribution. *Biotic Province:* Austroriparian. *Watershed:* Arkansas.

General Distribution. Southeastern United States (Florida, Alabama, Georgia, South Carolina), Oklahoma, and possibly south Texas southward through Mexico to Brazil.
Flight Season. Sep. (OK).
Identification. This is a largely brown dragonfly marked with pale green, and it is the only darner in our region with an all-brown thorax. The eyes are large and change from brown to green, a blue stripe across them, at maturity. The wings become heavily tinted with amber in older individuals. The abdomen is dull greenish brown and only slightly constricted at segment 3. The long cerci are often bro-

ken off in older females, and in males the epiproct is greater than 1/2 the length of the cerci.

Similar Species. Fawn Darner (*Boyeria vinosa*) has two pale yellow spots on each side of the thorax. The abdomen is strongly constricted at abdominal segment 3 in Bar-sided Darner (*G. mexicana*), which also has blue on abdominal segments 1–3, these characters not present in Twilight Darner. Bar-sided Darner also has a brown costal stripe on each wing.

Habitat. Fishless, ephemeral ponds with sufficient emergent vegetation and shade.

Discussion. A single male of this species was reported from the Ouachita National Forest, northwest of Page in Leflore County, Oklahoma, nearly 50 years ago. To my knowledge this species has not been confirmed in the south-central United States since, but it may occur in the lower Rio Grande Valley. It has also been taken at Victoria, in the Mexican state of Tamaulipas. As with other members of the genus, Twilight Darner may not be seen all day (usually hanging, at various heights, in shady forested areas), but will show itself in large numbers just after sunrise or before sunset, in large feeding swarms. discussed the possible importance of prey selection in visually acute predators like Twilight Darner, which would pressure potential prey items to mimic certain social wasps.

References. Kormondy (1960), O'Donnell (1996), Williams (1937).

Cyrano Darner
Genus *Nasiaeschna* Selys *in* Förster
(p. 155)

This is another monotypic genus, the members of its single species found widespread in the eastern portion of the south-central United States. The genus name derives from the uniquely protruding frons. The eyes are blue, and the vertex is bilobed. The body is brown, marked with green stripes, and the wings are clear. The abdomen is long and parallel-sided, not constricted behind the proximal abdominal segments. The caudal appendages in both sexes are distinctly short.

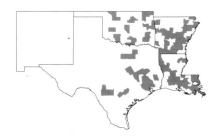

Cyrano Darner
Nasiaeschna pentacantha (Rambur)
(photo 30e)

Size. Total length: 62–73 mm; abdomen: 47–55 mm; hindwing: 45–50 mm.

Regional Distribution. *Biotic Provinces:* Austroriparian, Balconian, Carolinian, Kansan, Texan. *Watersheds:* Arkansas, Bayou Bartholomew, Brazos, Canadian, Cimarron, Colorado, Guadalupe, Mississippi, Neches, Nueces, Ouachita, Red, Sabine, San Antonio, San Jacinto, St. Francis, Trinity, White.

General Distribution. Widespread from southeastern Canada throughout the eastern United States to Texas and Oklahoma.

Flight Season. Mar. 30 (LA)–Oct. 12 (LA).

Identification. This is a smaller brown-and-green darner with a pronounced frons and brilliant blue eyes. It lacks a distinct black "T" spot. The thorax is brown with green stripes. The wings are clear, and the radial planate subtends a single row of cells. The abdomen of males has green lateral and middorsal stripes tapering posteriorly, these stripes generally interrupted throughout their length. The female cerci are short, scarcely longer than segment 10. The distinctive protruding frons is unique among North American dragonflies. Its common name derives from comparison of this feature with literatures' Cyrano de Bergerac.

Similar Species. Swamp Darner (*Epiaeschna heros*) and Regal Darner (*Coryphaeschna ingens*) are noticeably larger, with ringed abdomens. The latter also has a green thorax with brown stripes.

Habitat. Sheltered forest ponds, streams, and lake coves.

Discussion. This species seems never to stray far from the protection of wooded areas. It is often seen perching or flying along forest or path edges. It does not engage in feeding swarms, as many other darners do. Males have a distinctive patrol flight, continuously flicking their wings while flying slowly, darting at intruding dragonflies.

References. Dunkle (1985a, 2000).

Riffle Darner
Genus *Oplonaeschna* Selys
(p. 155)

This group of mosaic darner look-alikes comprises just two species, one in the southwest United States, the other (*Oplonaeschna magna* Gonzalez-S. and Novelo-G.) in central Mexico. Riffle darners are brown, with blue-and-black markings and distinctive wing venation.

Riffle Darner
Oplonaeschna armata (Hagen)
(photo 30f)

Size. Total length: 67–74 mm; abdomen: 50–57 mm; hindwing: 47–54 mm.
Regional Distribution. *Biotic Provinces:* Apachian, Chihuahuan, Navahonian. *Watersheds:* Colorado (NM), Pecos, Rio Grande.

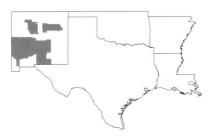

General Distribution. Arizona, New Mexico, and Utah; through Chihuahua, Mexico, and south to Guatemala.

Flight Season. Jun. 7 (NM)–Aug. 5 (NM).
Identification. Where Riffle Darner is found it can be rather common. The male's eyes are blue. The thorax is brown with two lateral thoracic stripes, the stripes blue above, fading to yellow below. The anterior lateral thoracic stripe is narrowed at its upper 1/3 to form two separated stripes. The abdomen in the male has small blue spots on the anterior segments and yellow spots on the posterior segments, 5–10. The cerci in the male are wedge-shaped, and those of the female are as long as segments 9 and 10, but break off after oviposition. Females may have blue, green, or yellow abdominal spots. Eyes of blue-form females are blue green, those of yellow-form females brownish yellow.
Similar Species. Some mosaic darners (*Aeshna*) are similar, but all have a complete anterior lateral thoracic stripe and larger pale spots on the abdomen. The radial planate is unforked in Riffle Darner and forked in mosaic darners.
Habitat. Rocky mountain streams in hardwood or pine forests subject to periodic flooding.
Discussion. Riffle Darners feed around the lower portions of trees in open forests. Males patrol most frequently in the morning, but will do so all day. One study found that adult emergence and flying season were related to annual air and water temperatures. Adults fly for about a month.

References. Dunkle (2000), Johnson (1968).

CLUBTAILS
(Family Gomphidae)

This is the second-largest dragonfly family, and consists of medium-sized distinctive greenish-yellow and brown species. They have widely separated eyes, as in the petaltails (Petaluridae), and a distinct, clublike widening of the posterior abdominal segments, usually more prominent in males. The wings are generally clear and lack both medial and radial planates, but there is a bracevein under the pterostigma. The legs vary in length, and there are usually numerous short spines on the femora. The caudal appendages are distinctive in most species, and the male epiproct is usually forked. In the tribe Gomphini, which includes the pond clubtails (*Arigomphus*), spinylegs (*Dromogomphus*) and common clubtails (*Gomphus*), the cerci are fused to the abdomen. Females lack an ovipositor, and are not accompanied by males during egg-laying.

To the general observer, these dragonflies will generally not be as obvious as those of other families. Most are found around streams and rivers and do not spend a great deal of time in flight. Most species are found resting on the ground, a rock, a leaf, or occasionally at the ends of twigs. They are often seen raising their abdomen when perched, a behavior known as obelisking. Carle (1986) revised the classification of this family, and I have largely followed his system.

References. Dunkle (1988), Garrison (1994b), Paulson (1983).

KEY TO THE GENERA OF CLUBTAILS (GOMPHIDAE)

1. Hind femora bearing several long spines intermingled with numerous smaller ones — **Spinyleg (*Dromogomphus*)**

1'. Hind femora without long spines intermingled with the usual numerous smaller ones — 2

2(1'). Triangles with 1 or more crossveins — 3

2'. Triangles without crossveins — 6

3(2). Body length greater than 70 mm; basal subcostal crossvein absent; subtriangles without crossveins — **Dragonhunter (*Hagenius brevistylus*)**

3'. Body length less than 70 mm; basal subcostal crossvein present; subtriangles of at least the forewings with crossveins — 4

4(3'). Supertriangles with 1 or more crossveins — 5

4'. Supertriangles without crossveins — **Sanddragon (*Progomphus*)**

5(4). Hindwing subtriangle with 2 or more cells; apicalmost spine on hind femora twice as long as preceding ones — **Leaftail (*Phyllogomphoides*)**

KEY TO THE GENERA OF CLUBTAILS (GOMPHIDAE) (*cont.*)

5'. Hindwing subtriangle usually 1-celled; apical-most spine on hind femora no longer, or only slightly so, than preceding ones — **Forceptail (*Aphylla*)**

6(2'). Hindwing with semicircular anal loop usually of 3 cells — **Snaketail (*Ophiogomphus*)**

6'. Hindwing *either* without semicircular anal loop *or* with loop of only 1 or 2 weakly bordered cells — 7

7(6'). Pterostigma of forewing short and thick, twice as long as wide, at its widest; hindwing with 5 paranal cells — **Ringtail (*Erpetogomphus*)**

7'. Pterostigma of forewing usually more elongate, 3 times as long as wide; hindwing with 4 or 5 paranal cells — 8

8(7'). Small, usually less than 40 mm in length; pterostigma less than 4 times as long as wide, more than twice as wide as the space behind its middle — **Least (*Stylogomphus albistylus*)**

8'. Larger, greater than 40 mm in length; pterostigma rarely less than 4 times as long as wide, less than twice as wide as the space behind its middle — 9

9(8'). Dark stripes on each side of pale middorsal thoracic carina faint or absent; thoracic stripes on side reduced — **Pond (*Arigomphus*)**

9'. Dark stripes on each side of middorsal carina conspicuous; at least some thoracic stripes on side fully developed — 10

10(9'). Top of frons 4 times as wide as long; long slender form — **Hanging (*Stylurus*)**

10'. Top of frons only 3 times as wide as long; form varied, but often more stocky — **Common (*Gomphus*)**

Forceptails
Genus *Aphylla* Selys
(p. 179)

Three species of this small, primarily Neotropical genus occur in North America, and all are present in the south-central United States. These are large dragonflies with relatively short legs and clear wings that superficially resemble leaftails (*Phyllogomphoides*). In forceptails the nodus is well beyond the midpoint of the forewing, and the hindwing subtriangle generally consists of a single cell.

The thorax and abdomen are red-brown marked with greenish-yellow stripes. The males have forceps-like cerci, as the common name implies.

The epiproct is reduced, appearing nearly absent, a situation not seen in any other North American dragonflies.

KEY TO THE SPECIES OF FORCEPTAILS (*APHYLLA*)

1. Sides of thorax with two broad pale-yellow-green stripes (females may occasionally have a small pale stripe between these) — **Two-striped (*williamsoni*)**

1'. Sides of thorax with three pale-yellow-green stripes — 2

2(1'). Pale lateral thoracic stripes all approximately same width; greatest width of lateral margin of segment 8 at least 0.8 mm in males and 0.5 mm in females — **Narrow-striped (*protracta*)**

2'. Pale midlateral thoracic stripe often narrower than others; greatest width of lateral margin of abdominal segment 8 less than 0.6 mm in males and 0.3 mm in females — **Broad-striped (*angustifolia*)**

Broad-striped Forceptail
Aphylla angustifolia Garrison
(p. 179; photos 31a, 31b)

Size. Total length: 62–68 mm; abdomen: 47–50 mm; hindwing: 36–40 mm.
Regional Distribution. *Biotic Provinces:* Austrorparian, Balconian, Tamaulipan, Texan. *Watersheds:* Brazos, Colorado, Guadalupe, Lower Rio Grande, Mississippi, Neches, Nueces, Red, San Antonio, Trinity.

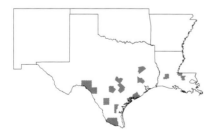

General Distribution. Coastal plain of Texas to Mississippi and south through Mexico to Belize and Guatemala.
Flight Season. May 11 (TX)–Oct. 17 (TX).

Identification. This species has three pale-greenish-yellow lateral stripes on the thorax. The middle stripe is generally narrower than the other two. The middorsal stripes are triangular. The abdomen is long, slender, and brown, marked with yellowish stripes. A wide lateral flange on segment 8 is more evident in males than in females. The mesoapical margin of the male cercus forms a ridge overlapping the appendage dorsally. The posterior margin of segment 10 is narrowly emarginate.
Similar Species. Narrow-Striped Forceptail (*A. protracta*) is similar, but all three of its pale lateral thoracic stripes are of the same width, and the flange on segment 8 is wider. Two-Striped Forceptail (*A. williamsoni*) has only two pale lateral thoracic stripes, and the pale humeral stripe is either lacking or only faintly visible. In leaftails (*Phyllogomphoides*), the abdomen is distinctly ringed, and the club is broader.
Habitat. Lakes, ponds, and pools of intermittent streams with muddy bottoms.
Discussion. Confusion between this species and the closely related Narrow-Striped Forceptail was clarified by Garrison's (1986) of Broad-striped description. He discussed the taxonomic history of

Forceptails (*Aphylla*)

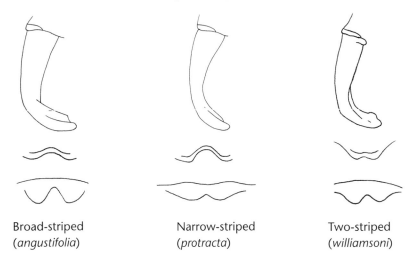

Broad-striped
(*angustifolia*)

Narrow-striped
(*protracta*)

Two-striped
(*williamsoni*)

Pond Clubtails (*Arigomphus*)

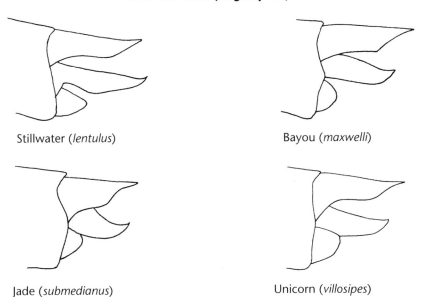

Stillwater (*lentulus*)

Bayou (*maxwelli*)

Jade (*submedianus*)

Unicorn (*villosipes*)

Fig. 27. Forceptails (*Aphylla*): from top to bottom, left cercus of male, dorsal view of posterior margin of abdominal segment 10 in male, ventral view of female vulvar lamina. Pond clubtails (*Arigomphus*): lateral views of male caudal appendages (pp. 178–183).

this species and gave characters to distinguish the two species. The key above is based, in part, on Garrison's work. The description given in Needham and Westfall (1955) for Narrow-Striped Forceptail was actually of Broad-Striped Forceptail. These two species occur together at several known localities, including Falcon Dam and the World Birding Center in the lower Rio Grande Valley of Texas. Broad-Striped Forceptail is generally much more common in the south-central United States than Narrow-Striped Forceptail. Early records of the latter should be suspect until they can be verified.

Similar Species. Narrow-Striped Forceptail is similar to both Broad-Striped and Two-Striped Forceptail (*A. williamsoni*). See *Similar Species* under the description of Broad-Striped Forceptail.
Habitat. Lakes, ponds, and pools of intermittent streams with muddy bottoms.
Discussion. This species has long been referred to in the literature as *Aphylla ambigua* (Selys). Little is known about the biology or behavior of this species, but males perch near water on sticks facing away and do not make regular patrols.

References. Dunkle (2000).

Narrow-striped Forceptail
Aphylla protracta (Hagen *in* Selys)
(p. 179; photo 31c)

Size. Total length: 64–66 mm; abdomen: 42–50 mm: hindwing: 35–49 mm.
Regional Distribution. *Biotic Provinces:* Austroriparian, Balconian, Tamaulipan, Texan. *Watersheds:* Brazos, Guadalupe, Lower Rio Grande, Nueces, Neches, Red, Sabine, San Antonio, San Jacinto, Trinity.

General Distribution. Central and southern Texas through Mexico to Costa Rica.
Flight Season. Apr. 27 (TX)–Nov. 15 (TX).
Identification. This Mexican species is less common in our region than Broad-Striped Forceptail (*A. angustifolia*). It has a greenish-yellow face and a thorax pattern similar to Broad-Striped, except that its three pale lateral thoracic stripes are all equally narrow. The abdomen is long, with pale markings, and the markings are less extensive than those in the Broad-Striped Forceptail. Segment 10 of males is deeply notched dorsally. The male cerci lack an overlapping ridge on their mesal margin dorsoapically.

Two-striped Forceptail
Aphylla williamsoni (Gloyd)
(p. 179; photo 31d)

Size. Total length: 71–76 mm; abdomen: 52–62 mm; hindwing; 37–43 mm.
Regional Distribution. *Biotic Province:* Austroriparian. *Watersheds:* Mississippi, Neches, Red, Sabine, Trinity.

General Distribution. Southeastern United States from Texas to Virginia.
Flight Season. Apr. 14 (LA)–Nov. 2 (LA).
Identification. This is the most distinctive and easily recognized of our forceptails. Two-Striped Forceptail generally has two broad, yellow lateral thoracic stripes. Occasionally, some females will have a thin pale stripe between these stripes, but the brown humeral and antehumeral stripes are always fused together, with only faint traces of pale color, if any. They have a wide lateral yellow-orange flange on abdominal segment 8.
Similar Species. The other two forceptail species in the region have three pale-yellow lateral thoracic stripes. Further differences among these species are given under the description of Broad-Striped Forceptail (*A. angustifolia*), above.

Habitat. Ponds, lakes, borrow pits, and sluggish streams.

Discussion. This is the best known of the three species in the region. It regularly forages in treetops. Males perch near the water on vegetation, occasion-ally making patrols along the shoreline. Females lay eggs late in the evening.

References. Bick and Aycock (1950), Dunkle (1989, 2000), Hornuff (1951).

Pond Clubtails
Genus *Arigomphus* Needham
(p. 179)

Pond clubtails are a distinctively pale group of seven eastern North American species, four of which occur in the south-central United States. The brown mid-dorsal and lateral thoracic stripes may be scarcely or entirely obscured on the greenish-gray thorax in this group. The hind femora in males are often clothed with many fine hairs. The wings are clear, with reduced venation, and the gaff is generally more than 1/2 as long as the inner side of the hindwing triangle. The terminal abdominal segments are only slightly enlarged laterally, in both sexes, and the cerci in the male are often forked apically. Females in this group are the only North American clubtails with a long vulvar lamina. Members of this group are unique among gomphids in regularly developing in semipermanent and artificial ponds and lakes.

KEY TO THE SPECIES OF POND CLUBTAILS (*ARIGOMPHUS*)

1. Middle of occiput with sharp elevation or spine, and usually edged by black	**Unicorn (*villosipes*)**
1'. Middle of occiput without sharp elevation or spine, usually not edged in black	2
2(1'). Lateral thoracic stripes well developed; small species, the hindwing 28-32 mm long	**Bayou (*maxwelli*)**
2'. Lateral thoracic stripes not well developed; larger species, the hindwing generally greater than 30 mm long	3
3(2'). Antehumeral and humeral dark stripes subequal in width	**Stillwater (*lentulus*)**
3'. Humeral stripe reduced to a line, much narrower than antehumeral stripe	**Jade (*submedianus*)**

Stillwater Clubtail
Arigomphus lentulus (Needham)
(p. 179; photo 31e)

Size. Total length: 48–57 mm; abdomen: 34–41 mm; hindwing: 30–37 mm.

Regional Distribution. *Biotic Provinces:* Austroriparian, Carolinian, Kansan, Navahonian, Texan. *Watersheds:* Arkansas, Brazos, Canadian, Cimarron, Colorado, Mississippi, Red, San Jacinto, St. Francis, Trinity.

General Distribution. Southeastern United States from Indiana to Texas.

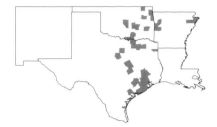

Flight Season. Apr. 25 (TX)–Jul. 17 (OK).

Identification. This is a smaller pale-greenish-gray species, with the thorax faintly marked in a fashion similar to that of Jade Clubtail (*A. submedianus*). The faint middorsal thoracic stripe is completely divided by a pale carina. The brownish humeral and antehumeral stripes are usually subequal in width, but the former is generally more developed and always separated from the latter by a thin stripe of greenish yellow. The lateral thoracic stripes are absent, visible only at their ends. The abdomen is rufous-brown, a pale middorsal stripe hardly visible. Segments 7 and 8 are black dorsally and segment 9 is pale brown. Segment 10 and the caudal appendages are yellowish.

Similar Species. Jade Clubtail is similar, but its humeral stripe is narrower than its antehumeral stripe, and segments 7–9 are uniformly dark. Bayou Clubtail (*A. maxwelli*) also has segments 7–9 uniformly dark, and it is smaller, with more well-developed middorsal and lateral thoracic stripes.

Habitat. Semipermanent and artificial ponds, lakes, and slow areas of streams with muddy bottoms.

Discussion. Although the Stillwater Clubtail has been reported from Louisiana, on the basis of sight records, the most recent study of the state revealed no confirmed records for this species in that state. It may turn up in the western portion of Louisiana, however. Little is known of its biology, but on all accounts it seems similar to other members in this group, the males typically perching on open shorelines.

References. Mauffray (1997), Needham and Westfall (1955), Vidrine et al. (1992a,b).

Bayou Clubtail
Arigomphus maxwelli (Ferguson)
(p. 179; photo 31f)

Size. Total length: 50–54 mm; abdomen: 35–40 mm; hindwing: 28–32 mm.

Regional Distribution. *Biotic Province:* Austroriparian. *Watersheds:* Bayou Bartholomew, Brazos, Mississippi, Neches, Ouachita, Red, Sabine, Trinity.

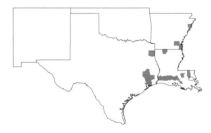

General Distribution. Southeastern United States from Florida to Texas.

Flight Season. May 2 (LA)–Jun. 20 (TX).

Identification. Bayou Clubtail is the most well-marked of the four pond clubtails in our region, with distinct brown stripes laterally on the thorax. It is generally uncommon, though it may be locally abundant where it is only occasionally taken in southeastern Texas and Louisiana. The face is yellowish and the thorax is olive-green. The middorsal thoracic stripe is entirely divided by a pale carina into a pair of widely separated brown stripes, each narrowing anteriorly. The antehumeral and humeral stripes are present and well developed; the former is slightly wider than the sinuate latter. The midlateral and third lateral stripes are generally present, but not as well developed, and visible only at their ends. The legs are pale basally, becoming black at the tibiae. The wings have a yellow costa and pale pterostigma. The abdomen is olive-green, but darker than the thorax. The middle segments have brown basal and apical rings. Segments 8 and 9 are reddish-brown, and segment 10 and the caudal appendages are yellow.

Similar Species. This species is most similar to Unicorn Clubtail (*A. villosipes*), which is larger and has a sharp elevation or spine on the occiput. Jade (*A. submedianus*) and Stillwater (*A. lentulus*) Clubtails are similar but larger, and have less well-defined dark thoracic markings.

Habitat. Ditches, bayous, and semipermanent lakes and ponds with muddy bottoms.

Discussion. This species was originally described from four males taken in Hardin County, in the Big Thicket area of southeastern Texas. It has since been taken in several other southeastern states. Males perch on open banks near the shoreline, where they face the water.

Jade Clubtail
Arigomphus submedianus (Williamson)
(p. 179; photo 32a)

Size. Total length: 51–55 mm; abdomen: 37–41 mm; hindwing: 34–36 mm.

Regional Distribution. *Biotic Provinces:* Austroriparian, Carolinian, Kansan, Tamaulipan, Texan. *Watersheds:* Arkansas, Bayou Bartholomew, Brazos, Canadian, Cimarron, Colorado, Guadalupe, Lower Rio Grande, Mississippi, Ouachita, Red, San Jacinto, St. Francis, Trinity, White.

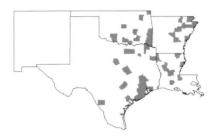

General Distribution. Great Plains from Nebraska and Wisconsin south to Texas.

Flight Season. April 4 (TX)–Aug. 5 (TX).

Identification. The face and thorax are olivaceous and pale. The faint brown middorsal stripe is entirely divided by the pale carina. There are both humeral and antehumeral brown stripes, but the former is usually less developed. The midlateral and third lateral thoracic stripes are evident only at their ends. The abdomen is greenish yellow, with brown basal and apical rings. Segments 7–9 are entirely rufous-brown. Segment 10 and the caudal appendages are pale yellow.

Similar Species. Stillwater Clubtail (*A. lentulus*) has well-developed dark humeral stripes, and abdominal segment 8 is generally darker than segments 7 or 9. The male cerci in Jade Clubtail are shorter and more compact than those in Stillwater Clubtail. Bayou Clubtail (*A. maxwelli*) is smaller and generally darker (brown, not reddish) with more well-defined thoracic stripes.

Habitat. Semipermanent and artificial ponds, lakes, and slow areas of streams with muddy bottoms.

Discussion. This species is sometimes locally abundant along the shores of ponds, small lakes, and borrow pits. It does not usually venture far from the water, resting on the ground at ponds' edge.

Unicorn Clubtail
Arigomphus villosipes (Selys)
(p. 179; photo 32b)

Size. Total length: 50–58 mm; abdomen: 37–41 mm; hindwing: 29–36 mm.

Regional Distribution. *Biotic Province:* Austroriparian. *Watershed:* Ouachita.

General Distribution. Generally Ontario southward to Arkansas.

Flight Season. Jun. 10 (AR).

Identification. This is a slightly larger, darker clubtail with a prominent sharp elevation or spine in the middle of the occiput, the spine usually edged by black. It is largely green with a relatively well-marked thorax and predominantly dark abdomen. The dark-brown middorsal thoracic stripe is divided by a pale carina, except at its upper end. The humeral and antehumeral stripes are both present and well developed, and separated by a green stripe of about equal width. The humeral stripe is abbreviated and free at its upper end. The remaining midlateral and third lateral stripes are visible only at their ends. The hind femora of the male are densely clothed with a series of hairs. The abdomen is largely black, especially distally. The pale-greenish middorsal stripe is interrupted on segments 3–7. Segments 8 and 9 are nearly all black, and 10 and the caudal appendages are yellowish brown.

Similar Species. The distribution of this species, and the prominent spine at the middle of the occiput, are unique among the pond clubtail species in the region.

Habitat. Semipermanent and artificial ponds, lakes, and slow areas of small streams with muddy bottoms.

Discussion. There is a single record of this eastern species in the region. It commonly rests on wet pond edges, rocks, and logs, where it can be extremely difficult to approach. It is apparently proficient at taking and feeding on smaller dragonflies.

References. Harp (1983).

Spinylegs
Genus *Dromogomphus* Selys

Spinylegs are a small genus of three North American species that are easily recognized by a row of 4–8 extremely long spines on the hind femora. All three species occur within the region; Southeastern Spinyleg (*D. armatus*), however, has been documented only as a larva in eastern Louisiana. The face of these dragonflies is yellowish green. The dark antehumeral and humeral stripes are variable but always prominent and often confluent. The remaining lateral thoracic stripes are variously present. The wings are clear, the costa yellow to dark brown. The yellowish-green abdomen is long and tapering, and the prominent club is more pronounced in males.

References. Westfall and Tennessen (1979).

KEY TO THE SPECIES OF SPINYLEGS (*DROMOGOMPHUS*)

1. Humeral and antehumeral dark stripes wide and confluent for at least most of their length, the pale stripe between them a hairline at most; midlateral thoracic stripe vestigial or absent ..... **Black-shouldered (*spinosus*)**

1'. Humeral and antehumeral dark stripes not wider than intervening pale stripe and not confluent throughout their length; midlateral thoracic stripe usually well marked 2

2(1'). Dark markings on abdominal segments 3–6 discontinuous dorsolaterally; caudal appendages yellow ..... **Flag-tailed Spinyleg (*spoliatus*)**

2'. Dark markings on segments 3–6 continuous dorsolaterally; caudal appendages brown or black ..... **Southeastern (*armatus*)**

Southeastern Spinyleg
Dromogomphus armatus Selys
(photo 32c)

Size. Total length: 60–68 mm; abdomen: 46–52 mm; hindwing: 36–42 mm.
Regional Distribution. *Biotic Province:* Austroriparian. *Watersheds:* Mississippi, Red.

General Distribution. Southeastern United States from Florida to Louisiana.
Flight Season. Dec. 15–Jan. 26 (LA, larval records).
Identification. This species just reaches the easternmost edge of our region, where it is uncommon. It is slightly larger than the other two species in the genus. Its face is green, with a prominent dark crossstripe. The top of the frons is yellowish and the vertex black. The thorax is green, and the brown antehumeral and humeral stripes are widely separated by a green stripe, the stripe at least as wide as each of them. The midlateral and third lateral thoracic stripes are both present and well developed. The wings have yellowish-brown costa. The abdomen is greenish, with an uninterrupted dark stripe dorsolaterally on segments 3–6, and segments 7–9 are orange-brown and expanded laterally.

Similar Species. Russet-tipped Clubtail (*Stylurus plagiatus*) has shorter legs without prominent spines on the hind femur. Flag-tailed Spinyleg (*D. spoliatus*) is paler and smaller, with pale rings on the middle abdominal segments. Black-shouldered Spinyleg (*D. spinosus*) has a broad black stripe on the shoulder.

Habitat. Small, sluggish coastal streams with relatively low turbidity, mucky bottoms, and emergent vegetation.

Discussion. Eastern Louisiana appears to be the western limit for this coastal species. Larvae have been reported from Iberia and Saint Tammany parishes. These are the only records of this species in the south-central United States.

References. Westfall and Tennessen (1979), Louton (1982).

Black-shouldered Spinyleg
Dromogomphus spinosus Selys
(photos 32d, 32e)

Size. Total length: 54–67 mm; abdomen: 42–45 mm; hindwing: 34–36 mm.

Regional Distribution. *Biotic Provinces:* Austroriparian, Balconian, Carolinian, Texan. *Watersheds:* Arkansas, Bayou Bartholomew, Brazos, Colorado, Guadalupe, Mississippi, Neches, Nueces, Ouachita, Red, Sabine, San Antonio, San Jacinto, St. Francis, Trinity, White.

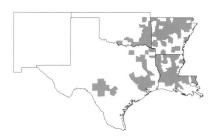

General Distribution. Throughout eastern United States and southeastern Canada.

Flight Season. May 25 (TX)–Nov. 11 (LA).

Identification. Black-shouldered Spinyleg is distinct from the other spinylegs in the region in having the antehumeral and humeral stripes fused for nearly their entire length. At most, there is a thin pale-green stripe between the two. The face is greenish and usually unmarked, but some individuals may have a dark cross-stripe. In the female, the top of the frons is green and the vertex is black, with a pair of distinct black spines. The midlateral stripe is generally reduced to a short stalk at its lower end, and the third lateral stripe is reduced to a thin line on the suture. The legs are black, and the wings have a dark costa. The abdomen is mostly black, with an interrupted greenish middorsal stripe. Segments 7–9 are dark brown to black and expanded laterally. The caudal appendages are black.

Similar Species. This is the only spinyleg and clubtail that has a broad, dark shoulder stripe formed from the fusion of the dark antehumeral and humeral stripes.

Habitat. Small to large streams and oxbows with slow to rapid flow and sandy or muddy bottoms.

Discussion. This species has a distinctive egg-laying behavior. Females fly quickly over the water, tapping the abdomen at regular intervals, depositing eggs. Pairs may stay in copula for some time, high in trees. I have seen this species abundant, perched on the ground and bridge guardrails near streams. One study documented an invasion of this species into a Tennessee lake that resulted in a dietary shift in co-existing larval baskettails (*Epitheca*).

References. Dunkle (1989), Kellicott (1899), Mahato and Johnson (1991).

Flag-tailed Spinyleg
Dromogomphus spoliatus (Hagen *in* Selys)
(photo 32f)

Size. Total length: 60–61 mm; abdomen: 43–46 mm; hindwing: 35–38 mm.

Regional Distribution. *Biotic Provinces:* Austroriparian, Balconian, Carolinian, Kansan, Navahonian, Tamaulipan, Texan. *Watersheds:* Arkansas, Bayou Bartholomew, Brazos, Canadian, Cimarron, Colorado, Guadalupe, Lower Rio Grande, Mississippi, Nueces, Ouachita, Pecos, Red, Sabine, San Antonio, San Jacinto, St. Francis, Trinity, White.

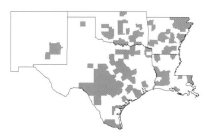

General Distribution. Southeastern United States and Canada, southwestward into Mexico.

Flight Season. May 20 (TX)–Sep. 29 (TX).

Identification. This species is lighter in color than Black-shouldered Spinyleg (*D. spinosus*), and is most similar to Southeastern Spinyleg (*D. armatus*), because both species have distinct dark antehumeral and humeral stripes. The face is yellow and never marked with a black cross-stripe. The vertex is dark brown and the females lack spines. The thorax is yellowish green. The middorsal thoracic stripe parallels the anterior collar, nearly contacting the antehumeral stripe at its lower end. The dark antehumeral and humeral stripes are separated by a pale yellowish-green stripe of equal or greater width. The midlateral stripe is more developed than in Black-shouldered Spinyleg (*D. spinosus*), but does not reach far beyond the spiracle. The third lateral stripe is hardly wider than the suture itself. The wings each have a yellow costa. The abdomen is brownish green, with an interrupted dark dorsolateral stripe on segments 3–6. Segments 7–9 are orange-brown and greatly expanded in males, more so than in other spinyleg species.

Similar Species. Southeastern Spinyleg is much less common and larger, and generally has darker markings. Southeastern Spinyleg also lacks the pale rings on the abdomen seen in Flag-tailed Spinyleg. Other clubtail species can be distinguished by having shorter legs that lack prominent spines.

Habitat. Small, clear sandy or mud-bottomed streams with a regular current.

Discussion. Mauffray (1997) reported that the six Louisiana parish records, cited in Bick (1957), were based on larval identifications and may be invalid. Those parishes (Bossier, East Feliciana, Saint Tammany, Tangipahoa, Washington, and Webster) have therefore not been incorporated in the distribution of this species. Flag-tailed Spinyleg can be common in the region. They may often be seen perching along the shore, or occasionally on vegetation. They regularly patrol over the water, where they may hover for extended periods, much as species of ringtails (*Erpetogomphus*) do.

Ringtails
Genus *Erpetogomphus* Hagen *in* Selys
(p. 190)

Ringtails are a group of 21 mainly Neotropical species, six of which occur primarily in the western portions of the region. They are medium-sized green dragonflies marked with brown or black. They have short legs and clear wings. The middorsal thoracic stripe, when present, is brown and widened anteriorly to form a triangle in our species. It is usually divided by the pale middorsal carina. The thorax is variously marked with brown lateral stripes. The abdomen is distinctly ringed, and the terminal segments are expanded to form a well-developed club. The caudal appendages and general maculation are the most useful characters for identifying males, and the vertex and occiput are distinctive in females. The cerci in males are variable, but never longer than segments 9 and 10 combined.

The genus can be divided it into five groups. All of the south-central United States species, except Blue-faced Ringtail (*E. eutainia*), belong to the *E. crotalinus* group and are characterized by two features: the upper surface of the male cerci is angulate, except in White-belted Ringtail (*E. compositus*), and the anterior hamule is divided distally.

References. Garrison (1994b).

KEY TO THE SPECIES OF RINGTAIL (*ERPETOGOMPHUS*)

MALES

1. Middorsal stripe absent; hind tibiae either entirely yellow externally or yellow with median longitudinal black line .... **Yellow-legged (*crotalinus*)**

KEY TO THE SPECIES OF RINGTAIL (*ERPETOGOMPHUS*) (*cont.*)

1'. Middorsal stripe present; hind tibiae entirely brown or black externally, lacking any yellow   2

2(1'). Dorsal surface of cerci distinctly angulate in lateral view   3

2'. Dorsal surface of cerci smoothly curved   5

3(2). Sides of thorax almost entirely green, with only a small, ill-defined dark humeral stripe   **Dashed (*heterodon*)**

3'. Sides of thorax green, but with the usual dark thoracic stripes   4

4(3'). Tips of cerci strongly acuminate; dark antehumeral stripe does not extend to the posterior edge of the thorax; base of wings with wash of amber; median area of occiput with a strongly raised tubercle   **Eastern (*designatus*)**

4'. Tips of cerci blunt, not acuminate; dark antehumeral stripe usually connected to humeral stripe; base of wings clear; median area of occiput only slightly raised   **Serpent (*lampropeltis*)**

5(2'). Cerci not uniform in color, the ventral carina at base of appendage black; this carina usually denticulate; thorax bluegreen in life; Texas Hill Country   **Blue-faced (*eutainia*)**

5'. Cerci of a uniform pale color, including the ventral carina at base of appendage; this carina smooth; thorax pale green in life   **White-belted (*compositus*)**

FEMALES

1. Middorsal stripe absent; hind tibiae either entirely yellow externally or yellow with median longitudinal black line   **Yellow-legged (*crotalinus*)**

1'. Middorsal stripe present; hind tibiae entirely brown or black externally, lacking any yellow   2

2(1'). Vulvar lamina on segment 9 followed by a distinct and prominent semicircular ridge, never with a posteriorly directed arm; Texas Hill Country   **Blue-faced (*eutainia*)**

2'. Vulvar lamina on segment 9 followed by a Y-shaped ridge   3

3(2'). Median surface of occiput with a strongly raised tubercle   **Eastern (*designatus*)**

3'. Median surface of occiput planar, or at most slightly raised   4

KEY TO THE SPECIES OF RINGTAIL (*ERPETOGOMPHUS*) (*cont.*)

4(3'). Thorax with a second complete dark lateral stripe ... 5

4'. Thorax without a second complete dark lateral stripe, or at most the stripe extending from base to just above metathoracic spiracle ... **Dashed (*heterodon*)**

5(4). Occiput in dorsal view narrow, its width less than the distance between median ocellus and occiput; base of wings with amber infusion ... **White-belted (*compositus*)**

5'. Occiput in dorsal view broad, its width almost equal to the distance between median ocellus and occiput; base of wings clear ... **Serpent (*lampropeltis*)**

White-belted Ringtail
Erpetogomphus compositus Hagen *in* Selys
(p. 190; photo 33a)

Size. Total length: 46–55 mm; abdomen: 31–39 mm; hindwing: 26–32 mm.

Regional Distribution. *Biotic Provinces:* Apachian, Balconian, Chihuahuan, Kansan, Navahonian, Tamaulipan. *Watersheds:* Colorado (NM), Nueces, Pecos, Upper Rio Grande.

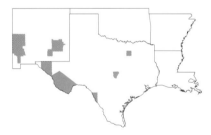

General Distribution. Western United States and northern Mexico.

Flight Season. Apr. 18 (TX)–Oct. 2 (TX).

Identification. This is one of the more distinctive of the six ringtail species in the region. Its face is nearly white, with only a few dark markings. The vertex is dark brown, often with a pale median spot extending posteriorly from the median ocellus. The thorax is pale green, more so in the front. The brown middorsal stripe is well defined and widens anteriorly to the collar. The antehumeral stripe widens early on, and is connected basally, but free at its upper end. The humeral stripe extends posteroventrally for some distance, but not as far as to connect with the midlateral stripe. The pale areas between these stripes are so pale as often to appear white. The hind femora are light pale green, the outer surfaces black and the tibiae mostly black. The wings are clear, with only a slight wash of yellow at their bases. The abdomen is pale gray, almost appearing white, for much of its length (segments 1–6), and strongly marked with black rings on the middle segments. In males, segment 7 is white dorsally on the anterior 1/2, becoming yellowish posteriorly. The remaining segments are yellowish brown, and generally darker in females. This is the only ringtail in the region where the male cerci are not strongly angulate.

Similar Species. Its white face and pale coloration make it distinctive among the other ringtail species in our region. Sulphur-tipped (*Gomphus militaris*) and Plains (*G. externus*) Clubtails have no abdominal rings.

Habitat. Desert streams, creeks, and irrigation ditches with wide sandy or rocky margins.

Discussion. This species is not usually common in the region, which reaches only to the eastern edge of its range, but it has been called one of the most conspicuous clubtails along desert streams and irrigation ditches in the southwestern United States. Although currently restricted to these desertlike streams, this species once ranged as far east as Dallas. Although often seen perched on sandbars of streams, it is readily found in shady, more protected areas in the late afternoon. Females lay eggs while hovering motionless over water and tapping their abdomens to the water surface.

References. Garrison (1994b), Hagen (1875).

Yellow-legged Ringtail
Erpetogomphus crotalinus Hagen *in* Selys
(p. 190; photo 33b)

Size. Total length: 45–49 mm; abdomen: 33–37 mm; hindwing: 29–35 mm.
Regional Distribution. *Biotic Province:* Navahonian. *Watershed:* Pecos.

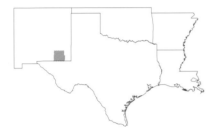

General Distribution. Arizona, New Mexico, and Mexico.
Flight Season. May–Oct.
Identification. This species is restricted to the higher elevations of central and western Mexico and the southwestern United States. Yellow-legged Ringtail is the only species in the region nearly lacking any sign of brown on a pale-yellow thorax. The abdomen is pale yellow, with interrupted dorsolateral black stripes. The male cerci are moderately to strongly angulate. Females are similar to males, but with a pair of pits on the frons anterolaterally and a medial notch in the occipital crest.
Similar Species. This species can be distinguished from other ringtails in the region by the yellow outer surfaces of the tibiae (those of the other ringtail species in the region are dark).
Habitat. Higher-elevation seasonal and permanent streams and creeks with wide sandy or rocky margins.
Discussion. This species can be common in certain habitats, but appears to be restricted primarily to central and western Mexico. Erroneous records of this species from Catron and Grant Counties in New Mexico are based on specimens of Dashed Ringtail (*E. heterodon*) (pers. comm. Rosser Garri-

son). The only valid record of this species in New Mexico is from Sitting Bull Falls in Eddy County, bordering Texas. This dragonfly has not, however, been found in Texas. These records, along with reared specimens from Arizona, constitute all known North American records. Yellow-legged Ringtail does fly alongside Eastern (*E. designatus*), Dashed, and Serpent (*E. lampropeltis*) Ringtails in Arizona and Mexico.

References. Evans (1995), Garrison (1994b), Novelo-G. and Gonzales-S. (1991).

Eastern Ringtail
Erpetogomphus designatus Hagen *in* Selys
(p. 190; photos 33c, 33d)

Size. Total length: 49–55 mm; abdomen: 34–37 mm; hindwing: 28–32 mm.
Regional Distribution. *Biotic Provinces:* Austroriparian, Balconian, Carolinian, Chihuahuan, Kansan, Navahonian, Tamaulipan, Texan. *Watersheds:* Arkansas, Bayou Bartholomew, Brazos, Canadian, Cimarron, Colorado, Guadalupe, Lower Rio Grande, Mississippi, Nueces, Ouachita, Pecos, Red, San Antonio, San Jacinto, St. Francis, Trinity, Upper Rio Grande, White.

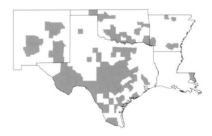

General Distribution. Widely distributed, including east of Texas, north to Montana, and southward into Mexico.
Flight Season. May 2 (TX)–Oct. 27 (TX).
Identification. This is the most widely distributed ringtail species in the region. The face of this yellowish-green species is pale green, and the vertex is dark brown. The green occiput is swollen medially in both sexes. The middorsal thoracic stripe widens anteriorly toward the collar. The brown antehumeral stripe tapers distally, and is free at both ends. The brown humeral stripe is complete and well developed, but does not extend considerably posteroventrally. The midlateral stripe is weakly developed and interrupted, often lacking at its upper end. The

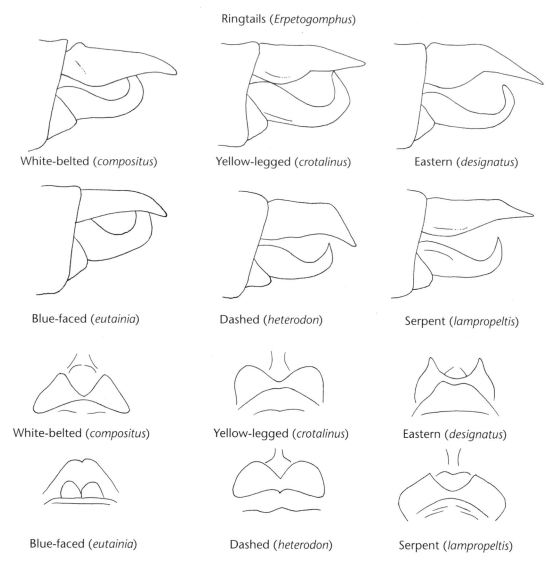

Ringtails (*Erpetogomphus*)

White-belted (*compositus*)   Yellow-legged (*crotalinus*)   Eastern (*designatus*)

Blue-faced (*eutainia*)   Dashed (*heterodon*)   Serpent (*lampropeltis*)

White-belted (*compositus*)   Yellow-legged (*crotalinus*)   Eastern (*designatus*)

Blue-faced (*eutainia*)   Dashed (*heterodon*)   Serpent (*lampropeltis*)

Fig. 28. Ringtails (*Erpetogomphus*): lateral views of male caudal appendages and ventral view of female vulvar lamina (pp. 188–192).

third lateral stripe is narrow, but complete. The femora are pale green, becoming darker distally, and the tibiae are dark brown and armed with black spines. The wings have a distinctive basal wash of yellow. The abdomen is pale green marked with reddish-brown rings and an interrupted dorsolateral stripe on segments 3–6. The remaining segments in the male are light yellowish brown and segments 7 and 8 in the female are darker dorsally.

Similar Species. This species can be distinguished from other ringtail species by a combination of characters, including the diffuse yellow or brown basally in the quadrangles of all wings, a dark brown (not black) pterostigma, and a medially swollen occiput in both sexes. The larger Flag-tailed Spinyleg (*Dromogomphus spoliatus*) has longer black legs and a distinctly larger club in both sexes.

Habitat. Clear streams and rivers of deciduous forests with moderate current.

Discussion. This is the best known and most widely distributed of the species in this group. It may be abundant, perched on the ground or on vegetation less than a meter above the ground. Emergent rocks surrounded by swift current along a stream margin are also favored perches. Females fly swiftly, occasionally hovering over the water, tapping their abdomens on the surface to lay eggs. Eastern Ringtail is partially sympatric with White-belted Ringtail (*E. compositus*). In central Texas it also flys with Blue-faced Ringtail (*E. eutainia*).

References. Ahrens (1938), Garrison (1994b), LaRivers (1938), Montgomery (1925, 1937), Williamson (1932).

Blue-faced Ringtail
Erpetogomphus eutainia Calvert
(p. 190; photos 33e, 33f)

Size. Total length: 47–51 mm; abdomen: 29–33 mm; hindwing: 23–28 mm.
Regional Distribution. *Biotic Province:* Texan. *Watershed:* Guadalupe.

General Distribution. Texas Hill Country south through Mexico to Costa Rica.
Flight Season. May 30 (TX)–Oct. 24 (TX).
Identification. This species is found in central Texas and in northern and central Mexico. It is our smallest ringtail, and has a distinctive blue face and blue dorsally on the thorax and basal abdominal segments. The occiput is slightly swollen medially. The thorax is blue dorsally and green laterally, with well-marked brown middorsal and lateral stripes, the former widening toward the collar. The antehumeral stripe is linear and free at its upper end, and the humeral stripe is long and complete. The midlateral and third lateral stripes are well-developed. The femora are dark brown to black, except for pale-

yellow areas midventrally; the tibiae are black. The wings are clear, with black pterostigmas. The abdomen is pale bluish green, with an interrupted dorsolateral brown stripe on segments 3–6, resulting in the appearance of basal and distal dark rings and white medial rings on each segment. Segment 7 is pale bluish green proximally and orange brown distally. Segments 8–10 are reddish-brown in the males and darker in females.

Similar Species. Similar species of ringtails and clubtails have pale rings basally on the abdominal segments, not in the middle of the segments as in Blue-faced Ringtail. This is our only clubtail that has blue on the face and thorax.

Habitat. Small rivulets and streams of central Texas, with swift current and cobble bottoms. Restricted at present to the San Marcos and Guadalupe Rivers.

Discussion. This species is found perching on bushes and grasses adjacent to the streams they patrol. For a clubtail, it has a very unusal flight pattern, almost appearing more like a damselfly, making only small movements from one perch to the next. It has been found only in Caldwell and Gonzales Counties in Central Texas. It has a long flight season, and can be quite common along the shorelines of both the San Marcos and Guadalupe Rivers, where it flies with Eastern Ringtail (*E. designatus*) as well as a number of other clubtail species.

Dashed Ringtail
Erpetogomphus heterodon Garrison
(p. 190; photo 34a)

Size. Total length: 50–53 mm; abdomen: 37–40 mm; hindwing: 33–36 mm.
Regional Distribution. *Biotic Provinces:* Apachian, Chihuahuan, Navahonian. *Watersheds:* Colorado (NM), Upper Rio Grande.

General Distribution. Southwestern United States and Chihuahua, Mexico.

Flight Season. Jun. 23 (TX)–Sep. 13 (TX).

Identification. This western species is most similar to Yellow-legged Ringtail (*E. crotalinus*), but is relatively well marked, with dark stripes on the thorax, and the outer surfaces of the tibiae are black, not yellow. The face and occiput are pale green, the occiput presenting only a slight medial swelling. The thorax is pale green, with a brown middorsal stripe widening anteriorly. There is an abbreviated antehumeral stripe, free at both ends. The humeral stripe is narrow, becoming more so at its lower end. The midlateral and third lateral stripes are only thinly visible at their lower and upper ends. The femora are pale green, and their outer surfaces and tibiae are black. The wings are clear, with a light-brown pterostigma. The abdomen is pale green with a dark-brown or black dorsolateral stripe interrupted anteriorly on segments 3–7. Segments 8–10 are predominantly orange-yellow, with a black dorsolateral stripe. The caudal appendages in the male are yellowish and strongly angulate.

Similar Species. The thoracic stripes on Eastern Ringtail (*E. designatus*) are more well developed, segments 1 and 2 are yellow dorsally, segment 8 lacks a dark stripe dorsolaterally, and there is a wash of yellow in the wings basally. Arizona Snaketail (*Ophiogomphus arizonicus*) lacks abdominal rings.

Habitat. Higher-altitude rivers and streams with swift current and rocky or cobble bottoms.

Discussion. I have seen this species with Serpent Ringtail (*E. lampropeltis*) in southwestern New Mexico, and it has been reported flying with Yellow-legged Ringtail in Rio Pacheco, in the Mexican state of Chihuahua. Females lay eggs, as do most other species in this group, by hovering over the water and tapping the abdomen to the water's surface. Males often perch on rocks, where they face the stream.

References. Garrison (1994b).

Serpent Ringtail
Erpetogomphus lampropeltis Kennedy
(p. 190; photo 34b)

Size. Total length: 49–53 mm; abdomen: 35–38 mm; hindwing: 27–31 mm.

Regional Distribution. *Biotic Provinces:* Apachian, Chihuahuan, Navahonian. *Watersheds:* Colorado (NM), Upper Rio Grande.

General Distribution. Southwestern United States and Mexico.

Flight Season. Aug. 4 (NM)–Oct. 8 (TX).

Identification. This southwestern species is most similar to White-belted Ringtail (*E. compositus*), but it is darker, lacks white abdominal rings, and differs in that the male has cerci distinctly angulate cerci in the male. The face has a slight blue cast and a pale brown cross-stripe. The occiput is dark brown. The thorax is darkly marked with brown. The middorsal stripe widens substantially toward the collar. The antehumeral stripe is long and parallel-sided for its entire length, but free at its upper end or only nearly joining the humeral stripe. The humeral stripe is long, well developed, and connected to the antehumeral stripe at its lower end, and runs posteroventrally a short distance. The midlateral stripe runs irregularly to nearly contact the third lateral stripe at its upper end. The legs are pale basally, and the outer surface of the femora and the entire tibiae are black. The wings are clear, with a dark pterostigma. The abdomen is pale green marked with dark brown. The brown dorsolateral stripe on segment 3 is interrupted medially and anteriorly. Segments 4–6 are only interrupted anteriorly by a pale greenish-yellow band. Segments 7–10 are largely rufous brown, becoming darker dorsally.

Similar Species. Eastern Ringtail (*E. designatus*) is similar, but the thorax is not as dark, and the antehumeral stripe is broken anteriorly. Differences with White-belted Ringtail are given above.

Habitat. Rivers and streams with swift current and rocky or cobble bottoms.

Discussion. Two subspecies are recognized. The nominate, or gray, form is restricted to four southern California counties. The green form found in our region, *E. lampropeltis natrix* Williamson and Williamson, is widely distributed throughout the southwestern United States and Mexico. This species has been taken with Yellow-legged Ringtail (*E. crotalinus*) in central Mexico, and I have taken it with Dashed Ringtail (*E. heterodon*) in western New Mexico, where males perched on exposed rocks. It has also been taken with White-belted Ringtail on occasion. Females lay eggs by flying rapidly over the water, periodically tapping the abdomen to the surface. Blue-faced Ringtail (*E. eutainia*) and this species seem to emerge later in the summer than the other ringtails.

References. Abbott and Stewart (1998), Garrison (1994b).

Common Clubtails
Genus *Gomphus* Leach
(pp. 196, 200)

In this large Holarctic complex of 38 North American species, relationships are in dispute, and the generic classification and placement are in turmoil. There are 12 largely eastern species found in the south-central United States. The face in common clubtails is usually greenish yellow, and may or may not have dark cross-stripes. The thorax is typically greenish or brown, and well marked with dark-brown or black middorsal and lateral stripes. The spines on the hind legs of females are oddly longer and more pronounced than those in males. These apparently help the females to pursue more efficient foraging. The abdomen is usually darker than the thorax, either brown or black, and striped with pale-yellowish-green longitudinal stripes. The classification of this group is controversial, but I recognize two subgroupings, presented below as subgenera. Many of the species in our region look similar, and it will often be necessary with this group to examine collected individuals critically and compare them with the key and the relevant figures before making a determination.

KEY TO THE SPECIES OF COMMON CLUBTAILS (*GOMPHUS*)

1. Vein A1 in hindwing runs straight or in an open curve from gaff to wing margin; front side of the forewing triangle is at least as long as the inner side (subgenus *Gomphus*)	2
1'. Vein A1 in hindwing is angulate or kinked at the outer end of the gaff; front side of the forewing triangle is no longer than the inner side	7
2(1). Gaff at least as long as the inner side of the triangle; short and stocky; eastern	**Banner (*apomyius*)**
2'. Gaff shorter than the inner side of the triangle	3
3(2'). Dark middorsal thoracic stripe parallel-sided	4
3'. Dark middorsal thoracic stripe widened downward, forming a triangle of brown	5
4(3). Tibia with a yellow line ending at tarsus	**Pronghorn (*graslinellus*)**
4'. Tibia with a yellow line running down onto tarsus	**Oklahoma (*oklahomensis*)**
5(3'). Yellow on outer surface of middle and hind tibiae	6

KEY TO THE SPECIES OF COMMON CLUBTAILS (*GOMPHUS*) (*cont.*)

5'. No yellow on outer surface of middle and hind tibiae **Rapids (*quadricolor*)**

6(5). Caudal appendages yellow; peduncle of penis warty externally **Sulphur-tipped (*militaris*)**

6'. Caudal appendages brown or black; peduncle smooth externally **Ashy (*lividus*)**

7(1'). Face lacking black horizontal stripe 8

7'. Face with prominent black horizontal stripe 11

8(7). Dorsum of segment 9 with area of pale yellow, which may be obscured in female 9

8'. Dorsum of segment 9 black, with little if any yellow **Cocoa (*hybridus*)**

9(8). Tibiae yellow on their external surfaces; female with slender yellow spine laterally on postocellar ridge **Plains (*externus*)**

9'. Tibiae entirely brown or black, without yellow on their external surfaces; female without spine laterally on postocellar ridge 10

10(9'). Male cerci strongly divergent; occipital crest of female straight to convex; no spine on postocellar ridge in female; South Texas and Mexico **Tamaulipan (*gonzalezi*)**

10'. Male cerci little divergent, if at all; occipital crest of female slightly biconvex; vestigial spine on postocellar ridge between lateral ocellus and compound eye in female; Arkansas and Oklahoma **Ozark (*ozarkensis*)**

11(7'). Humeral and antehumeral stripes separated their full lengths by long yellow stripe 2/3 or greater the width of the humeral **Gulf Coast (*modestus*)**

11'. Humeral and antehumeral stripes in contact near their upper ends, or separated by only a narrow yellow line **Cobra (*vastus*)**

SUBGENUS *GOMPHURUS* NEEDHAM
(p. 196)

Six of the 12 species occurring in the south-central United States fall into this group. They are of moderate to large size, and more robust than those in the subgenus *Gomphus*. The large size, combined with a smaller head and often large thorax and club, give this group a distinctive shape. The thorax is always well marked with middorsal and lateral dark stripes. The wings are clear, and the front side of the forewing triangle is no longer than the inner side. Vein A1 in the hindwing is angulate or kinked at the outer end of the gaff, before running to the wing margin. The abdomen is robust, with segments 7–9 greatly expanded laterally in both sexes. Each of these segments is progressively smaller than the former, and segment 10 is generally only 1/2 the length of segment 9. Segments 8 and 9 have pale-yellow or green markings laterally and dorsally. The caudal appendages in the male tend to be more slender than those in the subgenus *Gomphus*, and the larvae are generally broader in form.

Plains Clubtail
Gomphus externus Hagen *in* Selys
(p. 196; photos 34d, 34e)

Size. Total length: 52–60 mm; abdomen: 36–43 mm; hindwing: 30–35 mm.

Regional Distribution. *Biotic Provinces:* Austroriparian, Balconian, Carolinian, Chihuahuan, Kansan, Navahonian, Tamaulipan, Texan. *Watersheds:* Arkansas, Bayou Bartholomew, Brazos, Canadian, Cimarron, Colorado, Guadalupe, Mississippi, Neches, Nueces, Pecos, Red, Sabine, San Antonio, St. Francis, Trinity, Upper Rio Grande.

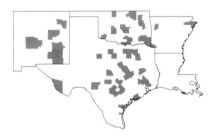

General Distribution. Utah south to Texas.

Flight Season. Mar. (TX)–July 26 (OK).

Identification. This species is widely distributed throughout most of the region, but has not as yet been reported from Louisiana. Its face is pale yellowish and devoid of dark marks. Females have a stout spine at each end of the vertex. The thorax is pale yellowish green, and the dark-brown middorsal stripe is widened slightly, to appear nearly parallel. This stripe is thinly divided by the pale middorsal carina. The antehumeral stripe is narrowly confluent with the humeral stripe at its upper end. There is a pale-yellowish stripe, no more than 1/2 their width, between these two stripes. The midlateral and third lateral stripes are well developed and only narrowly confluent at their lower ends. The paler-yellow stripe separates the two, but this may be-

come obscured in older individuals. The legs are dark brown or black, with a yellow stripe on the outer surface of the tibiae. The wings are clear, with a brown pterostigma. The abdomen is black, with interrupted pale middorsal and lateral stripes. Segments 7–9 are widely expanded, the middorsal stripe appearing as spearheads on segments 7 and 8. There is only a pale basal spot on segment 8 in females. Segment 9 shows a broad yellow stripe dorsally, and segment 10 is yellow. The male cerci are parallel, not divergent, when viewed dorsally.

Similar Species. This robust, medium-sized, yellowish species is most similar to Tamaulipan (*G. gonzalezi*) and Pronghorn (*G. graslinellus*) Clubtails. Tamaulipan Clubtail is generally larger and darker, with wider brown thoracic stripes, and the females lack the erect yellow spines at each end of the postocellar ridge. Pronghorn Clubtail is smaller, but may not always be reliably separated from Plains Clubtail. The epiproct of male Pronghorn Clubtails is barely wider than the cerci. Sulphur-tipped Clubtails (*G. militaris*) are smaller, and are yellow on the femora and more yellow on the club.

Habitat. Large muddy-bottomed rivers and streams with moderate flow.

Discussion. Females fly low over streams, tapping their abdomen to the water surface to oviposit. Adults emerge late at night and early in the morning on vegetation, logs, and artificial structures only a meter or so above the water. They generally perch on the ground, or just above, on low vegetation.

Tamaulipan Clubtail
Gomphus gonzalezi Dunkle
(p. 196; photo 34f)

Size. Total length: 47–50 mm; abdomen: 34–37 mm; hindwing: 27–31 mm.

Regional Distribution. *Biotic Province:* Tamaulipan. *Watershed:* Lower Rio Grande.

Common Clubtails (*Gomphus*)

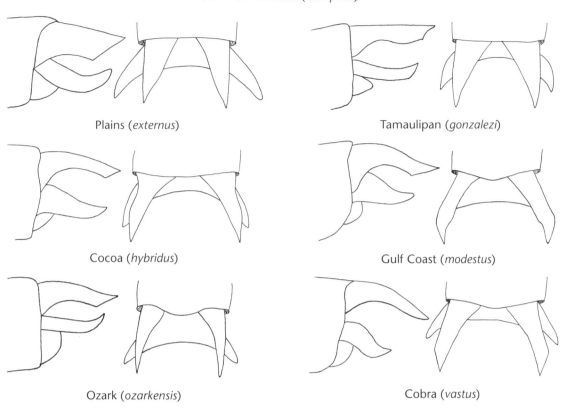

Plains (*externus*)

Tamaulipan (*gonzalezi*)

Cocoa (*hybridus*)

Gulf Coast (*modestus*)

Ozark (*ozarkensis*)

Cobra (*vastus*)

Fig. 29. Common Clubtails (*Gomphus*): lateral and dorsal views of male caudal appendages (pp. 195–198).

General Distribution. Texas and San Luis Potosí, Mexico.

Flight Season. Apr. 11 (TX)–May 8 (TX).

Identification. This recently described species is un-common and found only in far south Texas and northern Mexico. Its face is pale grayish green, with a green or brown vertex that lacks spines in fe-males. The thorax is pale green. The middorsal tho-racic stripe is parallel-sided. The antehumeral stripe is free at its upper end and widely separated from the humeral stripe by a pale area at least as wide as each dark stripe. The midlateral stripe is strongly developed. The third lateral stripe is long and con-fluent at its lower end with a fourth lateral stripe on the rear edge of the thorax. The legs are brown, and the wings are clear, with a dark-brown ptero-stigma. The abdomen is pale brown marked with grayish green on segments 1–6 and pale yellow on 7–10. The male cerci are brown and only slightly divergent dorsally, thus nearly parallel, with less of a ventral keel at 2/3 their length than in Plains Clubtail (*G. externus*).

Similar Species. This species is similar to Plains Clubtail, but smaller and paler, with narrower brown thoracic stripes and more yellow dorsally on segment 8 in males.

Habitat. Muddy canal-like channels and clear, spring-fed deep rivers.

Discussion. This is an early spring emerger. Males apparently wait for females during the middle of the day on overhanging vegetation or rock out-crops.

References. Dunkle (1992b).

Cocoa Clubtail
Gomphus hybridus Williamson
(p. 196; photos 35c, 35d)

Size. Total length: 48–53 mm; abdomen: 34–38 mm; hindwing: 27–32 mm.
Regional Distribution. *Biotic Provinces:* Austroriparian, Texan. *Watersheds:* Arkansas, Bayou Bartholomew, Mississippi, Neches, Red, Sabine.

General Distribution. Southeastern United States.
Flight Season. Mar. 30 (LA)–May 21 (AR).
Identification. Cocoa Clubtail is uncommon in the eastern portion of the region. Its face and thorax are greenish. The females have a short erect spine at the end of each postocellar ridge. A dark middorsal thoracic stripe widens toward the collar only slightly, and is divided by a pale carina. The antehumeral stripe is wide, often contacting the humeral stripe in one or more places, but becoming free at its upper end. The midlateral stripe is present but often interrupted above the spiracle, but the third lateral stripe is well developed. The legs are black, with only a pale-yellow line on the outer surface of the tibiae. The wings are clear, with a dark pterostigma. The abdomen is dark brown or black, with an interrupted middorsal stripe. The basal segments are yellowish green laterally. Segments 7–9, widened laterally, are darker than the preceding segments, often lacking any yellow dorsally on segment 9. Segment 10 either has a round yellow spot dorsally or is entirely black. The male cerci are divergent when viewed dorsally.
Similar Species. Abdominal segment 9 of Plains Clubtail (*G. externus*) has a dorsal yellow stripe, and segments 8 and 9 carry lateral yellow spots. Ozark Clubtail (*G. ozarkensis*) is larger and darker, and has a larger pale-yellow spot laterally on segment 8. Cobra Clubtail (*G. vastus*) is larger, and its dark thoracic stripes are separated.
Habitat. Large turbid rivers with moderate current and sandy bottoms.

Discussion. Males seem to prefer perching on the bank in semi-shaded areas. They are not active patrollers. Cobra Clubtail has been reported from Louisiana, but these records are most likely Cocoa Clubtail.

References. Mauffray (1997), Vidrine et al. (1992a,b).

Gulf Coast Clubtail
Gomphus modestus Needham
(p. 196; photo 36c)

Size. Total length: 55–63 mm; abdomen: 43–47 mm; hindwing: 34–38 mm.
Regional Distribution. *Biotic Provinces:* Austroriparian, Carolinian, Texan. *Watersheds:* Brazos, Mississippi, Ouachita, Red, Sabine, San Jacinto, Trinity.

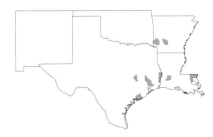

General Distribution. Southeastern United States from Mississippi to Texas.
Flight Season. May 8 (TX)–June 6 (LA).
Identification. The face is pale green, handsomely striped with black. The thorax is green, with a dark-brown middorsal stripe widening anteriorly and divided medially by a pale carina. The antehumeral stripe is wide and separated from the humeral stripe by a wide greenish-yellow stripe, but it does narrowly contact the middorsal stripe at its upper end. The midlateral stripe is well developed and broadly confluent with the humeral stripe at its lower end, appearing as a "U." The third lateral stripe is present and well developed. The wings sometimes have a hint of flavescence basally. The legs are black. The abdomen is black, except laterally on the basal segments. Segments 1–7 each have a pale hastate stripe middorsally, the stripe becoming shorter on the posterior segments. Segment 8 has a small basal yellow spot dorsally, and segments 9 and 10 are black dorsally. Segments 7–9 are yellow baslolaterally and broadly expanded in the male.
Similar Species. This is the largest of the *Gomphurus* occurring in the south-central United States. It is

closest to Cobra Clubtail (*G. vastus*) among the species in our region, but it is paler and has more complete lateral thoracic stripes. Cobra Clubtail also has a small pale spot laterally on segment 8, and wide black stripes on the face.

Habitat. Medium-sized coastal streams and rivers with muddy or sandy bottoms.

Discussion. Males often perch on the ground, facing the stream. Gulf Coast Clubtail is rarely seen in our region, a fact that may have more to do with males spending large amounts of time in trees rather than demonstrating their actual abundance. Males are only occasionally seen patrolling.

References. Westfall (1974).

Ozark Clubtail
Gomphus ozarkensis Westfall
(p. 196; photos 36f, 37a)

Size. Total length: 50–52 mm; abdomen: 35–37 mm; hindwing: 29–31 mm.

Regional Distribution. *Biotic Provinces:* Austroriparian, Carolinian, Kansan, Texan. *Watersheds:* Arkansas, Canadian, Cimarron, Ouachita, Red, White.

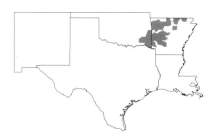

General Distribution. Southeastern United States; Ozarks.

Flight Season. Apr. 26 (TX)–June 14 (AR).

Identification. This uncommon species is largely restricted to the Interior Highlands of Arkansas and Oklahoma. Its eyes are greenish in tenerals, turning yellow in older individuals. The body is green and reddish brown in young individuals, becoming yellow and darker brown with age. The face is pale yellowish green and lacks any black stripes. The vertex is brown, and females bear a small spine at each end of the postocellar ridge. The thorax is greenish, with a dark middorsal stripe that widens only slightly anteriorly. The humeral and antehumeral stripes, and midlateral and third lateral thoracic stripes, are largely fused to form two broad dark

stripes, separated by at most a thin interrupted pale line. The wings are clear, becoming tinted with amber in older individuals. The abdomen is mostly dark brown, with an interrupted pale middorsal stripe. The basal segments are pale greenish laterally. Segments 7–9 are widely expanded, with yellow dorsally and laterally (the yellow on the dorsum of segment 9 may be obscured in older individuals). Segment 10 has a small yellow spot dorsally. The male cerci are nearly parallel, as in Plains Clubtail (*G. externus*).

Similar Species. Plains Clubtail is similar, but with broad yellow spots dorsally on segments 8–10. Pronghorn Clubtail (*G. graslinellus*) has a broad yellow spot dorsally on segment 9, and is generally smaller. The smaller Cocoa Clubtail (*G. hybridus*) is duller in color, with a faint hint of pale yellow laterally on segment 8 that is nearly isolated from the edge of the segment.

Habitat. Upland Interior Highland streams with moderate current.

Discussion. This species is considered rare because of its restricted range, little-known life history, and the threat that tourism poses to its Ozark upland stream habitats. A study in Arkansas found this species to have a short synchronous emergence period in the early spring. Maturation takes 18 and 25 days for males and females, respectively.

References. Bick (1983), Susanke and Harp (1991).

Cobra Clubtail
Gomphus vastus Walsh
(p. 196; photos 37d, 37e)

Size. Total length: 46–57 mm; abdomen: 33–42 mm; hindwing: 27–35 mm.

Regional Distribution. *Biotic Provinces:* Austroriparian, Balconian, Carolinian, Tamaulipan, Texan. *Watersheds:* Arkansas, Bayou Bartholomew, Brazos, Colorado, Guadalupe, Nueces, Ouachita, Red, Sabine, San Antonio, San Jacinto, St. Francis, Trinity, White.

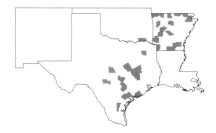

General Distribution. Eastern United States westward to Texas.

Flight Season. May 26 (TX)–Aug. 2 (TX).

Identification. This eastern species is widespread throughout the Austroriparian biotic province, except in Louisiana. The pale-green face is broadly striped with black, and the humeral and middorsal thoracic stripes are generally confluent at their upper ends, leaving the antehumeral stripe free at its upper end. The thorax is yellowish green, and the thin midlateral stripe is often interrupted above the spiracle. The third lateral stripe may be present, but is usually lacking. The abdomen is black, except for a thin, interrupted pale-yellow middorsal stripe and yellow on the basal segments. Segments 7–9 are broadly expanded, and there is only a small basal yellow spot laterally on 8 and a broad irregular lateral stripe on 9.

Similar Species. This species is closest to Gulf Coast Clubtail (*G. modestus*), but that species is generally paler, and has a narrow dark stripe on the face and distinct yellow spots laterally on segments 8 and 9.

Habitat. Medium-sized rivers or lakes with areas of alternating sand and gravel.

Discussion. Cobra Clubtails in Texas are larger and more brown in color than typical northeastern individuals. Cobra Clubtails are unusual among most clubtails, in being commonly found in both lakes and streams. They will perch on rocks along the margin of the rivers or lakes they inhabit. This species has been recorded in Louisiana on the basis of sight records, but further investigation suggests that these records are most likely attributed to Cocoa Clubtail. This species should, however, occur in Louisiana.

References. Kellicott (1899), Mauffray (1997), Vidrine et al. (1992a,b), Westfall (1974), Wilson (1909).

SUBGENUS *GOMPHUS* LEACH
(p. 200)

This group includes the remaining six species in the region. They are more slender than members of the subgenus *Gomphurus*, and males generally show much less lateral expansion of abdominal segments 7–9; females often show none. The face is pale yellowish or green and lacks black cross-stripes. The abdomen and thorax are generally duller in color, often brownish green, but with complete middorsal and lateral thoracic stripes. The wings are clear, and the front side of the forewing triangle is at least as long as the inner side. Vein A1 in the hindwing runs straight, or in an open curve, from the gaff to the wing margin, and the male caudal appendages are stockier on average than those in the subgenus *Gomphurus*.

Banner Clubtail
Gomphus apomyius Donnelly
(p. 200; photo 34c)

Size. Total length: 34–37 mm; abdomen: 26–29 mm; hindwing: 23–27 mm.

Regional Distribution. *Biotic Provinces:* Austroriparian, Carolinian. *Watersheds:* Mississippi, Ouachita, Trinity, White.

General Distribution. Southeastern United States from New Jersey to Texas.

Flight Season. Mar. 9 (TX)–May 24 (AR).

Identification. This is a small, uncommon species, originally described from the Sam Houston National Forest in Southeast Texas. Its face and occiput are yellowish, and the vertex is black. The thorax is pale yellow, with a dark, nearly parallel-sided middorsal stripe. The dark antehumeral stripe is wide and connects to the middorsal stripe at its upper end. The humeral stripe is widest at the top, where it is confluent with the antehumeral stripe (it also may come into contact with it at about 3/4 of its length). The thin pale stripe between the two stripes often disappears at its upper end to form a rounded triangular spot. The midlateral stripe is complete but thin, and the third lateral stripe is lacking or absent. The legs are black, except for the front and midfemora, which are brightly or obscurely yellow. The abdomen is dark brown, with yellow middorsally and laterally on segments 1–3. The spots are narrowed apically on segments 4–7, and the lateral expansions of segments 7–9 are yellow. The male cerci are black.

Common Clubtails (*Gomphus*)

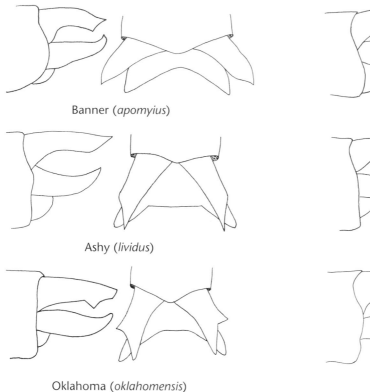

Banner (*apomyius*)

Ashy (*lividus*)

Oklahoma (*oklahomensis*)

Pronghorn (*graslinellus*)

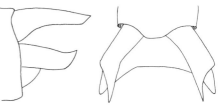

Sulphur-tipped (*militaris*)

Rapids (*quadricolor*)

Fig. 30. Common Clubtails (*Gomphus*): lateral and dorsal views of male caudal appendages (pp. 199–203).

Similar Species. This species is shorter and generally more robust than other common clubtails (*Gomphus*). Similar clubtails will have more extensive pale markings on the abdomen.

Habitat. Small, shaded streams with loose, flowing sand.

Discussion. To my knowledge, this species has not been seen in Texas, except from the type locality at Big Creek near Shepherd, in San Jacinto County, part of the Sam Houston National Forest. Little is known about the biology of this species, but it is known to forage at a height of almost 2 m along the forest edge, from morning to afternoon. It was described from mostly reared specimens, and the adult is rarely seen. Interestingly, it has not been found in southwest Louisiana.

References. Dunkle (2000).

Pronghorn Clubtail
Gomphus graslinellus (Walsh)
(p. 200; photos 35a, 35b)

Size. Total length: 44–54 mm; abdomen: 32–40 mm; hindwing: 28–35 mm.

Regional Distribution. *Biotic Provinces:* Austroriparian, Balconian, Carolinian, Kansan, Texan. *Watersheds:* Arkansas, Brazos, Canadian, Cimarron, Colorado, Guadalupe, Mississippi, Ouachita, Red, St. Francis, White.

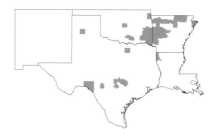

General Distribution. Widespread from southern British Columbia and Ontario southward to Texas.

Flight Season. Mar. 15 (TX)–July 13(OK).

Identification. This species has a largely northern distribution. Its face is greenish yellow, only scarcely marked with brown. The thorax is green, with well-defined dorsal and lateral stripes. The middorsal thoracic stripe is parallel-sided. The antehumeral stripe is largely fused with the humeral stripe, but diverging at its upper end and becoming free; only a thin pale line separates these two stripes medially. The remaining thoracic stripes are well separated. The legs are brown, becoming black distally, and there is a yellow stripe externally on the tibiae. The abdomen is dark brown or black, and has a pronounced club. A yellow middorsal stripe is interrupted, so as to appear as large spearheads on segments 1–7. Segment 8 has little yellow dorsally, but a wide stripe is present on segments 9 and 10. Laterally, segments 8–10 are all brightly marked with yellow. The male caudal appendages are dark brown.

Similar Species. Plains Clubtail (*G. externus*) is similar, and may not always be reliably separated. Some Plains Clubtails can be distinguished by the largely fused antehumeral and humeral stripes. The cerci in male Plains Clubtails lack teeth laterally. In Ozark Clubtail (*G. ozarkensis*) segment 9 is black dorsally or has at most a small pale-yellow spot. The paler Sulphur-tipped Clubtail (*G. militaris*) is yellow laterally on segment 7, and the pale yellow on segment 9 is only narrowly separated dorsally and laterally.

Habitat. Ponds, lakes, and slow reaches of small and large streams.

Discussion. Adults rest on the ground, on rocks, or on bushes near the water. Females lay eggs by flying low over the water and touching the abdomen to the surface every few meters. These clubtails emerge between daybreak and sunrise. Females fly by dipping and rising in a series of concave loops over the water when disturbed.

References. Needham and Hart (1901), Whitehouse (1941).

Ashy Clubtail
Gomphus lividus (Selys)
(p. 200; photos 35e, 35f)

Size. Total length: 46–57 mm; abdomen: 35–41 mm; hindwing: 28–35 mm.

Regional Distribution. *Biotic Provinces:* Austroriparian, Carolinian, Texan. *Watersheds:* Bayou Bartholo-

mew, Brazos, Mississippi, Neches, Ouachita, Red, Sabine, St. Francis, Trinity, White.

General Distribution. Eastern United States from Ontario to Florida and Texas.

Flight Season. Mar. 4 (TX)–June 6 (LA).

Identification. This early-spring species, found in the eastern part of the region, is darker than many of our other clubtails, with little color on the abdomen. The face is pale green without dark stripes, and the vertex is brown. The thorax is grayish green with a parallel-sided middorsal thoracic stripe. The antehumeral and humeral stripes are fused, the former sometimes free at its upper end; occasionally, a thin, interrupted yellowish line is visible between them. The midlateral and third lateral stripes are confluent, and a faint brown lateral stripe is often present at the rear edge of the thorax. The legs are brownish throughout, with a yellow line externally on the tibiae. The abdomen is largely black, with a thin pale-yellow, nearly continuous middorsal line on the middle segments. Segments 8 and 9 are only slightly enlarged in the male and have relatively little yellow dorsally. Segment 10 is brownish yellow, and the male caudal appendages are black.

Similar Species. This species is distinctive because of its slender form, pale color, and lack of contrasting markings.

Habitat. Sand- or mud-bottomed streams and rivers with moderate current; sheltered inlets and bays of lakes.

Discussion. This is often the most common clubtail in the early spring in the Big Thicket area of southeast Texas. Males tend to perch on the ground, or just above on vegetation. When disturbed, individuals will fly in a series of distinctive semicircles.

Sulphur-tipped Clubtail
Gomphus militaris Hagen *in* Selys
(p. 200; photos 36a, 36b)

Size. Total length: 47–54 mm; abdomen: 34–41 mm; hindwing: 28–35 mm.

Regional Distribution. *Biotic Provinces:* Austrororiparian, Balconian, Carolinian, Chihuahuan, Kansan, Navahonian, Tamaulipan, Texan. *Watersheds:* Arkansas, Bayou Bartholomew, Brazos, Canadian, Cimarron, Colorado, Guadalupe, Lower Rio Grande, Neches, Nueces, Pecos, Red, Sabine, San Antonio, San Jacinto, Trinity, Upper Rio Grande.

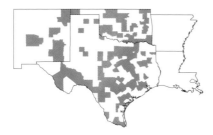

General Distribution. Southern Great Plains and Nuevo León, Mexico.

Flight Season. Apr. 1 (TX)–Aug. 14 (NM).

Identification. This is the most widespread common clubtail (*Gomphus*) in the region, although it is conspicuously absent in both Arkansas and Louisiana. Among the common clubtail species, it is marked with the most yellow and is the brightest-colored. The face is yellow, with no evidence of dark stripes. In females, there is a pair of minute spines on the vertex. The front of the thorax is distinctly more yellow than the darker sides. The brown middorsal thoracic stripe is narrow, but widens slightly toward the collar. The antehumeral stripe is well separated from the humeral stripe, and is free at its upper end. The pale stripe between them is 1/2 to 2/3 their width at its widest. The humeral stripe narrows at its lower end. The midlateral stripe is thin and always present, but generally interrupted above the spiracle. The third lateral stripe is also thin, but well developed. The legs have more yellow than those of other species. The predominantly black femora and tibiae both have yellow stripes on their outer surfaces. The black abdomen is conspicuously narrowed medially. There is a wide, pale-yellow middorsal stripe that is nearly continuous on the middle segments. Segments 7–9 are expanded laterally, less so in females, and segments 8–10 are diffusely yellow with brownish-yellow caudal appendages. Dorsal and lateral pale-yellow areas on segment 9 are narrowly separated. Males are distinctive because of the enlarged, distinctly warty peduncle.

Similar Species. This species is similar to Plains (*G.*

externus) and Pronghorn (*G. graslinellus*) Clubtail, but these species are darker, and the pale-yellow dorsal and lateral areas of segment 7 are widely separated. Oklahoma Clubtail (*G. oklahomensis*) differs similarly, and is much smaller. Flag-tailed Spinyleg (*Dromogomphus spoliatus*) is larger, with longer legs, pale basal rings on the middle abdominal segments, and distinct spines on the hind femora.

Habitat. Ponds, lakes, streams, and creeks with muddy bottoms.

Discussion. This species is found in a variety of habitats, often perching on the ground or on rocks adjoining the water. It can be equally as common away from the water, perching in open fields on vegetation roughly 1/2 a meter in height. Males patrol over water away from the bank.

Oklahoma Clubtail
Gomphus oklahomensis Pritchard
(p. 200; photos 36d, 36e)

Size. Total length: 44–49 mm; abdomen: 33–36 mm; hindwing: 24–30 mm.

Regional Distribution. *Biotic Provinces:* Austrororiparian, Carolinian, Kansan, Texan. *Watersheds:* Arkansas, Bayou Bartholomew, Brazos, Cimarron, Neches, Ouachita, Red, Sabine, San Jacinto, Trinity.

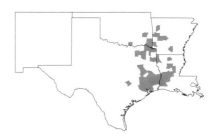

General Distribution. Restricted to Arkansas, Louisiana, Oklahoma, and Texas.

Flight Season. Mar. 23 (LA)–Aug. 31 (TX).

Identification. The face is pale green and the vertex dark brown. The thorax is grayish green, with a dark middorsal stripe that widens as it approaches the collar. The antehumeral and humeral stripes are separated by a pale stripe no more than 1/5 their width, except at their extreme lower ends, where they are confluent. The midlateral and third lateral stripes are diffusely joined together. The legs are brown, but pale on the under side of the femora and there is a yellowish-white stripe on the outer surface of the tibiae, extending down onto the tarsi. The abdomen is largely black, with a broad yellowish stripe ventrolaterally in females. The pale middorsal stripe is nearly continuous in females and tapers to a point on each segment in males. Segments 7–9 are only slightly expanded laterally in males and not at all in females. The middorsal stripe is reduced to a small basal spot on segment 8, or may appear as a disconnected thin line distally in females. There is a broad stripe on segment 9 and a conspicuous pale spot on segment 10. Segments 8 and 9 are yellow laterally, and the male caudal appendages are brown.

Similar Species. Pronghorn Clubtail (*G. graslinellus*) is larger and more brightly colored. In Ashy Clubtail (*G. lividus*) the antehumeral and humeral stripes are fused, and in Oklahoma Clubtail (*G. oklahomensis*) the two narrowly separated by a pale stripe.

Habitat. Small creeks and streams with moderate current and sand or mud bottoms.

Discussion. Females usually do not venture far from the bank, laying eggs from several centimeters above the water, by touching the abdomen to the surface and quickly rising again. Several dips are usually made before moving to a new location. Both sexes typically perch on the ground or low in vegetation.

Rapids Clubtail
Gomphus quadricolor Walsh
(p. 200; photos 37b, 37c)

Size. Total length: 42–45 mm; abdomen: 32–34 mm; hindwing: 25–27 mm.

Regional Distribution. *Biotic Province:* Carolinian. *Watershed:* White.

General Distribution. Southern Canada and eastern United States to Arkansas and Georgia.

Flight Season. May–Jul.

Identification. This small, greenish species is uncommon in the south-central United States. The middorsal thoracic stripe is brown, and widens anteriorly to form a triangle. The dark-brown humeral and antehumeral stripes are separated by a thin yellowish-green line. There are also dark midlateral and third lateral thoracic stripes. The legs are black. The abdomen is widened only slightly posteriorly. There is a pale-yellow middorsal abdominal stripe on segments 1–7, becoming narrower posteriorly to form separated triangles. Segments 8–10 are all black dorsally and yellow laterally. The caudal appendages are black.

Similar Species. This is the only common clubtail (*Gomphus*) in our region with no yellow dorsally on abdominal segments 8 or 9.

Habitat. Larger streams and rivers with cobble substrate.

Discussion. Little has been reported on this species, which barely ranges westward to northern Arkansas within the south-central United States. Males typically perch on the ground or on low vegetation, and are easily scared off.

Dragonhunter
Genus *Hagenius* Selys

The primitive-looking dragonfly of this monotypic genus is the largest North American clubtail. It is a widely distributed, common, voracious predator along many streams, but it is a relic species, its closest relatives being Palearctic and Asian in distribution. As its common name implies, it preys routinely on small and medium-sized insects, including other dragonflies. The black thorax is robust and striped with yellow. The legs are long, and the tibiae are armed with a row of strong spines that aid them in catching prey. The wings are clear and have a long, narrow pterostigma. The abdomen is long and robust, and gives hardly any indication of a club in either sex. The individual abdominal segments decrease in length from segment 6 on. The black caudal appendages are stout and shortened.

Dragonhunter
Hagenius brevistylus Selys
(photo 37f)

Size. Total length: 76–91 mm; abdomen: 53–65 mm; hindwing: 47–59 mm.

Regional Distribution. *Biotic Provinces:* Austroriparian, Balconian, Carolinian, Kansan, Texan. *Watersheds:* Arkansas, Bayou Bartholomew, Brazos, Cimarron, Colorado, Guadalupe, Mississippi, Neches, Ouachita, Red, Sabine, San Antonio, San Jacinto, St. Francis, Trinity, White.

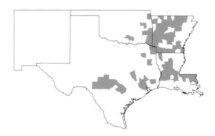

General Distribution. Widespread in eastern United States and Canada.

Flight Season. May 26 (LA)–Oct. 1 (TX).

Identification. This large, showy dragonfly is not likely to be confused with any other clubtail in the region. Its face is yellowish green and cross-striped with black. The thorax and abdomen are largely black. The thorax has three pairs of brilliant-yellow stripes, one middorsally, the others laterally. The legs are long and black. The wings are clear with black veins, but may become tinted with amber in older individuals. The abdomen is black, with a thin interrupted yellow middorsal stripe and a ventrolateral stripe running lengthwise to segment 8. The terminal segments are only slightly expanded, do not appear clubbed, and are characteristically tucked under during flight. (See also the generic discussion above.)

Similar Species. Though this species is unlikely to be confused with other clubtails, it is similar to some spiketails (*Cordulegaster*). The eyes of spiketails, however, touch at least partially on top of the head.

Habitat. Streams, rivers, and creeks with moderate to fast current and undercut banks.

Discussion. This dragonfly is often seen taking prey the size of midrange dragonflies and large swallowtail butterflies. One study indicated that the presence of this species resulted in aggregations of Ebony Jewelwing (*Calopteryx maculata*) wing clapping and stopping their feeding activities. When not foraging, this species will often perch on limbs of trees near water. The males can be remarkably bold, not easily scared off by human intruders when patrolling streams. Females usually lay eggs by regularly dropping down from a perch to water's edge and dipping their abdomen to the surface, but they will also fly back and forth in a small area periodically, tapping the abdomen to the water.

References. Erickson (1989).

Snaketails
Genus *Ophiogomphus* Selys

This is a relatively large group of 19 medium-sized North American species. Of these, there is a single Ozark species just entering the northeastern part of the region, and two species are found in the western edge of the area. In most species, the midlateral thoracic stripe is lacking. The legs are short and armed with stout spines. The wings are clear and have a distinct semicircular anal loop. The pale abdomen is usually stout and heavily marked. The cerci are short and stocky, generally only as long as segment 10. Females may have two pairs of spines on their heads, on the occiput and behind the eyes, but there is considerable variation in the presence or absence of these. Females may suffer severe head damage females during mating.

References. Carle (1981, 1992), Cook and Daigle (1985), Dunkel (1984a), Paulson (1998a).

KEY TO THE SPECIES OF SNAKETAILS (*OPHIOGOMPHUS*)

1. Dark antehumeral stripe completely absent; eastern	**Westfall's (*westfalli*)**
1' (2). Dark antehumeral stripe present, but reduced to small oval or elongated spot; western	2

KEY TO THE SPECIES OF SNAKETAILS (*OPHIOGOMPHUS*) (*cont.*)

2(1'). Male epiproct less than 1/2 the length of cerci; female often with stout occipital spine near each eye (may be lacking, however) and a strongly concave occipital crest — **Arizona (*arizonicus*)**

2'. Male epiproct longer than cerci; female lacking occipital spines and the occipital crest straight — **Pale (*severus*)**

Arizona Snaketail
Ophiogomphus arizonicus Kennedy
(photos 38a, 38b)

Size. Total length: 53–55 mm; abdomen: 40–42 mm; hindwing: 33–37 mm.
Regional Distribution. *Biotic Provinces:* Apachian, Navahonian. *Watershed:* Colorado (NM).

General Distribution. Restricted to Arizona and New Mexico.
Flight Season. Jun. 8 (NM)–Aug. 12 (NM).
Identification. This is a dull-green snaketail restricted to the desert streams in the western portions of our region. Its face is yellowish green and unmarked. Females often have a pair of spines on the strongly concave occipital ridge, but these may be lacking. The thorax has reduced markings including a faint middorsal stripe, an antehumeral stripe reduced to a spot at its posterior end, and a thin, but complete humeral stripe. The sides of the thorax are devoid of dark markings. The legs are pale basally with black tibiae. The abdomen is yellowish green with a continuous black lateral stripe down its length.
Similar Species. Pale Snaketail (*O. severus*) is most similar, but differs in range and the male having the epiproct only half as long as the cerci and spines on the occipital ridge of the female. It can be distinguished from Dashed Ringtail (*Erpetogomphus heterodon*) by the presence of pale abdominal rings

in that species. Pondhawks (*Erythemis*) have eyes touching on top of the head.
Habitat. Mountain streams with strong riffles and cobble substrate.
Discussion. Males are often seen perched on exposed rocks in the middle of the stream. They regularly patrol from these perches. Females may be scarce around the water when not laying eggs, but can be found perched on vegetation some distance from the stream.

Pale Snaketail
Ophiogomphus severus Hagen
(photos 38c, 38d)

Size. Total length: 50–52 mm; abdomen: 35–38 mm; hindwing: 29–34 mm.
Regional Distribution. *Biotic Provinces:* Chihuahuan, Coloradan, Kansan, Navahonian. *Watersheds:* Canadian, Pecos, Rio Grande.

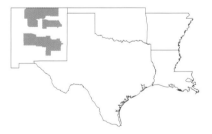

General Distribution. Western United States and Canada.
Flight Season. May 10 (NM)–Sep (NM).
Identification. This is another western species similar to Arizona Snaketail (*O. arizonicus*). See the description under that species for details. It differs in range and in the following respects: the epiproct of males is longer than the cerci; and females lack erect spines on the occipital ridge.

Similar Species. This species can be distinguished from Arizona Snaketail by the above characters. Other similar species are given under that species account.

Habitat. Primarily rivers and streams with a strong current and cobble bottom; occasionally larger lakes with mud substrate.

Discussion. Males often exhibit the characteristic body posture of this group, perching on exposed rocks with the tip of the abdomen curled downward in a snake-like position. They may forage and patrol from these rocks or from vegetation perches low to the ground. Pale Snaketail is found farther north than Arizona Snaketail.

Westfall's Snaketail
Ophiogomphus westfalli Cook & Daigle
(photos 38e, 38f)

Size. Total length: 48–50 mm; abdomen: 31–35 mm; hindwing: 29–31 mm.

Regional Distribution. *Biotic Provinces:* Austroriparian, Carolinian. *Watersheds:* Arkansas, Mississippi, Ouachita, White.

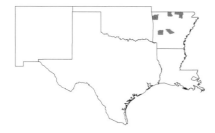

General Distribution. Ozark region of Arkansas, Kansas, and Missouri.

Flight Season. May 9 (AR)–July 18 (AR).

Identification. This species occurs in the Interior Highlands of the Arkansas Ozarks and Ouachita Mountains. It is green with conspicuously absent dark thoracic markings. Only a faint brown middorsal stripe is present. The female has prominent occipital spines and occasionally vestigial postoccipital spines. The wings have a yellow costa, and the basal 1/3 is flavescent. The pterostigma is black. The abdomen is pale yellow dorsally, with an interrupted dark-brown dorsolateral stripe on segments 1–6. Segments 7–9 are yellow dorsally, with black apically and a reddish-brown full-length dorsolateral stripe. Segment 10 is mostly yellow.

Similar Species. Pond clubtails (*Arigomphus*) that may be confused with this species have dark humeral and antehumeral stripes. Eastern Ringtail (*E. designatus*) is similar, but has a dark midfrontal thoracic stripe and distinct humeral and antehumeral stripes. There are also pale abdominal rings in that species.

Habitat. Clear forest mountain streams with strong riffles and cobble substrate.

Discussion. Rusty Snaketail, *O. rupinsulensis* (Walsh), was reported from Arkansas before this species was described. Those specimens have all been reexamined and found to be Westfall's Snaketail. Males regularly visit areas of streams with riffles, where they may perch on exposed rocks. They patrol around midday.

References. Cooke and Daigle (1985), Harp and Rickett (1977).

Leaftails
Genus *Phyllogomphoides* Belle

This primarily Neotropical group has two representatives in the southwestern United States. They are large, with long thin abdomens and short legs, reminiscent of the forceptails (*Aphylla*). They are greenish yellow in color, striped with brown, and the face is pale yellowish with brown cross-stripes. The thorax is yellowish green, with four or five brown or black lateral stripes. The legs are pale with dark outer surfaces, becoming entirely black distally. The spines of the hind femora are relatively stout, the apical spine twice as long as any other. The wings are clear, with a yellow to black costa and a dark pterostigma. The anal loop is generally well developed. The abdomen is largely brown, with wide pale-yellowish rings, and segments 8–10 are greatly expanded laterally. The male cerci are forcipate.

KEY TO THE SPECIES OF LEAFTAILS (*PHYLLOGOMPHOIDES*)

1. Third and fourth lateral stripes confluent at their lower ends — **Five-striped (*albrighti*)**

1'. Third and fourth lateral stripes not confluent, but distinctly separate, at their lower ends — **Four-striped (*stigmatus*)**

Five-striped Leaftail
Phyllogomphoides albrighti (Needham)
(photo 39a)

Size. Total length: 60–65 mm; abdomen: 45–49 mm; hindwing: 37–41 mm.
Regional Distribution. *Biotic Provinces:* Apachian, Austroriparian, Balconian, Chihuahuan, Navahonian, Tamaulipan, Texan. *Watersheds:* Brazos, Colorado, Colorado (NM), Guadalupe, Lower Rio Grande, Nueces, Red, San Antonio, Trinity, Upper Rio Grande.

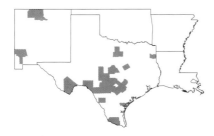

General Distribution. New Mexico, Texas, and Northern Mexico.
Flight Season. May 19 (TX)–Sep. 18 (TX).
Identification. The face is pale and the thorax yellowish green. The dark middorsal thoracic stripe is widest at its lower end, but not running onto the collar. The dark antehumeral stripe may be free at both ends or thinly confluent with the humeral stripe. The midlateral stripe is generally wider than the others and is usually confluent at least with the lower end of the humeral stripe. The third and fourth lateral stripes are broadly confluent at their lower ends. The wings are clear, with a dark pterostigma. The abdomen is blackish, becoming darker in older individuals, with broad yellowish anterior rings. Segments 8–10 are expanded considerably laterally in both sexes. The caudal appendages are yellow.
Similar Species. This species is similar to Four-striped Leaftail (*P. phyllogomphoides*) and Forceptail (*Aphylla*) species, but it can be distinguished from all these by the presence of a fourth lateral stripe at the rear margin of the thorax.
Habitat. Streams and rivers with swift current and cobble or muddy bottoms, emarginated by vegetation.
Discussion. This species was described from the San Antonio River near Berg's Mill in Bexar County, Texas, from specimens collected by Paul Albright, who stated, "They were quite wary, and were captured only by creeping up on them very slowly, and without any quick motions. They are not ordinarily very fast in flight." They inhabit streams and rivers, and are generally less abundant where they fly with Four-striped Leaftail. They can be common along streams, where they perch on vegetation, fences, or other structures a meter or so above the ground.

References. Needham (1950).

Four-striped Leaftail
Phyllogomphoides stigmatus (Say)
(p. 11; photo 39b)

Size. Total length: 65–70 mm; abdomen: 49–56 mm; hindwing: 39–44 mm.
Regional Distribution. *Biotic Provinces:* Balconian, Carolinian, Chihuahuan, Kansan, Tamaulipan, Texan. *Watersheds:* Arkansas, Brazos, Canadian, Cimarron, Colorado, Guadalupe, Nueces, Pecos, Red, San Antonio, San Jacinto, Trinity, Upper Rio Grande.

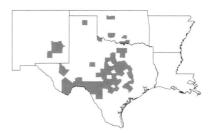

General Distribution. Southwestern United States and northeastern Mexico.

Flight Season. May 17 (TX)–Sep. 14 (TX).

Identification. The face and thorax are pale yellowish green. The thorax is like Five-striped Leaftail (*P. albrighti*), except that the lateral thoracic stripes are narrower. The third lateral stripe ends ventrally, and is never confluent with the midlateral stripe. The wings and abdomen are similar to those of Five-striped Leaftail except in females, which have a nearly cylindrical abdomen, with only a weak lateral expansion of the posterior segments.

Similar Species. This species can be distinguished from Five-striped Leaftail by the characters given above and under that species description. Forceptails (*Aphylla*) lack a 4th lateral thoracic stripe. Other similar clubtail species are much smaller.

Habitat. Ponds and slow reaches of streams with muddy bottom and heavy vegetation.

Discussion. The Four-striped Leaftail is more widely distributed than the Five-striped Leaftail, and is commonly found at livestock and artificial ponds, where males will perch high on grasses, facing the water. They are flighty, never staying perched for long, but often returning to their original perch after short feeding forays. Adults mature in open pastures of tall grass some distance from the water.

Sanddragons
Genus *Progomphus* Selys

This large New World genus is represented by four North American species, including two in the south-central United States. They are smaller clubtails, with short legs, and are generally grayish green to yellow marked with brown or black. The thorax is variously marked, and the wings have a touch of color basally. There is a basal subcostal crossvein in all wings, but no anal loop.

The anal triangle in males consists of three cells and does not extend to the hind margin of the wings, as it does in forceptails (*Aphylla*) and leaftails (*Phyllogomphoides*). The cerci are yellow, and there is only a slight swelling of abdominal segments 8–10.

References. Dunkle (1984b).

KEY TO THE SPECIES OF SANDDRAGONS (*PROGOMPHUS*)

1. Midlateral thoracic stripe absent above spiracle; thorax generally gray; larger species, greater than 56 mm in length   **Gray (*borealis*)**

1'. Midlateral thoracic stripe complete above spiracle; thorax not gray; smaller species, less than 55 mm in length   **Common (*obscurus*)**

Gray Sanddragon
Progomphus borealis McLachlan *in* Selys
(photos 39c, 39d)

Size. Total length: 56–62 mm; abdomen: 42–45 mm; hindwing: 33–36 mm.

Regional Distribution. *Biotic Provinces:* Apachian, Balconian, Chihuahuan, Kansan, Navahonian. *Watersheds:* Colorado, Pecos, Red, Upper Rio Grande.

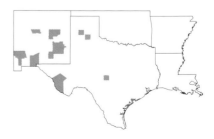

General Distribution. Southwestern United States and Mexico.

Flight Season. May 30 (TX)–Sep. 13 (TX).

Identification. This is the only western sanddragon in North America. It is easily distinguished from the eastern species by its larger size and the absence of a midlateral stripe. It is dull grayish green. The thorax is yellow in front and grayish green laterally. The antehumeral stripe is so angled at its upper end that it becomes nearly confluent with the middorsal stripe. It is confluent with the humeral stripe at its lower end, and again at about 2/3 its length, but the latter is free at its upper end. The midlateral stripe is entirely absent above the spiracle. The third lateral stripe is present and well developed. The wings are clear, with only a wash of brown at their extreme bases. The abdomen is largely black, ringed with yellow. Segments 8–10 are expanded slightly laterally, and the cerci are yellow. The male epiproct is black.

Similar Species. Sandragons are readily identifiable because of the bright-yellow cerci. The smaller Common Sanddragon (*P. obscurus*) has two dark lateral thoracic stripes and more brown basally in the wings.

Habitat. Shallow desert, sandy-bottomed streams.

Discussion. I have taken this species in Palo Duro Canyon State Park, Texas, on the Prairie Dog Town Fork of the Red River. This is apparently the easternmost locality for this species, and it was flying with Common Sanddragon. Gray Sanddragon flies along stream margins with heavy vegetation, occasionally resting on exposed sand banks in relative shade, where it can be inconspicuous. This species is often seen with its abdomen directed upward, in an obelisk position, so as to appear almost as if it is standing on its head. Females lay eggs while flying erratically, low over the water, and tapping their abdomens to the surface. Both sexes of this species have a straight, smooth, erect spine middorsally at the rear edge of the first abdominal tergite, apparently a persistent larval dorsal hook.

References. Dunkle (1984b).

Common Sanddragon
Progomphus obscurus (Rambur)
(photos 39e, 39f)

Size. Total length: 51–55 mm; abdomen: 39–43 mm; hindwing: 31–35 mm.

Regional Distribution. *Biotic Provinces:* Austroriparian, Balconian, Carolinian, Kansan, Navahonian, Tamaulipan, Texan. *Watersheds:* Arkansas, Bayou Bartholomew, Brazos, Canadian, Cimarron, Colorado, Colorado (NM), Guadalupe, Lower Rio Grande, Mississippi, Neches, Nueces, Ouachita, Pecos, Red, Sabine, San Antonio, San Jacinto, St. Francis, Trinity.

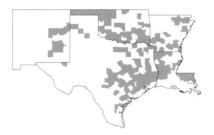

General Distribution. Eastern United States and southern Great Plains.

Flight Season. Apr. 11 (LA)–Sep. 9 (TX).

Identification. Common Sanddragon has a much wider range than Gray Sanddragon (*P. borealis*). It is slightly smaller, with a complete midlateral thoracic stripe and a strong patch of brown basally in each wing. The thorax is pale yellow, with a dark middorsal stripe widening anteriorly toward the collar and isolated from the antehumeral stripe at its upper end. The humeral and antehumeral stripes are normally confluent at both ends with the thin pale stripe between them. The midlateral and third lateral stripes are present and well developed. The legs are short, and darker than those in Gray Sanddragon. The basal brown spot in each wing extends out generally to the first antenodal crossvein.

Similar Species. Gray Sanddragon is larger, and lacks two complete lateral thoracic stripes. Eastern Ringtail (*Erpetogomphus designatus*) is similar, but the caudal appendages are both brown, not exhibiting the yellow cerci and the black epiproct of Common Sanddragon.

Habitat. Shallow streams and lakes with sandy bottoms.

Discussion. This eastern species was erroneously reported from Presidio County (the record is actually attributed to Gray Sanddragon). Common Sanddragon is the most abundant clubtail species of east Texas, found as larvae in many sandy-bottomed streams. Adults can be uncommon, taking refuge in wooded areas surrounding the streams they emerge from. It is not a particularly strong flier and is often seen obelisking. Females lay eggs by quickly flying low over the water and tapping their abdomens to the surface, and occasionally while hovering over a riffle. This is the only North American clubtail whose males guard females while they lay eggs. Males will remove sperm previously deposited by other males from females during mating, the first known instance in the family. In east Texas, at least, one study found that chironomid larvae and mayfly naiads make up the primary diet of the immatures. I have found this species flying alongside Gray Sanddragon as far west as southeastern New Mexico.

References. Brimley (1903), Byers (1925, 1930, 1939), Dunkle (1984b, 1989), Howe (1917), Montgomery (1925, 1933), Phillips (2001), Tinkham (1934), Williamson (1920, 1932), Wilson (1912).

Least Clubtail
Genus *Stylogomphus* Fraser

This is a small Asian group of clubtails with a single disjunct eastern North American species. This species has long been placed in the genus *Lanthus* (pygmy clubtails), but most recently it has been referred to *Stylogomphus*. The genus, which includes the smallest clubtail in North America, is a group of dark-green species with clear wings and an exceptionally slender abdomen in the male.

References. Carle (1980), Chao (1954).

Least Clubtail
Stylogomphus albistylus (Hagen *in* Selys)
(photo 40a)

Size. Total length: 31–36 mm; abdomen: 21–26 mm; hindwing: 20–23 mm.
Regional Distribution. *Biotic Provinces:* Austroriparian, Carolinian, Kansan, Texan. *Watersheds:* Arkansas, Cimarron, Mississippi, Ouachita, Red, St. Francis, White.

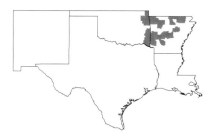

General Distribution. Eastern United States and Canada.
Flight Season. Jun. 21 (AR)–Jun. 26 (OK).

Identification. This species, the smallest clubtail in the region, is found in forest streams in eastern Oklahoma and western Arkansas. The body is dark green with distinct black crossstripes on the face. The thorax is green, the front appearing black except for a thin pale middorsal carina, two pale lateral stripes, and the anterior collar is green. The humeral and midlateral stripes are all dark and well developed. The 3rd lateral stripe is usually present, but may be lacking in some individuals. The wings are clear, with dark-brown pterostigma and sometimes a hint of flavescence basally. The abdomen is black, with narrow pale-yellowish-green basal rings. The caudal appendages are pale distally.
Similar Species. All other clubtails in our region are much larger.
Habitat. Shallow forest streams with moderate current.
Discussion. The larva of this species is much more easily found than the wary adults. Males will perch on emergent rocks and gravel bars in the sun, surrounded by swift current, and are quick to rapid flight if disturbed. Least Clubtails have been described as "wary and nervous, their flight impossible to follow." Females lay eggs in the afternoon by flying low to the water in a figure eight or similar pattern, tapping their abdomen and releasing eggs in areas of moderate riffle. One study found that Least Clubtail larvae had a significant preference for the interstitial spaces at the edge of riffles in Ozark streams.

References. Blust (1980), Kielb et al. (1996), Leonard (1940), Phillips (1996), Williamson (1932a).

Hanging Clubtails
Genus *Stylurus* Needham
(p. 212)

This group of 11 North American species is represented in our region by three eastern and one western species. The hanging clubtails are usually found perched on bushes or grasses, rarely on the ground. They are moderately sized yellowish to green dragonflies. The eyes in tenerals are gray, olive green, or brown, and become bright blue or green in adults. The top of the frons is distinctly narrow. The dark middorsal thoracic stripe is generally widened so much as to leave only a narrow isolated pale stripe on each side above the collar. The legs are relatively short, and the wings are clear, the front side of the forewing triangle distinctly longer than the inner side. The abdomen is long and slender, and males generally lack a well-developed club.

KEY TO THE SPECIES OF HANGING CLUBTAILS (*STYLURUS*)

1. Abdominal segment 9 noticeably longer than segment 8	**Arrow (*spiniceps*)**
1'. Segment 9 no longer than segment 8	2
2(1'). Lateral thoracic stripes vestigial or absent; western species	**Brimstone (*intricatus*)**
2'. Lateral thoracic stripes well developed; eastern species	3
3(2'). Sides of thorax yellowish; vertex entirely black; area between lateral thoracic stripes grayish	**Laura's (*laurae*)**
3'. Sides of thorax olivaceous; posterior half of vertex olivaceous green; area between lateral thoracic stripes olivaceous	**Russet-tipped (*plagiatus*)**

Brimstone Clubtail
Stylurus intricatus (Hagen *in* Selys)
(p. 212; photos 40b, 40c)

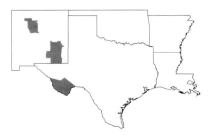

Size. Total length: 41–55 mm; abdomen: 32–43 mm; hindwing: 26–32 mm.
Regional Distribution. *Biotic Provinces:* Chihuahuan, Kansan, Navahonian. *Watersheds:* Pecos, Upper Rio Grande.
General Distribution. Western United States, Canada, and Central Great Plains, south to Texas.
Flight Season. Jun. 8 (TX)–Oct. 4 (TX).

Identification. This is the only western hanging clubtail in the region. The face is pale yellow green with little dark maculation on the head. The thorax

Hanging Clubtails (*Stylurus*)

Brimstone (*intricatus*)

Laura's (*laurae*)

Russet-tipped (*plagiatus*)

Arrow (*spiniceps*)

Fig. 31. Hanging Clubtails (*Stylurus*): lateral and dorsal views of male caudal appendages (pp. 211–214).

is pale green, with a dark-brown middorsal stripe interrupted medially by a pale carina. The narrow antehumeral stripe becomes exceedingly pale at its lower end. The remaining lateral stripes are obsolete or nearly so. The femora are yellow, except distally, where they become darker, running into black tibiae and tarsi. The abdomen is yellow green, with brown triangular lateral spots confluent dorsally. The caudal appendages are yellow, edged with black in the male.

Similar Species. Russet-tipped Clubtail (*S. plagiatus*) is larger and gray-green, rather than yellow green, and is well-marked on the thorax and abdomen. Eastern Ringtail (*Erpetogomphus designatus*) is bright green, and Flag-tailed Spinyleg (*Dromogomphus spoliatus*) has much longer legs and a rusty-brown club.

Habitat. Slow-flowing, open, desert streams and rivers.

Discussion. Kennedy (1917) described the behavior of this desert species, which ". . . spends much of its time seated on some bush or piece of driftwood, rarely alighting on the ground. However, when it is on the wing it is energetic, and the males fly rapidly back and forth in short beats, about 6 inches above the surface of the water. The females oviposit while flying in the same quick, nervous manner. . . . In copulation the male picks the female up, either

from over the water or from some bush, and after a very short nuptial flight settles for a very long period in copulation." This species was reported from Arkansas, but it was later determined that the record was based on misidentified material.

References. Harp (1983), Harp and Rickett (1977).

Laura's Clubtail
Stylurus laurae (Williamson)
(p. 212; photos 40d, 40e)

Size. Total length: 61–65 mm; abdomen: 42–48 mm; hindwing: 36–43 mm.

Regional Distribution. *Biotic Province:* Austroriparian. *Watersheds:* Mississippi, Ouachita, Red, Sabine, St. Francis, Trinity.

General Distribution. Eastern United States.

Flight Season. May 17 (AR)–Jul. 28 (LA).

Identification. This is the least common of the three eastern hanging clubtail species in the region. The head is greenish yellow, with a distinct black cross-stripe on the face. The dark middorsal thoracic stripe widens anteriorly, isolating a pale, smoothly rounded stripe that is nearly, but not quite, confluent with the pale collar. The antehumeral and humeral stripes are separated by a thin, often interrupted, pale line. The rest of the thorax is yellowish green, the remaining lateral stripes present and complete, although the midlateral stripe may be lacking at its lower end. The legs are pale basally, becoming black distally. There is a nearly complete yellow middorsal stripe on abdominal segments 1–7.

Similar Species. Russet-tipped Clubtail (*S. plagiatus*) lacks black on the terminal abdominal segments. Arrow Clubtail (*S. spiniceps*) has wider black lateral thoracic stripes, and Brimstone Clubtail (*S. intricatus*) is not found in the eastern portion of our area.

Habitat. Shallow, well-shaded, rivers and streams with cobble, sand, or mud substrate.

Discussion. Williamson (1932a) said that this species "almost invariably rested on leaves, 1–10 feet above the water. Two alighted on logs projecting from the water, but remained there only a few seconds. On leaves they were not wary and were easily approached and captured." This species was known in Texas only from larval collections taken in the Sam Houston National Forest, until a single adult female was photographed at Big Creek Scenic Area in June 1998. It is undoubtedly more common, but its secretive nature results in few adult sightings.

References. Donnelly (1978).

Russet-tipped Clubtail
Stylurus plagiatus (Selys)
(p. 212; photo 41a)

Size. Total length: 53–66 mm; abdomen: 38–50 mm; hindwing: 20–41 mm.

Regional Distribution. *Biotic Provinces:* Apachian, Austroriparian, Balconian, Carolinian, Kansan, Navahonian, Tamaulipan, Texan. *Watersheds:* Arkansas, Bayou Bartholomew, Brazos, Cimarron, Colorado, Colorado (NM), Guadalupe, Lower Rio Grande, Mississippi, Nueces, Ouachita, Pecos, Red, Sabine, San Antonio, San Jacinto, St. Francis, Trinity, White.

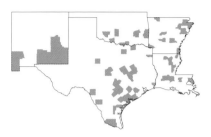

General Distribution. Eastern and southwestern United States to Nuevo León, Mexico.

Flight Season. May 25 (TX)–Nov. 7 (TX).

Identification. This is the most widespread of the three hanging clubtails in the region. It is darker than Laura's Clubtail (*S. laurae*) and has widely separated humeral and antehumeral stripes. The posterior half of the vertex is green, not black. The eyes are brilliant blue in mature adults. The brown middorsal stripe widens anteriorly, becoming confluent at both ends with the antehumeral stripe. The dark humeral and antehumeral stripes are widely separated by an olivaceous green stripe, as stated above. The midlateral stripe is thin, 1/2 the width of the humeral stripe and sinuate or interrupted at its upper end. The third lateral stripe is less developed and often interrupted in its lower half. The wings are clear, but may become amber in older females. The legs are as in Laura's Clubtail, pale basally, becoming black on the tibiae and tarsi. The pale middorsal abdominal stripe is nearly obsolete on the tawny-brown abdomen. Segments 7–10 are expanded laterally and orange-brown, as are the caudal appendages.

Similar Species. Two-striped (*Aphylla williamsoni*) and Narrow-striped (*A. angustifolia*) Forceptails are slightly larger, and the sides of their thorax is more brown than green. This is our only hanging clubtail that lacks dark markings on the terminal abdominal segments.

Habitat. Weedy rivers, streams, and lakes with moderate to little current.

Discussion. Pairs will often fly into the trees or bushes surrounding streams to mate. Females lay eggs in a fast, low, irregular flight, touching the water at intervals of several meters, or they may rest, perching between these flights. Interestingly, a female of this species has been captured in copula with Black-shouldered Spinyleg (*Dromogomphus spinosus*), and a male was found in copula with a female Southeastern Spinyleg (*D. armatus*). There is an apparent geographical variation in the eye color of this species: most individuals in the region change from gray or green to blue with age, but in Florida the eyes remain primarily green (Dunkle, pers. comm.).

References. Dunkle (1989), Williamson (1932).

Arrow Clubtail
Stylurus spiniceps (Walsh)
(p. 212; photo 40f)

Size. Total length: 58–67 mm; abdomen: 45–50 mm; hindwing: 35–38 mm.
Regional Distribution. *Biotic Provinces:* Austroriparian, Carolinian. *Watersheds:* Bayou Bartholomew, White.

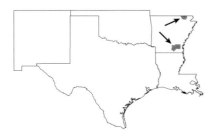

General Distribution. Southeastern Canada and eastern United States to Arkansas.
Flight Season. Jun.–Oct.

Identification. This brown clubtail has a distinctively long abdomen. Its face is dark, mostly black; the anterior half of the frons is yellowish, with short, black hairs. The thorax is brown, with two midfrontal, anteriorly diverging, yellow-green stripes and two narrow stripes perpendicular and anterior to those. Laterally, the thorax has broad dark-brown or black antehumeral and humeral stripes that are nearly fused throughout their length. The midlateral and 3rd lateral stripes are also fused or nearly so. The legs are black. The abdomen is long and largely black, with a small spot basally on each segment dorsally and a small patch of yellow laterally on 8 and 9. Segment 9 is noticeably longer than 8.
Similar Species. Laura's Clubtail (*S. laurae*) is smaller, and the dark lateral thoracic stripes are narrower. Abdominal segments 8 and 9 also show more yellow. Southeastern Spinyleg (*Dromogomphus armatus*) is larger, and has longer black legs equipped with long spines on the femora. Other similar clubtails like forceptails (*Aphylla*) have more green than brown laterally on the thorax and are generally more robust.
Habitat. Larger rivers, streams, and occasionally lakes with sandy bottoms.
Discussion. Males are most active from late afternoon to dark. They may perch on tree leaves like other hanging clubtails. Within our area, this species is known only from the extreme northeastern portions of Arkansas.

References. Dunkle (2000).

SPIKETAILS
(Family Cordulegastridae)

This family is represented in the United States by a single genus. Two of the eight North American species occur in the eastern limits of the region and two in the west. These dragonflies are similar to the petaltails in their large size and primitive appearance, but are generally black and yellow in coloration. The brilliant green or blue eyes meet at only a single point on top of the head, or may be barely separated. There is a medial cleft in the labium. The legs are short, the hind femora barely reaching the abdomen. The thorax has two or three lateral stripes. The wings are clear, and the anal triangle of the male is three or four cells. The anal loop is rather short and variable in number of cells, and the abdomen is distinctly cylindrical, lacking both dorsal and lateral carinae. The spike-like ovipositor of females is used to drive eggs into the bottom of typically small forest streams in a sewing-machine-like manner. The last two abdominal segments in females are soft and accordion-like, to aid in egg laying.

Spiketails
Genus *Cordulegaster* Leach

These large dark-brown or black dragonflies have distinctly brilliant green or blue eyes, and are vividly marked with yellow lateral thoracic stripes on the pterothorax and yellow spots on the abdomen. The clear wings lack a bracevein, and the triangles are usually two-celled. The well-developed anal loop is never foot-shaped and comprises two to ten cells. The abdomen is variably marked with yellow. Females have a spike-like ovipositor for depositing eggs in the substrate of shallow streams. The caudal appendages of the male are short, usually shorter than segment 10. The larvae can be locally abundant in narrow, gently flowing sandy- or muddy-bottomed streams of hardwood forests.

The taxonomy in this family has long been debated. Carle (1983) recognized two new genera among the North American species, and most recently Lohmann (1992) performed a cladistic analysis on the family that resulted in splitting the North American species into six genera. I do not recognize these genera, but rather follow Garrison (1997) in maintaining that all North American species should remain in a single genus. An undescribed species is known from Clark and Montgomery Counties, Arkansas, and is currently being described by Ken Tennessen.

KEY TO THE SPECIES OF SPIKETAILS (*CORDULEGASTER*)

1. Western species	2
1'. Eastern species	3
2 (1). Abdomen with elongated yellow spots	**Pacific (*dorsalis*)**
2'. Abdomen with bands or rings	**Apache (*diadema*)**
3(1'). Dorsal abdominal spots spear-shaped on middle abdominal segments, 2–7	**Arrowhead (*obliqua*)**
3'. Dorsal abdominal spots neither spear-shaped nor broadly interrupted medially	**Twin-spotted (*maculata*)**

Apache Spiketail
Cordulegaster diadema Selys
(photo 41b)

Measurements. Total length: 75–88 mm; abdomen: 57–66 mm; hindwing: 45–52 mm.
Regional Distribution. *Biotic Province:* Apachian. *Watershed:* Colorado (NM).

General Distribution. Arizona and western New Mexico south to Costa Rica.
Temporal Distribution. Aug. 5 (NM)–Nov.
Identification. This is one of two western spiketails occurring in our region. With its pale-yellow bands or rings across the middle of each abdominal segment, it is quite distinctive. The eyes are greenish in younger individuals, turning brilliant aquamarine blue. The thorax is brown, with a pair of broad yellow stripes laterally; a small pale stripe or spot may be visible between these. The dark-brown abdomen has pale medial bands on each segment (these lacking on the first segment in females).
Similar Species. The only other western spiketail, Pacific Spiketail (*C. dorsalis*), has elongated yellow spots on the abdomen. Bronzed River Cruiser (*Macromia annulata*) has a single yellow lateral thoracic stripe.
Habitat. Small, clear, mountain streams with areas of both flow and pools.
Discussion. Males are instantly noticed as they patrol along streams when the sun is out. When not flying they perch 2–3 m high on tree branches or shrubs, with their abdomen hanging vertically. This species is known only from Grant County, New Mexico, within our region.

Pacific Spiketail
Cordulegaster dorsalis Hagen in Selys
(photo 41c)

Size. Total length: 71–85 mm; abdomen: 55–64 mm; hindwing: 43–48 mm.

Regional Distribution. *Biotic Province:* Navahonian. *Watershed:* Upper Rio Grande.

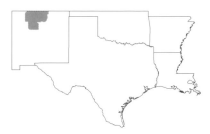

General Distribution. Western United States.
Temporal Distribution. Jul. (NM)–Aug. (NM).
Identification. This is the second of the two western spiketails occurring in our region. It is our only western species with elongated yellow median spots on each abdominal segment. The eyes are brownish initially, but become brilliant blue in mature individuals. The thorax is brown, with a pair of broad yellow stripes laterally. The ovipositor extends beyond the abdomen by at least the length of segment 10.
Similar Species. Apache Spiketail (*C. diadema*) has bands rather than spots on the abdomen. Bronzed River Cruiser (*Macromia annulata*) has a single yellow lateral thoracic stripe.
Habitat. Shady mountain foothill streams and possibly open desert springs.
Discussion. These strong fliers patrol 1–2 m off the ground in strong steady bursts, allowing at times for only quick glimpses. They are amazingly adept at darting in, out, and around bushes and thick riparian vegetation. Within our region, this species is known only from Rio Arriba, in Sandoval and Taos Counties in New Mexico. A paler form of this species, known from the Great Basin, may be a distinct species.

Twin-spotted Spiketail
Cordulegaster maculata Selys
(photo 41d)

Size. Total length: 65–76 mm; abdomen: 47–58 mm; hindwing: 38–50 mm.
Regional Distribution. *Biotic Province:* Austroriparian. *Watersheds:* Arkansas, Mississippi, Neches, Ouachita, Red, Sabine, St. Francis, Trinity.
General Distribution. Eastern United States and Canada.
Temporal Distribution. Mar. 2 (LA)–Apr. 18 (TX).

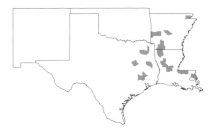

ron, Mississippi, Neches, Ouachita, Red, Sabine, St. Francis, Trinity, White.

Identification. This species is confined to the eastern Austroriparian biotic province within the region. Its eyes are brilliant blue or green in life. The pterothorax is dark brown, becoming paler below, where there are numerous gray hairs around the legs. There are two broad, pale stripes laterally on the thorax. The wings are clear, but sometimes become smoky. The abdomen is dark brown with a paired row of pale-yellow spots.

Similar Species. Arrowhead Spiketail (*C. obliqua*) has an arrowhead pattern dorsally on the abdomen. Fawn Darner (*Boyeria vinosa*) can be distinguished in the field by two lateral thoracic spots, rather than stripes. Springtime Darner (*Basiaeschna janata*) has basal brown spots in both wings and blue on the abdomen. Stream cruisers (*Didymops*) and river cruisers (*Macromia*) have a single lateral thoracic stripe.

Habitat. Small, rapidly flowing spring-fed forest streams and seepages with sandy or muck bottoms.

Discussion. This early spring species is widely distributed throughout the eastern half of the United States and Canada. Little has been published about its biology, but males are known to patrol long stretches of streams, sometimes just a few centimeters over the water. One study found that mayflies make up the majority (69%) of the larval diet. Mating lasts an average of 50 minutes.

References. Dunkle (1989), Johnson (1982).

Arrowhead Spiketail
Cordulegaster obliqua (Say)
(photo 41e)

Measurements. Total length: 72–88 mm; abdomen: 48–72 mm; hindwing: 41–60 mm.
Regional Distribution. *Biotic Provinces:* Austroriparian, Carolinian, Kansan, Texan. *Watersheds:* Arkansas, Bayou Bartholomew, Brazos, Canadian, Cimar-

General Distribution. Eastern United States and Canada.
Temporal Distribution. Apr. 17 (AR)–Jun. 7 (AR).
Identification. Like Twin-spotted Spiketail (*C. maculata*), this species is restricted to the eastern limits of the region. The eyes are aquamarine blue. There are two broad pale-yellow stripes laterally on the thorax. The wings are clear and rarely become smoky. The abdomen is dark red-brown, with yellow spear-shaped marks on segments 2–7.

Two forms, both of which occur in the region, are recognized. The northern form, *C. obliqua obliqua* (Say), is found in Texas and is generally smaller, at 72–80 mm. The forewing triangle usually consists of two cells, and the anal triangle of the male is three-celled. The southern form (*C. obliqua fasciata* Rambur), found in Arkansas and Louisiana, is generally larger, at 82–88 mm. The forewing triangle is generally three-celled, and the anal triangle in males consists of four-cells.

Similar Species. In addition to a later emergence period, Arrowhead Spiketail can be separated from Twin-spotted Spiketail by a row of spear-shaped pale-yellow dorsal markings on the abdomen. Arrowhead Spiketail is also darker than Twin-spotted Spiketail, often appearing nearly black. See also the discussion of *Similar species* for Twin-Spotted Spiketail.

Habitat. Small, rapidly flowing spring-fed forest streams and seepages with sandy or muck bottoms.

Discussion. This species emerges later and has a longer flight period than Twin-spotted Spiketail, but is less likely to be encountered. It is uncommon and elusive throughout its range. Little has been published about its behavior, but as far as is known, it is similar to Twin-Spotted Spiketail. It flies up and over treetops when disturbed.

References. Dunkle (2000), Needham (1905), Walker (1958).

CRUISERS & EMERALDS
(Family Corduliidae)

This family includes two distinct groups regarded as subfamilies here and described in greater detail below. Collectively, these dragonflies are medium-sized to large and often metallic in coloration. The compound eyes meet broadly on top of the head and generally bear a low tubercle on their rear margin. The thorax is often marked with pale stripes. The legs vary in length but are generally long, and may or may not have pronounced keels on the tibiae. The wings are generally clear or marked with brown, but with no bracevein under the pterostigma. The triangles in the forewing are twice, or more, as far from the arculus as those in the hindwing, and are generally elongate transversely. The anal loop can take one of two forms: foot-shaped, with little or no development of the toe; or semicircular. Males have an auricle on each side of abdominal segment 2.

References. May (1995c).

KEY TO THE GENERA OF CRUISERS & EMERALDS (CORDULIIDAE)

1. Single complete broad yellow stripe laterally on thorax	Subfamily Macromiinae 2
1'. Thorax without stripe, or if present, then interrupted	Subfamily Corduliinae 3
2(1). Body light brown; nodus of forewing about midway between base and apex of wing	**Stream Cruiser (*Didymops transversa*)**
2'. Body dark with metallic luster; nodus of forewing distinctly beyond middle of wing	**River Cruisers (*Macromia*)**
3(1'). Thorax with metallic-coppery reflections	**Striped Emeralds (*Somatochlora*)**
3'. Thorax without metallic-oppery reflections	4
4(3'). Thorax with a small yellow spot or interrupted stripe near spiracle	5
4'. Thorax without small yellow spot or stripe near spiracle	6
5(4). Yellow spot at level of thoracic spiracle	**Shadowdragons (*Neurocordulia*)**
5'. Yellow stripe, generally interrupted by spiracle, laterally on thorax	**Baskettails (*Epitheca*)**
6(4'). Wings with large brown spots at nodus and wingtips, and a large basal spot in hindwing	**Baskettails (*Epitheca*)**

6'. Wings with dark spot basally and sometimes small spots along antenodal crossveins, but never with large spots at nodus and wingtips   **Sundragons (*Helocordulia*)**

CRUISERS
Subfamily Macromiinae

This subfamily comprises a small group of fairly large individuals that fly swiftly along the margins of rivers and streams. They are primarily brown or black, and prominently marked with a yellow band encircling the thorax, like a belt. The otherwise dark thorax often appears to have a metallic luster. The eyes meet on top of the head for some distance, clearly separating the occiput and vertex. The legs are long and heavily spined. Males have a keel on the tibiae. The wings are narrow and venationally distinct. The hindwing triangle is remote from the arculus. The anal loop is as broad as it is long, appearing semicircular, and lacking a midrib. There is generally no radial or medial planate. The dark-brown to black abdomen lacks a ventrolateral carina. Females lack an ovipositor and release eggs by dragging their abdomens through the water while in flight.

References. Gloyd (1959).

Brown Cruisers
Genus *Didymops* Rambur

The two eastern North American species that make up this group are light or dull brown, lack any metallic luster, and are marked with yellow. A single widespread species occurs in our region; the other is restricted to Florida. The eyes become brilliant green at maturity and contact each other on top of the head, but for a relatively short distance, compared to river cruisers (*Macromia*). The front of the thorax lacks any stripes, but the side has the normal oblique lateral stripe. The wings are clear, and the costal margin is paler than the rest of the veins. The abdomen has pale rings on the middle segments.

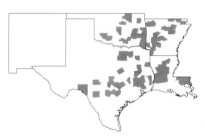

General Distribution. Eastern United States and Canada.

Flight Season. Mar. 12 (LA)–May 26 (AR).

Identification. This widely distributed species is distinctive among the cruiser species in the region. Its body is dull brown. The thorax is covered with whitish hairs and a single pale-yellowish stripe laterally. The wings each have a small brown spot basally. The abdomen has pale spots on segments 1–8 and is slightly clubbed. The caudal appendages are yellowish.

Similar Species. Bronzed River Cruiser (*Macromia annulata*) is dull, has yellow stripes on the front of the thorax, and lacks wings spots. Fawn Darner (*Boyeria vinosa*) has spots, not stripes, laterally on the thorax. The abdomen of the male is clubbed, which may re-

Stream Cruiser
Didymops transversa (Say)
(photo 41f)

Size. Total length: 56–60 mm; abdomen: 34–43 mm; hindwing: 34–48 mm.

Regional Distribution. *Biotic Provinces:* Austroriparian, Balconian, Carolinian, Kansan, Navahonian, Texan. *Watersheds:* Arkansas, Bayou Bartholomew, Brazos, Canadian, Cimarron, Colorado, Guadalupe, Mississippi, Neches, Nueces, Ouachita, Red, Sabine, San Antonio, San Jacinto, St. Francis, Trinity, White.

sult in confusion with some clubtails, but the eyes of Stream Cruiser touch on top of the head.

Habitat. Medium-sized to large streams and rivers.

Discussion. This is an early spring species that commonly perches obliquely on grasses and bushes. Younger individuals are often seen flying low, some distance from water, in open fields and along paths. Returning males patrol for long distances along the shoreline. Females may lay eggs over a long distance by tapping their abdomens to the water surface intermittently, or may choose to confine their egg-laying to a smaller area.

River Cruisers
Genus *Macromia* Rambur
(p. 13)

This large genus is represented in North America by seven species, all but two of which occur in our region. This group represents the larger members of the family. They are dark, generally have a metallic luster, and are marked with yellow. The eyes are brilliant green, and the vertex is distinctly bilobed and variously marked with yellow. The thorax and abdomen are dark brown or black. The former may or may not be marked dorsally with yellow, but always has a pale lateral stripe. The wings are clear, but may be heavily tinted in younger individuals. The legs are black, and the abdomen is long and robust, sometimes appearing club-shaped. Members of this genus are often found hanging inconspicuously, high in the branches of trees.

References. May (1997).

KEY TO THE SPECIES OF RIVER CRUISERS (*MACROMIA*)

1. Vertex dark, not marked with yellow	2
1'. Vertex pale, marked with yellow	4
2(1). Yellow antehumeral stripe well developed basally; yellow, when present, on abdominal segment 7 interrupted laterally	3
2'. Yellow antehumeral stripe, if present, vestigial; yellow on segment 7 encircling that segment	**Allegheny (*alleghaniensis*)**
3(2). Yellow ring on abdominal segment 2 widely interrupted middorsally; wings often flavescent in females	**Royal (*taeniolata*)**
3'. Yellow ring on segment 2 not interrupted, or only narrowly, middorsally; wings generally not flavescent in females	**Illinois (*illinoiensis*)**
4(1'). Vertex all pale; yellow spots on abdominal segments 3–6 entire or not completely divided by middorsal stripe; western species	**Bronzed (*annulata*)**
4'. Vertex with pale color restricted to marks on double cone-like summit; yellow spots on segments 3–6 narrowly interrupted by thin black middorsal stripe; eastern species	**Gilded (*pacifica*)**

Allegheny River Cruiser
Macromia alleghaniensis Williamson
(photo 42a)

Size. Total length: 65–72 mm; abdomen: 51–56 mm; hindwing: 45–50 mm.
Regional Distribution. *Biotic Province:* Austroriparian. *Watershed:* Ouachita.

General Distribution. Eastern United States.
Regional Flight Season. Jun. 19 (AR).
Identification. This species, with its entirely dark vertex, short tibial keels, and vestigial pale antehumeral stripe, is distinct in the region. It has a yellowish-brown face that becomes black dorsally on the frons, and there are two small yellow spots on the frons. The wings are clear, and the costa and pterostigma are dark. The keel on the middle tibia is 1/7 to 1/5 as long as the tibia itself. A yellow ring on abdominal segment 2 is interrupted middorsally. Segments 3–6 each bear a pair of spots dorsally. Segments 7 and 8 have a pale basal middorsal spot, which may be absent in females. Yellow on segment 7 encircles the abdomen, so that segments 7–9 are each yellow on the inferior basal margin.
Similar Species. Illinois River Cruiser (*M. illinoiensis*) is similar, but has partial frontal thoracic stripes and a complete band around segment 2. Royal River Cruiser (*M. taeniolata*) is noticeably larger, and generally bears a pair of yellow spots on segment 7.
Habitat. Cool upland streams.
Discussion. This species has been reported from southwestern Arkansas. Nothing has been published about the behavior of this species, but it is presumably similar to that of Illinois River Cruiser. In the original description, Williamson wrote, ". . . I recall nothing striking in its habits of life."

References. Harp and Harp (1996), Williamson (1909).

Bronzed River Cruiser
Macromia annulata Hagen
(photo 42b)

Size. Total length: 68–75 mm; abdomen: 52–57 mm; hindwing: 45–52 mm.
Regional Distribution. *Biotic Provinces:* Austroriparian, Balconian, Chihuahuan, Kansan, Navahonian, Tamaulipan, Texan. *Watersheds:* Brazos, Colorado, Guadalupe, Lower Rio Grande, Nueces, Pecos, San Antonio, Upper Rio Grande.

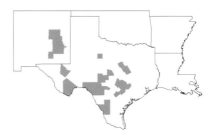

General Distribution. New Mexico, Texas, and northern Mexico.
Flight Season. Apr. 2 (TX)–Oct. 11 (TX).
Identification. This southwestern species is the only one with a whitish-yellow vertex and uninterrupted middorsal abdominal spots. The top of the frons is broadly marked with yellow and a pale narrow brown line medially on the sulcus. The thorax is dull to dark brown, with long well-developed antehumeral stripes. These stripes miss touching the alar crest by a distance equal to their width. The wings are clear, with a yellow costa and a hint of brown basally in females. The legs are brown, becoming black distally. The first abdominal segment bears a lateral yellow streak, unique among North American species. There is a wide, uninterrupted yellow ring on segment 2, and an uninterrupted row of pale dorsal spots middorsally on segments 3–8.
Similar Species. Stream Cruiser (*Didymops transversa*) is brown, emerges earlier in the spring, and lacks pale stripes on the front of the thorax. Gilded River Cruiser (*M. pacifica*) has brilliant green eyes, and the yellow color is much brighter.
Habitat. Large rivers and streams.
Discussion. This is the only western river cruiser occurring in the area. Nothing has been published about its biology, but it seems to be typical for the genus. Individuals, particularly males, fly fast over the water, some distance from shore. These cruisers can be abundant on some pristine western streams.

Illinois River Cruiser
Macromia illinoiensis Walsh
(photo 42c)

Size. Total length: 65–79 mm; abdomen: 47–56 mm; hindwing: 40–53 mm.

Regional Distribution. *Biotic Provinces:* Austroriparian, Balconian, Carolinian, Kansan, Tamaulipan, Texan. *Watersheds:* Arkansas, Bayou Bartholomew, Brazos, Canadian, Cimarron, Colorado, Guadalupe, Mississippi, Neches, Nueces, Ouachita, Red, Sabine, San Antonio, San Jacinto, St. Francis, Trinity, White.

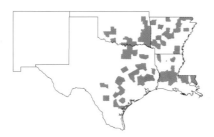

General Distribution. Eastern United States and Canada.

Flight Season. Apr. 27 (LA)–Sep. 28 (TX).

Identification. This is perhaps the most widespread and frequently encountered river cruiser in our area. The top of frons is dark, with two dorsal and two lateral pale spots. The vertex is entirely black. The antehumeral stripes on the dark thorax extend up 1/3 to 1/2 the height of the front of thorax. The wings have at most a hint of brown basally in the females. The legs are black and long. There is an uninterrupted yellow ring on abdominal segment 2. Segments 3–6 each have a dorsal spot that may be thinly interrupted. Segments 7 and 8 have a pale ring that is interrupted ventrolaterally, broadly so on 7.

Similar Species. Allegheny River Cruiser (*M. alleghaniensis*) is similar, but the yellow on abdominal segment 7 does not completely encircle that segment and there are no pale frontal thoracic stripes.

Habitat. Moderately large to large rivers and streams.

Discussion. This species is often seen flying along roads and forest paths. It is most active in the morning, when males patrol low above the water. There are two recognized subspecies. The nominate northern form is found north and west of the Appalachians, outside our region. The southern sub-species, *M. illinoiensis georgina* (Selys), which does occur in our region is distributed along the coastal plains and in the Mississippi Valley. Females lay eggs flying low over stream riffles, tapping their abdomens to the surface every few meters. Mating pairs may be found high in trees or lower to the ground in bushes or shrubs. River cruisers of this species are occasionally seen flying high in mixed feeding swarms.

References. Donnelly and Tennessen (1994).

Gilded River Cruiser
Macromia pacifica Hagen
(photo 42d)

Size. Total length: 62–70 mm; abdomen: 45–51 mm; hindwing: 40–46 mm.

Regional Distribution. *Biotic Provinces:* Austroriparian, Balconian, Carolinian, Kansan, Tamaulipan, Texan. *Watersheds:* Arkansas, Brazos, Canadian, Cimarron, Colorado, Guadalupe, Nueces, Ouachita, Red.

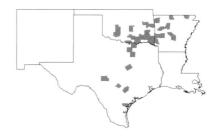

General Distribution. Eastern United States, Ohio to Texas.

Flight Season. May (TX)–Sep. 26 (TX).

Identification. Though uncommon, this is the brightest and most vividly marked river cruiser in our area. It has a yellow vertex and pale, narrowly interrupted abdominal spots. The frons is yellow on top, except for a broad medial dark stripe on the sulcus. The thorax is dark brown, with well-developed full-length antehumeral stripes. The wings are clear, though sometimes tinged with yellow, and have distinct basal brown spots. The legs are dark brown or black. The abdomen is also dark brown or black, strikingly marked with yellow on segments 2–9. These spots are often thinly interrupted middorsally on segments 3–6.

Similar Species. All other river cruisers in the area are more dull in coloration. Bronzed River Cruiser is much duller, lacks brilliant green eyes, and has medially connected spots on segments 2–6.

Habitat. Moderate-sized streams and rivers with pools and areas of slow flow.

Discussion. Individuals in the southern part of this species' range are on average smaller than more northern individuals. Williamson (1909) stated "The flight of *pacifica* is generally less swift than that of *taeniolata* and . . . [it] ranges less widely, patrolling possibly only one pool, while others return to the same point only after longer intervals." This species is not common in the south-central United States, but does apparently hybridize with Royal River Cruiser. These hybrids have been referred to as *M. wabashensis* Williamson and are known from Texas and Oklahoma.

References. Williams (1979b).

Royal River Cruiser
Macromia taeniolata Rambur
(photo 42e)

Size. Total length: 77–92 mm; abdomen: 53–68 mm; hindwing: 46–62 mm.

Regional Distribution. *Biotic Provinces:* Austroriparian, Carolinian, Texan. *Watersheds:* Arkansas, Bayou Bartholomew, Brazos, Mississippi, Neches, Nueces, Ouachita, Red, Sabine, San Jacinto, St. Francis, Trinity, White.

General Distribution. Eastern United States.

Flight Season. May 7 (LA)–Sep. (TX).

Identification. This is the largest and most robust of our river cruisers. It is easily recognized by its size and the metallic-blue color of its face, the top of the

frons, and the vertex. The thorax is dark brown, with pale-yellow antehumeral stripes extending 1/2 its length. The wings are clear, but deeply tinted amber in younger individuals. The legs are black. The abdomen is black, and not as vividly marked as in other species in the region. The yellow ring on segment 2 is narrowly interrupted dorsally. Pale spots on segments 3–6 are small and often obscured. Segment 7 generally has a pair of pale middorsal spots, and there is a small basal dorsolateral spot on segment 8. The remaining segments and caudal appendages are black.

Similar Species. Our other river cruisers are smaller and have a single pale spot dorsally on segment 7. Spiketails (*Cordulegaster*) have two lateral thoracic stripes.

Habitat. Rivers, streams, and lakes.

Discussion. Young individuals of this species, as with others, are often seen hanging obliquely from branches high in trees. Males patrol in a fashion similar to that of other species, but generally fly higher. Females lay eggs for about 2 minutes at a time, mostly in the afternoon.

References. Dunkle (1989).

EMERALDS
Subfamily Corduliinae

This subfamily consists mostly of slender, medium-sized dragonflies generally found breeding in streams or lakes with high oxygen content. Many are found in distinctive habitats and fly only early in the morning or late in the evening, which renders them less conspicuous to the casual observer. Most are metallically colored and have brilliant iridescent-green eyes at maturity. The eyes are prominent, and confluent for a long distance on top of the head. There is a tubercle on the rear margin, although it is generally not as prominent as in the cruisers (Macromiinae). The wings are either clear or with a brown basal spot that may extend variably out to the nodus. The anal loop is usually in the form of a foot or boot, but with little development of the toe region. In most species, there is a single bridge crossvein. The abdomen is generally longer than in the skimmers (Libellulidae), and there is a longitudinal ventrolateral carina on the middle segments in both sexes.

Baskettails
Genus *Epitheca*
(pp. 12, 13, 227)

This is a large group of similar-looking, medium-sized to large, nonmetallic-brown dragonflies. Their face is either yellow or brown. The eyes meet on top of the head for about the length of the occiput. The thorax is often thinly to diffusely covered with short black or white hairs, but usually a small yellow spot or interrupted stripe is visible laterally. The legs are long and the tibiae are keeled. The abdomen is dark brown or black, distinctly widened in some and constricted basally in others, usually not as much constricted in females. In both sexes, there is a pale-yellow lateral abdominal stripe.

Females in this group carry a large string of eggs with them, the string appearing as a pale-yellow or orange ball at the end of the abdomen. This egg mass is then released on a partially submerged piece of vegetation or debris, where it unravels. It has been noted that this tactic is advantageous, both so that females expose themselves only once to aquatic predators and because eggs placed near the water surface are exposed to higher oxygen concentrations and temperatures, speeding up the embryological development. North American species include two groups that historically have been recognized as separate genera (*Epicordulia* and *Tetragoneuria*), but more recently have been united with the Old World genus *Epitheca*, the placement I follow here. The *Epicordulia* subgroup includes a single distinctive eastern North American species that is the largest emerald in the region. The rest of the baskettails belong to the *Tetragoneuria* subgroup. This is a taxonomically difficult subgroup of medium-sized brown species, all of them similar. All have at least a hint of brown basally in the wings, and this may extend out to the level of the nodus in some. Six species occur in the south-central United States. Despite their having passed through four revisions, identifying these species is problematic at best. Many names have been synonymized, and still others probably should be. I have not studied this group in detail, and realize that the key and characters below will not be adequate to separate the species in many cases. For many of the species occurring together, the shape of the abdomen and the length of the male cerci seem to be the characters most likely to separate them.

References. Davis (1933), Donnelly (1992), Dunkle (1989), Kormondy (1959), May (1995d), Muttkowski (1911, 1915), Tennessen (1973), Walker (1966).

KEY TO THE SPECIES OF BASKETTAILS (*EPITHECA*)

1. Wings with large, but variable, brown spots basally, at nodus, and at wingtips	**Prince (*princeps*)**
1'. Wings clear, or at least never with a spot at wingtip, and generally only a trace of color basally in the hindwing	2
2(1'). Male cerci with a prominent anteapical tooth dorsally	**Robust (*spinosa*)**
2'. Male cerci without a dorsal anteapical tooth	3
3(2'). Middle abdominal segments wider than long; hindwing generally with brown extending from rear margin of wing outward nearly to nodus	**Mantled (*semiaquea*)**

3'. Middle abdominal segments narrow, longer than wide; hindwing with brown less extensive, generally not reaching rear margin of wing or nodus — 4

4(3'). Abdomen strongly constricted behind segment 3; male cerci in dorsal view nearly parallel; female caudal appendages as long as or longer than abdominal segments 9 and 10 combined — **Stripe-winged (*costalis*)**

4'. Abdomen not strongly constricted behind segment 3; male cerci divergent in dorsal view; female caudal appendages shorter, not as long as segments 9 and 10 combined — 5

5(4'). Pterostigma long, 3 mm; hindwing with basal brown spot not extending out to cover 1st antenodal crossvein — **Florida (*stella*)**

5'. Pterostigma shorter, 2 mm or less; hindwing with basal brown spot extending out to cover 1st antenodal crossvein — 6

6(5'). Abdomen parallel-sided; both fore- and hindwings with small transverse brown spot on nodus; hindwings generally with smaller spots on antenodal crossveins — **Dot-winged (*petechialis*)**

6'. Abdomen spindle-shaped, widening to segment 5 and narrowing to end; brown spot on nodus in neither fore- nor hindwings; hindwings without smaller spots on antenodal crossveins — **Common (*cynosura*)**

Stripe-winged Baskettail
Epitheca costalis (Selys)
(p. 227; photo 43a)

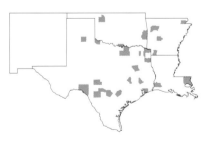

Size. Total length: 42–45 mm; abdomen: 28–32 mm; hindwing: 25–28 mm.

Regional Distribution. *Biotic Provinces:* Austroriparian, Balconian, Carolinian, Texan, Kansan. *Watersheds:* Arkansas, Brazos, Cimarron, Canadian, Colorado, Guadalupe, Mississippi, Nueces, Ouachita, Red, Sabine, San Antonio, San Jacinto, St. Francis, Trinity.

General Distribution. Eastern United States and southern Great Plains.

Flight Season. Mar. 25 (LA)–Jun. 27 (AR).

Identification. This is a typical baskettail with a hairy thorax and a spot of yellow on the side. The wings are marked with a basal spot of brown, or in some females there may be a brown stripe across the front edge of the wings. The males have a slender abdomen and long cerci (greater than 3.4 mm). These appendages are similar to those of the Common Baskettail (*E. cynosura*), but lack a ventral keel. The abdomen of the male is often strongly constricted behind segment 3. The abdomen has a yellow lateral stripe. The female caudal appendages are long, as long as segments 9 and 10 combined.

Similar Species. It is generally difficult to reliably identify males, except where they occur only with Common Baskettail (*E. cynosura*), which has a much stouter abdomen. Dot-winged Baskettail (*E. petechialis*) is in fact sometimes considered a form of Stripe-winged Baskettail. There are slight differences in the cerci of these two species, but some Dot-winged Baskettails have clear wings, and this character alone cannot reliably distinguish the two species. Sundragons (*Helocordulia*) have an orange ring encircling segment 3. Shadowdragons (*Neurocordulia*) lack a yellow spot or stripe laterally on the thorax.

Habitat. Lakes, ponds, and slow reaches of streams and rivers.

Discussion. Little has been published on the behavior of this species, but it seems to be similar to that of other, better-known baskettails. It may be seen in feeding swarms or perching on twigs and bushes in large numbers. Males patrol along shorelines for long distances. Mating pairs perch on stems at the water's edge. Larvae of this species and Common Baskettail have been found emerging on pine trees at unusually long distances and heights from the water. Exuviae were found 10.5 m from the water and at maximum heights of 5.5 m.

References. Tennessen (1979).

Common Baskettail
Epitheca cynosura (Say)
(p. 227; photos 43b, 43c)

Size. Total length: 36–44 mm; abdomen: 25–34 mm; hindwing: 26–30 mm.

Regional Distribution. *Biotic Provinces:* Austroriparian, Balconian, Carolinian, Kansan, Navahonian, Texan. *Watersheds:* Arkansas, Bayou Bartholomew, Brazos, Canadian, Cimarron, Colorado, Mississippi, Neches, Ouachita, Red, Sabine, San Jacinto, St. Francis, Trinity, White.

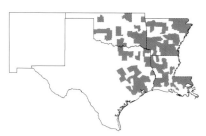

General Distribution. Eastern United States.

Flight Season. Feb. 17 (TX)–Jun. 18 (AR).

Identification. This species is brown, with a hairy thorax and a spot of yellow on each side. Some individuals have a basal triangular spot extending to the third antenodal crossvein in the hindwing. Others have only a basal spot of brown in each wing. The abdomen is broad and flattened in the middle segments, but it is not constricted, or only slightly so, behind the 3rd segment. The male cerci, when viewed laterally, have a ventral keel extending posteriorly from the ventral angle. The female caudal appendages are not longer than 2.25 mm.

Similar Species. Mantled Baskettail (*E. semiaquea*) is similar to those individuals of Common Baskettail, that have a brown triangle basally in the hindwing. In Common Baskettail this brown area is typically smaller, though, and does not extend to the hindwing margin or nodus. Shadowdragons lack a yellow spot or stripe laterally on the thorax. The abdomen is strongly constricted behind segment 3 in both Stripe-winged (*E. costalis*) and Dot-winged (*E. petechialis*) Baskettails.

Habitat. Almost any permanent or temporary, quiet water, including ponds, lakes, marshes, streams, and rivers, with submerged and emergent vegetation.

Discussion. Common Baskettail can be one of the most abundant early-spring, midsummer species. Larvae emerge on nearly any structure, natural or artificial, on which they can climb 1–3 m above the water, although they may emerge at distances much farther from the water. Adults may also venture some distance from water, and are commonly found along forest clearings and roads. Four types of flight have been recognized in this species: (1) a patrolling flight, consisting of extended periods of hovering, (2) a feeding flight, seen away from water and generally occurring during midmorning or early afternoon, (3) a copulatory flight, where both sexes mate in flight with no hovering and usually in a linear direction, often covering 1,300 m or more, and (4) a swarming flight involving both sexes and nearly always an additional species of baskettail. Females lay eggs in the fashion usual for this group, releasing a large string of eggs on partially submerged vegetation or debris.

References. Claus-Walker et al. (1997), Johnson (1986), Johnson et al. (1985), Kormondy (1959), Tennessen and Murray (1978).

Baskettails (*Epitheca*)

Stripe-winged (*costalis*)	Common (*cynosura*)	Dot-winged (*petechialis*)	Mantled (*semiaquea*)

Fig. 32. Baskettails (*Epitheca*): ventral views of basal abdominal segments in males (pp. 225–228).

(Redrawn with permission from original drawings by T. W. Donnelly.)

Dot-winged Baskettail
Epitheca petechialis (Muttkowski)
(p. 227; photo 43d)

Size. Total length: 41–43 mm; abdomen: 30–34 mm; hindwing: 27–31 mm.

Regional Distribution. *Biotic Provinces:* Austroriparian, Balconian, Chihuahuan, Kansan, Navahonian, Texan. *Watersheds:* Arkansas, Brazos, Canadian, Cimarron, Colorado, Guadalupe, Nueces, Pecos, Red, San Antonio, San Jacinto, Upper Rio Grande.

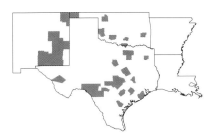

General Distribution. Colorado, Kansas, New Mexico, Oklahoma, and Texas.

Flight Season. Jan. (TX)–Jul. (TX).

Identification. This southern species has a relatively slender abdomen. Some individuals have a distinct row of brown spots on the antenodal crossveins in the hindwing, extending out to the nodus. These individuals are easy to identify. In others, these spots may not be as prominent. The slender abdomen is slightly constricted behind segment 3 in males. The male cerci are slightly divergent and lack sharp angulation downward, when viewed laterally. The female caudal appendages are approximately 2 mm long.

Similar Species. This species is nearly identical to Stripe-winged Baskettail (*E. costalis*), and is considered by some to be a form of that species. The clear-winged individuals may not be reliably distinguished from Stripe-winged Baskettail in the field. Differences from other species are given in the *Similar species* accounts under that species.

Habitat. Lakes, ponds, and slow reaches of streams and rivers.

Discussion. This species flies later in the year than the other species. It may be common along forest edges or in more open areas, where both clear and spotted-winged forms may be found.

Prince Baskettail
Epitheca princeps Hagen
(photo 43e)

Size. Total length: 58–68 mm; abdomen: 42–49 mm; hindwing: 38–43 mm.

Regional Distribution. *Biotic Provinces:* Austroriparian, Balconian, Carolinian, Kansan, Navahonian, Tamaulipan, Texan. *Watersheds:* Arkansas, Bayou Bartholomew, Brazos, Canadian, Cimarron, Colorado, Guadalupe, Lower Rio Grande, Mississippi, Neches, Nueces, Ouachita, Red, Sabine, San Antonio, San Jacinto, St. Francis, Trinity, White.

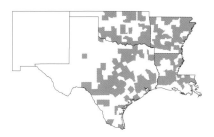

General Distribution. Eastern United States and Canada.

Flight Season. Mar. 31 (TX)–Sep. 1 (TX).

Identification. This is the most widespread and probably easiest emerald to recognize in the region. It is the only baskettail with brown wingtips. Its eyes are brown when young and brilliant green in older males. The thorax and abdomen are brown, and the latter has an obscured row of pale spots on each side. The wings are variably marked with brown basally, at the nodus, and apically.

Similar Species. Though not to be confused with other emeralds, Twelve-spotted Skimmer (*Libellula pulchella*) and female Common Whitetails (*Plathemis lydia*) have similar wing patterns. Prince Baskettail, though, has a much narrower abdomen than either of these species, and it lacks prominent yellow lateral thoracic stripes.

Habitat. Quiet reaches of streams, rivers, ponds, and lakes.

Discussion. This species may be mistaken for a patrolling darner because of its size and similar behavior. The more northern individuals are smaller and tend to have the brown markings on the wings more reduced. These baskettails fly high, often over the tree line, and may be seen mixed in with other species in feeding swarms. The males generally patrol long areas of shoreline, and fly at heights of a meter or two above the water. They will often perch vertically in trees with the abdomen conspicuously turned upward. Females lay eggs in a manner similar to that of other members of this genus, by depositing the egg mass on leaves or debris at the water surface. Adults may congregate in large numbers on the leeward side of bushes on windy days.

References. Robert (1963).

Mantled Baskettail
Epitheca semiaquea (Burmeister)
(p. 227; photo 43f)

Size. Total length: 34–38 mm; abdomen: 24–30 mm; hindwing: 24–31 mm.

Regional Distribution. *Biotic Provinces:* Austroriparian, Kansan, Texan. *Watersheds:* Brazos, Cimarron, Guadalupe, Neches, Red, San Jacinto, Trinity.

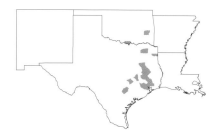

General Distribution. Eastern United States along coast and Oklahoma and Texas.

Flight Season. Mar. 25 (TX)–May 30 (TX).

Identification. The broad abdomen of this baskettail is never constricted behind segment 3, but tapers regularly after segment 6. The hindwings are often broadly colored with brown out past the fourth antenodal crossvein, to the level of the nodus or nearly so. This maculation is variable in specimens within the south-central United States, however, and these can be confused with maculated individuals of Common Baskettail (*E. cynosura*). The abdomen is the widest and shortest of all our Baskettails. Segments 4–6 are wider than long. The male cerci are widely divergent dorsally, and the female caudal appendages are short, between 1.2 and 1.5 mm long.

Similar Species. Common Baskettails with extensive markings on the hindwing are similar, but these markings do not reach the hind margin of the wing and rarely extend out to the nodus.

Habitat. Lakes and ponds with submerged and emergent vegetation.

Discussion. This species has only recently been discovered in Oklahoma. Little has been reported on the behavior of this species, but it is often found perched on twigs and bushes in open clearings, sometimes in large numbers. Eastern individuals have more extensive maculation in the hindwing.

References. Abbott (1996).

Robust Baskettail
Epitheca spinosa (**Hagen** *in* **Selys**)
(photo 44a)

Size. Total length: 42–45 mm; abdomen: 32–35 mm; hindwing: 29–33 mm.

Regional Distribution. *Biotic Provinces:* Austroriparian, Kansan. *Watersheds:* Mississippi, Ouachita.

General Distribution. Southeastern United States, from New Jersey to Oklahoma.

Flight Season. May 10 (AR).

Identification. Males of this southern coastal species have distinctive caudal appendages. The thorax is diffusely covered with white hairs, giving it a characteristically dull appearance. The hindwings may be variously marked with brown basally. The male cerci have a sharp dorsally projecting anteapical spine. The female caudal appendages are 2 mm long.

Similar Species. Other baskettails in the region are smaller and have narrower abdomens. This the only baskettail with a distinct dorsally projecting tubercle on the male cercus.

Habitat. Lakes, ponds, and wooded swamps with little flow.

Discussion. This species is uncommon and only sporadically reported in the south-central United States. Sid Dunkle (pers. comm.) has a single female in his collection from Washington Parish, Louisiana. A single male was reported by Harp and Harp (1996) from Clark County, Arkansas, and another by Bick and Bick (1957) from Latimer County, Oklahoma (previously reported by Bird (1932a) as *Epitheca canis* McLachlan, the Beaverpond Baskettail).

Within Texas, this species has also been documented only from Houston County.

Florida Baskettail
Epitheca stella (**Williamson** *in* **Muttkowski**)
(photo 44b)

Size. Total length: 44–47 mm; abdomen: 32–36 mm; hindwing: 27–32 mm.

Regional Distribution. *Biotic Province:* Austroriparian. *Watershed:* Bayou Bartholomew.

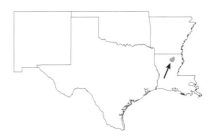

General Distribution. Southeastern coastal plain to Louisiana.

Flight Season. Apr. 14 (LA).

Identification. This species is similar to Common Baskettail (*Epitheca cynosura*), but it is generally larger. The frons is yellow, covered with numerous short black hairs above and white below. The pterothorax is densely clothed with white hairs. The wings are clear, with a touch of brown basally. The pterostigma is long. The abdomen is slender, regularly tapering beyond the swollen segment 3. Segments 1–3 are largely yellow. The pale lateral markings regularly diminish posteriorly, and segment 10 and the caudal appendages are black. The male caudal appendages are similar to those of Common Baskettail, but generally with the basal 1/3 straight, or nearly so.

Similar Species. Careful examination should be given to distinguish this species from Stripe-winged (*E. costalis*) and Common (*E. cynosura*) Baskettails. These two species are much more common in our region, and characters useful for separating them are given under each.

Habitat. Lakes and ponds.

Discussion. This species is sometimes considered endemic to Florida, but it has been infrequently taken in Georgia and Mississippi. A single male has been reported from Louisiana; the specimen was taken in Ouachita Parish near Monroe, Louisiana. Little has been reported on its behavior, which is apparently similar to that of Common Baskettail.

References. Dunkle (1989), Mulhern (1971), Paulson (1973).

Sundragons
Genus *Helocordulia* Needham

The two uncommon eastern species making up this small group are similar to baskettails (*Epitheca*). They differ from them in the structure of the male accessory genitalia and caudal appendages. The wings are clear, with a basal spot of brown, and in some individuals with spots on the antenodal crossveins. The venation in the genus is fairly plastic.

The male abdomen gradually widens after the basal constriction behind segment 3 and does not taper again after the middle segments. The sundragons have short flight seasons in the early spring, and are often seen along forest edges and roadsides. Females do not deposit eggs in a long gelatinous string as do baskettails.

KEY TO THE SPECIES OF SUNDRAGONS (*HELOCORDULIA*)

1. Hindwing with golden yellow spot in midst of basal brown spot; subgenital plate of female deeply bifid and greater than 1/2 as long as segment 9 — **Uhler's (*uhleri*)**

1'. Hindwing without yellow spot in midst of basal brown spot; subgenital plate of female emarginate and less than 1/3 as long as segment 9 — **Selys' (*selysii*)**

Selys' Sundragon
Helocordulia selysii (Hagen *in* Selys)
(photo 44c)

Size. Total length: 38–41 mm; abdomen: 29–31 mm; hindwing: 26–28 mm.
Regional Distribution. *Biotic Province:* Austroriparian. *Watersheds:* Mississippi, Ouachita, Red, St. Francis, Trinity.

General Distribution. Southeastern United States.
Flight Season. Mar. 8 (TX)–Apr. 10 (LA).

Identification. This is an uncommon dark species with clear wings. It has a pale-yellow face and a deep medial depression on the top of the frons. The thorax is brown, thickly clothed with silky hairs. The legs are black. Along the anterior margin of each wing there is a basal dark spot follwed by a series of smaller spots. The abdomen is dark, with pale spots laterally on the middle segments. Segment 3 has a nearly complete basal ring. The abdomen in males is slightly clubbed.
Similar Species. Uhler's Sundragon (*H. uhleri*) has a small amber spot next to the basal brown spot in each wing. Baskettails (*Epitheca*) all have a yellow spot or stripe laterally on the thorax, and shadowdragons (*Neurocordulia*) lack an orange ring on segment 3.
Habitat. Small, cool forest streams with sandy bottoms.
Discussion. Adults of this species prefer open sunny glades in woods and are much less common than Uhler's Sundragon, but at the western edge of their distribution, Selys' Sundragon seems more wide-

spread and more commonly encountered. Little has been reported on the biology of this species, but it may be seen foraging along forest edges. Males commonly hover while patrolling. It has recently been reported from Salado Creek in Independence County, Arkansas.

References. Harp and Harp (1996), Needham et al. (2000).

Uhler's Sundragon
Helocordulia uhleri Selys
(photo 44d)

Size. Total length: 41–46 mm; abdomen: 28–33 mm; hindwing: 25–30 mm.
Regional Distribution. *Biotic Provinces:* Austroriparian, Carolinian, Kansan. *Watersheds:* Ouachita, Red, White.

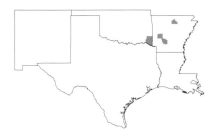

General Distribution. Widespread, eastern United States and Canada.
Flight Season. Apr. 12 (OK).
Identification. This uncommon species is similar to the preceding one, Uhler's Sundragon, but Uhler's has a small amber spot adjacent to the darker basal spot in each wing, and its range is more restricted within the south-central United States.
Similar Species. Differences between this species and Selys' Sundragon are given above and under the description for that species, along with differences between Selys' and other species.
Habitat. Small, rapid forest streams, often with impeded flow, and occasionally lakes.
Discussion. One author said this species ". . . is inconspicuous and flies with great speed, within two or three feet of the water, following the shore line closely." The species has only rarely been reported from Oklahoma and Arkansas. It was also reported from Louisiana, but it was later determined that all specimens were actually Selys' Sundragon, and the Louisiana records can probably be attributed to the past confusion in separating these two species.

References. Bick and Bick (1957), Harp and Rickett (1977), Mauffray (1997), Needham and Westfall (1955), Walker and Corbet (1975).

Shadowdragons
Genus *Neurocordulia* Selys

Of the seven inconspicuous medium-sized species of this group found in North America, all but two northern species occur in our area. The common name, shadowdragons, is derived from the fact that individuals of this group are seldom seen during the day, when they perch cryptically on twigs or bushes in the shade. They are crepuscular—typically active for only a short time at dusk and dawn. They are brownish, without iridescent eyes or metallic coloration. The wings are generally marked with brown spots, these often occurring on the antenodal crossveins and sometimes becoming diffuse. The frons is distinctly rounded. The erect, shelflike vertex often obstructs the view of the middle ocellus from above. There is a faint stripe or spot of yellow laterally on the thorax, around the spiracle. The legs are not especially long, and are generally pale. These are the only emeralds in the south-central United States having the veins M_4 and Cu_1 divergent to the wing margin. The abdomen is swollen basally, but becomes depressed in the middle segments.

References. Byers (1937), Davis (1929).

KEY TO THE SPECIES OF SHADOWDRAGONS (*NEUROCORDULIA*)

1. On each wing a dark basal spot, a dark spot at nodus,   2
   and row of spots on each antenodal crossvein

KEY TO THE SPECIES OF SHADOWDRAGONS (*NEUROCORDULIA*) (*cont.*)

1'. Each wing clear or nearly so, with at most small basal marking and/or light coloring on the antenodal crossveins	3
2(1). On each wing a row of amber dots along the full length of the front edge	**Orange (*xanthosoma*)**
2'. On each wing a row of brown dots extending only from base to nodus	**Umber (*obsoleta*)**
3(1'). Metathoracic trochanter, in males, with conspicuous truncate process on inner side, the process as long as segment is wide; caudal appendages of female long, 2.3 mm or greater	**Smoky (*molesta*)**
3'. Metathoracic trochanter, in males, lacking conspicuous truncate process on inner side; caudal appendages of female of normal length, less than 2.1 mm	4
4(3'). Keel on first tibia of male longer than width of tibia at its widest; caudal appendages of female shorter, less than 1.8 mm	**Alabama (*alabamensis*)**
4'. Keel on first tibia of male lacking or at most vestigial; caudal appendages of female 1.9–2.1 mm	**Cinnamon (*virginiensis*)**

Alabama Shadowdragon
Neurocordulia alabamensis Hodges *in*
Needham & Westfall
(photo 44e)

Size. Total length: 41–46 mm; abdomen: 29–32 mm; hindwing: 28–33 mm.
Regional Distribution. *Biotic Province:* Austroriparian. *Watersheds:* Mississippi, Ouachita, Trinity.

General Distribution. Southeastern United States.
Flight Season. May 17 (TX)–Jun. 19 (LA).

Identification. This is a coastal species seen only rarely as far west as Louisiana and Texas. Its body is pale brown with diffuse yellow on the sides of the thorax, the sides thus lighter than the front. The wings are clear or uniformly tinted with a row of brown spots along the entire costal margin of each wing. The legs are pale. The trochanter of the middle leg in the male lacks a distinct truncate process on its inner side. The distinct keel on the first tibia is about as long as the tibia is wide. The abdomen is pale brown, and the caudal appendages are pale yellowish-brown. The female caudal appendages are short, approximately 1.6 mm.
Similar Species. Orange (*N. xanthosoma*) and Umber (*N. obsoleta*) Shadowdragons have basal brown spots in the wings. Cinnamon (*N. virginiensis*) and Smoky (*N. molesta*) Shadowdragons lack spots on the front margins of their wings, and Coppery Emerald (*Somatochlora georgiana*) has clear wings.
Habitat. Small to medium-sized slow-flowing or spring-fed forest streams, frequently tannin-stained.

Discussion. All Louisiana records of this species are based on larvae. The only Texas records were taken from Hickman Branch, 3.2 km south of Coldspring in San Jacinto County, part of the Sam Houston National Forest. This species has been called one of the most elusive dragonflies in Florida, because of its crepuscular habits. Females lay eggs during a rapid crisscrossing flight in pools or areas of slow flow.

References. Donnelly (1978), Dunkle (1989), Mauffray (1997).

Smoky Shadowdragon
Neurocordulia molesta (Walsh)
(photo 44f)

Size. Total length: 45–53 mm; abdomen: 35–38 mm; hindwing: 33–38 mm.
Regional Distribution. *Biotic Provinces:* Austroriparian, Carolinian, Kansan, Texan. *Watersheds:* Arkansas, Bayou Bartholomew, Brazos, Canadian, Mississippi, Ouachita, Red, San Jacinto, St. Francis.

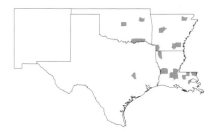

General Distribution. Eastern United States to Texas and Oklahoma.
Flight Season. May 21 (AR)–Jul. 23 (OK).
Identification. This is a larger species, second only to Orange Shadowdragon (*N. xanthosoma*), and widely distributed in the Southeast. Its face is olivaceous brown and the eyes are green. The thorax is brown in front, with only a pale-yellowish middorsal carina. The thorax becomes noticeably paler laterally. The midlateral stripe is diffuse yellow to the level of the spiracle, and then becomes obsolete except for a small yellowish spot. The legs are pale, and on the inner side of the middle trochanter of the male is a conspicuous truncate process. The wings are lightly tinted with amber and have brown spots on the antenodal crossveins and a larger spot at the nodus. The spots on the antenodal crossveins

are darkest on either side of the crossveins themselves. The abdomen is brown, and the caudal appendages are yellowish. The female appendages are long, approximately 2.4 mm.
Similar Species. The distinctive smoky wings and green eyes of this species will distinguish it from all other species in our area.
Habitat. Rivers and medium-sized streams with strong current.
Discussion. This uncommon species has been reported in Texas on the basis of a single male at the Little Brazos River in Brazos County. The biology and behavior of this species are poorly known. Males are active just before dusk and are unwary when patrolling near the shore.

References. Dunkle (2000), Louton (1982).

Umber Shadowdragon
Neurocordulia obsoleta (Say)
(photo 45a)

Size. Total length: 43–48 mm; abdomen: 33–35 mm; hindwing: 30–33 mm.
Regional Distribution. *Biotic Province:* Austroriparian. *Watershed:* Sabine.

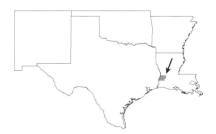

General Distribution. Eastern United States.
Flight Season. Jul. 1 (LA).
Identification. This brown, medium-sized species is similar to the others of the genus. The thorax is darkest on the front, but divided as usual by a pale middorsal carina. The wings are clear or lightly tinted with distinct maculation, including a large basal black spot on each wing, but generally larger on the hindwings. There is a row of smaller brown spots on the antenodal crossveins and a larger spot at the nodus. This is the only shadowdragon with a crossvein in the midbasal space of all wings. The legs are pale, and the abdomen is brown with dark caudal appendages.

Similar Species. Cinnamon (*N. virginiensis*) and Smoky (*N. molesta*) Shadowdragons lack spots on the front margins of their wing. Orange Shadowdragon (*N. xanthosoma*) has a different wing pattern. Similar-appearing baskettails (*Epitheca*) have a yellow lateral stripe on their abdomen.

Habitat. Rivers and lakes.

Discussion. A single male was reported from Beauregard Parish on the Texas–Louisiana border. It was also reported from Arkansas, but these records were later determined to be based on misidentified material. Nothing has been published about the behavior of this uncommon species, though like others in the group it flies out over riffles at dusk.

References. Harp (1983), Harp and Rickett (1977), Mauffray (1997).

yellowish brown. The female appendages are approximately 2 mm long.

Similar Species. Most other shadowdragons in the south-central United States have more coloration in the wings. Smoky Shadowdragon (*N. molesta*) has smoky wings and olive-green eyes. Similar-appearing baskettails (*Epitheca*) have a yellow lateral stripe on their abdomen.

Habitat. Medium-sized rivers with riffles.

Discussion. This species may fly well after sunset, or through forest understories during the day. Two stages of flight have been recognized in this species: the first involves an erratic flight 1–2 m above the water; the second involves males frequently skirmishing, followed by a lower flight 15 cm above the water.

References. Anadu et al. (1996), Dunkle (1989).

Cinnamon Shadowdragon
Neurocordulia virginiensis Davis
(photo 45b)

Size. Total length: 42–48mm; abdomen: 32–35 mm; hindwing: 29–32 mm.

Regional Distribution. *Biotic Provinces:* Austroriparian, Kansan. *Watersheds:* Cimarron, Arkansas, Mississippi, Ouachita.

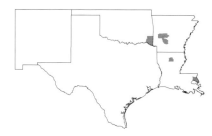

General Distribution. Southeastern United States.

Flight Season. May 21 (AR)–Jun. 18 (OK).

Identification. This is another medium-sized brown Shadowdragon, but its wings are the least marked of the species in our region. The thorax is darker in front and divided by a pale middorsal carina. The sides of the thorax are paler, with a diffuse yellow spot surrounding the spiracle. There are dark spots on the antenodal crossveins only before the triangles. The legs are pale, and the foretibiae in the males have only have a vestigial keel; there is no truncate process on the inner side of the middle trochanter. The abdomen is brown, and each segment has a dark transverse carina. The caudal appendages are

Orange Shadowdragon
Neurocordulia xanthosoma (Williamson)
(photo 45c)

Size. Total length: 48–51 mm; abdomen: 37–41 mm; hindwing: 36–42 mm.

Regional Distribution. *Biotic Provinces:* Austroriparian, Balconian, Carolinian, Kansan, Tamaulipan, Texan. *Watersheds:* Arkansas, Bayou Bartholomew, Brazos, Canadian, Cimarron, Colorado, Guadalupe, Lower Rio Grande, Ouachita, Pecos, Red, San Antonio, St. Francis, Trinity, White.

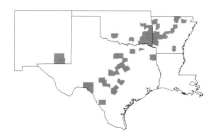

General Distribution. Southern Great Plains.

Flight Season. May 28 (AR)–Jul. 30 (TX).

Identification. This is the largest, most common, and most widely distributed of our shadowdragons. It is also the most distinctive. The thorax is pale yellowish brown, interrupted middorsally by a pale carina. There is only a hint of yellow laterally around each spiracle. The wings are lightly to strongly spotted along their entire costal margin. There is a darker black spot at the base of each hindwing. The wingtips are generally darkened, and the entire wing may become amber in males. The legs are pale. The abdomen is dark brown, with a carina outlined in black. The caudal appendages are pale, and the male cerci bear a ventrally projecting tooth, seen when viewed laterally.

Similar Species. The distinctive wing pattern will distinguish this species from our other shadowdragons. The spots along the front wing margin in Umber Shadowdragon (*N. obsoleta*) do not extend beyond the nodus.

Habitat. Medium-sized turbid rivers and streams with strong current.

Discussion. This species comes out at dusk or dawn, often in large numbers, to feed on smaller flying insects. Their pale color makes them diffi-cult to see against the sky, as they swarm 1.5 to 2.5 m off the ground. Mating pairs hang from branches or twigs in the shade. Egg-laying generally occurs 30 minutes to 1 hour after swarming. Females generally lay eggs alone, unguarded by males, by rapidly flying low over the water, occasionally pausing to release eggs at the surface, although there is at least one report of a female laying eggs while attended by the male. Larvae have the unusual behavior of feigning death. The westernmost record for this species is from the Black River in New Mexico.

References. Bird (1932b), Clark (1979), Harwell (1951), Williams (1979c).

Striped Emeralds
Genus *Somatochlora* Selys

This northern Holarctic genus is the largest group of emeralds and includes 26 North American species. Seven occur in the south-central United States. They are medium-sized, metallic-brown dragonflies that may be locally abundant, but are generally uncommon. Many fly at great heights, rarely coming low enough to view critically. The eyes become brilliant iridescent green in older adults. The front of the thorax is unmarked, but most have two variously shaped pale lateral stripes. The wings are clear, with a single bridge crossvein and generally a single crossvein in the triangles. There is a keel on the fore- and hind tibiae of males. The abdomen is generally dark, with a narrow pale ring and spots on the apical segments. The male caudal appendages are generally distinctive, and the females have a distinct spoutlike ovipositor.

References. Walker (1925).

KEY TO THE SPECIES OF STRIPED EMERALDS (*SOMATOCHLORA*)

---

1. Tibiae with yellow on their outer surface; coloration not so metallic; lateral thoracic stripes long and wide — **Coppery (*georgiana*)**

1'. Tibiae entirely black, without yellow on their outer surfaces; coloration strongly metallic; lateral thoracic stripes variable — 2

2(1'). Lateral thoracic stripes absent — **Mocha (*linearis*)**

2'. Pale lateral thoracic stripes present — 3

3(2'). Lateral thoracic stripes limited to two large oval spots — **Mountain (*semicircularis*)**

3'. Lateral thoracic stripes not abbreviated — 4

4(3'). Abdominal segment 2 with a single large spot of yellow, laterally, before the auricle; coloration generally pale — **Clamp-tipped (*tenebrosa*)**

KEY TO THE SPECIES OF STRIPED EMERALDS (*SOMATOCHLORA*) (*cont.*)

4'. Segment 2 with more than 1 spot of yellow, laterally, before the auricle; coloration generally more vivid	5
5(4'). First lateral thoracic stripe narrower than the second, and angulated or interrupted at its middle	**Fine-lined (*filosa*)**
5'. First lateral thoracic stripe running straight and uninterrupted for its full length, and generally not distinctly narrower than the second	6
6(5'). Obscure pale basal markings on abdominal segments 4–8	**Texas (*margarita*)**
6'. No obscure pale basal markings on segments 4–8	**Ozark (*ozarkensis*)**

Fine-lined Emerald
Somatochlora filosa Hagen
(photo 45d)

Size. Total length: 52–69 mm; abdomen: 41–54 mm; hindwing: 35–46 mm.
Regional Distribution. *Biotic Province:* Austroriparian. *Watersheds:* Mississippi, Neches, Red, St. Francis.

General Distribution. Southeastern United States, New Jersey to Texas.
Flight Season. Jul. 8 (LA)–Sep. 24 (LA).
Identification. This is an uncommon species with brilliant iridescent-green eyes, a dark metallic-green thorax, and two pale lateral stripes. It is the only species in the region with the first thoracic stripe interrupted or distinctly angulated medially. The second stripe is wider. Its face is pale in front and metallic blue on top. The wings in young females often become amber apically. The legs are black. The abdomen is dark metallic brown or black, the basal segments marked with three pale stripes laterally. There are narrow white basal rings on segments 8–10. The female has a grooved ovipositor.
Similar Species. The distinctive pale lateral stripes

will differentiate this species from the other striped emeralds in the region.
Habitat. Probably spring-fed seeps and forest streams.
Discussion. This species has been reported from a limited number of counties in Louisiana and Arkansas, and in east Texas it is known only from a single female. This species is typical of the genus, usually seen flying high over paths, trails, and roads in the early morning and in the late afternoon and evening. Males feed in forest clearings.

References. Abbott and Stewart (1998), Barr (1981), Mauffray (1997).

Coppery Emerald
Somatochlora georgiana Walker
(photo 45e)

Size. Total length: 47–50 mm; abdomen: 34–38 mm; hindwing: 32–34 mm.
Regional Distribution. *Biotic Province:* Austroriparian. *Watershed:* Red.

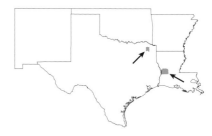

General Distribution. Southeast United States; coastal states from New Jersey to Texas.

Flight Season. Jun. (TX)–Aug. 6 (LA).

Identification. This southeastern coastal species has only rarely been reported from as far west as Louisiana and Texas. It is distinctive among the striped emeralds in the region by having yellow on the outer surface of the otherwise brown tibiae. It is generally duller and not as metallic as our other species. The thorax is brown, with two well-developed pale lateral stripes, the posterior much wider than the anterior. The wings are clear, and the abdomen is dark brown, with the usual yellow basal stripes. The female ovipositor is short, triangular, and directed ventrally.

Similar Species. No other striped emerald in our area has yellow on the tibiae. Shadowdragons (*Neurocordulia*) and baskettails (*Epitheca*) lack distinctive thoracic stripes.

Habitat. Pools and slow-flowing, tannin-stained forest streams.

Discussion. This species has been reported from a single parish/county in both Louisiana and Texas. Both specimens were collected in the early 1950's and are in the G.H. Beatty collection at the Pennsylvania State University. No additional collections of this species have been made in the region. Adults are generally encountered in the early or late afternoon, feeding 10–20 m above dirt roads and forest clearings. They perch on branches of trees. Egg-laying usually occurs at midstream, the females flying erratically less than 1 m above the water.

References. Daigle (1994).

Mocha Emerald
Somatochlora linearis (Hagen)
(photo 45f)

Size. Total length: 56–70 mm; abdomen: 42–56 mm; hindwing: 38–50 mm.

Regional Distribution. *Biotic Provinces:* Austroriparian, Carolinian, Kansan, Texan. *Watersheds:* Arkansas, Bayou Bartholomew, Brazos, Cimarron, Mississippi, Neches, Ouachita, Red, Sabine, St. Francis, Trinity, White.

General Distribution. Eastern United States and Canada.

Flight Season. Jun. 12 (LA)–Sep. 20 (TX).

Identification. This distinctive species is the largest, most widespread, and probably most frequently encountered of the striped emeralds in the region. It is the only one in the region lacking pale-yellow lateral thoracic stripes. Its face is pale brown, and the top of the head is metallic blue. The eyes become iridescent green in older individuals. The thorax is brown and metallic green without pale stripes. The wings are clear, but may be tinted amber in older individuals. The legs are black. The abdomen is dark brown with metallic reflections. Segment 2 has a large pale spot basally. Orange laterobasal spots on segments 3–8 have fade with age. The male cerci bifurcate apically, and the female has a triangular-shaped ventrally projecting ovipositor.

Similar Species. The lack of lateral thoracic stripes on these emeralds distinguishes them from the other striped emeralds.

Habitat. Permanent and temporary forest streams.

Discussion. This species is most often seen flying early in the morning and late in the afternoon, high over the trees (it only occasionally flies low to the ground). Between those times, adults perch on twigs in full shade. Males patrol low to the water, at heights of a meter or less, making frequent hovering stops. Females lay eggs unaccompanied by the male in sand or mud at the water's edge, by stabbing the ovipositor into the substrate.

References. Dunkle (1989), Williamson (1922a).

Texas Emerald
Somatochlora margarita Donnelly
(photo 46a)

Size. Total length: 50–54 mm; abdomen: 32–41 mm; hindwing: 32–37 mm.

Regional Distribution. *Biotic Province:* Austroriparian. *Watersheds:* Neches, Red, Sabine, Trinity.

General Distribution. Piney woods of East Texas and Louisiana.

Flight Season. May 27 (TX)–July 2 (TX).

Identification. This species has a largely pale face, and the top of the head is metallic blue. The eyes are brilliant iridescent green in older individuals. The thorax is brown with bluish-green reflections and two well-developed pale lateral stripes. The first is slightly narrower than the second. The legs are entirely black, and the wings are clear. The abdomen is dark brown with metallic reflections. The basal segments are marked with pale yellow. Segment 3 has a conspicuous pale triangular spot anterolaterally. Segments 4–8 each have obscure pale spots anterolaterally.

Similar Species. This species is closely related to Ozark (*S. ozarkensis*) and Fine-lined (*S. filosa*) Emeralds, but it has obscure pale basal spots on abdominal segments 4–8.

Habitat. Small, sandy forest streams with moderate current.

Discussion. This species is endemic to the longleaf and loblolly pine forests of southeastern Texas, but it is uncommon and until recently had not been seen more than a few kilometers from its type locality, in the Sam Houston National Forest. By now, its initial range has been expanded to a now estimated 16,000 km² area in eastern Texas and western Louisiana. I reported it from Louisiana in 1996 and expanded its range north and west in 1997 with a single female found at Engeling Wildlife Management Area in Anderson County, Texas. Little is known, and nothing apart from the original description has been published on the behavior of this species. It flies high, at treetop level, along dirt roads. Females occasionally come down into open fields or forest clearings to feed. They fly in the early morning and in late afternoon to dusk, as is usual for the genus.

References. Price et al. (1989).

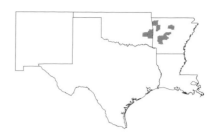

General Distribution. Ozark region of Arkansas, Missouri, and Oklahoma.

Flight Season. Jun. 10 (OK)–Sep. 7 (AR).

Identification. This species is apparently endemic to the Interior Highlands of Arkansas, Oklahoma, and Missouri. Its face is pale, and the top of the head is metallic greenish blue. Its eyes become iridescent green in older individuals. The thorax is brown, with metallic reflections and two well-developed pale stripes laterally. The posterior stripe is slightly wider than the anterior one. The wings are clear, occasionally becoming amber. The legs are largely black, and the abdomen is dark brown. The basal segments of the abdomen are marked with pale yellow, including a subtriangular spot, dorsolaterally on segment 3. There are no basal spots on segments 4–8.

Similar Species. Texas Emerald (*S. margarita*) has basal spots on abdominal segments 4–8. Fine-lined Emerald (*S. filosa*) has narrower lateral thoracic stripes.

Habitat. Forest streams with moderate riffles.

Discussion. This species was described from Cunneotubby Creek near Wilburton in Latimer County, Oklahoma. The flight of Ozark Emerald has been referred to as irregular. It is crepuscular, flying from just after daybreak to early morning or during the late evening, but is rarely seen during the day. No photo is given for Ozark Emerald, since it looks essentially the same as Texas Emerald.

References. Pritchard (1936).

Ozark Emerald
Somatochlora ozarkensis Bird

Size. Total length: 50–56 mm; abdomen: 37–44 mm; hindwing: 33–40 mm.

Regional Distribution. *Biotic Provinces:* Austroriparian, Carolinian, Kansan. *Watersheds:* Arkansas, Ouachita, Red, White.

Mountain Emerald
Somatochlora semicircularis (Selys)
(photo 46b)

Size. Total length: 48–51 mm; abdomen: 35–40 mm; hindwing: 27–32 mm.

Regional Distribution. *Biotic Provinces:* Coloradan, Navahonian. *Watersheds:* Pecos, Rio Grande.

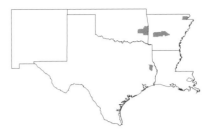

General Distribution. Western United States and Canada south to New Mexico.

Flight Season. Jun.–Oct.

Identification. In our region, this northern species just enters New Mexico. It is a shiny metallic green emerald. Its face is dark, with many hairs. The top of the frons and vertex are metallic green. The thorax is metallic green, with two pale oval spots laterally, and is thickly covered with fine hairs. The legs are black. The wings are clear, but may appear smoky. The abdomen is black. On segment 2 there is a pair of small pale spots and an interrupted apical band. There is a pale basal triangle on segment 3. Pale abdominal markings become obscured in older individuals.

Similar Species. No other striped emerald is found in the western portion of the region.

Habitat. Mountain ponds, including those that are ephemeral.

Discussion. May be common in clearings on sunny days.

Clamp-tipped Emerald
Somatochlora tenebrosa (Say)
(photos 46c, 46d)

Size. Total length: 48–65 mm; abdomen: 33–46 mm; hindwing: 33–42 mm.

Regional Distribution. *Biotic Provinces:* Austroriparian, Kansan. *Watersheds:* Arkansas, Ouachita, Red, Sabine, St. Francis.

General Distribution. Widespread throughout eastern United States and Canada.

Flight Season. Jul. 17 (AR).

Identification. This common eastern species has rarely been seen as far west as Arkansas or Oklahoma. It is distinct, among the species in the region, in having a single large circular pale spot laterally on abdominal segment 2. It has an orange-yellow face, and the top of the head is deep metallic blue or black. The thorax is brown and metallic green, with well-developed pale stripes laterally. The wings are clear, but occasionally become tinted with amber. The legs are black. The abdomen is brown, with metallic reflections. Segment 2 is as above. Segment 3 has a short dorsolateral spot and a larger ventral spot. The large angulate cerci of the male are distinctive. The female has a long, compressed ventrally projecting ovipositor.

Similar Species. No other Emerald in our region has a single large circular spot laterally on the second abdominal segment, and the male's cerci are unique.

Habitat. Small forested streams with intermittent riffles and pools.

Discussion. This species is typical of the genus, flying in the early morning or in late afternoon and evening. It is known to have a preference for shade. Wilson (1912) observed this species ". . . patrolling back and forth just after sunset in one corner of an old pasture near a small brook at the foot of the mountains. They were strong and rapid fliers. . . ." Males will often hover over the creeks and streams they patrol.

Reference. Walker and Corbet (1975).

SKIMMERS
(Family Libellulidae)

This is the largest family of Odonata and includes nearly a third of the species in the south-central United States. Members of the family are worldwide in distribution and among some of the most common and recognizable species. They generally inhabit ponds, lakes, and marshes, where they are often seen flying or perched atop twigs and bushes. Many are vividly colored and have distinctive wing markings. In most species the males accompany the females during egg-laying.

The eyes are large, and meet on top of the head for a considerable distance, but they lack the tubercles on their hind margin found in the emeralds (Corduliidae). The triangles in the fore- and hindwings are of different shapes, and males lack an anal triangle, with the result that the wings in both sexes are rounded basally. The radial and median planates are well developed. The anal loop is distinctively shaped as a foot, with a well-developed ankle and toe region. The abdomen is generally shorter and more depressed than in the emeralds, and males lack auricles on abdominal segment 2.

KEY TO THE GENERA OF SKIMMERS (LIBELLULIDAE)

1. Hindwings and usually forewings with brownish-yellow spot at nodus; toe of anal loop resting on posterior wing margin	**Evening Skimmer (*Tholymis citrina*)**
1'. Hindwings and forewings without spot at nodus, or, if present, then other markings in the wings as well; toe of anal loop not resting on posterior wing margin	2
2(1'). Antenodal crossveins of both wings with row of black spots	**Filigree Skimmer (*Pseudoleon superbus*)**
2'. Antenodal crossveins of both wings without row of spots	3
3(2'). Vein M_2 wavy (only slightly so in clubskimmers and sylphs)	4
3'. Vein M_2 smoothly curved	11
4(3). Hindwing with 2 cubito-anal crossveins and vein Cu1 arising from outer side of triangle; hindwing narrow at base; south Texas and Mexico	**Gray-waisted Skimmer (*Cannaphila insularis*)**
4'. Hindwing without 2 cubito-anal crossveins and vein Cu1 not arising from outer side of triangle; hindwing wider at base	5

KEY TO THE GENERA OF SKIMMERS (LIBELLULIDAE) (*cont.*)

5(4'). Wings with more than 1 crossvein | 22

5'. Wings with a single bridge crossvein | 6

6(5'). Pterostigma extremely long, surmounting 5 or 6 crossveins | **Tropical King Skimmers (*Orthemis*)**

6'. Pterostigma moderately long, surmounting no more than 4 crossveins | 7

7(6'). Hindwing with 2 cubito-anal crossveins | **Rainpool Gliders (*Pantala*)**

7'. Hindwing with a single cubito-anal crossvein | 8

8(7'). Forewing with 2 rows of cells beyond triangle | 9

8'. Forewing with 3 rows of cells beyond triangle | 10

9(8). Forewing subtriangle with 1 or 2 cells; median planate absent in forewing | **Sylphs (*Macrothemis*)**

9'. Forewing subtriangle with 3 cells; median planate present in forewing | **Clubskimmers (*Brechmorhoga mendax*)**

10(8'). 2 to 4 parallel rows of cells between vein A_2 and marginal row at hind angle of hindwing | **Setwings (*Dythemis*)**

10'. 4 or 5 irregular rows of cells between vein A_2 and marginal row at hind angle of hindwing | **Red Rock Skimmer (*Paltothemis lineatipes*)**

11(3'). Midrib of anal loop nearly straight, or only slightly kinked at ankle | 12

11'. Midrib of anal loop angulate | 14

12(11). Forewing triangle of 2–4 cells, usually with 3 or 4 cells beyond in trigonal interspace; radial planate often subtending 2 rows of cells | **Small Pennants (*Celithemis*)**

12'. Forewing triangle of 1 cell, with 2 rows of cells beyond in trigonal interspace; radial planate subtending a single row of cells | 13

13(12'). Inner side of forewing triangle about as long as front side; more than a single bridge crossvein | **Amberwings (*Perithemis*)**

13'. Inner side of forewing triangle much longer than front side; generally only a single bridge crossvein | **Marl Pennant (*Macrodiplax balteata*)**

14(11'). Wings with more than a single bridge crossvein | **Tropical Dashers (*Micrathyria*)**

14'. Wings with a single bridge crossvein | 15

KEY TO THE GENERA OF SKIMMERS (LIBELLULIDAE) (cont.)

15(14'). Wings with a triple-length vacant space before single crossvein, the crossvein either under the distal end of pterostigma or just beyond it — **Blue Dasher (*Pachydiplax longipennis*)**

15'. Wings with 1 or more crossveins under the pterostigma and without a triple-length vacant space — 16

16(15'). Wings with a single crossvein under pterostigma; no dark band across entire base of hindwing — **Meadowhawks (*Sympetrum*)**

16'. Wings generally with 2 or more crossveins under pterostigma; often a dark band across entire base of hindwing — 17

17(16'). Hindwing with 2 paranal cells before anal loop — **Tropical Pennants (*Brachymesia*)**

17'. Hindwing with 3 paranal cells before anal loop — 18

18(17'). Pterostigma trapezoidal, front side distinctly longer than rear; some double-length cells above apical planate reaching from planate to M_1 — 19

18'. Pterostigma not trapezoidal, front and rear sides equal in length; apical planate poorly developed, no double-length cells reaching to M_1 — 21

19(18). All of the cells above apical planate double-length and in a single row; 1 crossvein under pterostigma — **Hyacinth Glider (*Miathyria marcella*)**

19'. 1/2 of the cells above apical planate in a single row, followed by a double row; 2 crossveins under pterostigma — 20

20(19'). Forewing with 3 rows of cells in trigonal interspace — **Aztec Glider (*Tauriphila azteca*)**

20'. Forewing with 4 rows of cells in trigonal interspace — **Saddlebags (*Tramea*)**

21(18'). Spines on outer angle of hind femur gradually increasing in length distally — **Dragonlets (*Erythrodiplax*)**

21'. Spines on basal 1/2 to 2/3 of outer angle of hind femur short and of nearly equal length, with 2–4 large spines on distal 1/2 to 1/3 — **Pondhawks (*Erythemis*)**

22(5). Forewing with pair of brown stripes at midbasal area — **Blue Corporal (*Ladona deplanata*)**

22'. Forewing variously marked, but not as above — 23

KEY TO THE GENERA OF SKIMMERS (LIBELLULIDAE) (*cont.*)

23 (22'). First abdominal sternite of male with pair of large, conspicuous processes; middle abdominal segments of female with side margins parallel, not tapering **Whitetails (*Plathemis*)**

23'. First abdominal sternite of male without conspicuous processes; middle abdominal segments of female with side margins tapering posteriorly **King Skimmers (*Libellula*)**

Tropical Pennants

Genus *Brachymesia* Kirby

This is a small genus of three species, two of them Neotropical. All three occur in our region. They are medium-sized with a brown thorax, and have a red, brown, or black abdomen. The wings are clear, smoky, or adorned with a large dark spot beyond the nodus. The apical planate subtends 2 or 3 rows of cells. The abdomen is swollen basally and strongly compressed to the end. These dragonflies are found around lakes, ponds, marshes, and ditches, often perched high on the riparian vegetation.

References. Byers (1936).

KEY TO THE SPECIES OF TROPICAL PENNANTS (*BRACHYMESIA*)

1. Abdomen red or brown, without black middorsal stripe **Red-tailed (*furcata*)**

1'. Abdomen with black middorsal stripe or entirely dark blue-black 2

2(1'). Face black with light markings; pterostigma white **Four-spotted (*gravida*)**

2'. Face pale; pterostigma buff-colored **Tawny (*herbida*)**

Red-tailed Pennant
Brachymesia furcata (Hagen)
(photos 46e, 46f)

Size. Total length: 38–46 mm; abdomen: 23–30 mm; hindwing: 30–36 mm.
Regional Distribution. *Biotic Provinces:* Austroriparian, Balconian, Chihuahuan, Tamaulipan, Texan. *Watersheds:* Colorado, Guadalupe, Lower Rio Grande, Nueces, San Jacinto, Upper Rio Grande.

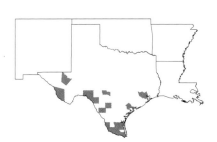

General Distribution. Southwestern United States and southern Florida through Central America to Chile and Argentina.

Flight Season. May 13 (TX)–Oct. 28 (TX).

Identification. This species is found in the southern portions of Texas and can be distinguished from the other two tropical pennants by having a relatively short abdomen, distinctly shorter than the wings, that becomes brilliant red in mature individuals. The face is yellow or red, and the prothorax is bilobed and covered by a dense fringe of long hairs. The pterothorax is brown, unmarked, and densely clothed with short hairs. The wings are clear, with a hint of yellow basally in the forewing and a larger spot in the hindwing. The legs are brown, becoming black distally. The abdomen is swollen basally and strongly compressed, tapering toward the tip. It is bright red in older males and some females, but usually yellowish-brown in the latter.

Similar Species. Mayan Setwing (*Dythemis maya*) has a longer abdomen, and the basal fifth of the hindwing is orange. Flame (*Libellula saturata*) and Neon (*L. croceipennis*) Skimmers are larger and have extensive brown basally on the wings. Tropical king skimmers (*Orthemis*) are much larger and have no color at their wing bases.

Habitat. Ponds, lakes, and ditches with permanent or semipermanent water, including brackish waters.

Discussion. This species frequently forages from atop bushes and tall grasses. Males, when not patrolling, perch on twigs or other vegetation extending out over the water. The species may be common around cattle tanks and resacas of south Texas. Pairs mate for a short period, just 15 seconds, and the male then guards the female as she dips the eggs in the water along the shoreline.

References. Dunkle (1989, 2000).

Four-spotted Pennant
Brachymesia gravida (Calvert)
(photos 47a, 47b)

Size. Total length: 47–55 mm; abdomen: 30–40 mm; hindwing: 32–42 mm.

Regional Distribution. *Biotic Provinces:* Austroriparian, Balconian, Chihuahuan, Kansan, Navahonian, Tamaulipan, Texan. *Watersheds:* Arkansas, Bayou Bartholomew, Brazos, Canadian, Cimarron, Colorado, Guadalupe, Lower Rio Grande, Mississippi, Neches, Nueces, Ouachita, Red, Sabine, San Antonio, San Jacinto, Trinity, Upper Rio Grande, White.

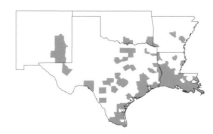

General Distribution. Found largely along coastal areas from New Jersey to Texas and on to Arizona.

Flight Season. Apr. 27 (LA)–Oct. 27 (TX).

Identification. This is the most widespread and most commonly encountered tropical pennant in our area. The face is black and white, becoming entirely black, along with the top of the head, in older individuals. The thorax is brown, becoming dark bluish black with age, but less so in females. The wings have a distinct white pterostigma. There are four rows of cells beyond the forewing triangle. Males and mature females develop a dark-brown spot between the nodus and pterostigma in all wings. The legs are black. The abdomen is brown, with a dark interrupted longitudinal middorsal stripe in tenerals and females. This stripe becomes entirely black in males. The abdomen is swollen basally and compressed for a short distance thereafter, but never tapers to the end.

Similar Species. Young individuals are similar to Tawny Pennant (*B. herbida*), but the latter have a tan face and pterostigma and they lack prominent wing spots. Band-winged Dragonlet (*Erythrodiplax umbrata*) is similar, but has complete bands in the wings rather than spots, and has pale lateral spots on the middle abdominal segments.

Habitat. Ponds, lakes, and roadside ditches, including brackish waters.

Discussion. This species is typical of the genus, often perching high on twigs, stems, and bush tops. It may be abundant on fence wire and telephone lines. After a brief mating, females oviposit in a fashion similar to that of Red-tailed Pennant (*B. furcata*), dipping eggs into the water, sometimes guarded by the male.

Tawny Pennant
Brachymesia herbida (Gundlach)
(photo 47c)

Size. Total length: 43–48 mm; abdomen: 32–36 mm; hindwing: 33–38 mm.

Regional Distribution. *Biotic Provinces:* Austroriparian, Balconian, Chihuahuan, Tamaulipan, Texan. *Watersheds:* Colorado, Guadalupe, Lower Rio Grande, Neches, San Jacinto, Upper Rio Grande.

General Distribution. Southern Florida and Texas through Central America south to Argentina.
Flight Season. May 19 (TX)–Aug. 21 (TX).
Identification. This species is rare in the south-central United States, and may be confused with teneral individuals of Four-spotted Pennant (*B. gravida*). The pterostigma is tan, not white, and the wings lack a dark spot, although they may be smoky. The face is tan or brown, and the thorax is brown and unmarked. The wings may be amber or smoky, especially along the front border, but they never have a dark spot beyond the nodus. The abdomen has a black middorsal stripe and a yellow lateral stripe.
Similar Species. See *Similar Species* under Four-spotted Pennant, above.
Habitat. Ponds, lakes, marshes, and roadside ditches, including brackish waters.
Discussion. Nothing has been published on the biology of this tropical species, but it seems most similar to Four-spotted Pennant in that respect. It occurs year-round farther south in its established range.

Clubskimmers
Genus *Brechmorhoga* Kirby

This is a Neotropical genus of moderate size, a single species of which occurs in the western part of our region (Masked Clubskimmer (*B. pertinax*) occurs in Arizona). Most species are blue-gray in color. The wings are clear or smoky in some females, generally have two crossveins in the forewing triangle, and lack an apical planate. The combination of their color and the slightly expanded terminal abdominal segments may lead to the illusion of a clubtail (Gomphidae) dragonfly and is the basis for the name "clubskimmer."

Pale-faced Clubskimmer
Brechmorhoga mendax (Hagen)
(photo 47d)

Size. Total length: 52–64 mm; abdomen: 34–46 mm; hindwing: 32–44 mm.
Regional Distribution. *Biotic Provinces:* Apachian, Austroriparian, Balconian, Chihuahuan, Kansan, Navahonian, Tamaulipan, Texan. *Watersheds:* Arkansas, Brazos, Canadian, Colorado, Colorado (NM), Lower Rio Grande, Guadalupe, Nueces, Ouachita, Pecos, Red, San Antonio, Upper Rio Grande.
General Distribution. Southwestern United States and Great Plains.

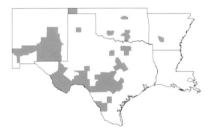

Flight Season. Mar. 17 (TX)–Nov. 11 (TX).
Identification. This species is widespread throughout the western portion of the region. It has a pale face and bluish-gray thorax. There are two broadly confluent stripes middorsally on the thorax for 2/3 to 3/4 of its length. The broad humeral stripe is brown. Two additional lateral stripes are confluent above. The wings, which sometimes become amber in females, have a dark basal spot, the spot more prominent in the hindwing. The legs are brown, becoming darker distally. The abdomen, largely black with pale basal segments, is slender basally, then slightly expanded at segments 7–9 in males. There are paired bluish-gray spots dorsally on segments 3–6, but these quickly become obscured with age. Segment 7 bears a pair of large pale dorsal spots that are never obscured.

Similar Species. The stout shape, clubbed abdomen, color, and markings of this species may lead to its misidentification as a clubtail (Gomphidae), but the eyes are in contact on top of the head. Tropical dashers (*Micrathyria*) are much smaller and behave differently.

Habitat. Sand and cobble streams and rivers.

Discussion. Males of this species typically have small territories that they patrol less than a meter over the water, and can be elusive. These clubskim-mers have been described as the most graceful on the wing of any odonate. Females lay eggs by making short, straight or figure-eight runs low over the water and dipping eggs at the surface. Adults are active all day, but may retreat to shaded areas of the stream in the heat of the day. They are often abundant at dusk in clearings near streams, where they feed on emerging mayflies and caddisflies.

References. Kennedy (1917).

Narrow-winged Skimmers
Genus *Cannaphila* Kirby

This is a small Neotropical genus of three species, a single one of which occurs as far north as southern Texas. They are rather nondescript as a group, except for having the basal portion of the hindwings only slightly wider than the forewings. The toe of the anal loop is poorly developed for a skimmer. There are two cubito-anal crossveins, and vein Cu1 arises from the outer side of the hindwing triangle.

Gray-waisted Skimmer
Cannaphila insularis Kirby
(photo 47e)

Size. Total length: 36–39 mm; abdomen: 24–26 mm; hindwing: 29–32 mm.

Regional Distribution. *Biotic Provinces:* Balconian, Tamaulipan, Texan. *Watersheds:* Guadalupe, Lower Rio Grande, San Antonio.

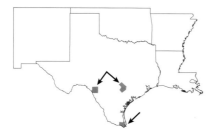

General Distribution. Southern Texas through Mexico to Panama.

Flight Season. Jun. 27 (TX)–Sep. 8 (TX).

Identification. This uncommon nondescript species is only rarely reported from Texas. The best diagnostic character is the unusually narrow base of the hindwing. The body is brown, and the face a pale yellow or white. The top of the head becomes metallic green in older males. The eyes are reddish, turning metallic green in older individuals. The thorax is brown, darker in front, with streaks of yellow becoming obscured with age. The wings are unusually narrow, as described above, and have dark wingtips and a brown pterostigma. The legs are brown, turning black distally. The abdomen is brown, edged by a black carina, with some pale yellow dorsally and basally on segments 1–8. The female has a broad flange on segment 8. Older individuals (males and most females) become entirely black, with a gray pruinescence on the abdomen.

Similar Species. Young Blue Dashers (*Pachydiplax longipennis*) have an interrupted dorsolateral abdominal stripe. Older Slaty Skimmers (*Libellula incesta*) and Black Setwings (*Dythemis nigrescens*) have dark or black faces.

Habitat. Marshy ponds, lakes, and streams.

Discussion. Gray-waisted Skimmer forages from the tips of vegetation just above ground level to 3 m high. Females lay eggs in areas of dense vegetation.

References. Dunkle (2000).

Small Pennants
Genus *Celithemis* Hagen

This is a genus of eight usually colorful eastern species. All but a single northern species occur in the south-central United States. Members of this group are commonly seen perching on grass, bushes, or tall twigs in open fields or around ponds, lakes, or marshes. The face is pale yellow, often turning brilliant red in older individuals. The thorax is pale, with dark stripes that become obscured with age. The humeral and antehumeral stripes generally form a single broad dark shoulder stripe. The wings are generally colorful and variously marked, with at least a basal bicolored spot in the hindwing. The color pattern of the wings is the best basis for identification. The legs are usually black, with a touch of pale color on the underside of the femora. The midrib of the anal loop is nearly straight. The abdomen is slender and only slightly compressed along the middle segments.

References. Williamson (1922b).

KEY TO THE SPECIES OF SMALL PENNANTS (*CELITHEMIS*)

1. Wings clear or with only a touch of color basally — 2

1'. Wings with large dark spots at or near the nodus — 5

2(1). Basal color in hindwing extending 1/2 of the way to the nodus — 3

2'. Basal color in hindwing extending less than 1/3 the distance to the nodus — 4

3(2). Thorax unmarked laterally — **Amanda's (*amanda*)**

3'. Thorax with dark lateral stripes — **Faded (*ornata*)**

4(2'). Wing veins red — **Red-veined (*bertha*)**

4'. Wing veins not red — **Double-ringed (*verna*)**

5(1'). Wings yellow to orange, with dark broad stripe just before pterostigma extending entire width of wing — **Halloween (*eponina*)**

5'. Wings clear with dark markings, but never with dark broad stripe just before pterostigma extending across width of wing — 6

6(5'). Basal dark spot touching nodus of wing; round spot between nodus and pterostigma large, touching costa — **Banded (*fasciata*)**

6'. Basal dark spot not touching nodus of wing; round spot between nodus and pterostigma small, not touching costa — **Calico (*elisa*)**

Amanda's Pennant
Celithemis amanda (Hagen)
(photo 47f)

Size. Total length: 24–31 mm; abdomen: 16–22 mm; hindwing: 21–27 mm.

Regional Distribution. *Biotic Province:* Austroriparian. *Watersheds:* Mississippi, Neches, Ouachita, Red, Sabine, Trinity.

General Distribution. Southeastern United States, generally along coast, from Florida to Texas.

Flight Season. Jun. 17 (LA)–Sep. 9 (LA).

Identification. This is a small species, uncommon in the south-central United States, can be recognized by the extent of the wing markings and the unmarked thorax. Its face is yellow, becoming brown or red in older males. The thorax is yellow, turning brown with age, and largely unmarked. The forewings are clear, but have red veins. The hindwings have a large basal amber or brown spot extending approximately 1/4 the length of each wing. The spot generally contains two anterior black stripes and a single posterior black stripe. The abdomen is largely dark brown or black, with pale dorsal spots on segments 1–7. Segment 3 and the basal portion of 4 are pale yellow, turning red in older individuals.

Similar Species. On Faded Pennant (*C. ornata*) two or three dark stripes run through a basal spot in the hindwing. Other similar small pennants have larger spots in the hindwing.

Habitat. Calm lakes, ponds, and marshes with emergent vegetation.

Discussion. This species is uncommon west of the Mississippi River. As is typical for the genus, it is often seen perching on tall grasses or bushes around the water. Pairs mate while perched on emergent vegetation. Females thereafter lay eggs along the shore, accompanied by the male in a fashion similar to that of Faded Pennant.

Red-veined Pennant
Celithemis bertha Williamson
(photo 48a)

Size. Total length: 26–37 mm; abdomen: 16–23 mm; hindwing: 23–28 mm.

Regional Distribution. *Biotic Province:* Austroriparian. *Watersheds:* Mississippi, Red.

General Distribution. Southeastern United States from North Carolina to Louisiana.

Flight Season. May 28 (LA)–Sep. 23 (LA).

Identification. This is another, smaller uncommon species found west of the Mississippi River that within our region is known only from Louisiana. Males become bright red, including the wing veins. There is a small basal amber or black spot in the hindwing. The face, thorax, and pale abdominal markings are yellow in younger individuals but become red with age. The wide, black humeral stripe and 3rd lateral stripe are not connected. The mid-lateral thoracic stripe is broad and fully developed below the spiracle. The face, thorax, and wing veins remain yellow in females. The abdomen is slender and mostly black, with pale basal markings laterally. There are pale dorsal spots on segments 3–7.

Similar Species. This species can be distinguished from other small pennants by its thoracic markings, the red veins in mature males, and the relative absence of markings in the hindwing. The lateral thoracic markings in Faded Pennant (*C. ornata*) are connected dorsally and the hindwing spot is larger. In female and young male Seaside Dragonlets (*Erythrodiplax berenice*), the thorax is entirely black or pale with numerous black stripes.

Habitat. Lakes, ponds, pools, roadside ditches, and borrow pits with emergent vegetation.

Discussion. This species was erroneously reported from the Jones State Forest in Montgomery County, Texas. The specimen in question was actually Faded Pennant (R. Orr, pers. comm.). Red-veined Pennant does not appear to range as far west as the Piney Woods of east Texas, though it has been noted that the species has an apparent prefrence for pine trees as roosting sites. This species is similar to Amanda's Pennant (*C. amanda*) in its habits and behavior. Females lay eggs accompanied by males.

References. Dunkle (1989), Price et al. (1989).

Calico Pennant
Celithemis elisa (Hagen)
(photos 48b, 48c)

Size. Total length: 24–34 mm; abdomen: 16–22 mm; hindwing: 25–30 mm.
Regional Distribution. *Biotic Provinces:* Austroriparian, Balconian, Carolinian, Kansan, Texan. *Watersheds:* Arkansas, Bayou Bartholomew, Brazos, Canadian, Cimarron, Guadalupe, Mississippi, Neches, Nueces, Ouachita, Red, Sabine, San Antonio, San Jacinto, St. Francis, Trinity, White.

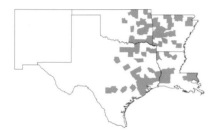

General Distribution. Widespread throughout eastern United States and Canada.
Flight Season. Mar. 9 (LA)–Sep. 25 (LA).
Identification. This is one of the more common and widespread small pennants in our area. It is easily distinguished by the wing markings, which include a brown spot basally, beyond the nodus and at the tips. The basal spot in the hindwing is large, occupying 1/4 or more of the wing area, and it is usually bicolored, encompassing a central amber area. The yellow pterostigma becomes red with age. The face is yellow, but becomes bright red in older males. The thorax is yellowish brown, with a typical dark middorsal stripe and diffuse brown stripes laterally on each suture. The abdomen is dark brown or black, with basal pale markings laterally on segments 1–4 and dorsally on segments 3–7.

Similar Species. Halloween Pennant (*C. eponina*) is larger and has orange wings. Banded Pennant (*C. fasicata*) has black bands in the wings, the basal band extending to the nodus.
Habitat. Lakes, ponds, and borrow pits with calm, clear waters and emergent vegetation.
Discussion. This species perches on top of tall grasses and weeds in open fields and surrounding water. Males are not territorial and perch facing away from the water, apparently to intercept females as they approach the water. Mating lasts an average of 5 minutes, and egg-laying, accompanied by the male or alone, requires 3–5 minutes.

References. Dunkle (1989), Waage (1986).

Halloween Pennant
Celithemis eponina (Drury)
(photo 48d)

Size. Total length: 30–42 mm; abdomen: 20–30 mm; hindwing: 27–35 mm.
Regional Distribution. *Biotic Provinces:* Austroriparian, Balconian, Carolinian, Kansan, Navahonian, Tamaulipan, Texan. *Watersheds:* Arkansas, Bayou Bartholomew, Brazos, Canadian, Cimarron, Colorado, Guadalupe, Lower Rio Grande, Mississippi, Neches, Nueces, Ouachita, Pecos, Red, Sabine, San Antonio, San Jacinto, St. Francis, Trinity, White.

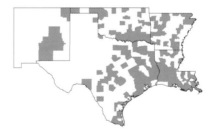

General Distribution. Widespread throughout eastern United States and Canada.
Flight Season. May 16 (OK)–Oct. 15 (TX).
Identification. This is the largest, most widely distributed, and perhaps most colorful small pennant species in the region. Its common name derives from its distinctive orange and brown or black wings. The face is yellowish or olivaceous, becoming darker with age, and red in males. The thorax is yellowish green, with a dark middorsal stripe and

narrow lateral stripes on the sutures; the midlateral stripe is usually not continuous after the spiracle. This is the only species with completely yellowish-orange wings marked with broad dark-brown or black stripes and a red pterostigma. The abdomen is slender, with pale yellow dorsal spots on segments 3–7 that become red with age.

Similar Species. This is a very recognizable species. Banded Pennant (*C. fasciata*) has clear wings with black bands, and Calico Pennant (*C. elisa*) is smaller, with patterning on otherwise clear wings.

Habitat. Lakes, ponds, borrow pits, and marshes with emergent vegetation.

Discussion. This species may be locally abundant. It forages from atop tall grasses, weeds, and stems in open fields, some distance from the water. It perches in highly unusual fashion, with the fore- and hindwings in different planes, the forewings held somewhat vertically and the hindwings held horizontally. Most activity takes place in the early to midmorning hours. Females lay eggs manner similar to that of other species, accompanied by males. Halloween Pennant's fluttering flight has been compared to that of a butterfly.

References. Dunkle (1989), Miller (1982), Needham (1901), Whedon (1914).

Banded Pennant
Celithemis fasciata Kirby
(photos 48e, 48f)

Size. Total length: 28–38 mm; abdomen: 17–26 mm; hindwing: 24–32 mm.

Regional Distribution. *Biotic Provinces:* Austroriparian, Balconian, Carolinian, Kansan, Texan. *Watersheds:* Arkansas, Bayou Bartholomew, Brazos, Canadian, Cimarron, Colorado, Guadalupe, Mississippi, Neches, Nueces, Ouachita, Pecos, Red, Sabine, San Antonio, San Jacinto, St. Francis, Trinity, White.

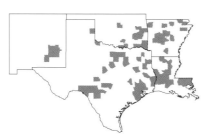

General Distribution. Eastern United States from New York to Texas.

Flight Season. May 17 (LA)–Sep. 18 (TX).

Identification. This species is easily identified by its dark color and clear wings, the wings bearing large black spots. Its face and body are bright yellow, but quickly become black or dark blue in males. The thorax is yellow, striped with black on the humeral, midlateral, and third lateral sutures, but becomes entirely black at maturity. The large basal black spot on the wings extends out to the nodus and encompasses an amber central area in the hindwing. There is a dark spot in the outer half of each wing, and the wingtips are black. The abdomen is black, with pale-yellowish dorsal markings on segments 5–7 that quickly become obscured in older individuals.

Similar Species. Halloween Pennant (*C. eponina*) has orange wings. The basal markings in the hindwing of Calico Pennant (*C. elisa*) are smaller and do not extend out to the level of the nodus.

Habitat. Permanent lakes, ponds, and borrow pits with emergent vegetation.

Discussion. Banded Pennant prefers protected areas of ponds and lakes with thick growths of trees or bushes. They forage, as do other species in this group, from tall grasses or stems. They often perch, in a fashion similar to that of Halloween Pennant, with the forewings elevated above the hindwings. Females lay eggs accompanied by males or alone.

Faded Pennant
Celithemis ornata (Rambur)
(photos 49a, 49b)

Size. Total length: 31–36 mm; abdomen: 21–26 mm; hindwing: 21–28 mm.

Regional Distribution. *Biotic Province:* Austroriparian. *Watersheds:* Mississippi, Red, Sabine, San Jacinto, Trinity.

General Distribution. Eastern U.S. coast from New Jersey and Florida to Texas.

Flight Season. Mar. 12 (LA)–Sep. 9 (LA).

Identification. This small reddish species is similar to Amanda's Pennant (*C. amanda*), but has a smaller spot in the hindwing and the thorax is marked as below. The face and thorax are pale olivaceous. The middorsal thoracic stripe is wide and brown. There is a broad humeral, a midlateral, and third lateral stripe on each side of the thorax. The last two stripes are confluent above. The basal 1/5 of the hindwing is amber or brown. The abdomen is black with pale yellow spots dorsally on segments 1–7, the spots becoming red in older males.

Similar Species. This species can be distinguished from Amanda's Pennant by the character described above. Mature Red-veined Pennants (*C. bertha*) are brighter red, and their lateral thoracic markings are separated.

Habitat. Lakes, ponds, and pools with calm waters, and slow reaches of streams, all with emergent vegetation.

Discussion. This dainty species behaves much like the other small pennants discussed above. It may be seen perched high on tall grasses and stems when foraging in open fields. It is more common in the eastern portion of its range. Females lay eggs accompanied by males, along the shorelines of ponds, lakes, and marshes.

Double-ringed Pennant
Celithemis verna Pritchard
(photos 49c, 49d)

Size. Total length: 31–36 mm; abdomen: 20–24 mm; hindwing: 25–29 mm.

Regional Distribution. *Biotic Provinces:* Austroriparian, Kansan, Texan. *Watersheds:* Bayou Bartholomew, Canadian, Cimarron, Mississippi, Ouachita, Red, Trinity.

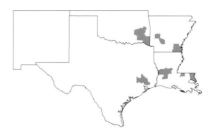

General Distribution. Southeastern United States from New Jersey to Florida and Texas.

Flight Season. Apr. 12 (LA)–Jun. 12 (LA).

Identification. This species differs from the other small pennants in the south-central United States in lacking pale spots dorsally on abdominal segments 5–7, and in having only a small black basal spot on the hindwing. Its face and thorax are yellowish, becoming black in older males. The black middorsal and humeral stripes are confluent at their lower ends. The midlateral and third lateral stripes are interrupted and irregular. The abdomen is black except for pale basal segments in tenerals and females.

Similar Species. Seaside Dragonlet (*Erythrodiplax berenice*) lacks a dark basal spot in the hindwing.

Habitat. Newly formed lakes and ponds with emergent vegetation.

Discussion. This species was originally described from Quinton, Oklahoma, where Pritchard stated that he ". . . had very little success . . . in finding *verna* in the daytime. At daybreak, however, . . . numbers were found emerging among water lilies. During the day, several teneral specimens which were flushed from the vegetation bordering the lake fluttered to the tree tops." Minter Westfall noted this species as locally common near Hendersonville, North Carolina, but difficult to capture. It is an active species, often found out over the water just beyond reach. Westfall used a slingshot to capture single males. Females lay eggs accompanied by the male, dipping the abdomen in the water along the shore at frequent intervals.

References. Pritchard (1935).

Setwings
Genus *Dythemis* Hagen

This genus comprises seven New World species, four of which occur in the south-central United States. Members of this group are commonly seen perching atop grasses and other vegetation with both wings depressed downward and the abdomen raised above the rest of the body. These "setwings" are often found near streams and rivers of moderate current and lakes and ponds with emergent vegetation.

They are medium-sized, grayish blue to yellow or red, some species becoming a heavily pruinose dark blue with age. The wings may have large basal spots or only a hint of color. The anal loop in the hind wing is a well-developed foot and is strongly angulate at the ankle. There are generally two to four parallel rows of cells between vein A2 and the hindwing margin.

KEY TO THE SPECIES OF SETWINGS (*DYTHEMIS*)

1. Wings with a wide basal crossband of red or brown out to the hindwing triangle	2
1'. Wings without a wide basal crossband of color	3
2(1). Thorax and abdomen bright red; basal color of wings orange-red	**Mayan (*maya*)**
2'. Thorax and abdomen brown or black; basal color of wings dark brown	**Checkered (*fugax*)**
3(1'). Top of frons metallic purple; female with pale lateral streaks on abdominal segments 4–7	**Black (*nigrescens*)**
3'. Top of frons not metallic; female without pale lateral streaks on abdomen	**Swift (*velox*)**

Checkered Setwing
Dythemis fugax Hagen
(photo 50a)

Size. Total length: 42–51 mm; abdomen: 30–35 mm; hindwing: 35–40 mm.

Regional Distribution. *Biotic Provinces:* Apachian, Austroriparian, Balconian, Carolinian, Chihuahuan, Kansan, Navahonian, Tamaulipan, Texan. *Watersheds:* Arkansas, Brazos, Canadian, Cimarron, Colorado, Colorado (NM), Guadalupe, Lower Rio Grande, Mississippi, Neches, Nueces, Pecos, Red, Sabine, San Antonio, San Jacinto, Trinity, Upper Rio Grande.

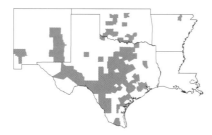

General Distribution. South-central United States.

Flight Season. Apr. 25 (TX)–Dec. 26 (TX).

Identification. This is primarily a south-central United States species, easily recognized by its dark color and the large basal brown spots that extend out not quite to 1/4 the length of the wings. It is the only dark setwing in the region, and has a broad dark basal spot in both pairs of wings. The face is olivaceous in young males and females, and bright red in older males. The brown middorsal and antehumeral stripes are broadly confluent above, exposing only a small isolated pale area. The side of the thorax is bluish gray or olivaceous, with four brown diffuse lateral stripes. The legs are black, and the wings are as above, with a smaller spot of brown at the nodus and generally with dark wingtips. The abdomen is bluish gray to yellow basally and dark brown to black for most of its length. There are pale spots laterally on segments 4–9 and middorsally on 4–7, these becoming large and most conspicuous on 7, giving it a distinctive checkered pattern. The remaining segments and caudal appendages are dark brown or black.

Similar Species. No other setwing in the region has broad basal markings in both wings and a checkered abdomen. Marl Pennant (*Macrodiplax balteata*) does not have the characteristic checkered pattern on the abdomen, and lacks a dark spot in the basal portion of the forewing.

Habitat. Ponds and lakes with emergent vegetation.

Discussion. This species may be abundant, perching on tall vegetation surrounding ponds and lakes, or in open fields away from water. Males perch at the tips of vegetation and along fence lines with both pairs of wings depressed downward. When disturbed they usually do not fly far, returning quickly to their perch. The species was only recently reported in Arkansas for the first time.

References. Harp and Harp (1996).

Mayan Setwing
Dythemis maya Calvert
(photos 49e, 49f)

Size. Total length: 37–41 mm; abdomen: 23–26 mm; hindwing: 30–33 mm.

Regional Distribution. *Biotic Provinces:* Chihuahuan, Tamaulipan. *Watershed:* Upper Rio Grande.

General Distribution. Southeast Arizona and west Texas (Big Bend) south through Mexico.

Flight Season. Jul. 28 (TX)–Oct. 3 (TX).

Identification. This is an uncommon brilliant-red species known only from isolated localities throughout its range. Its face, thorax, and abdomen are unmarked and all red, brilliantly so in males. Young individuals are brown. Both wings have a broad basal crossband of deep amber for 1/4 or more of their length. Females may have dark wingtips. The legs are brown with black tarsi. The abdomen is noticeably wider than those in other setwing species.

Similar Species. This the only red setwing in the area, but it may be confused with other red skim-

mers that it flies with. Red Rock Skimmer (*Paltothemis lineatipes*) is profusely marked with black on the thorax and abdomen and perches on rocks and on the ground. Neon (*Libellula croceipennis*) and Flame (*L. saturata*) Skimmers are larger and have more extensive color in the wings. Red-tailed Pennant (*Brachymesia furcata*) lacks the broad amber patches in the hindwing and has thin black rings around each segment.

Habitat. Small arid streams with moderate to swift current.

Discussion. This species was first collected in the United States in Big Bend Ranch State Natural Area in Presidio County, Texas, where it can be locally common. At that site it flies with a number of other similar red species, namely Flame, Neon, and Red Rock Skimmers. (Differences among these species are given above.) Mayan Setwing flies from midmorning well into the afternoon. Males patrol short territories scarcely more than 10 m in breadth, with regular attentiveness. When not patrolling, they perch on vegetation overhanging the stream.

References. Abbott (1996).

Black Setwing
Dythemis nigrescens Calvert
(photos 50b, 50c)

Size. Total length: 42–50 mm; abdomen: 26–32 mm; hindwing: 31–34 mm.

Regional Distribution. *Biotic Provinces:* Balconian, Chihuahuan, Kansan, Tamaulipan, Texan. *Watersheds:* Brazos, Colorado, Guadalupe, Lower Rio Grande, Nueces, Pecos, San Antonio, Upper Rio Grande.

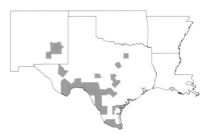

General Distribution. Arizona, New Mexico, and Texas south through Mexico.

Flight Season. Apr. 11 (TX)–Dec. 26 (TX).

Identification. This dark Mexican species, similar to Swift Setwing (*D. velox*), is found in the southwestern part of the region. The top of the frons is metallic purple, and older individuals become entirely pruinose dark blue. The face and thorax of tenerals and females are olivaceous. A broad, dark middorsal stripe covers nearly the entire front of the thorax. Diffuse lateral stripes form a "HII" or "HIY" pattern on the side of the thorax. The entire thorax becomes obscured by a dark blue pruinosity in older individuals. The wings are clear, with at most a slight spot of dark brown at the extreme base of both wings and generally at the extreme tips. With age the wings become amber throughout. Generally there are four rows of cells between vein A2 and the hindwing margin. The legs are brown basally, becoming black on the tarsi and tibiae. The abdomen in the male is slender beyond the swollen, pale basal segments. There are paired, pale, middorsal spots on segments 4–7, those on 7 conspicuously enlarged. The remaining segments and caudal appendages are dark.

Similar Species. The vertex in Swift Setwing is not metallic, there are no pale lateral spots on segments 4–7, and the lateral thoracic stripes appear as "YIY." Marl Pennant (*Macrodiplax balteata*) is more robust, and has a distinct dark basal spot in the hindwing. Slaty Skimmer (*Libellula incesta*) is much larger.

Habitat. Creeks, streams, and rivers with moderate current.

Discussion. Individuals perch atop twigs near the water, but generally in open areas. The species exhibits behavior similar to that of Swift Setwing.

Swift Setwing
Dythemis velox Hagen
(photo 50d)

Size. Total length: 42–50 mm; abdomen: 25–32 mm; hindwing: 30–36 mm.

Regional Distribution. *Biotic Provinces:* Austroriparian, Balconian, Carolinian, Chihuahuan, Kansan, Navahonian, Tamaulipan, Texan. *Watersheds:* Arkansas, Bayou Bartholomew, Brazos, Canadian, Cimarron, Colorado, Guadalupe, Lower Rio Grande, Mississippi, Neches, Nueces, Ouachita, Pecos, Red, Sabine, San Antonio, San Jacinto, St. Francis, Trinity, Upper Rio Grande, White.

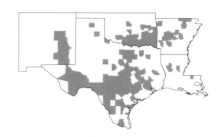

General Distribution. Southern United States from Florida to Arizona, and Mexico; also Cuba.

Flight Season. Mar. 17 (TX)–Oct. 14 (TX).

Identification. This clear-winged species is similar to Black Setwing (*D. nigrescens*), but the top of the frons is dull and never metallic. Its face is olivaceous in tenerals and females and dark brown in older males. The lateral thoracic pattern of dark stripes is "YIY." This pattern is nearly always visible, since older males generally lack heavy pruinescence. The wingtips are dark and usually have three rows of cells between vein A2 and the hindwing margin. The wings become amber in older individuals and have a spot of brown at the extreme base in both wings. The legs are black. The abdomen has pale yellow-greenish dorsolateral spots on segments 3–7, the spots becoming most conspicuous on segment 7 and sometimes lacking on 5 and 6.

Similar Species. This species is closest to Black Setwing, but Swift Setwing has a "YIY" lateral thoracic pattern, not "HII" or "HIY." It also has darker wingtips and lacks a metallic vertex. (See *Similar Species* under Black Setwing for details on separating this species from others.)

Habitat. Lakes, ponds, and borrow pits, as well as creeks, streams, and rivers with moderate current.

Discussion. This is the most widespread of the North American setwings, and is apparently continuing to expand its range. Bick (1957) did not find it in Louisiana, but Mauffray (1997), 40 years later, reported that it was common in the northern part of the state, west of Baton Rouge. It is often found along ponds, borrow pits, streams, creeks, and rivers, where it perches high on tall grasses and weeds with the wings depressed downward and the abdomen held above the rest of the body, sometimes considerably so. Males patrol small areas along the stream and creek edges, where they may be flighty.

Pondhawk
Genus *Erythemis* Hagen

This Neotropical genus numbers ten medium-sized species. The five occurring in the south-central United States are green, red, blue, or black, and members of this group often change considerably in color with age. They are stocky, with long legs, the legs well-armed with black spines. The wings are clear, or with color only basally. The abdomen is narrowed at its middle segments and distinctly triangular in cross section.

References. Kennedy (1923a), Williamson (1923b).

KEY TO THE SPECIES OF PONDHAWKS (*ERYTHEMIS*)

1. Total length greater than 55 mm; green	**Great (*vesiculosa*)**
1'. Total length less than 50 mm; green, red, blue, or black	2
2(1). Thorax black, brown, or yellowish	3
2'. Thorax green (or pruinose blue in older males)	4
3(2). Abdomen broad, segments 4–6 no more than twice as long as wide	**Flame-tailed (*peruviana*)**
3'. Abdomen long and narrow, segments 4–6 at least four times as long ventrally as wide	**Pin-tailed (*plebeja*)**
4(2'). Face all green; caudal appendages yellow	**Eastern (*simplicicollis*)**
4'. Face with black across frons; caudal appendages blackish	**Western (*collocata*)**

Western Pondhawk
Erythemis collocata (Hagen)
(photos 50e, 50f)

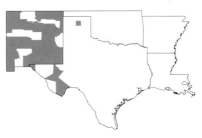

Size. Total length: 39–42 mm; abdomen: 23–30 mm; hindwing: 30–33 mm.
Regional Distribution. *Biotic Provinces:* Apachian, Chihuahuan, Coloradan, Kansan, Navahonian. *Watersheds:* Canadian, Colorado (NM), Pecos, Upper Rio Grande.
General Distribution. Western United States and Mexico.
Flight Season. May 26 (TX)–Oct. 4 (TX).
Identification. This common, widespread, western species is similar to Eastern Pondhawk (*E. simplici-* *collis*). Its face and thorax are bright green in tenerals and females. Its green eyes become blue in older individuals (more so in males). The thorax and abdomen in older males turns completely powder blue or black with a waxy pruinescence. The wings are clear. In young males and females, abdominal seg-

ments 1–3 are green and 4–10 are paler yellowish green with a black middorsal stripe through segments 2–9. The middle abdominal segments are broader than long, and the caudal appendages are black. The female has a scooplike ovipositor that projects ventrally.

Similar Species. Eastern Pondhawk is similar, but older males are paler blue, with yellow or white cerci. Females have a more spotted abdomen, and the middle abdominal segments in both sexes are narrow, longer than their respective widths. Great Pondhawk (*E. vesiculosa*) is larger and has complete black stripes or bands dorsally on abdominal segments 4–6, rather than spots. Snaketails (*Ophiogomphus*) and Ringtails (*Erpetogomphus*) all have the eyes separated on top of the head.

Habitat. Ponds, lakes, and slow-flowing waters of streams and creeks.

Discussion. This species was long considered a variant or subspecies of Eastern Pondhawk, and some evidence suggests that they may intergrade. It is common in parts of western Texas, where it does occur with Eastern Pondhawk. Nothing has been written on what mechanisms may allow these nearly identical species to distinguish each other where they occur together. Western Pondhawk is often seen perching on the ground, in the manner of clubtails, for which they may be mistaken, owing to their green color. They are capable and fierce predators in the air, taking small to large prey at will.

References. Gloyd (1958), Harrison and Lighton (1998).

General Distribution. Texas Hill Country south through Latin America to Argentina.

Flight Season. Jul. (TX).

Identification. Mature males of this distinctive pondhawk have bright-red abdomens contrasting with a darker black thorax. The face is black and the top of the frons yellow in young males and females. The thorax is dark in young males and females, with a broad stripe of yellow dorsally. The legs are black. The wings are clear, but have a small, narrow spot at the base of the hindwing. Abdominal segments 1–3 are swollen, and the rest of the abdomen tapers gradually to 9. The abdomen becomes brilliant red, more so dorsally, in older males. The cerci are red.

Similar Species. No other dragonfly in the region has the distinctive red abdomen contrasting with a black thorax that we see in this species.

Habitat. Ponds, lakes, ditches, and slow reaches of rivers and streams.

Discussion. Within North America, this species is known from only a single locality in the Texas Hill Country, where it was only recently discovered (summer 2001). The presence of multiple individuals there suggests a breeding population, though it must have only recently moved into the area. It is a common, even ubiquitous species in the tropics, and may become well-established in southern Texas. This pondhawk behaves like the other pondhawks in the south-central United States. The bright-red abdomen of males contrasts with the green vegetation where they occasionally perch.

Flame-tailed Pondhawk
Erythemis peruviana (Rambur)
(photo 51a)

Size. Total length: 37–43 mm; abdomen: 23–28 mm; hindwing: 29–34 mm.

Regional Distribution. *Biotic Province:* Balconian. *Watershed:* Colorado.

Pin-tailed Pondhawk
Erythemis plebeja (Burmeister)
(photos 51b, 51c)

Size. Total length: 41–49 mm; abdomen: 30–39 mm; hindwing: 30–37 mm.

Regional Distribution. *Biotic Provinces:* Balconian, Tamaulipan, Texan. *Watersheds:* Colorado, Guadalupe, Lower Rio Grande, San Antonio.

General Distribution. Southern Florida and Texas south through Central America to Argentina.

Flight Season. Apr. 11 (TX)–Oct. 15 (TX).

Identification. This is the only black pondhawk in the south-central United States. Its face and thorax are brownish in young males and females and black in older males. The hindwing of both sexes has a small black spot basally. Abdominal segments 1–3 are greatly swollen. The remaining segments are thin. Abdominal segments 1–3 are light brown, and segments 4–7 have light-brown rings. The entire abdomen, except for the brown cerci, becomes black in older males. The female has a ventrally projecting spoutlike ovipositor on segment 9.

Similar Species. This species is easily distinguished from other pondhawks by the combination of its dark color, the small basal black spot in the hindwing, and the extremely thin abdomen. Marl Pennant (*Macrodiplax balteata*) is larger, with more black basally in the hindwing, and the abdomen is not noticeably narrowed. The young male and female Band-winged (*Erythrodiplax umbrata*) and Black-winged (*E. funerea*) Dragonlets lack the prominent swollen basal abdominal segments. Other skimmers like Black Setwing (*Dythemis nigrescens*) and Gray-waisted Skimmer (*Cannaphila insularis*) have much smaller dark spots in the hindwing, spots that may be entirely absent.

Habitat. Ponds, lakes, ditches, and slow reaches of rivers and streams.

Discussion. This species, an extremely active, aggressive flier, reaches north only as far as southern Texas, southern Florida, and northeastern Mexico. Males perch on vegetation low over the water. Mating is reported to take an average of 40 seconds. Other aspects of its reproduction are presumed similar to that of the better-known Eastern Pondhawk. It probably flies year-round in extreme southern Texas.

References. Dunkle (1989).

Eastern Pondhawk
Erythemis simplicicollis (Say)
(photos 51d, 51e)

Size. Total length: 36–48 mm; abdomen: 24–30 mm; hindwing: 30–34 mm.

Regional Distribution. *Biotic Provinces:* Austroriparian, Balconian, Carolinian, Chihuahuan, Kansan, Navahonian, Tamaulipan, Texan. *Watersheds:* Arkansas, Bayou Bartholomew, Brazos, Canadian, Cimarron, Colorado, Guadalupe, Lower Rio Grande, Mississippi, Neches, Nueces, Ouachita, Pecos, Red, Sabine, San Antonio, San Jacinto, St. Francis, Trinity, Upper Rio Grande, White.

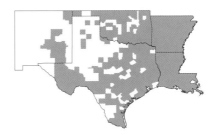

General Distribution. Throughout central and eastern United States and Canada, east of the Rockies.

Flight Season. Year-round.

Identification. This is one of the most widely distributed species in the region and in all of the eastern United States. It has a green face. Young males and females are bright green, becoming powder blue in older males, starting basally on the abdomen and on the front of the thorax. The wings are clear. The abdomen, before becoming obscured in older males, is black with green dorsolateral spots on segments 4–6 and yellow or pale cerci. Females have a ventrally projecting spoutlike ovipositor below segment 9. It is not unusual to see green males with varying degrees of pruinosity.

Similar Species. This species most closely resembles Western Pondhawk (*E. collocata*), which is darker overall, the abdomen is broader, the males have black cerci, and young males and females have a dark dorsal abdominal stripe. Great Pondhawk (*E. vesiculosa*) is larger, with a more slender abdomen and dark rings on segments 4–7. Other similar skimmers can be distinguished by the face color and/or markings in the wings.

Habitat. Ponds, lakes, ditches, and slow-moving creeks, streams, and rivers.

Discussion. This species is known from every county and parish in Arkansas and Louisiana. Its green color and common habit of resting on the ground, trash, logs, or other objects may result in its initially being confused with clubtails, but the eyes are widely joined on top of the head. It regularly preys on a variety of small and large insects, up to and including other Eastern Pondhawks. Although Eastern Pondhawks inhabit almost any slow-moving body of water, they are often found around plants on the water surface, such as water lilies, lotus, and duckweed, where males patrol their territories. Males of this species display a unique "leap frogging" behavior when defending territories. A male chasing another male will suddenly move under the male in front. This swapping of positions will often occur repeatedly. The change in males, from a green coloration to the pruinose blue over the entire thorax and the first seven abdominal segments occurs through a predictable progression of color patterns. Over a period of 2–3 weeks, this species acquires 17 different color patterns. The rate of color change significantly decreases with both decreasing food consumption and declining air temperature. Two studies found that sperm from the most recent mating competes for fertilizations with sperm stored from previous matings, but only if the female oviposits on the following day without re-mating. Sperm mixing in the bursa of females took 24 to 48 hours, at which time the last male to mate had replaced an average of 57–75% of the sperm stored by females from previous matings.

References. Bell and Whitcomb (1961), Dunkle (1989), McVey (1981, 1985, 1988), McVey and Smittle (1984), Sanborn (1996), Waage (1986), Williamson (1900b).

Great Pondhawk
Erythemis vesiculosa (Fabricius)
(photo 51f)

Size. Total length: 55–65 mm; abdomen: 40–48 mm; hindwing: 38–45 mm.
Regional Distribution. *Biotic Provinces:* Austroriparian, Balconian, Carolinian, Kansan, Navahonian, Tamaulipan, Texan. *Watersheds:* Arkansas, Brazos, Canadian, Cimarron, Colorado, Guadalupe, Lower Rio Grande, Neches, Nueces, Pecos, Red, San Antonio, San Jacinto, Trinity.
General Distribution. Southern United States from Arizona to Florida and throughout Central America south to Argentina.

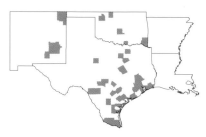

Flight Season. Apr. (TX)–Sep. 20 (TX); all year farther south.
Identification. This species is called Great Pondhawk because of its great size. It has dull-gray or brown eyes, and the face and thorax are bright green and unmarked. The legs are greenish basally, but black for most of their length, with large prominent femoral and tibial spines. The wings are clear, with a green pterostigma in young individuals. The abdomen is largely green. Segments 1–3 are swollen, 4–7 have dark rings, and 8–10 are black. The cerci are pale yellow or green. The ovipositor is short, inconspicuous, and spoutlike in the female.
Similar Species. The size of Great Pondhawk and the distinctive wide dark bands on abdominal segments 4–7 will distinguish it from other pondhawks, including the smaller green Eastern (*E. simplicicollis*) and Western (*E. collocata*) Pondhawks which have interrupted rings or spots on abdominal segments 4–7. Similar-looking clubtails will all have the eyes separated on top of the head.
Habitat. Ponds, lakes, ditches, and slow-moving creeks, streams, and rivers.
Discussion. Like other pondhawks in the region, this species is a strong flier and skillful hunter, taking other flying insects, such as horseflies, butterflies, and other dragonflies up to its own size. It commonly rests on the ground or just above, on objects or vegetation. It is generally wary of movement. Mating occurs while perched, and females will lay eggs in nearly any standing or slow-moving body of water.

Dragonlets
Genus *Erythrodiplax* Brauer

This is another large, primarily Neotropical genus. The group includes six North American species, all of which occur in the south-central United States. The species in our area are all brown, blue, or black, with either minimal basal wing markings or wide bands of black in the outer half of the wings. Members of this group may change considerably in color with age, and females of some species occur in different color forms. These species are generally pale as tenerals, but the face and top of the head may become metallic blue, black, or red in older males. The abdomen is generally robust and often gently tapering beyond the compressed basal segments. Females and tenerals often have pale dorsolateral spots on segments 3–7, the spots almost always obscured in older males.

References. Borror (1942), Paulson (2002).

KEY TO THE SPECIES OF DRAGONLETS (*ERYTHRODIPLAX*)

1. Larger, total length greater than 38 mm; wingtips generally dark — 2

1'. Smaller, total length less than 35 mm; wingtips clear — 3

2(1). Males with broad black band extending from wing base to between nodus and pterostigma; median planate generally subtends 2 rows of cells — **Black-winged (*funerea*)**

2'. Males with broad black band extending only from nodus to pterostigma; median planate generally subtends a single row of cells — **Band-winged (*umbrata*)**

3(1'). Hindwing with small dark spot basally; vein Cu1 in hindwing arises from anal angle of triangle — 4

3'. Hindwing without small dark spot basally; vein Cu1 in hindwing distinctly separated from anal angle of triangle — **Seaside (*berenice*)**

4(3). Frons bluish black; basal spot in hindwing variable, usually dark brown, but never red or reddish brown — 5

4'. Frons red or reddish brown; basal spot in hindwing dark red or reddish brown — **Red-faced (*fusca*)**

5(4). Hindwing less than 21 mm long; basal spot in hindwing small, extending at most to cubital crossvein — **Little Blue (*minuscula*)**

5'. Hindwing 21 mm long or more; basal spot in hindwing larger, often extending to base of A_2 or beyond — **Plateau (*basifusca*)**

Plateau Dragonlet
Erythrodiplax basifusca (Calvert)
(photos 52a, 52b)

Size. Total length: 26–30 mm; abdomen: 18–24 mm; hindwing: 21–26 mm.

Regional Distribution. *Biotic Provinces:* Apachian, Chihuahuan, Kansan. *Watersheds:* Colorado (NM), Pecos, Upper Rio Grande.

General Distribution. Southwestern United States from Texas to Arizona, south through Mexico to Argentina and Chile.

Flight Season. Jul. 13 (TX)–Oct. 22 (TX).

Identification. This is a small brown southwestern species. Its body is olivaceous or brownish, but quickly becomes dark blue or black in older males. The top of the head changes to metallic blue or black in these individuals. The thorax is largely unmarked, but darker brown in front and more olivaceous on the sides, before the pruinescence sets in. The legs are completely black in males, brown with black externally in females. The wings are clear, or have at most a dark amber spot, basally on the front portion of the hindwing, that extends no farther than the first antenodal crossvein. The abdomen is stout, with swollen basal segments. The middle segments are flattened dorsoventrally. The abdomen is brown in females and tenerals, and bluish black in older males. The cerci are black.

Similar Species. Little Blue Dragonlet (*E. minuscula*) has a smaller basal hindwing spot (the spot may be lacking) and white cerci in the males. Older Red-faced Dragonlets (*E. fusca*) have a red face and a brown thorax. Similar pondhawks (*Erythemis*) are larger and paler blue, and have a green face. Band-winged Meadowhawk (*Sympetrum semicinctum*) has black lateral thoracic markings.

Habitat. Marshy creeks, streams, and ponds.

Discussion. This species was known as *E. connata* until Needham et al. (2000) recognized it as a different species, which was later discussed by Paulson (2002), who diagnosed it. Males of this dragonfly vigorously defend their relatively small territories, chasing away even slightly larger dragonflies. They perch low on vegetation surrounding ponds, or low on stems overhanging the water. They often perch in a fashion similar to that of setwings, with both wings moderately depressed. Pairs may mate on the wing or perched, and males generally guard females as they lay eggs.

Seaside Dragonlet
Erythrodiplax berenice (Drury)
(photos 52c, 52d)

Size. Total length: 28–35 mm; abdomen: 15–23 mm; hindwing: 18–26 mm.

Regional Distribution. *Biotic Provinces:* Austroriparian, Balconian, Chihuahuan, Kansan, Navahonian, Tamaulipan, Texan. *Watersheds:* Brazos, Colorado, Lower Rio Grande, Mississippi, Neches, Pecos, Red, Sabine, San Antonio, San Jacinto, Trinity, Upper Rio Grande.

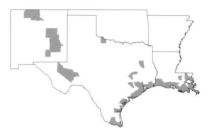

General Distribution. Atlantic Coast from southern Canada to Texas and New Mexico; West Indies and Central America south to Venezuela and Trinidad.

Flight Season. Jan. 16 (TX)–Nov. 4 (LA).

Identification. This is a small, common, primarily coastal species. Young males and females have a pale face with broad black stripes. The thorax is pale yellowish, heavily striped with black on the front and sides, the stripes broadly confluent at their lower ends. The face and thorax quickly become entirely dark blue or black in males and nearly so in females. The top of the frons becomes metallic blue. The wings are clear, or have a hint of amber basally and at the nodus in females. The legs are black. The abdomen is largely black, with broad orange or yellow spots dorsally on segments 1–7. These spots quickly become obscured in males, more slowly in females. Females are variable during this transition time: the thorax turns black before the abdomen in some individuals, and there may or may not be an amber spot near the nodus in each wing. Females have a triangular-shaped, scooped-out, ovipositor projecting ventrally from segment 9.

Similar Species. Similar dragonflies will generally not be found so closely tied to coastal areas. Gray-waisted Skimmer (*Cannaphila insularis*) is larger, and the hindwing bases are narrow. Similar pondhawks (*Erythemis*) have the basal abdominal segments distinctly swollen. Female small pennants (*Celithemis*) lack ovipositors, and those males that have a dark spot in the basal area of the hindwing will also have a different lateral thoracic color pattern.

Habitat. Salt marshes, estuaries, bays, and occasionally inland lakes high in salinity.

Discussion. Seaside Dragonlet is the closest thing we have to a marine dragonfly in North America, capable of breeding in waters with high salt concentrations. It is limited to salt marshes and estuaries along the coast and certain inland saline lakes. They may be locally abundant, perched in great numbers on bushes, stems, or on the ground. Populations of males are often found defending relatively small territories around isolated pools. The female, accompanied by the male, lays eggs in these pools by dipping her abdomen numerous times while hovering in place. This species probably occurs year-round in the southern limits of the region.

References. Dunson (1980).

Black-winged Dragonlet
Erythrodiplax funerea (Hagen)
(photo 52e)

Size. Total length: 38–42 mm; abdomen: 20–33 mm; hindwing: 25–34 mm.
Regional Distribution. *Biotic Provinces:* Balconian, Chihuahuan, Tamaulipan. *Watersheds:* Rio Grande, San Antonio.

General Distribution. Arizona, Texas, and California; through Central America south to Colombia and Ecuador.
Flight Season. Jun. (TX).
Identification. This species is an uncommon stray into the south-central United States from Mexico and Central America. It is a large, dark species with variable wings that are generally extensively marked with brown or black. In mature males and andromorphic females the black in the wings extends basally out beyond the nodus, halfway to the pterostigma and beyond. Young individuals and gynomorphic females are most similar to Band-winged Dragonlet (*E. umbrata*), with either a basal spot or a wide, isolated dark band in the middle of the wing. The wingtips are usually marked with brown or black. The face is black. The thorax is brownish in young individuals, but well-marked with black

stripes, the stripes becoming diffuse with age, so that the entire thorax becomes black in males. The thorax of the mature female is generally violet. The abdomen is brown, with pale lateral rectangular spots on segments 5–7.
Similar Species. Band-winged Dragonlets are similar, and young males and females of the two species may be difficult to differentiate. Generally, the wing markings are not as extensive basally, and Band-winged Dragonlet is far more common in the region. Female Band-winged Dragonlets generally have an olivaceous thorax and abdomen. The dark markings in male Filigree Skimmers (*Pseudoleon superbus*) extend to the pterostigma, and the wings of females are much more spotted. Male Marl Pennants (*Macrodiplax balteata*) are similar to females but with reduced markings, and their abdomen tapers distinctly toward its tip. Female Marl Pennants have shorter abdomens with distinct yellow spots. The base of the abdomen in Pin-tailed Pondhawk (*Erythemis plebeja*) is distinctly swollen.
Habitat. Open, temporary pools and ponds.
Discussion. This tropical species' range just barely reaches northward into our area. It is found primarily along the Pacific coast. Its similar eastern counterpart, Band-winged Dragonlet, is found along the Atlantic seaboard. Other records outside the region are from Mesa, Arizona, and Allende, Nuevo León, Mexico. This species aestivates, as adults, in forests before rains, where it then turns black within three days, moving to open ponds and pools. Males and pairs in copula perch on vegetation around these ponds, while females lay eggs, occasionally dipping their abdomens among the emergent vegetation (pers. comm., Sidney Dunkle).

Red-faced Dragonlet
Erythrodiplax fusca (Rambur)
(photo 52f)

Size. Total length: 24–28 mm; abdomen: 16–22 mm; hindwing: 19–28 mm.

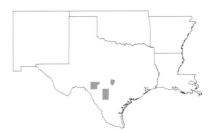

Regional Distribution. *Biotic Provinces:* Balconian, Tamaulipan. *Watersheds:* Colorado, Guadalupe, Nueces, Rio Grande.

General Distribution. South Texas and Mexico south to Argentina.

Flight Season. Jul. 14 (TX)–Aug. 12 (TX).

Identification. This small species is closely related to Plateau (*E. basifusca*) and Little Blue (*E. minuscula*) Dragonlets. It can generally be distinguished from these in the field by its predominantly reddish color, including the head, thorax, and basal wing spot. The basal wing spot is more prominent in males than in females, and generally varies from a faint-yellowish wash to a distinct dark-reddish spot extending out to the triangle in the hindwing. Young males are yellowish brown with a dark antehumeral stripe on the thorax and a brown lateral stripe on the abdomen. With age, in both sexes, the yellowish color quickly becomes bright red or brownish. The abdomen, especially in older males, becomes dark with a light powder-blue pruinescence. The cerci are brown. Older males are distinct, since they are the only dragonfly with a bright-red face and powder-blue abdomen in our area.

Similar Species. Mature male Little Blue Dragonlets have a completely blue thorax and abdomen. Blue Corporal (*Ladona deplanata*) is larger, with brown basal stripes in both wings. Similar meadowhawks (*Sympetrum*) are generally larger and lack dark abdominal stripes.

Habitat. Marshy swamps, pools, lakes, and streams with moderate current and periodic pools.

Discussion. This species is uncommon in the Hill Country of central Texas and southward. It may be seen foraging in open fields or several meters from water. Males perch low on the stems overhanging the water.

Little Blue Dragonlet
Erythrodiplax minuscula (Rambur)
(photo 53a)

Size. Total length: 22–27 mm; abdomen: 14–17 mm; hindwing: 15–21 mm.

Regional Distribution. *Biotic Provinces:* Austroriparian, Balconian, Carolinian, Tamaulipan, Texan. *Watersheds:* Arkansas, Bayou Bartholomew, Brazos, Mississippi, Neches, Ouachita, Red, Sabine, San Antonio, San Jacinto, St. Francis, Trinity, White.

General Distribution. Eastern United States from New Jersey to Texas.

Flight Season. Apr. 3 (LA)–Dec. 9 (LA).

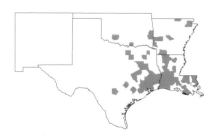

Identification. This is the most widespread of the small dragonlet species in the region. It is similar to its southwestern counterparts, Plateau (*E. basifusca*) and Red-faced (*E. fusca*) Dragonlets. It can generally be distinguished in the field from either of these by its eastern distribution and the distinctive powder-blue pruinose appearance of older individuals. Young males and females are greenish brown or olivaceous. The face is olivaceous, and the front of the thorax is devoid of stripes and darker than the sides. The wings are clear, or with only a small basal spot in the hindwing. There are a black middorsal stripe and a pair of interrupted lateral stripes on the abdomen. Mature individuals develop a powder-blue pruinose cast that envelops the body from the thorax posteriorly and the terminal abdominal segments anteriorly. Abdominal segments 7–10 become entirely black. The male cerci are pale, nearly white, and the female has short but distinct, triangular, spoutlike ovipositor below segment 9.

Similar Species. Blue Corporal (*Ladona deplanata*) is larger, with brown basal stripes in both wings. Similar meadowhawks (*Sympetrum*) are generally larger and lack dark abdominal stripes.

Habitat. Marshy ponds, pools, lakes, and slow-moving streams.

Discussion. This dragonlet commonly perches low on grasses or other ground cover. They generally do not travel far, even when disturbed. Males patrol and defend small territories close to the water's edge, where competition is minimal and they have an easier time escaping larger dragonflies. Mating occurs quickly, generally in less than 20 seconds, and the females then lay eggs among emergent plants, with the males guarding.

Band-winged Dragonlet
Erythrodiplax umbrata (Linnaeus)
(photos 53b, 53c)

Size. Total length: 38–47 mm; abdomen: 23–34 mm; hindwing: 25–34 mm.

Regional Distribution. *Biotic Provinces:* Austroriparian, Balconian, Chihuahuan, Kansan, Navahonian, Tamaulipan, Texan. *Watersheds:* Arkansas, Bayou Bartholomew, Brazos, Canadian, Cimarron, Colorado, Guadalupe, Lower Rio Grande, Mississippi, Neches, Nueces, Red, Sabine, San Antonio, San Jacinto, St. Francis, Trinity, Upper Rio Grande.

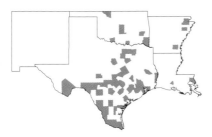

General Distribution. Florida and Texas south throughout Central America and south to Argentina; occasional stray to eastern United States.
Flight Season. Year-round in south Texas.
Identification. This is a larger dragonlet, second in size only to the similar Black-winged Dragonlet (*E. funerea*). Mature males are olivaceous, with a broad stripe in the wings extending between the nodus and pterostigma. Males become pruinose, but the cerci remain pale. The wing band becomes progressively darker with age. Females may be similar to the male, but with a paler, reduced stripe in each wing that does not reach the pterostigma. Other, generally more common, females lack a prominent wing band, but have dark wingtips. Young individuals of both sexes have pale rectangular spots laterally on the abdomen. Occasionally, the hindwing may be amber or brown basally.
Similar Species. The similar Black-winged Dragonlet is much less common, and characters are given under that species to separate it from Band-winged. Other similar species include Filigree Skimmer (*Pseudoleon superbus*), which has much heavier maculation of the wings, and Great Pondhawk (*Erythemis vesiculosa*), which may be confused with young male and female Band-wingeds with unmarked wings. Great Pondhawk, however, is bright green, and with the abdomen is well marked with black. The face and thorax of Band-winged Dragonlet are olivaceous or greenish-brown in young individuals of both sexes.
Habitat. Permanent and temporary marshy ponds, pools, and lakes.
Discussion. This species will occasionally roost in large numbers on the branches of trees with their wings characteristically depressed below the body. Males guard females during egg-laying, like other members of this genus, and will patrol around ponds.

Corporals
Genus *Ladona* Needham

This genus includes six species worldwide, of which three are found in North America. These species are sometimes included with the King Skimmers (*Libellula*). One of the North American species occurs in the south-central United States.

References. Bennefield (1965), May (1992).

Blue Corporal
Ladona deplanata (Rambur)
(photos 53d, 53e)

Size. Total length: 31–35 mm; abdomen: 19–24 mm; hindwing: 22–26 mm.
Regional Distribution. *Biotic Provinces:* Austroriparian, Carolinian, Kansan, Texan. *Watersheds:* Arkansas, Bayou Bartholomew, Brazos, Canadian, Cimarron, Mississippi, Neches, Ouachita, Red, Sabine, San Jacinto, St. Francis, Trinity, White.

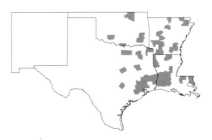

General Distribution. Eastern United States from Maine to Texas.
Flight Season. Feb. 25 (TX)–May 21 (AR).

Identification. This is a brown, moderate-sized southeastern species with a notable clear streak running through the basal brown spot in all wings. Its face is light tan and darkens with age. The vertex is black and shallowly emarginate. The thorax is brown, with two distinct white stripes in front resembling a corporal's stripes. The wings are clear, with only the basal brown area described above and brown pterostigmata. The legs are brown. The abdomen is strongly depressed and brown, with a black middorsal stripe and carinae outlined in black. The front of the thorax and the basal area of the abdomen develop a pruinose appearance, becoming steel blue, in mature males.

Similar Species. Little Blue Dragonlet (*Erythrodiplax minuscula*) is much smaller and lacks a dark basal area in the wings and stripes on the front of the thorax. Mature Eastern Pondhawks (*Erythemis simplicicollis*) have a green face and lack wing markings and frontal thoracic stripes.

Habitat. Sloughs, ponds, lakes, borrow pits, and open areas of slow streams often with sandy bottoms.

Discussion. This is one of the few skimmers that may be commonly seen perching on the ground with the wings depressed. They will also perch vertically on trees exposed to sunlight in the late afternoon, probably using the depressed abdomen as a heat collector. In this and other behaviors, it is similar to Whitetails (*Plathemis*) in habit. It emerges in the early spring for a relatively short time. Males patrol the edges of ponds and lakes, sometimes resting on floating debris or low on vegetation. They have a low fluttering flight occasionally interrupted by hovering. Mating occurs on the wing, and females lay eggs immediately after, while guarded by the male. They lay eggs by short dips of the abdomen to the water. Females are not often encountered near the water, except to mate or lay eggs. This species is unusual among most skimmers in the south in overwintering as a final instar larva.

King Skimmers
Genus *Libellula* Linnaeus

This large Holarctic group of cosmopolitan species generally constitutes the dominant dragonflies around ponds, pools, and lakes. King skimmers are well represented in the region, 17 of the 18 North American species occurring in the south-central United States. They are often brightly colored, and their distinct wing maculation is useful for field identifications. They are generally stocky dragonflies with a robust thorax that is usually unmarked and densely clothed with hair. The wings are often conspicuously colored with stripes or spots, sometimes with white. The arculus is closer to the second antenodal crossvein than to the first. The forewing triangle may comprise two to five cells, but generally no more than three or four, while the hindwing triangle nearly always comprises only two cells. Vein R_3 is usually strongly undulate or waved. The broad, robust abdomen is shorter than the wings and regularly tapers rearward, but in some females segment 8 is widened laterally. This expansion is used to throw water along with eggs onto the shore, where they may experience less mortality.

This is a relatively well-known group, but there are differences of opinion on how they should be classified. Four distinct subgroupings are recognized: (1) *Plathemis* (Whitetails), two species with broad abdomens, white pruinescence on male abdomens, and a distinct habit of perching on the ground; (2) *Ladona* (Corporals), small species with short, broad abdomens and the habit of perching on the ground; (3) *Belonia* (Flame and Neon Skimmers), two brilliant orange-red species; and (4) *Libellula*, containing the majority of species. Current evidence suggests that the first two groups should each be elevated to its own genus, as I have done, though I am retaining *Belonia* as a subgenus of *Libellula*.

References. Artiss et al. (2001), Carle and Kjer (2002), Kennedy (1922a,b), Ris (1910).

KEY TO THE SPECIES OF KING SKIMMERS (*LIBELLULA*)

1. Abdomen bright red, orange, or tan, without black middorsal stripe; wings with amber basally, out to nodus or beyond — 2

1'. Abdomen variously colored, but *if* red or orange, *then* with black middorsal stripe; wings without basal amber area extending out to nodus — 3

2(1). Wings with reddish-amber color extending outward to pterostigma; hindwing with dark red-brown basal area — **Flame (*saturata*)**

2'. Wings with reddish-amber color extending outward only to nodus; hindwing without dark red-brown basal area — **Neon (*croceipennis*)**

3(1'). Basal 1/3 of both fore- and hindwings covered full width by blackish band; wingtips clear — **Widow (*luctuosa*)**

3'. Basal 1/3 of wings without black markings *or*, if present, *then* not completely covering basal 1/3 of wing; wingtips variable — 4

4(3'). Anterior 1/4 of fore- and hindwings covered full-length by diffuse brownish band strongly tinged with amber yellow; pterostigma white, heavily outlined in black — **Yellow-sided (*flavida*)**

4'. Anterior 1/4 of wings not covered by diffuse brownish band; pterostigma variable — 5

5(4'). Wings throughout yellow, orange, or red, but with no dark marking in a definite color pattern — 6

5'. Wings clear, or with a more or less definite pattern in brown — 7

6(5). Hind tibiae light brown, with black spines; costa darker before nodus, yellow beyond; found along coast — **Needham's (*needhami*)**

6'. Hind tibiae black, with black spines; costa not bicolored, uniform light brown; found more inland — **Golden-winged (*auripennis*)**

7(5'). Costa distinctly white out to pterostigma; short band of yellow across wing bases; western — **Bleached (*composita*)**

7'. Costa variable, but never white; no band of yellow across wing bases — 8

8(7'). Pterostigma bicolored, brown and yellow-white — 9

KEY TO THE SPECIES OF KING SKIMMERS (*LIBELLULA*) *(cont.)*

8'. Pterostigma uniformly colored	10
9(8). Face white; hindwing greater than 35 mm long; western	**Comanche (*comanche*)**
9'. Face dark blue; hindwing less than 35 mm long; eastern	**Spangled (*cyanea*)**
10(8'). Wings with wide crossband of brown at nodus	11
10'. Wings without wide crossband of brown at nodus	13
11(10). Hindwing less than 37 mm long; wings with light-brown spots bordered with yellow	**Painted (*semifasciata*)**
11'. Hindwing greater than 37 mm long; wings with dark-brown or black spots, not bordered with yellow	12
12(11'). Wingtips entirely brown beyond pterostigma	**Twelve-spotted (*pulchella*)**
12'. Wingtips largely clear beyond pterostigma	**Eight-spotted (*forensis*)**
13(10'). Hindwing with large brown basal spot traversed by white crossveins and connected by streak of yellow in membrane to small brown spot at nodus; western	**Four-spotted (*quadrimaculata*)**
13'. Hindwing without large brown basal spot	14
14(13'). Both fore- and hindwings with conspicuous basal black spots reaching out to triangles; western	**Hoary (*nodisticta*)**
14'. Markings in fore- and hindwings not reaching triangles	16
15(14'). Face white; sides of thorax pale without a brown triangle near base of forewing	**Great Blue (*vibrans*)**
15'. Face not white, in most cases brown or black; sides of thorax with a brown triangle near base of forewing	16
16(15'). Both sexes with a black basal streak in each wing; frons of female with a sharply defined black triangle; labrum black	**Bar-winged (*axilena*)**
16'. Males and usually females lacking a black basal streak in each wing; frons of female brown to diffusely black; labrum pale	**Slaty (*incesta*)**

Golden-winged Skimmer
Libellula auripennis Burmeister
(photo 53f)

Size. Total length: 45–58 mm; abdomen: 32–40 mm; hindwing: 35–45 mm.

Regional Distribution. *Biotic Provinces:* Austroriparian, Tamaulipan, Texan. *Watersheds:* Arkansas, Bayou Bartholomew, Brazos, Mississippi, Neches, Nueces, Ouachita, Red, Sabine, San Jacinto, Trinity.

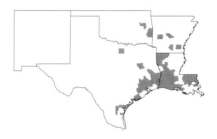

General Distribution. Eastern United States to Texas.

Flight Season. Jun. 4 (LA)–Sep. 9 (LA).

Identification. This relatively large, beautiful, red species is found throughout the eastern portions of our region. It is similar to the more coastally distributed Needham's Skimmer (*L. needhami*). The face is brown in young males and females, becoming bright red in older males. The thorax is brown with two diffuse pale stripes laterally. The wings have a yellow pterostigma and a white costal vein. The hind tibiae are reddish brown. The abdomen is yellow, with a black middorsal stripe. The front of the thorax in mature males is rusty red, and the pterostigma and abdomen are bright red. The wing veins are reddish orange throughout.

Similar Species. Needham's Skimmer has black veins over most areas of its wings, the hind tibiae are bicolored, males have a redder face and body, and the costal vein is bicolored. Young males and all female Needham's Skimmers have an unmarked thorax. Young Yellow-sided Skimmers (*L. flavida*) have shorter abdomens, their wingtips are dark to the pterostigma, the costa is generally dark out at least to the nodus, and the thorax is more robust, with a prominent pale middorsal stripe.

Habitat. Ponds, pools, ditches, lakes, and occasionally slow-flowing streams.

Discussion. This species is common around open ponds and lakes, where males actively defend their territories, but becomes much less common as it approaches coastal waters, where it is replaced by Needham's Skimmer. These two species were long confused with one another, and literature records like those of Wright (1943a,b), published before Westfall's (1943) clarification of the two species, undoubtedly include a mix of records. Males may be exceedingly wary of intruders, but often return to the top of a favored twig or branch. Females perch high in trees or lower to the ground on vegetation some distance from the water. They lack distinct lateral flanges on abdominal segment 8, and therefore lay eggs by dipping the abdomen to the water's surface, usually doing so while guarded by the male. This species is a voracious predator, taking damselflies, horseflies, butterflies, and other small insects readily. They are also victims of other predatory insects, however, such as robber flies.

References. Paulson (1966).

Bar-winged Skimmer
Libellula axilena Westwood
(photo 54a)

Size. Total length: 50–62 mm; abdomen: 37–42 mm; hindwing: 41–49 mm.

Regional Distribution. *Biotic Province:* Austroriparian. *Watersheds:* Mississippi, Neches, Red, Sabine, St. Francis, Trinity.

General Distribution. Eastern United States from New York to Texas.

Flight Season. May 23 (LA)–Sep. 9 (LA).

Identification. This large species is similar to Slaty (*L. incesta*) and Great Blue (*L. vibrans*) Skimmers. It has a pale face that darkens with age to black, with a metallic-purple luster dorsally. The thorax is brown and yellow in females and young males. Mature males become dark, with a pale-gray pruinescece starting at the front of the thorax, between the wings and progressing posteriorly. The wings are generally clear, with a touch of white pruinosity basally in the hindwings and a dark bar on the anterior edge in the basal 1/5 of all wings. There is also a black bar between the nodus and pterostigma, and the tips of each wing are black.

Similar Species. Females and young males are most similar to Slaty and Great Blue Skimmers, but the darker face with contrasting pale lateral borders will distinguish them. These species also lack a black bar between the nodus and pterostigma. Slaty Skimmer has clear wings, or if black streaks are present they are not as extensive, and Great Blue Skimmer has a white face. Other similar king skimmers (*Libellula*) lack the white pruinosity at the base of the hindwing.

Habitat. Forest ponds, pools, and ditches.

Discussion. This species has been found in several Louisiana Parishes and recently in southeast Texas, which is the westernmost recorded locality for this species. The reproductive behavior of this species is similar to that of Slaty Skimmer. Pairs mate while perched on stems or vegetation, and females lay eggs guarded by, but unaccompanied by, the male.

Comanche Skimmer
Libellula comanche Calvert
(photo 54b, 54c)

Size. Total length: 45–57 mm; abdomen: 30–36 mm; hindwing: 35–46 mm.

Regional Distribution. *Biotic Provinces:* Austroriparian, Balconian, Carolinian, Chihuahuan, Kansan, Navahonian, Tamaulipan, Texan. *Watersheds:* Arkansas, Brazos, Canadian, Cimarron, Colorado, Guadalupe, Lower Rio Grande, Nueces, Pecos, Red, San Antonio, Trinity, Upper Rio Grande.

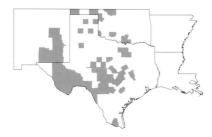

General Distribution. Western United States from California and Montana to Texas and northern Mexico.

Flight Season. May 31 (TX)–Sep. 13 (TX).

Identification. This is a handsome southwestern species commonly found from the Texas Hill Country westward. It is one of two skimmers in the region with a distinctly bicolored pterostigma, half white and half brown. It has a white face with a dark stripe across the labrum. The thorax is brown in front, divided by a broad pale middorsal stripe. The sides are pale or cream-colored, with a distinct brown stripe on the third lateral suture connected by a line above and below that extends to the front of the thorax. The line below diverges to cover the spiracle and encompass a pale spot below it. The wings are clear, with at most a yellowish patch of color along the front margin and brown at the extreme wingtips. The pterostigmata are pale yellow-white proximally and black in their distal half. The legs are black. The abdomen is broad, never narrowed basally, but tapers gradually rearward. It is brownish, with darker middorsal and lateral stripes. The thorax and abdomen in mature males become covered with dark-blue pruinescence, so that only the epiproct remains pale.

Similar Species. The only other dragonfly with a bicolored pterostigma is Spangled Skimmer (*L. cyanea*), which is smaller and has a black face. It is found in the eastern part of the region, though the ranges of the two narrowly overlap. Yellow-sided Skimmer (*L. flavida*) is also similar, but it has a dark face and lacks the bicolored pterostigma.

Habitat. Ponds, lakes, and sluggish streams.

Discussion. This species is an active flier around its usual weedy pond and lake haunts. It perches atop grasses, bent stems, and other vegetation near or overhanging the water. It is replaced in the southeastern United States by the similar Spangled Skimmer.

Bleached Skimmer
Libellula composita (Hagen)
(photos 54d, 54e)

Size. Total length: 42–48 mm; abdomen: 28–33 mm; hindwing: 33–37 mm.

Regional Distribution. *Biotic Provinces:* Chihuahuan, Kansan, Navahonian, Texan. *Watersheds:* Canadian, Pecos, Red, Upper Rio Grande.

General Distribution. Western United States, including Big Bend of Texas.

Flight Season. Jun. 11 (NM)–Aug. 27 (NM).

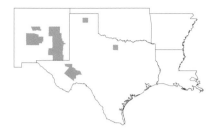

Identification. This western desert species is uncommon in the region. It is distinct in life, with its white face and unusual white eyes. The top of the vertex and occiput are white, so that the head is predominantly pale in color. The thorax is brown in front, with a broad, pale middorsal stripe. The sides are pale white or olivaceous, with two black stripes on the humeral and third lateral sutures. The stripes are thinly connected above and below by an irregular line. The legs are distinctly pale basally, but become black beyond. The wings are marked with brown basally, extending out to the arculus, and there is a smaller spot on the nodus. The costa are bright yellow, the other veins black. The abdomen is dark, with broad middorsal and lateral stripes and a black band posteriorly on each segment, giving it the appearance of a series of large quadrate spots, each becoming smaller on subsequent segments. The latter 1/2 of the abdomen in the male is entirely black. The is a slight expansion laterally on segment 8 in females.

Similar Species. Marl Pennant (*Macrodiplax balteata*) is smaller, lacks wing spot at the nodus, and does not have whitish pruinosity. Segments 1–6 of Four-spotted Skimmer (*L. quadrimaculata*) are brown dorsally.

Habitat. Alkaline desert aponds and lakes, often associated with springs.

Discussion. This species is uncommon and largely restricted to habitats with alkaline waters. Some of the species often associated with it include Desert Whitetail (*Plathemis subornata*), Eight-spotted Skimmer (*L. forensis*), and Flame Skimmer (*L. saturata*). Little is known of the biology of this species, but females lay eggs in tandem. Males are active patrollers, flying both near the shoreline and out over the middle of the water.

Neon Skimmer
Libellula croceipennis Selys
(photos 54f, 55a)

Size. Total length: 54–59 mm; abdomen: 32–39 mm; hindwing: 35–47 mm.

Regional Distribution. *Biotic Provinces:* Austroriparian, Balconian, Chihuahuan, Kansan, Navahonian, Tamaulipan, Texan. *Watersheds:* Arkansas, Brazos, Canadian, Cimarron, Colorado, Guadalupe, Lower Rio Grande, Neches, Nueces, Pecos, Red, San Antonio, San Jacinto, Trinity, Upper Rio Grande.

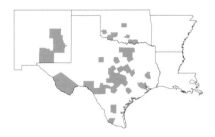

General Distribution. Southwestern United States through Central America south to Colombia.

Flight Season. May 25 (TX)–Oct. 18 (TX).

Identification. This is a brilliant robust, bright-red species. Its face, front of the thorax, and entire abdomen, including caudal appendages, are all brilliant red in mature individuals. The sides of the thorax are reddish brown and unmarked. The wings have a diffuse amber-yellow area basally that extends out to the triangle, where it narrows toward the costal margin, terminating near the nodus. The pterostigma are brown and generally longer (6 mm) than in Flame Skimmer (*L. saturata*, less than 5 mm). Females may have clear wings. The legs are brown and armed with black spines. Abdominal segment 8 in females is broadly expanded laterally. The thorax and abdomen in young individuals are reddish brown with a pale-yellowish middorsal stripe.

Similar Species. The amber color in the wings of Flame Skimmer (*L. saturata*) is more extensive, extending beyond the nodus, and there is a darker band basally in the wing. Needham's (*L. needhami*) and Golden-winged (*L. auripennis*) Skimmer both have a black middorsal stripe down the abdomen. The Mayan Setwing (*Dythemis maya*) has a much more slender thorax and abdomen.

Habitat. Ponds, lakes, and sluggish streams.

Discussion. This is one of the most conspicuous visitors to lakes and ponds in central Texas. Its bright-red color and erratic movements rarely let it go unnoticed. Males may be seen perched on top of tall grasses and weeds, but when females are present, they are generally seen chasing them in attempts to mate. Unusually among dragonflies, Neon Skimmers exhibit courtship behavior. Males typically approach females only when the are laying eggs. The male approaches a female with its abdomen raised and clearly visible to the female. The female then leaves, or the male makes sudden quick advances toward her until she does flee, whereupon he attempts to seize her. Males apparently also exhibit threat displays to other males by lowering the abdomen. Copulation typically occurs while the pair is perched on limbs or twigs near the water, but it may take place in flight. The entire process usually does not take longer than 30 seconds. Females then lay eggs guarded by the male for only an initial short time. Oviposition typically occurs at midday, the female flying swiftly forward, dipping the abdomen in the water, and subsequently throwing eggs with droplets of water on shore. They then fly up and back again, repeating this sequence several times.

References. Williams (1977).

Spangled Skimmer
Libellula cyanea Fabricius
(photo 55b, 55c)

Size. Total length: 40–48 mm; abdomen: 29–34 mm; hindwing: 31–37 mm.

Regional Distribution. *Biotic Provinces:* Austroriparian, Balconian, Carolinian, Kansan, Texan. *Watersheds:* Arkansas, Bayou Bartholomew, Brazos, Canadian, Cimarron, Colorado, Mississippi, Neches, Ouachita, Red, Sabine, San Antonio, St. Francis, Trinity, White.

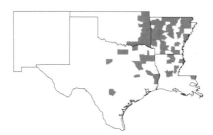

General Distribution. Eastern United States from Maine to Texas.

Flight Season. May 3 (LA)–Jul. 28 (LA).

Identification. Like the similar Comanche Skimmer (*L. comanche*), this eastern species has a bicolored pterostigma, but it is smaller and has a dark-black face. The front of the thorax is brown, with a broad pale-yellow or white middorsal stripe. The sides are pale, with a distinct brown third lateral stripe isolating the two large pale areas. The wings are clear, with a white-and-brown pterostigma, a distinct dark stripe on the costal area that does not extend beyond the triangle, and dark-colored wingtips. The front area of the wings beyond the nodus is tinged with amber or yellow, particularly in tenerals and females. The legs are brown basally, becoming black beyond. The abdomen is short, broad, and pale yellow, with distinct dark middorsal and lateral stripes. Segment 8 is only slightly expanded laterally in the female. The entire thorax and abdomen, including caudal appendages, become pruinose dark steel blue in mature males.

Similar Species. Comanche Skimmer is larger and has a white face. Yellow-sided Skimmer (*L. flavida*) lacks a distinctly bicolored pterostigma, and the wingtips may be brown or clear. Slaty Skimmer (*L. incesta*) is much larger and also lacks a bicolored pterostigma. Bar-winged Skimmer (*L. axilena*) has dark markings on the anterior edge of the wings, is larger, and lacks the bicolored pterostigma.

Habitat. Marshy ponds, pools, and lakes.

Discussion. This species is common around farm stock ponds and around waters dammed by beavers. Adults perch on top of grasses surrounding their usually marshy habitat. Aspects of their reproductive behavior and biology that I have witnessed do not appear especially different from those of other members in the genus. Females lay eggs alone, but guarded by the male.

References. Calvert (1907).

Yellow-sided Skimmer
Libellula flavida Rambur
(photos 55d, 55e)

Size. Total length: 47–52 mm; abdomen: 31–36 mm; hindwing: 36–42 mm.

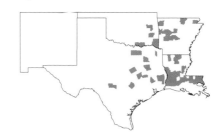

Regional Distribution. *Biotic Provinces:* Austroripari-an, Balconian, Carolinian, Kansan, Texan. *Watersheds:* Arkansas, Brazos, Canadian, Cimarron, Colorado, Guadalupe, Mississippi, Neches, Ouachita, Red, Sabine, St. Francis, Trinity, White.

General Distribution. Eastern United States from New York to Florida and Texas.

Flight Season. Apr. 24 (TX)–Sep. 6 (LA).

Identification. This is a moderately sized, widespread, but fairly uncommon species of the south-central United States. Its face is pale, but quickly turns black in males. The thorax is brown in front, with a pale-cream-colored middorsal stripe. This area becomes pruinose blue in mature males. The sides are cream-colored and divided by a brown stripe on the third lateral suture. The wings are deeply tinged with amber along the front 1/5 of each wing, especially toward the tips, and there is a dark-brown stripe on either side of the midbasal space. The legs are brown basally and become black beyond. The abdomen is brown, with a dark mid-dorsal stripe, and the middle segments are strongly depressed. The abdomen becomes dark with a powder-blue covering dorsally at maturity.

Similar Species. Female and teneral Golden-winged (*L. auripennis*) and Needham's (*L. needhami*) Skimmers have longer abdomens, and the sides of their thorax are light brown. Young Spangled Skimmers (*L. cyanea*) are smaller and have a bicolored pterostigma. Comanche Skimmer (*L. comanche*) also has a bicolored pterostigma and white face.

Habitat. Marshy ponds, lakes, borrow pits, and slow-flowing streams.

Discussion. This species was considered a variation of Spangled Skimmer until detailed morphological differences between the two species were described. Little has been published on its behavior, but it is a wary species. Males are typically found patrolling in small numbers around ponds and borrow pits. Often, only one or two individuals will be present around what would seem suitable habitat.

References. Calvert (1907).

Eight-spotted Skimmer
Libellula forensis Hagen
(photo 55f)

Size. Total length: 44–50 mm; abdomen: 27–32 mm; hindwing: 35–41 mm.

Regional Distribution. *Biotic Provinces:* Chihuahuan, Coloradan, Kansan, Navahonian. *Watersheds:* Canadian, Colorado (NM), Pecos, Upper Rio Grande.

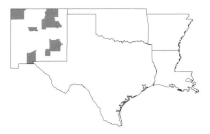

General Distribution. West of the Rockies in the United States and Canada.

Flight Season. Jun. 28 (NM).

Identification. This western species is similar to the more widely distributed Twelve-spotted Skimmer (*L. pulchella*). Its face is pale yellow in females and dark blue or black in males. The thorax is brown, with a pair of lateral yellow stripes, one on each side of the spiracle. There is a dark-brown spot basally in each wing, broadly extending out to the triangle, and a second, more diffuse spot extending from the nodus toward the pterostigma, but not reaching it. Mature males and some females develop white opaque spots between these darker areas. The legs are black. The stout abdomen is dark brown, with a pale-yellow lateral stripe interrupted along the posterior segments, resulting in a series of spots. Segment 8 in females is slightly expanded laterally. The abdomen in older individuals develops a gray-pruinose appearance.

Similar Species. Twelve-spotted Skimmer has dark wingtips. Hoary Skimmer (*L. nodisticta*) has a brown spot in the wing at the nodus, rather than a broad band. Male Common (*Plathemis lydia*) and Desert (*P. subornata*) Whitetails have shorter abdomens and a brown band in the wing that reaches from the nodus to the pterostigma.

Habitat. Muck-bottomed ponds, lakes, and sloughs.

Discussion. This species is found only in New Mexico, within our region, where it is often seen flying alongside the similar Twelve-spotted Skimmer, and they may indistinguishable in flight, though Eight-spotted Skimmers often have a larger white spot near the pterostigma. Females lay eggs, unattended by the male, by tapping their abdomens to the water's surface along the shoreline.

Slaty Skimmer
Libellula incesta Hagen
(photos 56a, 56b)

Size. Total length: 45–56 mm; abdomen: 30–36 mm; hindwing: 35–43 mm.

Regional Distribution. *Biotic Provinces:* Austroriparian, Balconian, Carolinian, Kansan, Texan. *Watersheds:* Arkansas, Bayou Bartholomew, Brazos, Canadian, Cimarron, Colorado, Guadalupe, Mississippi, Neches, Nueces, Ouachita, Red, Sabine, San Jacinto, St. Francis, Trinity, White.

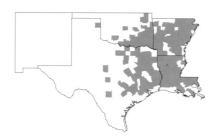

General Distribution. Widespread throughout eastern United States and Canada.

Flight Season. Jun. 17 (LA)–Sep. 25 (LA).

Identification. This species has a tan or brown face that quickly darkens, and the top of the head becomes metallic black in both sexes. The front of the thorax is brown, lacking a definite middorsal stripe. The sides of the thorax lack defined lateral stripes, having only a short dark triangular spot below the forewing, and the upper part of the third lateral suture is outlined in brown. The wings are clear, with dark wingtips and an occasional small dark spot at the nodus. The legs are black, with brown only at their extreme bases. The abdomen is slender, yellow, and slightly depressed, tapering regularly rearward, and in young individuals has the usual dark lateral and middorsal stripes. Segment 8 in the female is widened laterally. The thorax and entire abdomen develop a deep-steel-blue pruinescence in both sexes.

Similar Species. Yellow-sided Skimmer (*L. flavida*) generally shows amber along the front margin of the wings. Great Blue Skimmer (*L. vibrans*) is slightly larger, has a white face, basal wing streaks, and a dark spot at the nodus. Bar-winged Skimmer (*L. axilena*) has white basally in the hindwing, pruinescence on the thorax, and a paler abdomen. Gray-waisted Skimmer (*Cannaphila insularis*) has a pale face, and the bases of the hindwing are distinctly narrowed.

Habitat. Marshy ponds, lakes, and slow-flowing forest streams with muck bottoms.

Discussion. This species may be one of the most common dragonflies at a forest pond or other quiet waters. Like many other skimmers, males perch atop of tall grasses and weeds in sunlit areas. Females are seldom seen around water except to mate, which takes an average of 30 seconds, followed by egg-laying. Females deposit their eggs alone, but guarded by males. They use their abdomens to throw the eggs, along with water droplets, to the shoreline or to open water. Females are sexually mature and can mate while still showing their pale nonmaturated coloration.

References. Dunkle (1985b, 1989).

Widow Skimmer
Libellula luctuosa Burmeister
(photos 56c, 56d)

Size. Total length: 38–50 mm; abdomen: 24–32 mm; hindwing: 33–41 mm.

Regional Distribution. *Biotic Provinces:* Apachian, Austroriparian, Balconian, Carolinian, Chihuahuan, Coloradan, Kansan, Navahonian, Tamaulipan, Texan. *Watersheds:* Arkansas, Bayou Bartholomew, Brazos, Canadian, Cimarron, Colorado, Colorado (NM), Guadalupe, Mississippi, Neches, Nueces, Ouachita, Pecos, Red, Sabine, San Antonio, San Jacinto, St. Francis, Trinity, Upper Rio Grande, White.

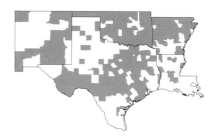

General Distribution. Widespread across United States and southeastern Canada; absent from Great Plains.

Flight Season. May 10 (LA)–Sep. 13 (TX).

Identification. This is one the most widespread and easily recognized dragonflies in the region. Its face is pale yellow or brown in females and young males, but darkens to black along with the top of the head in mature males. The pterothorax is dark brown, with a pale-yellow middorsal stripe that extends onto the prothorax. This area becomes obscured with brown color in females and black in mature males, ultimately turning powder blue. The sides are pale yellow, with an ill-defined dark stripe on the third lateral suture. This area becomes obscured in females and turns a dark brassy brown in older males. There are large black bands on the wings extending basally to the nodus in both sexes. The wingtips are occasionally darkened, especially in females and western individuals. Mature males develop a white area beyond the dark basal stripe, which extends to the wing apex. The legs are black. The abdomen is only moderately depressed and tapers rearward. It is pale yellow, with broad black middorsal and lateral stripes. The yellow is interrupted only by a black carina. Segment 8 in females is slightly expanded laterally. The caudal appendages are black. The color of the abdomen darkens in both sexes and becomes powder-blue pruinose in males.

Similar Species. No other skimmer has broad wing-bands on both the fore- and hindwings.

Habitat. Still bodies of water, including marshy ponds, lakes, and borrow pits.

Discussion. This widespread species is found nearly everywhere in North America except along the Gulf Coast of the southeastern United States and the Great Basin. It is an active flier around nearly any still body of water, or creek or stream, where males may be seen regularly battling over territories. Females rhythmically dip their abdomens to the water while flying just above the surface, unaccompanied by, but occasionally guarded by, the male. Three different variations of this species have been recognized, including the paler *odiosa* form that intergrades with the darker nominate form in the Hill Country of Texas. The density of males increases dramatically during the breeding season, with two or more males simultaneously defending a given territory.

References. Campanella (1975), Ferguson (1940), Garrison (1976), Moore (1987, 1989, 1990).

Needham's Skimmer
Libellula needhami Westfall
(photo 56e)

Size. Total length: 45–57 mm; abdomen: 32–39 mm; hindwing: 35–45 mm.

Regional Distribution. *Biotic Provinces:* Austroriparian, Tamaulipan, Texan. *Watersheds:* Bayou Bartholomew, Brazos, Lower Rio Grande, Mississippi, Neches, Nueces, Ouachita, Red, Sabine, San Antonio, San Jacinto, Trinity.

General Distribution. Atlantic seaboard and Gulf coastal plain.

Flight Season. Apr. 24 (LA)–Sep. 10 (LA).

Identification. This species is largely restricted to the coastline. Its face, thorax, and abdomen are all yellowish-brown in young individuals and females, with all but the sides of the thorax becoming vivid red. The thorax lacks lateral stripes. The legs are brown, with black only on the spines. The wings are amber or orange on the front 1/2, with a yellow-orange pterostigma, and have black veins in the posterior 2/3 of each wing. The costa is bicolored, being dark basally to the nodus and lighter beyond to the pterostigma. There is a black middorsal stripe on the abdomen.

Similar Species. See Golden-winged Skimmer (*L. auripennis*) for differences from that species. Young Yellow-striped Skimmers (*L. flavida*) have pale lateral thoracic stripes.

Habitat. Marshy ponds and lakes, including brackish waters.

Discussion. This species, although reported as far inland as just north of the Arkansas–Louisiana border, is much more common along the coastal areas, where it replaces Golden-winged Skimmer. It may be one of the most abundant species, along with Four-spotted Pennant (*Brachymesia gravida*) and Seaside Dragonlet (*Erythrodiplax berenice*), at brackish waters, where it typically perches low on vegetation surrounding or overhanging the water. Females are often encountered only some distance from the water when mating. Pairs mate while perched, and females lay eggs, guarded or unaccompanied by males, by vigorously tapping their abdomens to the water surface.

Hoary Skimmer
Libellula nodisticta Hagen
(photo 56f)

Size. Total length: 46–52 mm; abdomen: 32–35 mm; hindwing: 37–42 mm.

Regional Distribution. *Biotic Provinces:* Apachian, Chihuahuan, Navahonian. *Watersheds:* Canadian, Colorado (NM), Pecos, Upper Rio Grande.

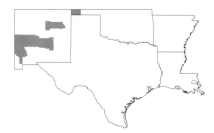

General Distribution. Western United States, west of the Rocky Mountains and south into Mexico.

Flight Season. Jun. 9 (NM)–Jul. (NM).

Identification. This is a distinct, but uncommon, western species that just enters our region. It face is yellow, with a black crossstripe, and the top of the frons is furrowed and black. The thorax is brown, with a pale middorsal stripe that becomes obscured with age. The lateral sutures on the sides of the thorax are outlined in black, and there are four yellow spots. The wings have a brown basal stripe extending out to the level of the triangle, and there is a smaller spot on the nodus. The area behind and below the basal stripe becomes pruinose with age. The veins and pterostigma are black, and the legs are black. The abdomen is brown or black, becoming darker posteriorly, and has an interrupted pale lateral stripe. The caudal appendages are black. Males develop pruinescence over the front of the thorax and the entire abdomen.

Similar Species. Bleached Skimmer (*L. composita*) has a white face, amber basal wing spots, and no thoracic spots. Eight-spotted Skimmer (*L. forensis*) has a broad crossband of brown at the wing nodus. Four-spotted Skimmer (*L. quadrimaculata*) lacks the basal brown stripe on the forewing.

Habitat. Ponds, lakes, and streams with little flow.

Discussion. This species is not found east of the Navahonian biotic province. Despite its local abundance in certain western localities, little has been documented about its behavior or habits.

Twelve-spotted Skimmer
Libellula pulchella Drury
(photo 57a)

Size. Total length: 51–58 mm; abdomen: 32–36 mm; hindwing: 42–48 mm.

Regional Distribution. *Biotic Provinces:* Apachian, Austroriparian, Balconian, Carolinian, Chihuahuan, Coloradan, Kansan, Navahonian, Tamaulipan, Texan. *Watersheds:* Arkansas, Bayou Bartholomew, Brazos, Canadian, Cimarron, Colorado, Colorado (NM), Guadalupe, Mississippi, Neches, Nueces, Ouachita, Pecos, Red, Sabine, San Antonio, San Jacinto, St. Francis, Trinity, Upper Rio Grande, White.

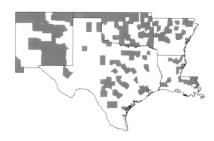

General Distribution. Throughout the contiguous United States and southern Canada.

Flight Season. Mar. 28 (LA)–Nov. 7 (LA).

Identification. This large, handsome, brown skimmer is found in all 48 contiguous United States. Its distinct wing pattern of dark brown or black wing spots, basally, at the nodus, and at the wingtips, will readily distinguish it from most dragonflies in the region. The face is dull yellowish brown. The thorax is brown, and lacks a middorsal stripe; laterally there, are a pair of pale-yellowish stripes. The wings are spotted with dark brown or black bands, as mentioned above. The mature males develop two white spots in each forewing and three in each hindwing, resulting in their traditional common name, "ten spot." The species has more recently been given the name "twelve spot" to represent the more conspicuous brown spots of each wing. The legs are brown at their extreme bases and black beyond. The abdomen is brown, with a broad pale-yellow uninterrupted stripe on each side and a narrower one along the middorsal carina. The caudal appendages are brown, darkening with age. Females have a slight lateral expansion of abdominal segment 8.

Similar Species. Eight-spotted Skimmer (*L. forensis*) lacks dark wingtips. Female Common Whitetails

(*Plathemis lydia*) are smaller, and have pale legs and a white zigzag lateral abdominal stripe. Prince Basket-tail (*Epitheca princeps*) has green eyes, a long, slender abdomen, and no thoracic or abdominal stripes.

Habitat. Shallow ponds, lakes, marshes, and slow streams.

Discussion. Twelve-spotted Skimmer tends to prefer open pond and lake shores well exposed to sunlight. It is an aggressive, strong flier, entering into numerous skirmishes with other males and intruders, and rarely being displaced, which often makes it difficult to catch. Territories are established in areas over the water that are free of surface vegetation. Mature males seldom perch, but when they do, they can be found on top of tall grasses and bushes surrounding the water. The female deposits eggs along the shoreline of shoals and bays by regularly tapping her abdomen to the water surface, unattended by the male. The similar Eight-spotted Skimmer is almost always found alongside Twelve-spotted Skimmer where their ranges overlap.

References. Fitzhugh and Marden (1997), Marden (1995), Pezalla (1979).

Four-spotted Skimmer
Libellula quadrimaculata Linnaeus
(photo 57b)

Size. Total length: 41–45 mm; abdomen: 25–30 mm; hindwing: 31–38 mm.

Regional Distribution. *Biotic Provinces:* Chihuahuan, Coloradan, Navahonian. *Watersheds:* Colorado (NM), Pecos, Upper Rio Grande.

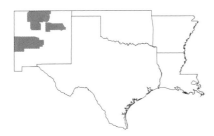

General Distribution. Canada, northern United States, and northeastern Europe and Asia.

Flight Season. Jul. 12 (NM).

Identification. This well-patterned species' range just enters the western portions of the south-central United States. Its has a yellow face and a dull-brown thorax thickly covered with hairs. Abdominal segments 1–6 are dull brown, 7–10 are black with a pale yellow stripe laterally. The wings have amber basal streaks and a small black spot at the nodus. The hindwings have a black triangular spot basally. Segment 8 in females is expanded slightly or not at all.

Similar Species. The wing pattern is distinctive among species in our region. Hoary Skimmer (*L. nodisticta*) has a basal brown stripe in the forewing.

Habitat. Marshy bogs, ponds, and lakes, especially peaty waters.

Discussion. Though this species has been called the most common skimmer in Canada, in the south-central United States it has been reported only from Arkansas and several north-central New Mexico counties. Because of its more northern distribution, the casual observer in the region is unlikely to come across this species. In much of its typical northern range, however, it is one of the earliest dragonflies to emerge in the spring. It is commonly seen in open fields and along forest margins, where it perches low on vegetation or on the ground, similar to the preferences of Common Whitetail (*Plathemis lydia*). This species is known to form large aggregations and migrate. As males mature they patrol their territories vigorously, around nearly any standing body of water. They readily take other dragonflies their size and smaller as prey. Mating takes place in flight, generally lasting only a few seconds, but sometimes as long as a minute. The female deposits her eggs unaccompanied, but guarded by the male, by regularly dipping her abdomen to the water surface.

References. Burton (1996), Convey (1990), Evans (1995), Muttkowski (1911), Schiemenz (1953), Walker and Corbet (1975), Whitehouse (1941).

Flame Skimmer
Libellula saturata Uhler
(photo 57c)

Size. Total length: 52–60 mm; abdomen: 32–40 mm; hindwing: 41–45 mm.

Regional Distribution. *Biotic Provinces:* Apachian, Austroriparian, Balconian, Chihuahuan, Coloradan, Kansan, Navahonian, Tamaulipan, Texan. *Watersheds:* Brazos, Canadian, Colorado, Colorado (NM),

Guadalupe, Nueces, Pecos, Red, Sabine, San Antonio, San Jacinto, Upper Rio Grande.

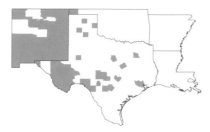

General Distribution. Western United States and Mexico.

Flight Season. Apr. 29 (TX)–Sep. 27 (TX).

Identification. This is a dull-orange southwestern species. The light-brown face of young individuals quickly becomes bright red with age. The stocky thorax and abdomen are brick red and lack lateral stripes. Tenerals and females have a tan thorax and abdomen. The wings have red veins and an amber patch that extends out to the pterostigma. There is a darker brown stripe midbasally in the hindwing. The legs are red with black spurs. The caudal appendages are red, and segment 8 in the female is expanded laterally.

Similar Species. Neon Skimmer (*L. croceipennis*) is similar, but the amber color in the wings is not as extensive in that sepcies and does not extend beyond the nodus. Neon Skimmer also lacks the darker-brown stripe covering the midbasal space in the hindwing. Mayan Setwing (*Dythemis maya*) has a much more slender body. Golden-winged (*L. auripennis*) and Needham's (*L. needhami*) Skimmers have a black middorsal stripe running down the abdomen. Female and young Roseate (*Orthemis ferruginea*) and Orange-bellied (*O. discolor*) Skimmers both have lateral thoracic markings.

Habitat. Ponds, lakes, and slow streams, including artificial ponds.

Discussion. This conspicuous dragonfly commands the notice of even the most casual observer. Males are found searching long stretches of streams for potential mates, or they are seen perched on tall vegetation near the ponds and pools used by females for egg-laying. Males will warn off intruders by flying toward them and then along with them, in an ascending flight, with only one male returning to the perch. Females lay eggs in a manner similar to that of Neon Skimmer, by throwing water along with the eggs toward the shore. Males will guard females from a perch for only a short time after mating in flight. Males tend to occur at areas along streams where receptive females are likely to visit, both seasonally and during the course of the day. Both of these observations indicate that male mate-searching patterns in this species are sexually selected. The small disjunct population in Houston, Texas, represents the easternmost record for this species, likely accidentally introduced as larvae with aquatic plants (Robert Honig, pers. comm.).

References. Alcock (1989a), DeBano (1993, 1996).

Painted Skimmer
Libellula semifasciata Burmeister
(photo 57d)

Size. Total length: 39–48 mm; abdomen: 25–31 mm; hindwing: 31–38 mm.

Regional Distribution. *Biotic Provinces:* Austroriparian, Carolinian, Kansan, Texan. *Watersheds:* Arkansas, Bayou Bartholomew, Brazos, Mississippi, Neches, Ouachita, Red, Sabine, San Jacinto, St. Francis, Trinity.

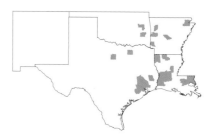

General Distribution. Eastern United States and Canada from Florida to Texas.

Flight Season. Feb. 24 (LA)–Aug. 25 (LA).

Identification. This is a moderately sized yellow-and-brown species found in the eastern portions of the region. Its wings are distinctive, with a basal yellow wash of color that extends along the costal area to the nodus. There is a dark-brown stripe beyond the midbasal space in the forewing and below in the hindwing, as well as a brown spot at the nodus and apically on both wings. The face is olivaceous, turning red in mature males. The thorax is tawny brown, thinly beset with short brown hairs, and lacks a middorsal stripe. The sides each have an oblique yellowish-white stripe behind the darker first and third lateral stripes. The legs are pale brown basally and black beyond. The abdomen is brown, and tapers strongly posteriorly. It is pale-yellowish-brown basally, often appearing translucent until segment 6. A pale lateral stripe is present on each side. Segments 7–10 and the caudal appendages are black dorsally. The lateral margins of segment 8 are only narrowly expanded laterally.

Similar Species. Similar small pennants (*Celithemis*) are smaller and have more extensive wing markings basally. King skimmers (*Libellula*) have similar markings and lack dark wingtips.

Habitat. Marshy forest seepages, ponds, and slow streams.

Discussion. Flying casually around the forest ponds it patrols, this species can be quite inconspicuous. It seldom occurs in large numbers. Its flight is usually swift, with sweeping curves.

Great Blue Skimmer
Libellula vibrans Fabricius
(photos 57e, 57f)

Size. Total length: 50–63 mm; abdomen: 37–43 mm; hindwing: 46–52 mm.

Regional Distribution. *Biotic Provinces:* Austroriparian, Carolinian, Kansan, Texan. *Watersheds:* Arkansas, Bayou Bartholomew, Brazos, Mississippi, Neches, Ouachita, Red, Sabine, San Jacinto, St. Francis, Trinity, White.

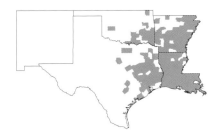

General Distribution. Eastern United States.

Flight Season. Mar. 26 (LA)–Sep. 25 (LA).

Identification. This is the largest of our skimmers, and it is found throughout the eastern portions of our region. Its face is white. The thorax is brown with a narrow white middorsal stripe, and the sides are pale grayish white, with a dark stripe along the third lateral suture, the stripe obsolete at its lower end. The wings are clear, with a narrow dark stripe basally and a small spot at the nodus. The wingtips are dark. The femora are pale over their basal half, but their remaining length, the tibiae, and the tarsi are black. The abdomen is yellow, with a black middorsal stripe. The mature males develop a pruinose pale-blue color first on the front of the thorax, then on the abdomen. The yellow on the abdomen of mature females becomes brown. Segment 8 in females is broadly expanded laterally.

Similar Species. Slaty Skimmer (*L. incesta*) is slightly smaller, and has a dark face and clear wings. Barwinged Skimmer (*L. axilena*) has white basally in the hindwing, pruinescence on the thorax, and a paler abdomen. Yellow-sided Skimmer (*L. flavida*) generally has amber along the front margin of the wings. Gray-waisted Skimmer (*Cannaphila insularis*) has a pale face, and the bases of the hindwing are distinctly narrowed.

Habitat. Swampy ponds, lakes, and slow forest streams.

Discussion. This large handsome dragonfly is common around forest ponds and sloughs during the summer, where it perches for lengthy periods. It is remarkably approachable at its shady perches. Mating occurs while pairs are perched, and generally takes less than 30 seconds. Females then lay eggs by tossing them along with water onto the shoreline. Males may mate before they are fully pruinose.

References. Dunkle (1985b, 1989).

Marl Pennants
Genus *Macrodiplax* Brauer

This is a tropical genus of just two species, one of them, *Macrodiplax cora* (Kaup), found in the Old World and a single species found in the southern limits of our region. The group is recognizable by the deeply notched frons and broad hindwings. In the forewings, there are large paranal cells adjacent to smaller marginal cells. The forewing triangle is unusually broad and typically devoid of crossveins. A radial planate encompasses five cells in both fore- and hindwings. The single New World species is found in the southern United States, and is generally associated with large ponds and lakes, often with brackish water.

Marl Pennant
Macrodiplax balteata (Hagen)
(photos 58a, 58b)

Size. Total length: 35–42 mm; abdomen: 25–30 mm; hindwing: 31–35 mm.

Regional Distribution. *Biotic Provinces:* Austroriparian, Balconian, Chihuahuan, Kansan, Navahonian, Tamaulipan, Texan. *Watersheds:* Colorado, Lower Rio Grande, Mississippi, Nueces, Pecos, Red, Sabine, San Antonio, San Jacinto, Trinity, Upper Rio Grande.

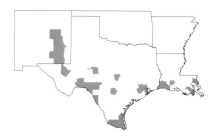

General Distribution. Southern United States, primarily coastal, and Central America, West Indies, and Venezuela.

Flight Season. May 7 (TX)–Aug. 4 (TX).

Identification. This widely distributed species is recognizable by the large black spot at the base of the hindwing and the much smaller, but similar, spot in the forewing. The face and thorax in young individuals and females are white and gray, respectively. The lateral sutures on the side of the thorax are irregularly outlined by black and connected below, to appear as "W." The wings are as above, with a narrow brown pterostigma. The legs are pale at their base and black throughout the rest of their length. The abdomen is pale yellow, outlined laterally, middorsally, and on the carinae by black. Segments 8–10 are entirely black. The mature male darkens extensively, his face and entire body turning black.

Similar Species. Pin-tailed Pondhawk (*Erythemis plebeja*) has a much narrower abdomen, a pale face in both sexes, and a larger spot in the hindwing. The similar Small Pennants (*Celithemis*), are all smaller, with a lighter stripe through the dark basal patch in each hindwing. Spot-winged Glider (*Pantala hymenaea*) is larger and golden brown. Saddlebag gliders (*Tramea*) have either a narrow or a broad band, but never a spot, basally in the hindwing.

Habitat. Large brackish ponds and lakes.

Discussion. This widely distributed species is found across the southern half of the region. It is usually found associated with large brackish bodies of water, but not always. Males and females of this species are found in equal abundance around the water. They will perch on the tips of vegetation at varying heights, with both wings and abdomen elevated. Occasionally, they may be found in large feeding swarms with Hyacinth Glider (*Miathyria marcella*), rainpool gliders (*Pantala*), and saddlebag gliders (*Tramea*). Males patrol for some distance over open water and along the shoreline. Mating takes place in flight and generally over the open water of ponds and lakes. Males accompany females during egg-laying, which takes place in open waters or along the shoreline, the female tapping the abdomen to the surface during long regular approaches to the water. This species is found yearround throughout most of its range.

Sylphs
Genus *Macrothemis* Hagen

Just three species of this large Neotropical genus occur in North America, all of them found in the southern limits of our region. The species in this group are slender, with small heads, seemingly delicate thoraxes, and narrowed abdomens. A unique characteristic of all but two species (*Macrothemis inequiunguis* Calvert and *M. aurimaculata* Donnelly) is the unusually elongate inner tooth of the tarsal claw. Species in our region are brown or black, with pale-cream-colored thoracic stripes and clear wings. The frons and vertex are a metallic color in males.

References. Calvert (1898b), Donnelly (1984), May (1998a).

KEY TO THE SPECIES OF SYLPHS (*MACROTHEMIS*)

1. Sides of thorax with only pale spots, not full-length stripes	**Straw-colored (*inacuta*)**
1'. Sides of thorax with full-length stripes, the spots present or not	2
2(1'). Sides of thorax with complete pale anterior stripe and 2 or 3 posterior spots	**Ivory-striped (*imitans*)**
2'. Sides of thorax with dark stripes in a "IY" or "YY" pattern	**Jade-striped (*inequiunguis*)**

Ivory-striped Sylph
Macrothemis imitans Karsch
(photo 58c)

Size. Total length: 35–37 mm; abdomen: 25–27 mm; hindwing: 25–29 mm.

Regional Distribution. *Biotic Provinces:* Balconian, Tamaulipan, Texan. *Watersheds:* Guadalupe, Lower Rio Grande, Nueces, San Antonio.

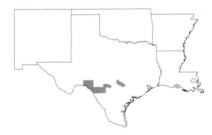

General Distribution. Central Texas and Mexico south to Argentina.

Flight Season. Jul. 1 (TX)–Oct. 24 (TX).

Identification. This is one of the two smaller sylphs that occur in the region. The top of the head is distinctively metallic blue in males and brown in females. The eyes of males are deep aquamarine blue. The thorax is dark brown or black, with a pair of pale abbreviated middorsal stripes and a whitish-green anterior lateral stripe followed by two or three pale spots. The middorsal thoracic stripes are scant or entirely absent in females. The wings are clear, with a touch of yellow flavescence throughout. The pale middorsal stripe on the abdomen is interrupted, and segments 7–9 are broadly expanded in the male. The male cerci are upturned to appear as a high-heeled shoe, when viewed laterally.

Similar Species. Jade-striped Sylph (*M. inequiunguis*) can be distinguished by its nearly complete middorsal thoracic stripes and broader dark lateral stripes, the latter forming a "IY" or "YY" pattern. Straw-colored Sylph (*M. inacuta*) is larger, and the pale anterior lateral thoracic stripe is interrupted. Blue Dasher (*Pachydiplax longipennis*) lacks a clubbed abdomen. Thornbush Dasher (*Micrathyria hagenii*) has a "IYI" lateral thoracic pattern.

Habitat. Rocky streams and rivers.

Discussion. This species feeds in sustained flights, and males fly back and forth, low over shallow riffles. It is known from only a few counties in central Texas west to the Devils River.

Straw-colored Sylph
Macrothemis inacuta Calvert
(photo 58d)

Size. Total length: 39–42 mm; abdomen: 25–29 mm; hindwing: 27–31 mm.

Regional Distribution. *Biotic Provinces:* Balconian, Tamaulipan. *Watersheds:* Lower Rio Grande, Nueces.

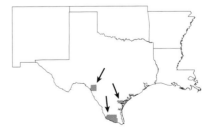

General Distribution. Southern Arizona and Texas; Mexico through Central America south to Venezuela and Brazil.

Flight Season. May 11 (TX)–Oct. 17 (TX).

Identification. This species is the largest of the three sylphs occurring in our region. The pale thoracic stripes are not as prominent in this species as they are in the other two. The face is hairy and olivaceous, becoming metallic blue along the top of the head in mature males. The eyes are dark brown, becoming aquamarine in males. The thorax is brown, with pale-cream stripes on the front of the thorax and on either side of the middorsal carina, thus appearing as inverted L's. The lateral sutures are thinly outlined in black. The pale anterior thoracic stripe is interrupted, and is followed posteriorly by three pale spots. The wings are clear, but often develop a wash of amber, and each has a basal brown spot that may extend out as far as the triangle in the hindwing. The legs are brown. The abdomen is brown, long, and slender, with black carinae.

Similar Species. This is the only sylph in our region without a complete lateral thoracic stripe. No other species in our area are brown with blue eyes.

Habitat. Clear rocky streams and rivers.

Discussion. This species perches vertically on tree branches from 1.5 to several meters high. Males will feed in sustained flights. Within our region, this species is found primarily in the lower Rio Grande Valley.

References. Calvert (1899).

Jade-striped Sylph
Macrothemis inequiunguis Calvert
(photo 58e)

Size. Total length: 32–36 mm; abdomen: 25–28 mm; hindwing: 24–30 mm.

Regional Distribution. *Biotic Province:* Balconian. *Watersheds:* Guadalupe, San Antonio.

General Distribution. Central Texas, Mexico, and Central America south to Venezuela.

Flight Season. May 31 (TX)–Nov. 13 (TX).

Identification. This is a smaller brown sylph with a pale-green thorax and black lateral thoracic stripes forming a "IY" or "YY" pattern. Its face is brown in females and young males, but becomes metallic green in mature males. The femora are brown, becoming black distally along with the remainder of the leg. The wings are clear or with a wash of amber throughout. There is a dark spot basally in each wing. The abdomen is brown, darker on top, with a pale interrupted lateral stripe on segments 1–5. There is a pair of pale oval spots dorsally on segment 7.

Similar Species. Ivory-striped Sylph (*M. imitans*) has prominent pale frontal thoracic markings in the male and no frontal thoracic markings in the female. Clubskimmers (*Brechmorhoga*) are much larger and greener. Blue Dasher (*Pachydiplax longipennis*) lacks a clubbed abdomen. Thornbush Dasher (*Micrathyria hagenii*) has a "IYI" lateral thoracic pattern. Three-striped Dasher (*M. didyma*) has a "III" lateral thoracic pattern, and Spot-tailed Dasher (*M. aequalis*) either is pruinose gray or has a "WII" lateral thoracic pattern.

Habitat. Rocky streams and rivers.

Discussion. This species is very uncommon within our range. It is known only from a handful of counties in central Texas.

Hyacinth Gliders
Genus *Miathyria* Kirby

Two of the species of this Neotropical genus are associated with water hyacinth (*Eichhornia*) and water lettuce (*Pistia*). Only one of these is found in the south-central United States, and it occurs primarily along the southern coastal regions of Texas and Louisiana. These are brownish, moderately sized dragonflies with distinctly broad hindwings that taper apically. The hindwings are marked with a long, narrow basal stripe similar to saddlebag gliders (*Tramea*). There are few antenodal crossveins in the hindwing, generally four in *Miathyria marcella*, and the apical planate subtends a single row of double-height cells. The abdomen in this group narrows apically after the swollen basal segments. The caudal appendages of the male are distinctly sigmoid when viewed laterally.

Hyacinth Glider
Miathyria marcella (Selys *in* Sagra)
(photo 58f)

Size. Total length: 35–41 mm; abdomen: 21–27 mm; hindwing: 27–34 mm.

Regional Distribution. *Biotic Provinces:* Austroriparian, Balconian, Tamaulipan, Texan. *Watersheds:* Arkansas, Brazos, Colorado, Guadalupe, Lower Rio Grande, Mississippi, Neches, Nueces, Red, Sabine, San Jacinto, St. Francis, Trinity.

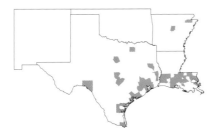

General Distribution. Southern United States from Virginia to Texas; also Mexico and West Indies south to Argentina.

Flight Season. Apr. 13 (LA)–Nov. 24 (LA).

Identification. This beautiful golden-brown species is found all along the coastal areas of southern Texas and Louisiana, as well as some distance inland. Its face is pale, the top of the frons becoming metallic violet in older males. The thorax is tawny brown with a pair of cream-colored oblique lateral stripes. These stripes may become obscured with age in males as the thorax becomes progressively darker violet in color, starting anteriorly and between the wings. The wings themselves are clear, with light-brown or reddish veins and a dark basal band in each hindwing. There are relatively few antenodal crossveins compared to other dragonflies of this size, with seven and four in the fore- and hindwings, respectively. The legs are dark brown with paler bases. The abdomen is orange-brown with a black middorsal stripe.

Similar Species. Saddlebag gliders (*Tramea*) are similar, but all are larger, with the black on the abdomen restricted to segments 8–10, and all but Striped Saddlebags (*T. calverti*) lack thoracic stripes.

Habitat. Marshy ponds and lakes, including brackish waters, with water hyacinth.

Discussion. This species is found throughout Mexico and Central and South America, but was not reported from the United States until 1950. It has since been reported as far northward as Virginia. Its spread is correlated with the introduction and spread of water hyacinth (*Eichhornia*) into this country. Individuals are commonly seen in large feeding swarms away from water, often along roadsides. They are active fliers, rarely resting, but when at rest they perch vertically on twigs, stems, or other vegetation low to the ground, with their abdomens pointed downward (in a fashion similar to that of the darners). Males patrol territories low over the water, often hovering extensively. Territories in this species are established by display of the orange abdomen by one male to intruding males. Mating occurs in flight. The male may accompany the female as she lays eggs, or she may do this alone or guarded by the male. Eggs are deposited at the base of water hyacinth (*Eichhornia*), water lettuce (*Pistia*), and other floating aquatic vegetation by rapid descents from 1–2 m above the water.

References. Bick et al. (1950), Dunkle (1989), Paulson (1966).

Tropical Dashers
Genus *Micrathyria* Kirby

This is another large group of Neotropical species, represented by only a few species in North America. Three species occur in central and southern Texas. They are all small, dark dragonflies with predominantly pale faces, a greenish thorax, and spotted abdomens. The hind lobe of the prothorax is distinctly enlarged and bilobed in some species, with a fringe of long hairs. The wings are clear and highly variable in venation, but always have two or three bridge crossveins. The abdomen is generally swollen both basally and distally, with slender middle segments. Abdominal segment 10 is considerably shortened.

Members of this genus are closely associated with small ponds and swamps, where adults perch on the tips of stems and branches. This group is of evolutionary significance because it contains species that oviposit exophytically by recognizing and choosing individual plant species, although the success of the larvae is not dependent upon the selection.

References. May (1977, 1980), Needham (1943), Paulson (1966, 1969).

KEY TO THE SPECIES OF TROPICAL DASHERS (*MICRATHYRIA*)

1. Total length less than 32 mm; hindwing less than 25 mm long	**Spot-tailed (*aequalis*)**
1'. Total length greater than 33 mm; hindwing greater than 25 mm long	2
2(1). Thorax with 3 unbranched lateral stripes	**Three-striped (*didyma*)**
2'. Thorax with a single midlateral stripe, forked at its upper end	**Thornbush (*hagenii*)**

Spot-tailed Dasher
Micrathyria aequalis (Hagen)
(photo 59a)

Size. Total length: 26–34 mm; abdomen: 15–24 mm; hindwing: 20–26 mm.
Regional Distribution. *Biotic Province:* Tamaulipan. *Watershed:* Lower Rio Grande.

General Distribution. Southern Florida, Texas, West Indies, and Central America south to Ecuador.
Flight Season. May 7 (TX)–Oct. 18 (TX).
Identification. This small species occurs in the lower Rio Grande Valley of southernmost Texas. Its face is nearly white, with grey eyes that become brilliant green, and the top of head is metallic green in mature males. The thorax is pale yellowish green in females and young males. The front of the thorax has a pair of yellowish stripes, and each side has three diffuse brown stripes along the lateral sutures, confluent below to form a "WII" pattern. The wings are clear, and the legs are black. The slender abdomen is brown, with a pair of interrupted pale stripes dorsally, ending on segment 7 or 8. Older males become heavily pruinose, obscuring the thoracic and abdominal pattern with gray. Females usually retain this pattern, but colors may be dull and the wingtips may darken.
Similar Species. Blue Dasher (*Pachydiplax longipennis*), Three-striped Dasher (*M. didyma*), and Thornbush Dasher (*M. hagenii*) are all similar, but all have three distinct straight black stripes or a "IYI pattern" laterally on the thorax. Females of Seaside Dragonlet (*Erythrodiplax berenice*) are easily distinguished by their prominent ventrally projecting ovipositor. Swift (*Dythemis velox*) and Black (*D. nigrescens*) Setwings are larger, and have dark wingtips and a different lateral thoracic pattern. Pale-faced Clubskimmer (*Brechmorhoga mendax*) is much larger.
Habitat. Permanent and temporary ponds, sloughs, and lakes.
Discussion. This species barely enters the southern tip of Texas, where it can nonetheless be locally common. Individuals may fly furiously along the edges of their pond habitats. Males will perch at varying heights up to 2 m over the water on twigs and branches, usually exposed to sunlight. Females tend to remain farther back from water when not mating or laying eggs. On warmer days, both sexes will adopt a typical obelisk position. Females lay eggs alone, but are often interrupted by males. Females land on floating leaves, below which they in-

vert the end of the abdomen to deposit eggs on the underside. An estimated 2,000 eggs have been documented in a 2.5 cm² area, though these are probably not all from the same female.

References. May (1977, 1980), Needham (1943).

Three-striped Dasher
Micrathyria didyma (Selys)
(photo 59b)

Size. Total length: 32–40 mm; abdomen: 22–28 mm; hindwing: 25–32 mm.
Regional Distribution. *Biotic Province:* Tamaulipan. *Watershed:* Lower Rio Grande.

General Distribution. Southern Florida and Texas, south through Mexico and the West Indies to Ecuador.
Flight Season. Year-round (Mexico).
Identification. Like Spot-tailed Dasher (*M. aequalis*), Three-striped Dasher reaches just into the lower Rio Grande Valley of Texas. Its face is white, the top of the head becoming metallic green in mature males and staying brown in females. The green thorax is marked with brown in front and has three dark oblique stripes laterally. The wings are clear, generally with two cells instead of one in the forewing triangle. The legs are black. The abdomen is slender and black, with pale-green trapezoidal spots on segments 2–7, the most prominent of which is on segment 7. Segments 8–10 and the caudal appendages are black.
Similar Species. Spot-tailed Dasher (*M. aequalis*) is smaller, and has a pattern of stripes that appears as "WII" on the sides of the thorax and triangular spots dorsally on segment 7. Older males of Thornbush Dasher (*M. hagenii*) become heavily pruinose, and the midlateral thoracic stripe is forked at its upper end, forming a "IYI" pattern. Blue Dasher (*Pachydiplax longipennis*) has pale streaks on seg-

ment 7 and spots on abdominal segment 8 dorsally. Female Seaside Dragonlets (*Erythrodiplax berenice*) have a prominent ventrally projecting ovipositor. Swift (*Dythemis velox*) and Black (*D. nigrescens*) Setwings are larger, and have dark wingtips and a different lateral thoracic pattern. Pale-faced Clubskimmer (*Brechmorhoga mendax*) is much larger.
Habitat. Weedy pools, ponds, brooks, and ditches in the shade.
Discussion. The only record of this species in Texas is a single male taken at Bentsen State Park in Hidalgo County. This species is typically found along forest clearings, where it feeds. Little has been written on its biology or behavior. One author described the habitat where he collected larvae as a "little, weedy, spring-fed brook." It is found year-round throughout most if not all of its range.

References. Abbott (1996), Needham (1943).

Thornbush Dasher
Micrathyria hagenii Kirby
(photo 59c)

Size. Total length: 30–36 mm; abdomen: 18–25 mm; hindwing: 24–30 mm.
Regional Distribution. *Biotic Provinces:* Austroriparian, Balconian, Carolinian, Chihuahuan, Kansan, Tamaulipan, Texan. *Watersheds:* Arkansas, Brazos, Colorado, Lower Rio Grande, San Antonio, San Jacinto, St. Francis, Upper Rio Grande.

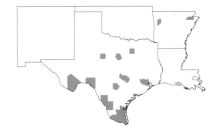

General Distribution. Arkansas and Texas, south through Mexico and the West Indies to Panama.
Flight Season. May. 7 (TX)–Dec. 23 (TX).
Identification. This is the most widespread tropical dasher in North America. Its face is pale yellow. The top of the frons is brown in females and young males, but becomes metallic blue in mature males. The front of the thorax is brown, with two pale-green stripes that do not reach the alar carina.

The sides are green, with three brown oblique stripes. The middle one is usually forked at its upper end, often joining with the stripes on either side to form a "IYI" pattern. The underside of the front femur is green in females and young males, and the rest of the leg is black. The wings are clear, with only a flavescent tinge basally and a small brown basal spot in the hindwing. The black abdomen is slender, but widens slightly at segments 7–9. There is a row of greenish spots dorsolaterally on segments 1–7, those on segment 7 the most pronounced. The remaining abdominal segments and caudal appendages are black. Mature males develop a pruinose grayish color on the thorax and abdomen. Females are similar to males, but lack the pruinosity, and their abdominal spots are larger throughout.

Similar Species. Three-striped Dasher (*M. didyma*) and Blue Dasher (*Pachydiplax longipennis*) are generally smaller and have three distinct lateral thoracic stripes forming a "III" pattern. The smaller female Spot-tailed Dasher (*M. aequalis*) has a "WII" lateral thoracic pattern and prominent triangular spots dorsally on abdominal segment 7. Swift (*Dythemis velox*) and Black (*D. nigrescens*) Setwings are larger, have dark wingtips, and have a different lateral thoracic pattern. Pale-faced Clubskimmer (*Brechmorhoga mendax*) is much larger. Female Seaside Dragonlets (*Erythrodiplax berenice*) have a prominent ventrally projecting ovipositor.

Habitat. Heavily vegetated ponds and lakes.

Discussion. This species has been reported as far north as Franklin County, Arkansas. Although it is widely distributed, it becomes locally common only from Central Texas southward, where it breeds. One study documented females ovipositing as they hover low over the water, at 0.25 m, extruding egg masses approximately 2 mm in diameter and flicking their abdomen upward to release them. They will also extrude eggs on floating vegetation. The species is often found perched on thick vegetation surrounding the ponds it haunts. It is often recognizable because of its common habit of raising its abdomen, with its pale white spots, nearly vertically over its brilliant iridescent-green eyes. This species is found year-round throughout much of its range, and the limited flight season stated above may not be completely representative.

References. Harp and Rickett (1977), Paulson (1969).

Tropical King Skimmers
Genus *Orthemis* Hagen

This is a moderately diverse Neotropical genus of some 18 species, two of which range northward into our region. They are large, robust dragonflies similar in size to king skimmers (*Libellula*), but males are predominantly brown in color, with red or purple abdomens. The top of the frons becomes metallic violet in mature males. The wings are clear, with a long brown or black pterostigma that surmounts five or six crossveins. The abdomen tapers strongly rearward and appears triangular in cross section. Abdominal segment 8 is pronounced, and expanded laterally in females.

KEY TO THE SPECIES OF TROPICAL KING SKIMMERS (*ORTHEMIS*)

1. Face dark, not red; distinct dark markings on ventrolateral portion of metepisternum; second lateral thoracic stripe distinct, dark on lower part of mesepimeron; the two halves of the vulvar lamina, in females, meet evenly, forming a wide "U" — **Roseate (*ferruginea*)**

1'. Face red; no dark markings on venter of thorax; second lateral thoracic stripe nearly absent; the two halves of the vulvar lamina, in females, meet at a slight angle, forming a shallow "V" — **Orange-bellied (*discolor*)**

Orange-bellied Skimmer
Orthemis discolor (Burmeister)
(photo 59d)

Size. Total length: 47–56 mm; abdomen: 33–39 mm; hindwing: 35–44 mm.
Regional Distribution. *Biotic Provinces:* Tamaulipan, Texan. *Watersheds:* Brazos, Guadalupe, Rio Grande.
General Distribution. Central and south Texas south through Central America.

Flight Season. Aug. 17 (TX)–Oct. 20 (TX).
Identification. This species has a red face and orange labrum. Young individuals are brown, with an unmarked, ventrally orange-yellow thorax. The top of the head turns dark metallic violet in mature males, and the thorax and abdomen become reddish with a uniform purple pruinose cast in both sexes. The wings are clear, with dark wing veins. There is a wide, dark, lateral flange on segment 8 in females. The black vulvar laminae meet at a slight angle, forming a shallow "V."
Similar Species. The only species likely to be confused with this one is the similar and more widespread Roseate Skimmer (*O. ferruginea*), which is more magenta or purple. The ventral side of its thorax is grayish yellow and often has distinct dark markings on the ventrolateral portion of the metepisternum that are generally not present on the Orange-bellied Skimmer. The sides of the thorax in the Roseate Skimmer have a distinct pale-yellow stripe along the ventral suture. The wing veins are more orange, and the lateral flange of abdominal segment 8 in females is pale. The vulvar laminae in the Roseate Skimmer are parallel, forming a short-armed "U." Flame (*Libellula saturata*) and Neon (*L. croceipennis*) Skimmers both lack lateral thoracic markings and have much stockier bodies. Red-tailed Pennant (*Brachymesia furcata*) is smaller and has amber basally in the hindwing.
Habitat. Temporary and permanent ponds, lakes, ditches, and slow streams.

Discussion. This species was long considered a southern form of, and synonym of, the similar Roseate Skimmer until recently. Both tropical king skimmers in the region can occur together. It is unclear exactly what behavioral differences there may be, but early observations suggest that Orange-bellied Skimmer may prefer shady areas. Where both occur together, Orange-bellied Skimmer seems to be more abundant. Novelo's (1981) behavioral study of Roseate Skimmer may have included both species. The early and late flight season dates listed above reflect relatively few observations in our region. Orange-bellied is found year-round throughout most of its range. In Texas, this species can be locally abundant at locations in the lower Rio Grande Valley and on the King Ranch in Kleburg County.

References. DeMarmels (1988), Donnelly (1995), Dunkle (1998), Paulson (1998b).

Roseate Skimmer
Orthemis ferruginea (Fabricius)
(photos 59e, 59f)

Size. Total length: 46–55 mm; abdomen: 33–39 mm; hindwing: 35–44 mm.
Regional Distribution. *Biotic Provinces:* Apachian, Austroriparian, Balconian, Carolinian, Chihuahuan, Kansan, Navahonian, Tamaulipan, Texan. *Watersheds:* Arkansas, Brazos, Canadian, Cimarron, Colorado, Colorado (NM), Guadalupe, Lower Rio Grande, Mississippi, Neches, Nueces, Ouachita, Pecos, Red, Sabine, San Antonio, San Jacinto, Trinity, Upper Rio Grande.

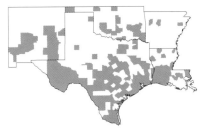

General Distribution. Southern United States, Mexico, West Indies, and Central America south to Chile.
Flight Season. Apr. 27 (LA)–Dec. 1 (LA).
Identification. This handsome and widespread species is found throughout the south-central United States. It is brown initially in both sexes, with pale stripes on the thorax forming an irregular "HII" pattern. The abdomen is uniform brown in young indi-

viduals. Mature adults develop a pale-bluish thorax and a bright-pinkish or -purple abdomen. The wings are clear with orangish veins. The lateral flanges of abdominal segment 8 in females are generally pale.

Similar Species. For additional characteristics and differences, see the discussion for the similar Orange-bellied Skimmer, above.

Habitat. Temporary and permanent ponds, lakes, ditches, and slow streams.

Discussion. This is a widespread species that seems to invade new habitats and is capable of readily expanding its range. It is found throughout the New World tropics, including the Bahamas, West Indies, and Hawaii. It behaves much like many king skimmers (*Libellula*), foraging from the top of tall vegetation. It is an aggressive predator, taking insects only slightly smaller than itself. Males will regularly and vigorously patrol territories averaging 10 m in length. Males use their abdomens to ward off intruding males, by bending the tip downward. They pursue females in flight, where mating takes place for an average of 10 seconds. Oviposition by females takes an average of 1–3 minutes and is done by flicking the eggs along with water droplets toward the shoreline. The male guards the female during this time, often hovering close to her and bending the abdomen down, almost at a right angle, when numerous competing males are present. Two emergence peaks have been reported in Louisiana, one in the spring and a second one in the late summer to early fall.

References. Harvey and Hubbard (1987), Mauffray (1997), Novelo-G. (1981), Novelo-G. and Gonzalez-S. (1984), Young (1980).

Blue Dasher
Genus *Pachydiplax* Brauer

This genus includes a single species of variably sized individuals found throughout North America and southward to Belize and the Bahamas. It is recognizable by the striped thorax and pruinose blue abdomen of mature males and older females. The wings have only a single crossvein under the pterostigma, with a long vacant space proximal to it. The abdomen is depressed and short, especially in females.

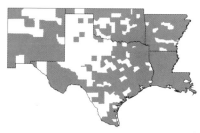

General Distribution. Throughout southern Canada and United States, except Great Basin; also Mexico, Bermuda, Bahamas, and Belize.

Flight Season. Year-round (TX).

Identification. This rather distinctive species is found throughout our area. It has a white face, and the top of the frons is metallic blue in mature individuals of both sexes. The eyes are brilliant blue or green in males and reddish brown in females. The front of the thorax is brown, with a thin pale carina medially and a wider pale stripe on either side. The sides are pale green, with three full-length brown stripes. The wings are typically clear, but may be flavescent, and have a dark-brown stripe on either side of the midbasal space in males. The legs are black and heavily armed. The abdomen is black, with a pair of pale-yellow stripes dorsally, interrupt-

Blue Dasher
Pachydiplax longipennis (Burmeister)
(p. 5; photo 60a)

Size. Total length: 28–45 mm; abdomen: 23–35 mm; hindwing: 30–43 mm.

Regional Distribution. *Biotic Provinces:* Apachian, Austroriparian, Balconian, Carolinian, Chihuahuan, Kansan, Navahonian, Tamaulipan, Texan. *Watersheds:* Arkansas, Bayou Bartholomew, Brazos, Canadian, Cimarron, Colorado, Colorado (NM), Guadalupe, Lower Rio Grande, Mississippi, Neches, Nueces, Ouachita, Pecos, Red, Sabine, San Antonio, San Jacinto, St. Francis, Trinity, Upper Rio Grande, White.

ed on segments 3–8 to appear as dashes. Segment 9 and the caudal appendages are black, and 10 is pale. The abdomen is considerably shorter than in females. Older males develop a pale pruinose blue color dorsally, and more slowly laterally, on the thorax and over the entire abdomen. Females become pruinose, but much more slowly than males.

Similar Species. Mature males resemble mature males of Eastern Pondhawk (*Erythemis simplicicollis*) and Western Pondhawk (*E. collocata*), but the pondhawks have an unmarked thorax. Female Seaside Dragonlets (*Erythrodiplax berenice*) have a prominent ventrally projecting ovipositor. Species of tropical dashers (*Micrathyria*) are similar, but in all except Three-striped Dasher (*M. didyma*) the lateral thoracic markings are branched. All also have pronounced greenish spots dorsally on abdominal segment 7.

Habitat. Ponds, lakes, marshes, ditches, slow streams, and other quiet bodies of water.

Discussion. This species is found around nearly any standing body of water, where it is often the most common and abundant species. It is known from every county in Arkansas and every parish in Louisiana. As a result, it ranks as one of the most well-studied dragonflies in North America. The Blue Dasher is often seen perched vertically on twigs and branches at a variety of heights, from just above ground level to the treetops, with the wings depressed downward. On warmer days individuals will raise the abdomen in an obelisk position, reducing heat absorbance. They are aggressive predators, regularly taking over 10% of their body weight in prey daily. Adults roost in trees and are occasionally attracted to lights at night. This species will defend favored feeding sites for several days in a row. Breeding territories are established along the shoreline, where males will investigate all intruders, defending against other males and chasing them out by raising their pruinose blue abdomens. Multiple territories may be established in a single day. Mating takes place while in flight or perched, and may last from 1/2 to 2 minutes. The male will guard the female from a nearby perch while she deposits eggs by flying low over the water and repeatedly tapping the abdomen to the surface, but never bobbing the entire body up and down. She may lay 300–700 eggs in only 35 seconds, usually in a heavily vegetated pond margin. Females remain farther back from the water when not laying eggs or mating.

The tremendous variation in size within this species is generally correlated with the appearance of larger individuals during the spring months and progressively smaller ones in the summer and fall. One study found that the larvae of this species had a strong preference for the leaf axil area of the aquatic plants they were associated with. The axil provides them protection and makes them less susceptible to fish predation.

References. Baird and May (1997), Bick (1950, 1957), Byers (1930), Fried and May (1983), Frost (1971), Johnson (1962c), Mackinnon and May (1994), Mauffray (1997), May (1984), Needham (1946), Paulson (1966), Penn (1951), Robey (1975), Root (1924), Sherman (1983), Wellborn and Robinson (1987), Wright (1943b).

Rock Skimmers
Genus *Paltothemis* Karsch

This group contains a Mexican species and a single species found in the southwestern United States that occurs southward to Venezuela. Members of the group are closely related to the setwings (*Dythemis*) but differ in several points. Individuals have broad wings with a narrow, black, and relatively short pterostigma. The forewing triangle points inward, and the cubital vein is strongly angulate. The anal area in the hindwing contains four or five irregular double rows of cells. They get their name from the unusual habit among skimmers of perching horizontally or vertically on rocks and bridge pillars.

References. Garrison (1982).

Red Rock Skimmer
Paltothemis lineatipes Karsch
(photo 60b)

Size. Total length: 44–53 mm; abdomen: 29–36 mm; hindwing: 41–46 mm.

Regional Distribution. *Biotic Provinces:* Apachian, Austroriparian, Balconian, Chihuahuan, Kansan, Navahonian, Tamaulipan. *Watersheds:* Arkansas, Colorado (NM), Pecos, Red, Upper Rio Grande.

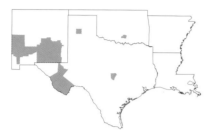

General Distribution. Southwestern United States and Mexico south to Costa Rica.

Flight Season. Apr. 23 (TX)–Oct. 13 (NM).

Identification. This moderate-sized dragonfly is found throughout the southwestern parts of the region. Young males and females have a pale face, which becomes bright red along with the vertex in older males. The thorax and abdomen are brownish gray heavily marked with black. The front of the thorax has a dark rectangle anteriorly. Two large dark spots, one in front of the humeral suture and one below it, are visible from the side. The midlateral and third lateral sutures, along with the rear margin of the pterothorax, are each marked with a heavy, but irregular, black stripe. The wings each have a broad area of flavescence basally, the effect is more pronounced in males, and a darker stripe on either side of the midbasal space. The legs are pale externally and dark on the inside. The abdomen is marked with black on the carinae and irregularly so on the rest of each segment. In mature males, the front of the thorax and the entire abdomen are red.

Similar Species. This species may be confused with several other red species with which it occurs. Red Rock Skimmer, however, is darker and more heavily striped then these others. Mayan Setwing (*Dythemis maya*), Neon (*Libellula croceipennis*) and Flame (*L. saturata*) Skimmers all have an unmarked thorax and abdomen.

Habitat. Small, sunlit, rocky, forest streams.

Discussion. Red Rock Skimmer is unique among skimmers in combining the broad hindwing of glider dragonflies (*Pantala* and *Tramea*) with the behavior of perchers and fliers. Adults make a habit of gliding during both feeding and patrolling flights. When perched, they nearly always do so horizontally on rocks or vertically on bridge pillars. Males patrol small 15-m sections of streams where the water trickles through the rocks early in the morning, rarely being seen after midday. Mating takes place in flight, and subsequent egg-laying by females occurs alone or guarded by the male. Females drop rapidly to the water from a height of 12 cm, dipping the abdomen in regular 1-second intervals.

One study found that in the absence of intraspecific competition, males defended territories more than twice as large as those defended during a high-density year, as defined by a high rate of male-male interactions, and regular raiding of neighboring territories to steal females. Further studies showed that males engage in meandering searching flights to locate potential egg deposition sites in their territories. This is followed by an inspection flight of a suitable place that will be displayed to the female after she has been captured.

References. Alcock (1987b, 1989b, 1990), Dunkle (1978).

Rainpool Gliders
Genus *Pantala* Hagen

Two medium-sized brownish-yellow dragonflies constitute this genus. They have pale faces that become red with maturity, and large wings that are broad basally, allowing them more sustained flight. Both species in this genus are found widespread throughout North America, where they are often seen soaring for hours in open fields. Wandering Glider (*P. flavescens*) is a well-known migratory species with a circumtropical distribution. The two species are among the first of the dragonflies to colonize a newly formed habitat, which may include nearly any standing body of water, such as temporary ponds, pools, and watering troughs. They get their name from their unusual tactic of breeding in these temporary pools, where the larvae are fast-growing.

KEY TO THE SPECIES OF RAINPOOL GLIDERS (*PANTALA*)

1. Hindwing with a large round brown spot in the anal area	**Spot-winged (*hymenaea*)**
1'. Hindwing without a brown spot in the anal area, but some yellow flavescence may be present	**Wandering (*flavescens*)**

Wandering Glider
Pantala flavescens (Fabricius)
(photo 60c)

Size. Total length: 44–51 mm; abdomen: 25–34 mm; hindwing: 35–42 mm.

Regional Distribution. *Biotic Provinces:* Apachian, Austroriparian, Balconian, Carolinian, Chihuahuan, Kansan, Navahonian, Tamaulipan, Texan. *Watersheds:* Arkansas, Bayou Bartholomew, Brazos, Canadian, Cimarron, Colorado, Colorado (NM), Guadalupe, Lower Rio Grande, Mississippi, Neches, Nueces, Ouachita, Pecos, Red, Sabine, San Antonio, San Jacinto, St. Francis, Trinity, Upper Rio Grande, White.

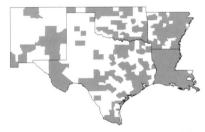

General Distribution. Throughout United States and southern Canada; also West Indies and Central America south to Chile and Argentina; found on all continents but Europe and Antarctica.

Flight Season. Year-round (TX).

Identification. This is the most widespread and most cosmopolitan dragonfly in the region. With its predominantly yellow color, it is distinctive. It has a pale-yellow face that becomes reddish in older males. The thorax is olivaceous brown and largely unmarked. The wings are clear, with brown apices in the males. The legs are pale basally, becoming black for most of their length. The stout, tapered abdomen is yellow, with black lateral stripes on the swollen basal segments. There is a thin, dark middorsal abdominal stripe that widens and becomes noticeably darker on segments 8–10. The pale caudal appendages are more or less bicolored in males, darkening in the outer half.

Similar Species. Spot-winged Glider (*P. hymenaea*) has a distinct brown spot basally in the hindwing and is generally darker in color. Similar meadowhawks (*Sympetrum*) have normal-shaped wings and parallel-sided abdomens. Saddlebag gliders (*Tramea*) all have wide crossbands in the hindwings.

Habitat. Permanent and temporary ponds, pools, and other water bodies, including brackish ones.

Discussion. Few if any dragonfly species have been accorded a more appropriate common name. It is circumtropical in distribution, found in nearly every contiguous state, extreme southern Canada, southward throughout Central and South America, the Bahamas, West Indies, and Hawaii, and throughout the Eastern Hemisphere, except for Europe. It is a strong flier, regularly encountered by ocean freighters, and a well-known migratory species. Because of its ability to drift with the wind, feeding on aerial plankton, until it finally encounters a rain pool in which to breed, it has been called "the world's most evolved dragonfly." It is generally more abundant in the fall, when offspring from earlier matings in the spring migrate southward.

Wandering Gliders are often encountered in large mixed feeding swarms, along with saddlebag gliders (*Tramea*), where they prey upon small flying insects. Males patrol territories of varying breadths 1–2 m above the water. Mating takes place in flight and lasts from 30 seconds to an unusually long 5 minutes. Females lay eggs in temporary ponds or rainpools by tapping their abdomen to the water surface, alone or accompanied by the male. It is not unusual, however, to see females attempting to lay eggs on automobile rooftops, asphalt roads, or other shiny structures that they mistake for water. They usually perch vertically on low stems and twigs, but sometimes they perch horizontally with the abdomen depressed below the rest of the body. Eggs are deposited in ponds while flying in tandem straight over the water. Feeding swarms consist of both males and females in equal numbers, over land, and can occur at any time from dawn to dusk. The larva can complete its larval cycle in as little as five weeks, but early instars of the larvae are extremely tolerant to drought, living several months in dry mud when necessary.

References. Corbet (1963), Dunkle (1989), McLachlan (1896), Reichholf (1973), Svihla (1961), Trottier (1967), Van Damme and Dumont (1999), Wakana (1959), Warren (1915).

Spot-winged Glider
Pantala hymenaea (Say)
(photo 60d)

Size. Total length: 43–51 mm; abdomen: 29–35 mm; hindwing: 39–45 mm.

Regional Distribution. *Biotic Provinces:* Apachian, Austroriparian, Balconian, Carolinian, Chihuahuan, Kansan, Navahonian, Tamaulipan, Texan. *Watersheds:* Arkansas, Bayou Bartholomew, Brazos, Canadian, Cimarron, Colorado, Colorado (NM), Guadalupe, Lower Rio Grande, Mississippi, Neches, Nueces, Ouachita, Pecos, Red, Sabine, San Antonio, San Jacinto, St. Francis, Trinity, Upper Rio Grande, White.

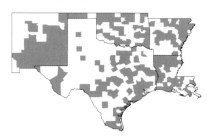

General Distribution. Throughout southern Canada and United States; also West Indies and Central America south to Argentina and Chile.

Flight Season. Year-round (TX).

Identification. This species is similar to Wandering Glider (*P. flavescens*) and equally cosmopolitan in the south-central United States. The face and thorax are essentially as in that species. Each hindwing, however, has a distinct round dark spot basally. The abdomen is darker and mottled.

Similar Species. Wandering Glider lacks the brown spot basally in the hindwing. All saddlebag gliders (*Tramea*) in the region have a basal band rather than a spot in the hindwing. The Hyacinth Glider (*Miathyria marcella*) has a narrow band in the hindwing rather than a spot. Otherwise similar skimmers lack the hindwing spot.

Habitat. Open, temporary and artificial ponds and pools, including brackish waters.

Discussion. This species, although not found globally, is as widely distributed throughout North America as the Wandering Glider, and its behavior is much the same as that of that species. It is a strong flier, generally only taking a perch to roost at night. It is an early colonizer of the temporary and artificial ponds where it breeds. Males patrol larger more linear territories than do Wandering Glider. Females lay eggs by tapping the abdomen to the water while flying quickly over the water or while hovering, either accompanied by the male or alone. This species has been slow to colonize along the West coast of the United States, but there is an apparent increase over much of that region. One study showed that larvae of this species in Oklahoma could complete development in less than five weeks during the summer months.

References. Bick (1951), Dunkle (1989), Paulson and Garrison (1977).

Amberwings
Genus *Perithemis* Hagen

This primarily Neotropical group contains a few species of small, robust, brown-and-yellow dragonflies. All are readily recognized, as a group, by their small size, the solid amber wings of the male, and the spotted wings of the female. Many members of this genus have complex courtship displays. The thorax is olivaceous green-brown, and the abdomen is narrowed at both ends in each sex. The abdomen is shorter than the wings. The forewing triangle is unusual, the inner and anterior sides being equal, or nearly so, in length. The anal loop forms a foot that is hardly bent, if at all, at the ankle. Eastern Amberwing (*P. tenera*) is the only species widely distributed across North America, but all three North American species are found within our region.

References. Ris (1930).

KEY TO THE SPECIES OF AMBERWINGS (*PERITHEMIS*)

1. Thorax without definite stripes	**Mexican (*intensa*)**
1'. Thorax with a pair of greenish stripes	2
2(1'). Legs mostly pale; pale dorsal markings on the abdomen take the form of chevrons; generally, triangles and subtriangles without crossveins	**Eastern (*tenera*)**
2'. Legs mostly dark; pale dorsal markings on the abdomen form a straight line; generally, some triangles or subtriangles divided by crossveins	**Slough (*domitia*)**

Slough Amberwing
Perithemis domitia (Drury)
(photo 60e)

Size. Total length: 21–25 mm; abdomen: 12–16 mm; hindwing: 16–20 mm.
Regional Distribution. *Biotic Provinces:* Chihuahuan, Tamaulipan. *Watersheds:* Lower Rio Grande, Upper Rio Grande.

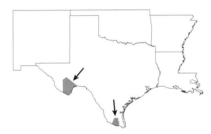

General Distribution. Southern Arizona and Texas south through Mexico to Ecuador and Brazil.
Flight Season. May 9 (TX)–Oct. 19 (TX).
Identification. This is a small Latin American species whose range just reaches into the lower Rio Grande Valley. Its face is yellow, and the vertex and occiput are brown. The thorax is brown, with two wide olivaceous stripes that become obscured with age. The male's wings are amber, with dark-red venation and pterostigma. The female's wings are amber out to the nodus, with dark brown spots. The legs are brown with black joints. The brown, short abdomen has a narrow waist basally, widening medially and narrowing again apically, thus appearing spindle-shaped. A series of pale stripes forms an interrupted, but straight, line on either side of the midline of the abdomen.

Similar Species. This species could easily be confused with the similar Eastern Amberwing (*P. tenera*), but that species has chevrons dorsally on the abdomen, not stripes. Slough Amberwing also tends to prefer shady areas, rather than the open sunny fields and meadows preferred by Eastern Amberwing. All other similar skimmers with amber in their wings are much larger.
Habitat. Shaded sloughs, ponds, pools, roadside ditches, and other still waters.
Discussion. Though widespread farther south, this species has been found at only a few localites in Texas, including the lower Rio Grande Valley and Big Bend National Park. Breeding populations are known at these localities. Slough Amberwings are often found taking cover in shady areas, unlike both Mexican (*P. intensa*) and Eastern Amberwings. Needham et al. (2000) reported that "Adults fly low over water, never departing far from it. They dart about very swiftly and perch frequently on emergent twigs or grass stems. Males on meeting face to face in flight may dart upward to considerable heights, threatening each other, but return at once to low-level perches."

Mexican Amberwing
Perithemis intensa Kirby
(photo 60f)

Size. Total length: 24–26 mm; abdomen: 15–16 mm; hindwing: 20–22 mm.
Regional Distribution. *Biotic Province:* Apachian. *Watershed:* Colorado (NM).

General Distribution. Southwestern United States and Mexico.

Flight Season. Jul. 26 (NM)–Sep. 6(NM).

Identification. This Mexican species is found only in far western New Mexico within our region. It has a pale yellow face, and the thorax is light brown, generally lacking stripes, but occasionally bearing darker midfrontal and humeral stripes. The abdomen is light brown, with a faint, narrow and waved dorsolateral brown stripe, the stripe often lacking completely in males. The wings in males are orange with yellow veins, a darker-brown spot halfway between the base and nodus, and a dark-red pterostigma. The female's wings are clear, with amber bands between the base and nodus and between the nodus and pterostigma. Though variable, there are generally also dark-brown spots in the basal 1/4 of the wing and at the nodus. The wing patterning is variable in both sexes.

Similar Species. With its general lack of thoracic and abdominal stripes this species is distinctive. Slough Amberwing (*P. domitia*) has dark-brown stripes dorsally on the abdomen, and Eastern Amberwing (*P. tenera*) has a row of chevrons dorsally on the abdomen. Other skimmers with amber in the wings are significantly larger.

Habitat. Open sloughs, ponds, pools, roadside ditches, and other still waters.

Discussion. Little has been reported about the behavior of this small Mexican species. From what is known, it appears to behave similarly to Eastern Amberwing.

Eastern Amberwing
Perithemis tenera (Say)
(photo 61a)

Size. Total length: 19–25 mm; abdomen: 12–16 mm; hindwing: 16–21 mm.

Regional Distribution. *Biotic Provinces:* Apachian, Austroriparian, Balconian, Carolinian, Chihuahuan, Kansan, Navahonian, Tamaulipan, Texan. *Watersheds:* Arkansas, Bayou Bartholomew, Brazos, Canadian, Cimarron, Colorado, Colorado (NM), Guadalupe, Lower Rio Grande, Mississippi, Neches, Nueces, Ouachita, Pecos, Red, Sabine, San Antonio, San Jacinto, St. Francis, Trinity, Upper Rio Grande, White.

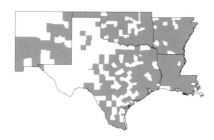

General Distribution. Eastern and central United States, southeastern Canada, and Mexico.

Flight Season. Feb. 1 (LA)–Nov. 21 (LA).

Identification. This is a small species with a brown thorax. There is a pair of wide greenish stripes laterally and middorsally on the thorax. Males have orange or amber wings that usually develop a brown spot above the triangles. Females have variously shaped brown spots or stripes running through the amber areas. Both sexes have red pterostigmata. The abdomen is narrowed basally, but thick thereafter, and brown, with a row of dark chevrons dorsally.

Similar Species. Slough Amberwing (*P. domitia*) has dark-brown stripes, not chevrons, dorsally on the abdomen. Other skimmers with amber in the wings are significantly larger.

Habitat. Open sloughs, ponds, pools, roadside ditches, and other still waters.

Discussion. This small dragonfly has been well studied. It has an elaborate courtship behavior. Males come to the water's edge early in the morning in search of a territory. They then patrol and defend these territories as potential egg-laying sites, where they regularly perch on emergent sticks or twigs. These small territories, less than 5 m², are accepted by the male only if he is not disturbed and there is no competition from other males. A female appears and is courted by the male. He will fly out to her and lead her back to his prospective oviposition site, hovering with his abdomen turned up. When he is accepted by the female, which is signaled by a slower wing beat, the pair perch on a twig and mate, taking 20–30 seconds. Females then lay eggs either accompanied by the male or alone and guarded. Females tap the abdomen against sticks or twigs within the oviposition area, attaching to it a gelatinous mass just above the waterline. The clumps of eggs thus released into the water seem to explode into individual eggs as the clump drifts downward through the water.

Both sexes of this group mimic wasps by perching at the ends of grasses or weeds and, while beating their wings, pumping the abdomen up and down. Females also fly with the hindwings held together vertically and the abdomen bent up. Females with more darkly pigmented wings tend to select the more favorable oviposition sites, such as logs or sticks, and the lighter-pigmented females select less favorable patches of floating vegetation. One study found that males that were prevented from mating were much more likely to change potential oviposition sites the following day than males that were allowed to mate, possibly implying that males use their reproductive success to determine the quality of oviposition sites. Andromorphic females, with diffusely amber wings, are occasionally reported.

Reference. Dunkle (1989), Hardy (1966), Jacobs (1955), Montgomery (1937), Shiffer (1968), Switzer (1977a,b).

Whitetails
Genus *Plathemis* Hagen

This genus contains two North American species, both of which occur in the south-central United States. Their common name derives from the distinctive white pruinosity the mature males develop on their abdomens. They are closely related to king skimmers (*Libellula*) and are sometimes included in that group, though they are distinctive in many respects.

KEY TO THE SPECIES OF WHITETAILS (*PLATHEMIS*)

1. Middle band of wings in males a uniform dark brown; wingtips in female brown; widespread — **Common Whitetail (*lydia*)**

1'. Middle band of wings in males divided by a stripe of pale color; wingtips in female clear — **Desert Whitetail (*subornata*)**

Common Whitetail
Plathemis lydia (Drury)
(photos 61b, 61c)

Size. Total length: 38–48 mm; abdomen: 23–29 mm; hindwing: 29–35 mm.

Regional Distribution. *Biotic Provinces:* Apachian, Austroriparian, Balconian, Carolinian, Chihuahuan, Coloradan, Kansan, Navahonian, Tamaulipan, Texan. *Watersheds:* Arkansas, Bayou Bartholomew, Brazos, Canadian, Cimarron, Colorado, Colorado (NM), Guadalupe, Lower Rio Grande, Mississippi, Neches, Nueces, Ouachita, Pecos, Red, Sabine, San Antonio, San Jacinto, St. Francis, Trinity, Upper Rio Grande, White.

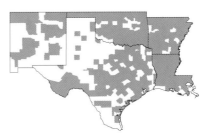

General Distribution. Throughout United States and southern Canada.

Flight Season. Mar. 3 (LA)–Nov. 24 (LA).

Identification. This is a ubiquitous species, commonly seen at almost any standing body of water during the summer. It is moderate-sized and stout, with a distinct dimorphism between the sexes. Males have large, broad brown or black bands in the outer portion of each wing, whereas the females wings are less maculated and bear three spots, one basally, another at the nodus, and the third apically, thus appearing as a smaller version of Twelve-spotted Skimmer (*Libellula pulchella*). Its face is yellowish-brown initially, but becomes noticeably darker in both sexes. The top of the head is deep brown. The robust thorax is brown, unmarked in front, and has two yellowish lateral stripes giving way to white at their upper ends. The wings in the male are as above, with a small white spot below the basal dark area in the hindwing. In both sexes, the legs are brown. In males the abdomen is broad, appearing triangular in cross section as it tapers apically. The female abdomen is strongly depressed. In both sexes the abdomen is brown, with an interrupted white line laterally, appearing as a series of individual stripes. The thorax in mature males becomes darker and the lateral stripes are obscured. The most noticeable change, however, is the total envelopment of the male's abdomen by a white pruinescence.

Similar Species. Female Twelve-spotted Skimmer (*L. pulchella*) is similar, but larger, and has yellow dorsolateral stripes on the abdomen. Prince Baskettail (*Epitheca princeps*) is larger, with green eyes, a long slender abdomen, and a lack of any white in the wings. Desert Whitetail (*P. subornata*) has a pale window within the brown wing band at the nodus, and has distinct white areas basally. Eight-spotted Skimmer (*L. forensis*) has brown wing bands at the nodus that do not reach the pterostigma. Other similar banded dragonflies like Four-spotted Pennant (*Brachymesia gravida*) and Band-winged Dragonlet (*Erythrodiplax umbrata*) lack basal wing markings.

Habitat. Nearly any pool, pond, lake, or quiet stream.

Discussion. This is one of the most familiar dragonflies to the casual observer, as well as one of the most studied. It has been collected in every county in Arkansas (Harp, pers. comm.). The distinct white abdomen of mature males is used in displays to threaten other males. They elevate the abdomen above the rest of the body and fly toward an intruder. Males patrol moderate-sized habitats of 15–30 m breadth around the shores of ponds, lakes, and occasionally streams. They will often venture some distance from their breeding sites, and may commonly be seen along roadsides and path margins perching on the ground, logs, or low vegetation. Adults mature after an average of two weeks, after which they return to bodies of water to breed. Males are aggressive, often stealing females from other males. Mating is quick, occurring as the pair hovers over the water, usually no longer than 3 seconds. Males will often attempt to guard more than one female as they lay some 1,000 eggs, by tapping the tip of the abdomen to the water at regular intervals. Recently, several male specimens having dark wingtips were reported from Oregon. All the males I have seen from our region have had clear wingtips.

References. Campanella and Wolf (1974), Dickerson et al. (1982), Jacobs (1955), Koenig (1990, 1991), Koenig and Albano (1985, 1987a,b), McMillan (1984), Savard and Girard (1996), Valley (1998), Waltz (1982).

Desert Whitetail
Plathemis subornata Hagen
(photos 61d, 61e)

Size. Total length: 41–52 mm; abdomen: 22–31 mm; hindwing: 31–38 mm.

Regional Distribution. *Biotic Provinces:* Apachian, Austroriparian, Chihuahuan, Coloradan, Kansan, Navahonian. *Watersheds:* Canadian, Colorado (NM), Pecos, Red, Upper Rio Grande.

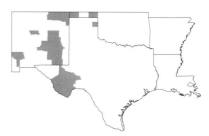

General Distribution. Western United States, southwest Canada, and northern Mexico.

Flight Season. Jul. 25 (TX)–Aug. 5 (OK).

Identification. This western desert species is similar to Common Whitetail (*P. lydia*). Females are easily distinguished by the lack of color apically in the wings. The males typically have a more pronounced clear streak through the dark midbasal area of all wings, but this becomes obscured in older males. The area between the dark spots becomes entirely pruinose in the Desert Whitetail, and the ventral tubercle on the first abdominal segment is shallowly bifurcate, appearing "V" shaped. The face is yellowish, the black median stripe becoming obscured as the face darkens with age. The thorax is dark brown, with two pale-yellow oblique lateral stripes that also are obscured with age. The wings in both sexes have a dark basal area, as noted above, extending out to the triangle, and a second broad stripe in the outer half of the wing that is significantly lighter in its middle 1/3, sometimes appearing clear. The abdomen, stocky and dark brown, has a series of interrupted yellow stripes laterally. Segment 8 is not expanded laterally in females.

Similar Species. Differences from Common Whitetail are given above, under that species. Twelve-spotted Skimmer (*Libellula pulchella*) has brown wingtips, and Hoary Skimmer (*L. nodisticta*) has a brown spot, rather than a broad stripe, at the nodus.

Habitat. Desert pools, ponds, and slow streams with thick emergent vegetation and mud bottoms.

Discussion. This species is strictly western, found in semi- and full-desert environments, often alongside Common Whitetail. The larvae often transform just above the water on thick clusters or reeds and grasses. The species was reported among the most common in alkaline lakes and ponds of Nevada; ". . . it was seemingly restricted to the pond areas, where it beat over the water in regular circuits. . . ." Desert Whitetail, Eastern Pondhawk (*Erythemis simplicicollis*), and Variegated Meadowhawk (*Sympetrum corruptum*) were the only inhabitants of the smaller springs, often consisting of only muddy seeps. This desert species has been reported from as far distant as Nanaimo, British Columbia, which constitutes a considerable range extension northward, not to mention a dramatic shift in ecological conditions.

References. Cannings (1983), Gloyd (1958), La-Rivers (1946), Needham et al. (2000), Williamson (1906b).

Filigree Skimmer
Genus *Pseudoleon* Kirby

This genus contains a single unmistakable black, ornately patterned species found in the southwestern United States southward through Mexico to Costa Rica. The eyes are strikingly patterned with long stripes that often remain after preservation. The wings are variably and ornately patterned with black. The foot in the hindwing venation is broad, usually with several ankle cells. The forewing triangle is unusually narrow, and the triangle in the hindwing is concave on its outer side.

Filigree Skimmer
Pseudoleon superbus (Hagen)
(photo 61f)

Size. Total length: 34–45 mm; abdomen: 21–29 mm; hindwing: 30–36 mm.

Regional Distribution. *Biotic Provinces:* Apachian, Balconian, Chihuahuan, Kansan, Navahonian, Tamaulipan. *Watersheds:* Colorado (NM), Lower Rio Grande, Nueces, Pecos, San Antonio, Upper Rio Grande.

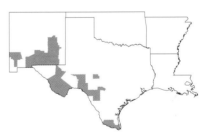

General Distribution. Southwestern United States and Mexico south to Costa Rica.

Flight Season. Jun. 11 (TX)–Jul. 17 (TX).

Identification. This dark species is found throughout the southwestern parts of Texas and New Mexico. Young individuals have a pale face that quickly becomes black with age. The eyes are light, with dark longitudinal stripes. The thorax is initially tan, with numerous darker-brown stripes, becoming diffusely black with age. The wings are variably patterned with black, but always with a wide dark band at the nodus. Males usually have more black in their wings, the hindwing often becoming entirely black, except for the clear apical tip, and the forewing black except for the basal half. The legs are brown, but darken with age. The abdomen is brown, and is marked with a series of "V" marks outlined internally by a thinner pale line on segments 3–7. This pattern becomes obscured by an almost iridescent-black color in males.

Similar Species. Few other dragonflies in our region are as dark. Mature males may be entirely black, except for the clear area basally in the forewing and the clear wingtips. Black-winged Dragonlet (*Erythrodiplax funerea*) is more slender, and the black in the wings extends out to halfway between the nodus and the pterostigma. Band-winged Dragonlet (*E. umbrata*) has clear basal areas in both wings.

Habitat. Desert ponds and slow streams.

Discussion. Filigree Skimmer typically perches on the ground or on rocks, with the wings characteristically depressed below the rest of the body. They can be wary, however, and can rapidly take flight from this position, especially on hot days. Males are territorial, and have numerous midair skirmishes. Females are commonly seen fluttering low around grasses and roots floating in ponds. They lay eggs by hovering over the water and then "thrusting" the abdomen into algal mats floating on the surface. Females are commonly interrupted by pursuing males during oviposition, whereupon they immediately stop laying, flee the area, and lay eggs somewhere else.

References. Needham et al. (2000).

Meadowhawks
Genus *Sympetrum* Newman

This is a widely distributed group of nearly 60 species found predominantly in the Northern Hemisphere. Thirteen species occur in North America and, of those, nine are found in the south-central United States. Two are widespread within that region, one is predominantly eastern, and the other six are western species. They are small to medium-sized yellow or red dragonflies that are generally seen flying in meadows and swamps. They are weak fliers, usually abundant in the fall, seen resting on the tips of twigs, branches, or tall grasses. On cooler days some species will rest on the ground or rocks for maximum sun exposure. The head is rounded, and has a low frons and a well-developed furrow. The thorax is usually moderately hairy. The wings

generally have some degree of flavescence. The abdomen is slender, compressed vertically on the basal segments, especially in the male, and more parallel-sided beyond. Mating occupies relatively lengthy periods, up to 30 minutes in some cases. Because of certain differing characters in the genitalia and wing venation, two of the species have historically been placed in a separate genus, *Tarnetrum*. Various authors have since disagreed with this separation, and I follow them in not giving *Tarnetrum* generic status.

References. Carle (1993), Walker and Corbet (1975), Garrison (1997), Gloyd and Wright (1959), Kormondy (1958, 1960), Walker (1917).

KEY TO THE SPECIES OF MEADOWHAWKS (*SYMPETRUM*)

1. Largely black or, if not, then the sides of the thorax with black ladderlike markings enclosing 2 or 3 pale spots	**Black (*danae*)**
1'. Body not black, and thorax lacking such markings	2
2(1') Face white and top of head bluish green	**Blue-faced (*ambiguum*)**
2'. Face pale, but not white, and top of head not bluish green	3

KEY TO THE SPECIES OF MEADOWHAWKS (*SYMPETRUM*) (*cont.*)

3(2'). Thorax with 2 pale lateral stripes	4
3'. Thorax without 2 pale lateral stripes	6
4(3). Basal 1/4 of wings amber	**Cardinal (*illotum*)**
4'. Wings amber at most at extreme base	5
5(4'). Black middorsal strip on abdominal segments 8 and 9	**Variegated (*corruptum*)**
5'. Abdominal segments 8 and 9 without black middorsal stripe	**Striped (*pallipes*)**
6(2'). Wings with amber or orange band in basal 1/2	**Band-winged (*semicinctum*)**
6'. Wings without amber or orange band in basal 1/2	7
7(6'). Legs mostly yellow	**Yellow-legged (*vicinum*)**
7'. Legs black	8
8(7'). Yellow or amber band behind front margin of each wing	**Saffron-winged (*costiferum*)**
8'. No band of color behind front margin of wings	**Cherry-faced (*internum*)**

Blue-faced Meadowhawk
Sympetrum ambiguum (Rambur)
(photo 62a)

Size. Total length: 31–38 mm; abdomen: 22–25 mm; hindwing: 26–28 mm.

Regional Distribution. *Biotic Provinces:* Austroriparian, Carolinian, Kansan, Texan. *Watersheds:* Arkansas, Bayou Bartholomew, Brazos, Canadian, Cimarron, Mississippi, Neches, Ouachita, Red, Sabine, San Jacinto, St. Francis, Trinity, White.

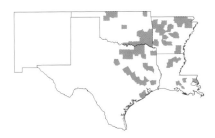

General Distribution. Southeastern Canada and eastern United States west to Texas.

Flight Season. Jun. 13 (AR)–Nov. 27 (LA).

Identification. This is the only predominantly eastern species of the genus in our region. It has a white face, bluish above. The thorax is grayish brown or olivaceous, with lateral sutures outlined by thin brown stripes. The wings are clear, with only a small spot of flavescence at their extreme bases. The costa is yellow, and the pterostigma is brown, with yellow around its outer edges. The legs are pale brown but darker at the joints. The abdomen is brown, with diffuse black rings apically around segments 4–9 in young males and females. The abdomen turns red in mature males.

Similar Species. The combination of pale legs and blue on top of the frons makes this species distinct in the region. Yellow-legged Meadowhawk (*S. vicinum*) lacks black rings around the abdomen.

Habitat. Partially shaded temporary and permanent ponds, pools, marshes, swamps, and sloughs.

Discussion. This species is partial to shaded areas and forest edges. It is typical of the group in perching at the tips of twigs, stems, and grasses, but it often does so at greater heights than other species. It will sometimes perch with its abdomen raised above the rest of the body in an obelisk position, like many other meadowhawks. Males bring females down low to weeds, stems, and other perches, and pairs are even occasionally seen mating on the ground. The female lays eggs alone, but is guarded by the male, as she extrudes eggs along the shore or over a dry pond or pool, where they remain undeveloped until the pond fills again.

Variegated Meadowhawk
Sympetrum corruptum (Hagen)
(photos 62b, 62c)

Size. Total length: 33–43 mm; abdomen: 23–29 mm; hindwing: 27–33 mm.

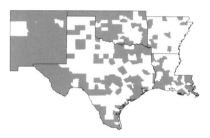

Regional Distribution. *Biotic Provinces:* Apachian, Austroriparian, Balconian, Carolinian, Chihuahuan, Coloradan, Kansan, Navahonian, Tamaulipan, Texan. *Watersheds:* Arkansas, Bayou Bartholomew, Brazos, Canadian, Cimarron, Colorado, Colorado (NM), Guadalupe, Lower Rio Grande, Mississippi, Neches, Nueces, Ouachita, Pecos, Red, Sabine, San Antonio, San Jacinto, St. Francis, Trinity, Upper Rio Grande, White.
General Distribution. Throughout United States and southern Canada; also Mexico south to Belize and Honduras.
Flight Season. Year-round (TX).
Identification. This is the most widespread meadowhawk in the region. It is largely tan or gray, its face tan in young males and females but becoming red in mature males. The thorax has two oblique lateral white stripes, each with a distinct round yellow spot at its lower end. The yellow always remains visible, but the white becomes obscured in mature males. The wings are clear, with yellow veins in the costal and subcostal areas. The pterostigma is tan, bordered by yellow and red. The legs are dark brown, except on their outer surfaces. The abdomen is grayish, with a yellowish-orange middorsal stripe and orange rings apically on segments 3–7. A row of white spots are present laterally on segments 2–8,

and segments 8 and 9 each have a large black spot dorsally. The orange color of the abdomen turns red in older males.
Similar Species. Striped Meadowhawk (*S. pallipes*) and Cardinal Meadowhawk (*S. illotum*) both lack black dorsally on abdominal segments 8 and 9. Striped Meadowhawk also lacks yellow spots laterally on the thorax.
Habitat. Ponds and slow streams, preferably with sandy or cobble bottoms, but occasionally including brackish waters.
Discussion. Variegated Meadowhawks may be seen on the ground more often than other meadowhawks. They will also perch readily on the tips of grass stems and tree branches. They can be numerous flying over roads, lawns, meadows, marshes, and ponds. They are more abundant in the early spring and late fall months, but have been taken every month in Texas. Variegated Meadowhawk has been described as very adaptable, ". . . found in a greater variety of environments than any other." Mating occurs while perched on twigs, stems, or other vegetation. Females lay eggs accompanied by males in the open water of ponds and lakes. Mass movements of this species have been reported on several occasions.

References. Arnaud (1972), Kennedy (1915a), Opler (1971), Turner (1965).

Saffron-winged Meadowhawk
Sympetrum costiferum (Hagen)
(photo 62d)

Size. Total length: 31–37 mm; abdomen: 22–26 mm; hindwing: 25–28 mm.
Regional Distribution. *Biotic Province:* Kansan. *Watershed:* Pecos.

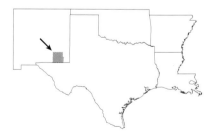

General Distribution. Canada and northern United States south to New Mexico.

Flight Season. Jul. (NM)–Oct. (NM).

Identification. This primarily northern, reddish species has been reported as far south as southeastern New Mexico. Young males and females are yellow. The face is hairy and pale white or olivaceous, with a darker, green labrum. The thorax is brownish red in older males. The dark markings seen on the sutures of young individuals become obscured with age. Young males and females have golden yellow diffused along the costal margin of each wing, the yellow covering otherwise red veins. This color lessens or disappears in mature males, but may become diffuse over nearly all of the hindwing in females. The legs are pale yellow externally and dark medially. The abdomen is distinctly spindle-shaped, narrowed basally and then widening slightly at segments 5 and 6 and narrowing again distally. The abdomen is red in mature adults, with a prominent black stripe on each lateral carina. Segments 8 and 9 are black dorsally.

Similar Species. Yellow-legged Meadowhawk (*S. vicinum*) has entirely yellow legs, is more slender-bodied, and has amber restricted to the extreme wing bases.

Habitat. Shallow marshes, bays, and lagoons of lakes and reservoirs, including saline ones.

Discussion. This is a late-fall species. The only record of this species in the south-central United States is from Eddy County, New Mexico. It is apparently more tolerant of saline waters than other meadowhawks are. This species is another one that commonly perches on the ground. Females generally lay eggs accompanied by the male. One author reported seeing thousands of this species perched on telephone wires, and I have seen the same behavior in other meadowhawks.

Reference. Evans (1995), Kennedy (1915b), Walker (1917), Walker and Corbet (1975).

Black Meadowhawk
Sympetrum danae (Sulzer)
(photos 62e, 62f)

Size. Total length: 21–23 mm; abdomen: 19–24 mm; hindwing: 21–26 mm.

Regional Distribution. *Biotic Provinces:* Coloradan, Navahonian. *Watersheds:* Canadian, Colorado (NM), Pecos, Upper Rio Grande.

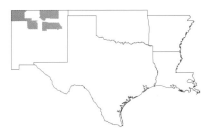

General Distribution. Canada and northern United States south to New Mexico; also Europe to Japan.

Flight Season. Jun. 15 (NM)–Nov. 15 (NM).

Identification. This small, black meadowhawk is uncommon in northern New Mexico. The face is pale yellow in young males and females, turning black in mature males. The thorax is yellowish, with two or three small pale spots enveloped by ladderlike black markings laterally. The wings are clear, with a pale costal vein that darkens with age. There is a small spot of amber basally and sometimes one at the nodus. The pterostigma and legs are black. Young individuals have a yellow abdomen with a black lateral stripe. The abdominal segments quickly (more so in males) become black with a row of pale dorsolateral spots. Females have a short, spoutlike ovipositor.

Similar Species. This is the only dark meadowhawk in our region, and the only one with a black face (in older males). Female Seaside Dragonlet (*Erythrodiplax berenice*) has dark markings on a pale face, and has a distinctly longer ovipositor. The similar small pennants (*Celithemis*) have more extensive basal, amber-colored wing markings.

Habitat. Marshy areas, ponds, lakes, and sometimes saline ponds.

Discussion. This ubiquitous northern species is uncommon in New Mexico. It behaves like other meadowhawks, perching lower on vegetation. It forages in open fields, and males are not territorial when away from the water.

Cardinal Meadowhawk
Sympetrum illotum (Hagen)
(photo 63a)

Size. Total length: 36–40 mm; abdomen: 23–26 mm; hindwing: 26–29 mm.

Regional Distribution. *Biotic Provinces:* Apachian, Chihuahuan, Kansan, Navahonian. *Watersheds:* Colorado (NM), Pecos, Upper Rio Grande.

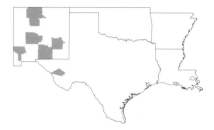

General Distribution. Western United States, West Indies, and Central America south to Chile and Argentina.

Flight Season. May 22 (TX).

Identification. The face is pale brown but becomes bright red in front and on top in mature males. The thorax is brown, with two oblique, abbreviated or interrupted, white spots laterally. In mature adults only the rounded lower end of these spots remains conspicuous. The wings are diffused with yellow out to the level of the nodus, and there are one or two darker, brown streaks extending at least to the first antenodal crossvein in the subcostal and/or cubital areas of the wings. The legs are reddish brown. The abdomen is dark brownish red and parallel-sided for most of its length. The caudal appendages are red. The subgenital plate of the female is emarginate and extends beyond the posterior margin of segment 8 by 1/2 the length of that segment.

Similar Species. Variegated Meadowhawk (*S. corruptum*) has a black-and-white pattern on the abdomen. Striped Meadowhawk (*S. pallipes*) has distinct stripes, rather than spots, laterally on the thorax. Both of these species lack amber in the wings.

Habitat. Small ponds and slow streams.

Discussion. This species and Variegated Meadowhawk are the only two species in the region that belong to the more robust-bodied *Tarnetrum* group. Within our region this species occurs in southern New Mexico and west Texas. It perches on the tips of twigs, grasses, and other vegetation, with its wings depressed below the abdomen. Mating is initiated in flight or on a twig or branch and requires about 30 seconds. The male then generally accompanies the female as she lays eggs by making numerous dips to the surface with the abdomen.

References. Abbott (1996), Evans (1995), Kennedy (1917).

Cherry-faced Meadowhawk
Sympetrum internum Montgomery
(photo 63b)

Size. Total length: 27–34 mm; abdomen: 20–23 mm; hindwing: 25–27 mm.

Regional Distribution. *Biotic Provinces:* Navahonian, Texan. *Watersheds:* Arkansas, Canadian, Rio Grande.

General Distribution. Canada and northern United States south to Oklahoma and New Mexico.

Flight Season. Oct. 8 (OK).

Identification. This species is found only as far south as north-central Oklahoma. It is a handsome species with a cherry-red face at maturity. The thorax is reddish brown and thickly clothed with hairs of the same color, but it is unmarked both on the front and on its sides. The wings are clear, with only a hint of yellow flavescence at their extreme base. The femora are pale beneath, and the rest of the leg is black. The abdomen is cherry-red with large black subequal triangles laterally on segments 4–8 and sometimes 9. Segment 10 and the caudal appendages are yellow, with black on the apices of the cerci.

Similar Species. Saffron-winged Meadowhawk (*S. costiferum*) has more yellow on the legs, and has a narrower black lateral abdominal stripe. Band-winged Meadowhawk (*S. semicinctum*) has black lateral thoracic markings.

Habitat. Ponds, pools, and slow shady streams.

Discussion. This species is similar to the larger and more northern Ruby Meadowhawk (*S. rubicundulum* (Say)), which does not occur in the south-central United States. This species is known from Cleveland and Payne Counties in Oklahoma, which represent the most southern extent of its range. It has also been reported from Arkansas and Texas, but all these records are undoubtedly in error.

References. Bick and Bick (1957), Montgomery (1943), Needham and Westfall (1955).

Striped Meadowhawk
Sympetrum pallipes (Hagen)
(photo 63c)

Size. Total length: 34–38 mm; abdomen: 22–26 mm; hindwing: 26–28 mm.

Regional Distribution. *Biotic Provinces:* Coloradan, Kansan, Navahonian. *Watersheds:* Canadian, Colorado (NM), Pecos, Rio Grande.

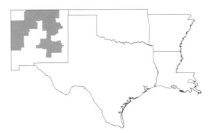

General Distribution. Western United States and Canada.

Flight Season. Jun.–Aug. 7 (NM).

Identification. This species is distinctive among the slender-bodied meadowhawks in the region because of its pale cream-colored lateral thoracic stripes. Its face and thorax are otherwise brown. The legs are light brown in young individuals and turn black with age. The wings are clear, with a small touch of amber basally. The abdomen is initially yellow-brown but becomes pale red in older individuals. There are few to no dark markings on the abdomen.

Similar Species. Cardinal Meadowhawk (*S. illotum*) has extensive amber in the basal areas of the wings, and is generally more robust. Variegated Meadowhawk (*S. corruptum*) has two yellow spots laterally on the thorax.

Habitat. Permanent marshy ponds, pools, and streams with thick riparian vegetation.

Discussion. Nothing has been reported on the biology of this western species, but females lay eggs in tandem.

Band-winged Meadowhawk
Sympetrum semicinctum (Say)
(photo 63d)

Size. Total length: 30–38 mm; abdomen: 18–24 mm; hindwing: 23–27 mm.

Regional Distribution. *Biotic Provinces:* Chihuahuan, Coloradan, Kansan, Navahonian. *Watersheds:* Canadian, Cimarron, Colorado (NM), Pecos, Red, Rio Grande.

General Distribution. Western United States and Canada.

Flight Season. Aug. 1 (NM)–Oct. 21 (NM).

Identification. This is a brownish western species with distinctive amber-colored wings. Its face is pale yellow. The thorax is dark olivaceous in front and pale yellowish green laterally, with distinct black stripes, usually confluent at their lower end. In older individuals the pale color becomes dull greenish or gray, and the dark stripes become obscured. The oblique black stripe anterior to the spiracle always remains visible. The wings are distinctive among the species in our region. They have a broad yellow flavescence or amber band that extends out to the nodus, and a darker, brown band covering the outer half of that stripe. The legs are black. The abdomen is yellowish dorsally, with a black ventrolateral stripe on either side. Segments 8 and 9 are black dorsally, and the caudal appendages are yellow.

Similar Species. Female Black Meadowhawks (*S. danae*) and other similar meadowhawks have much less color basally in the wings. Cardinal Meadowhawk (*S. illotum*) has two pale lateral thoracic spots. Amberwings (*Perithemis*) are much smaller, and the wings of their females are completely amber.

Habitat. Marshy, sometimes spring-fed, muddy-bottomed ponds, sloughs, and swamps.

Discussion. Considerable confusion has existed between the Western (*S. occidentale* Bartenev) and Band-winged (*S. semicinctum*) Meadowhawks. Some scientists consider these two variable entities to be distinct, but I consider the Western Meadowhawk, which is found in California and Nevada, outside our region, to be a synonym of the Band-winged Meadowhawk here. Three forms (subspecies) of *S.*

occidentale have been described. I have seen members of this species perched on fence lines and telephone lines by the thousands.

References. Walker (1951).

Yellow-legged Meadowhawk
Sympetrum vicinum (Hagen)
(photo 63e)

Size. Total length: 26–35 mm; abdomen: 18–23 mm; hindwing: 20–25 mm.
Regional Distribution. *Biotic Provinces:* Austroriparian, Balconian, Carolinian, Chihuahuan, Kansan, Navahonian, Texan. *Watersheds:* Arkansas, Bayou Bartholomew, Brazos, Canadian, Cimarron, Colorado, Guadalupe, Nueces, Ouachita, Pecos, Red, San Antonio, San Jacinto, St. Francis, Upper Rio Grande, White.

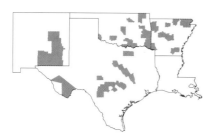

General Distribution. Southern Canada and throughout the United States, except the extreme Southwest.
Flight Season. Jun. 11 (TX)–Nov. 9 (OK).
Identification. This is a smaller, more delicate, but widely distributed meadowhawk. Its face is yellowish, becoming red in older males. The thorax is darker in front and greenish brown laterally, with no markings. The wings are clear, with a slight hint of amber at the extreme base in each wing. The legs are pale yellow with no black. The abdomen is uniform brown, but becomes red, along with the front of the thorax, in mature males. The female has a ventrally projecting scoop-shaped subgenital plate that becomes more pronounced after laying eggs.
Similar Species. This is the only meadowhawk in our region with a pale face and legs and clear wings. Blue-faced Meadowhawk (*S. ambiguum*) has blue on top of the head and black markings on the thorax and legs. Saffron-winged Meadowhawk (*S. costiferum*) is larger, and has more amber in the wings and black on the legs. Young male and female Plateau Dragonlets (*Erythrodiplax basifusca*) are smaller, and the females have stockier abdomens.
Habitat. Permanent ponds and slow-flowing streams.
Discussion. This thin-legged species flies in the late summer and early fall, and is found throughout the region. Carle (1993) included Louisiana in his distribution for the species, but could not recall the source of his data. Although it is to be expected in the northern part of that state, Mauffray (1997) listed it as doubtful from Louisiana. Individuals of Yellow-legged Meadowhawk tend to perch higher up on vegetation, such as bushes and grasses, than many other meadowhawks. They tend to breed only in permanent waters, including slow streams and ponds. Females lay eggs in tandem along the bank by tapping the abdomen alternately against the water and then the bank.

References. Calvert (1926), May (1998b).

Pasture Gliders
Genus *Tauriphila* Kirby

This is a group of five Neotropical species, two of which are found in North America. Garnet Glider (*T. australis* (Hagen)) is found in Florida and Aztec Glider (*T. azteca* Calvert) ranges into southern Texas. Members of the group are similar in appearance to saddlebag gliders (*Tramea*). They are generally smaller than members of that genus, however, and are unique in several venational characters.

The forewing triangle is composed of two cells, and the pterostigma is of equal length in the fore- and hindwings (in saddlebag gliders the forewing pterostigma is distinctly longer). The radial planate subtends a single row of cells. The caudal appendages are of normal length and do not appear unusually long, as they do in the saddlebag gliders.

Aztec Glider
Tauriphila azteca Calvert
(photo 63f)

Size. Total length: 40–44 mm; abdomen: 29–35 mm; hindwing: 34–38 mm.
Regional Distribution. *Biotic Province:* Tamaulipan. *Watershed:* Lower Rio Grande.

General Distribution. South Texas, Mexico, West Indies, Guatemala, and Costa Rica.
Flight Season. Jun. 8 (TX).
Identification. The species has a yellow abdomen marked with distinct dark bands at each segment. These contrasting colors along with the dark basal spot in each hindwing will readily distinguish this species from others in our region. Males have a brown face with a metallic-violet luster. The thorax is brown and largely unmarked. There is a black middorsal stripe on abdominal segments 8–10, and the caudal appendages are black.
Similar Species. Hyacinth Glider (*Miathyria marcella*) is smaller, and has pale lateral thoracic stripes and a middorsal black stripe on the abdomen. Marl Pennant (*Macrodiplax balteata*) has a larger round spot basally in the hindwing.
Habitat. Slow, calm waters with emergent or floating vegetation.
Discussion. The only record of this species in our region is of a single male specimen in the Florida State Collection of Arthropods that had been collected in Kingsville, Texas. As its common name implies, this species has a distinct gliding flight, ranging widely when feeding. Males patrol territories over floating plants, where females lay eggs at their bases (Dunkle, pers. comm.).

References. Abbott (1996).

Evening Skimmers
Genus *Tholymis* Hagen

This is a tropical genus of few species, with a single inhabitant in the New World. It is a vagrant in southern Florida and Texas. Members of this genus have a broadly rounded heel in the anal loop and an open toe, with veins A1 and A2 ending at the hindwing margin. Also characteristic of this genus is a second row of cells inserted medially in the radial and median planates. This species has an amber spot below the nodus.

Evening Skimmer
Tholymis citrina Hagen
(photo 64a)

Size. Total length: 48–53 mm; abdomen: 32–40 mm; hindwing: 36–39 mm.
Regional Distribution. *Biotic Province:* Tamaulipan. *Watershed:* Lower Rio Grande.

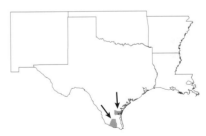

General Distribution. Mexico, West Indies, and Central America south to Brazil and Chile; stray to south Florida and Texas.
Flight Season. Sep. 16 (TX).
Identification. This is a medium-sized dragonfly capable of strong, efficient flight. Its face is pale yellow, but in mature males it darkens with age and becomes metallic blue along with the top of the frons and vertex. The thorax is olivaceous brown and unmarked, but the front and sides become bluish black in older

males. The wings have a spot of amber below the nodus, brown crossveins, and a tawny pterostigma. The spot is ill-defined and more diffuse in the forewing. Generally, there are three paranal cells before the anal loop. The legs are pale and armed with dark spines. The abdomen is pale brown, with a mid-dorsal stripe that is darkest on segment 9. The caudal appendages and segment 10 are pale in young individuals, but they darken with age in both sexes.

Similar Species. No other skimmer in the region has only a single amber spot below the nodus. Some shadowdragons (*Neurocordulia*) may look superficially similar, but they have heavier markings in the wings. (Check the ranges of the species as well.)

Habitat. Vegetated ponds and lakes.

Discussion. This species has been reported in North America on only three occasions, once from the southern tip of Florida, and the other two times from Texas. A single female was collected by Smith and Hodges from the Rio Grande Valley (Hidalgo County), in 1950, and in 2003 T. Langschied found a male on the King Ranch. This species' normal distribution is southern Mexico southward to Brazil. It is a crepuscular flier, which may in part explain the paucity of records.

References. Barber and Elia (1994).

Saddlebags
Genus *Tramea* Hagen

This is a globally distributed group of moderate-sized to large dragonflies. Five of the seven North American species occur in the south-central United States. They are typically red, brown, or black, with large heads and a long eye seam. They have a broad saddlebag-like crossband basally in the hindwing, giving them their characteristic appearance and common name. They have broad hindwings that taper to a point, permitting them to remain in flight for extended periods of time. The pterostigma is trapezoidal and distinctly longer in the forewing than in the hindwing. The broad basal wing bands are used to shade the depressed abdomen from the sun on hot days. The forewing triangle is generally two or three cells. The apical planate subtends two rows of cells for most of its length. The inside of the foot in the anal loop is composed of numerous branching veins, making it difficult to follow. The abdomen is long and slender.

Members of this group have an unusual egg-laying behavior in which the male and female start out in tandem, the male releases the female while she drops to the water and dips her abdomen to the surface, and she then returns and is grasped by the male as they travel to another ovipositing site. The eggs are in thin sticky strings that attach to sub-

KEY TO THE SPECIES OF SADDLEBAGS (*TRAMEA*)

1. Sides of thorax with pale-yellow stripes	**Striped (*calverti*)**
1'. Sides of thorax without pale-yellow stripes	2
2(1'). Base of hindwing with narrow crossband of dark color extending outward to level of anal crossing	**Antillean (*insularis*)**
2'. Base of hindwing with broad crossband of dark color extending to distal angle of triangle	3
3(2'). Basal color of hindwing red or brown	4
3'. Basal color of hindwing black	**Black (*lacerata*)**
4(3). Top of frons metallic violet	**Carolina (*carolina*)**
4'. Top of frons variable, but never metallic violet	**Red (*onusta*)**

merged vegetation. This genus is similar to rainpool gliders (*Pantala*) in its habit of continual sustained flight, only occasionally perching horizontally on the tips of tall grasses or other vegetation.

Striped Saddlebags
Tramea calverti Muttkowski
(photo 64b)

Size. Total length: 44–49 mm; abdomen: 30–33 mm; hindwing: 37–42 mm.
Regional Distribution. *Biotic Provinces:* Balconian, Tamaulipan, Texan. *Watersheds:* Brazos, Colorado, Guadalupe, Lower Rio Grande, Nueces, Trinity.

General Distribution. Southern Arizona, central and south Texas, straying up along East Coast; also Mexico, West Indies, and Central America south to Argentina.
Flight Season. Apr. (TX)–Aug. 4 (TX).
Identification. This tropical species has extended its northern limits into Texas. It has a pale yellow face that turns red in mature males, and the vertex is metallic violet. The thorax is brown, with two pale oblique lateral stripes. The wings are clear, with a broad brown basal stripe in the hindwing. The legs are pale basally and darker beyond. The abdomen is yellowish in females and red in mature males. Segments 8–10 are black dorsally.
Similar Species. This is the only saddlebag glider in the region with a striped thorax. It is most similar to the smaller Hyacinth Glider (*Miathyria marcella*), but it lacks the complete middorsal black stripe on the abdomen of that species.
Habitat. Temporary and permanent ponds and slow streams.
Discussion. This species has been found as far north as New York and Massachusetts. It is widely distributed in Texas, but is most commonly encountered in the southern portions of the state. It is typical of the genus in having a strong, steady, high flight, often not descending below 2 m. They

will perch horizontally at the tips of tall grasses and branches with a clear view of intruders in open fields and around ponds. The female lays eggs in the manner typical of the genus, as described above.

Carolina Saddlebags
Tramea carolina (Linnaeus)
(photo 64c)

Size. Total length: 45–54 mm; abdomen: 30–36 mm; hindwing: 41–46 mm.
Regional Distribution. *Biotic Provinces:* Austroriparian, Balconian, Carolinian, Kansan, Texan. *Watersheds:* Arkansas, Bayou Bartholomew, Brazos, Canadian, Cimarron, Mississippi, Neches, Ouachita, Red, Sabine, San Antonio, San Jacinto, St. Francis, Trinity, White.

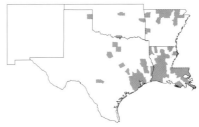

General Distribution. Southern Canada and eastern United States southward to Texas.
Flight Season. Year-round throughout the southern parts of its range.
Identification. This is a large, handsome red species found predominantly throughout the eastern portion of the region. Its face is initially pale but becomes red, and the top of the head, including the vertex, turns metallic violet in older individuals. The thorax is reddish brown and unmarked. The wings have red veins along the front margin. The hindwing has a nearly solid dark-reddish-brown band that extends out to the outer side of the anal loop. The veins in this area are red. The legs are brown basally and turn darker for most of the rest of their length. The abdomen is brownish red, becoming bright red in mature males. Segments 8 and 9 are largely black.
Similar Species. This species is similar to Red Saddlebags (*T. onusta*), especially in flight, but that species lacks the violet color on top of the head. Segments 8 and 9 are pale laterally, and the crossband in the hindwing is smaller and generally interrupted by a larger clear stripe medially. Gloyd (1958) provided further distinctions between these two species.

Habitat. Ponds, lakes, and slow streams with thick emergent vegetation.

Discussion. This species is generally not seen in the large feeding swarms in which Red and Black (*T. lacerata*) Saddlebags take part. Males fly feeding and patrolling territories nearly all day. They will perch, as usual, horizontally on the tops of tall vegetation, giving them a clear view of their territory. Pairs mate while perched in vegetation or high in trees, and remain there for some time. Females typically lay eggs in the manner described for the genus, but they may also oviposit alone. Females that lay eggs alone do so at a rate nearly ten times faster than those in tandem.

References. Carpenter (1991), Davis (1898), Dunkle (1989), Sherman (1983).

cept for segments 8, 9, and sometimes 10, which are black dorsally.

Similar Species. Hyacinth Glider (*Miathyria marcella*) is smaller and has a dark middorsal stripe running the length of the abdomen. Aztec Glider (*Tauriphila azteca*) has a dark thorax and a yellow abdomen.

Habitat. Ponds, lakes, and slow streams.

Discussion. In our region, this species is known from only a couple of localities in west Texas, including Big Bend National Park, the Devils River, and Ft. Clark Springs near Brackettville. There are apparent breeding populations in both localities, and I suspect this species may become more common in Texas as it establishes itself further. Individuals can be seen flying quickly at great heights.

References. Abbott (1996).

Antillean Saddlebags
Tramea insularis Hagen
(photo 64d)

Size. Total length: 41–48 mm; abdomen: 26–31 mm; hindwing: 36–40 mm.

Regional Distribution. *Biotic Provinces:* Balconian, Chihuahuan. *Watershed:* Upper Rio Grande.

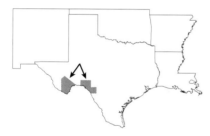

General Distribution. Florida and southern and western Texas; also Mexico, Bahamas, and Greater Antilles.

Flight Season. May 23 (TX).

Identification. This species is found predominantly in the Bahamas and Greater Antilles, but is fairly common in southeastern Florida and ranges into Mexico and Texas. The face is brown in females and young males but becomes black in mature males. The top of the head is metallic violet. The thorax is brown and unmarked. The wings have predominantly red veins anteriorly, and the hindwings each have a brown basal crossband. The legs are black except at their extreme bases. The abdomen is red, ex-

Black Saddlebags
Tramea lacerata Hagen
(photo 64e)

Size. Total length: 47–55 mm; abdomen: 31–38 mm; hindwing: 40–48 mm.

Regional Distribution. *Biotic Provinces:* Apachian, Austroriparian, Balconian, Carolinian, Chihuahuan, Coloradan, Kansan, Navahonian, Tamaulipan, Texan. *Watersheds:* Arkansas, Bayou Bartholomew, Brazos, Canadian, Cimarron, Colorado, Colorado (NM), Guadalupe, Lower Rio Grande, Mississippi, Neches, Nueces, Ouachita, Pecos, Red, Sabine, San Antonio, San Jacinto, St. Francis, Trinity, Upper Rio Grande, White.

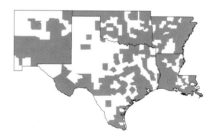

General Distribution. Southern Canada and throughout United States.

Flight Season. Mar. 30 (LA)–Nov. 5 (LA).

Identification. This large dragonfly is the only black saddlebags glider in the region. It has a yellowish face in young males and females that becomes entirely black in mature males. The top of the head, including the vertex, is deep metallic violet. The thorax is brown, with black iridescence on the sides. The wings are clear, except for a broad black crossband covering the basal 1/4 of the hindwing. There is a large prominent clear spot medially in this area, extending to the inner wing margin. The legs are black. The abdomen is black, with a pair of yellowish spots dorsally on the middle segments that become obscured, except on segment 7.

Similar Species. Marl Pennant (*Macrodiplax balteata*) has a round spot rather than saddlebag marking in the basal hindwing area. Azetec Glider (*Tauriphila azteca*) has a yellow abdomen.

Habitat. Marshy ponds, lakes, ditches, and slow streams.

Discussion. This species migrates northward in the spring. Males are often seen in large feeding swarms throughout the day. Females lay eggs as described for the genus, but may do so alone or without being released from the male's grasp. This species probably occurs year round in the southern areas of its range.

References. Beatty (1946), Borror (1953).

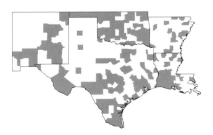

Red Saddlebags
Tramea onusta Hagen
(photo 64f)

Size. Total length: 41–48 mm; abdomen: 28–34 mm; hindwing: 37–42 mm.

Regional Distribution. *Biotic Provinces:* Apachian, Austroriparian, Balconian, Carolinian, Chihuahuan, Kansan, Navahonian, Tamaulipan, Texan. *Watersheds:* Arkansas, Bayou Bartholomew, Brazos, Canadian, Cimarron, Colorado, Colorado (NM), Guadalupe, Lower Rio Grande, Mississippi, Neches, Nueces, Ouachita, Pecos, Red, Sabine, San Antonio, San Jacinto, St. Francis, Trinity, Upper Rio Grande, White.

General Distribution. Southern United States, becoming more common in the Southwest; also Mexico, West Indies, and Central America south to Venezuela.

Flight Season. Feb. 20 (TX)–Nov. (TX).

Identification. This is another red species that is widely distributed across the region. Its face is pale brown initially but turns red in mature males. The thorax is brown and unmarked. The wings have reddish-brown veins anteriorly, and the hindwing has a large brown basal crossband that generally does not extend beyond the midrib of the anal loop. Usually, there is a large central clear spot in this crossband. The legs are pale, turning black more distally. The abdomen is yellowish brown in females, but turns red in mature males. Segments 8–10 are black dorsally and pale laterally.

Similar Species. This species is most similar to Carolina Saddlebags (*T. carolina*), but that species has more extensive markings in the hindwing and a metallic-purple vertex, and abdominal segments 8–10 are nearly all black dorsally.

Habitat. Permanent and temporary ponds, lakes, and slow streams.

Discussion. This species is commonly seen throughout the region feeding over large fields, meadows, and roadways. Males patrol large territories, often flying at great heights. Mating occurs while the pair is perched high in bushes and trees. Females will lay eggs alone or in tandem, and generally deposit them on algal mats at the water surface.

ADDITIONAL SPECIES

As this book was going to press, four new species were discovered in Texas raising the total number for that state to 217. This is a testament to the growing popularity of odonates and the number of people now watching them. As well, it is undoubtedly a sign of the many discoveries still waiting to be made. Though these species were not able to be included in the text, I have summarized the new species and records below in an effort to be as complete as possible.

Fiery-eyed Dancer
(*Argia oenea*) (Coenagrionidae)

A population of this species was discovered for the first time in Texas at Chinati Hot Springs (Presidio Co.) by Robert Tizard in early May 2004. Fiery-eyed Dancer is a tropical species that extends from Panama to the southwestern United States. This species is very similar to the Coppery Dancer (*A. cuprea*) located in the Texas Hill country. Like that species, it can be recognized by the brilliant red eyes and metallic coppery red to brown thorax. Segment 8 of males is blue or violet while it is largely black in Coppery Dancer.

Dark-tailed Forceptail
(*Phyllocycla breviphylla*) (Gomphidae)

Two female specimens of this species were collected in late May 2004 by Omar Boconegra at Anauac Wildlife Management Area in Cameron Co., Texas. This represents the first confirmation of this species and genus in the United States (a photograph had been taken earlier of this species, but the identification could not be confirmed). This species may be a synonym of *P. elongata*. It is very similar to our other Forceptails (*Aphylla*) and Leaftails (*Phyllogomphoides*). Leaftails lack a differentiated posterodorsal rim on abdominal segment 10 and the hindwing subtriangle is composed of two or more cells. There are numerous teeth (their length no more than 1/6 of the femur width) on the distal half of the hind femur in *Aphylla*, while there are fewer, but larger teeth (their length at least 1/4 the femur width) on the hind femur in *Phyllocycla*.

Claret Pondhawk
(*Erythemis mithroides*) (Libellulidae)

A single male of this species was photographed at Santa Ana National Wildlife Refuge (Hidalgo Co., Texas) in early May 2004 by Martin Reid. This represents the first record for this species in the United States. This species is tropical ranging southward to Brazil and Paraguay. There is evidence that the Mexican populations of this taxon actually represent an undescribed species (pers. comm. Dennis Paulson). It is similar to the other pondhawks in that it regularly perches on the ground. It is uniformly red over its head, thorax and abdomen. There is a red to brown spot basally in the hindwing that extends out to the anal crossing.

Leptobasis melinogaster (Coenagrionidae)

In late June of 2004, Tom Langschied and Jim Sinclair discovered a population of this Mexican species in Kleberg County, Texas. It represents a new genus and species for the United States. It is like no other damselfly in our region, with a greenish thorax and long thin abdomen (32 mm) with segments 7–10 cream colored. No common name has been established for this species.

All of these finds indicate the tremendous opportunity for discovery in the Odonata. It is truly amazing that in a two month period four species have been discovered for the first time in Texas. All but one, Fiery-eyed Dancer, is new for the United States! With the growing popularity of Odonata and the advent of digital cameras, many more discoveries are waiting to be made. I would strongly encourage individuals who think they have found something new for an area to collect specimens as they are still essential in many cases for the identifications of this group.

The Dragonfly Society of the Americas Common Names Committee has also made some name changes that will affect the south-central U.S. fauna. I direct readers to the following url, where the changes should be reflected: http://www.ups.edu/biology/museum/NAdragons.html. I also direct readers to Odonata Central, http://www.odonatacentral.com, for up-to-date information about the south-central U.S. Odonata fauna.

John C. Abbott
July 15, 2004

Checklist of South-Central United States Odonata

BROAD-WINGED DAMSELS (CALOPTERYGIDAE)
- ☐ Sparkling Jewelwing (*Calopteryx dimidiata*)
- ☐ Ebony Jewelwing (*Calopteryx maculata*)

- ☐ American Rubyspot (*Hetaerina americana*)
- ☐ Smoky Rubyspot (*Hetaerina titia*)
- ☐ Canyon Rubyspot (*Hetaerina vulnerata*)

SPREADWINGS (LESTIDAE)
- ☐ Great Spreadwing (*Archilestes grandis*)

- ☐ Plateau Spreadwing (*Lestes alacer*)
- ☐ Spotted Spreadwing (*Lestes congener*)
- ☐ Common Spreadwing (*Lestes disjunctus*)
- ☐ Emerald Spreadwing (*Lestes dryas*)
- ☐ Rainpool Spreadwing (*Lestes forficula*)
- ☐ Elegant Spreadwing (*Lestes inaequalis*)
- ☐ Slender Spreadwing (*Lestes rectangularis*)
- ☐ Chalky Spreadwing (*Lestes sigma*)
- ☐ Lyre-tipped Spreadwing (*Lestes unguiculatus*)
- ☐ Swamp Spreadwing (*Lestes vigilax*)

THREADTAILS (PROTONEURIDAE)
- ☐ Coral-fronted Threadtail (*Neoneura aaroni*)
- ☐ Amelia's Threadtail (*Neoneura amelia*)

- ☐ Orange-striped Threadtail (*Protoneura cara*)

POND DAMSELS (COENAGRIONIDAE)
- ☐ Mexican Wedgetail (*Acanthagrion quadratum*)

- ☐ Western Red Damsel (*Amphiagrion abbreviatum*)

- ☐ Paiute Dancer (*Argia alberta*)
- ☐ Blue-fronted Dancer (*Argia apicalis*)
- ☐ Comanche Dancer (*Argia barretti*)
- ☐ Seepage Dancer (*Argia bipunctulata*)
- ☐ Coppery Dancer (*Argia cuprea*)
- ☐ Variable Dancer (*Argia fumipennis*)
- ☐ Lavender Dancer (*Argia hinei*)
- ☐ Kiowa Dancer (*Argia immunda*)
- ☐ Leonora's Dancer (*Argia leonorae*)
- ☐ Sooty Dancer (*Argia lugens*)

- ☐ Powdered Dancer (*Argia moesta*)
- ☐ Apache Dancer (*Argia munda*)
- ☐ Aztec Dancer (*Argia nahuana*)
- ☐ Amethyst Dancer (*Argia pallens*)
- ☐ Springwater Dancer (*Argia plana*)
- ☐ Golden-winged Dancer (*Argia rhoadsi*)
- ☐ Blue-ringed Dancer (*Argia sedula*)
- ☐ Tezpi Dancer (*Argia tezpi*)
- ☐ Blue-tipped Dancer (*Argia tibialis*)
- ☐ Tonto Dancer (*Argia tonto*)
- ☐ Dusky Dancer (*Argia translata*)
- ☐ Vivid Dancer (*Argia vivida*)

- ☐ Aurora Damsel (*Chromagrion conditum*)

- ☐ River Bluet (*Enallagma anna*)
- ☐ Rainbow Bluet (*Enallagma antennatum*)
- ☐ Azure Bluet (*Enallagma aspersum*)
- ☐ Double-striped Bluet (*Enallagma basidens*)
- ☐ Boreal Bluet (*Enallagma boreale*)
- ☐ Tule Bluet (*Enallagma carunculatum*)
- ☐ Familiar Bluet (*Enallagma civile*)
- ☐ Alkali Bluet (*Enallagma clausum*)
- ☐ Cherry Bluet (*Enallagma concisum*)
- ☐ Northern Bluet (*Enallagma cyathigerum*)
- ☐ Attenuated Bluet (*Enallagma daeckii*)
- ☐ Turquoise Bluet (*Enallagma divagans*)
- ☐ Atlantic Bluet (*Enallagma doubledayi*)
- ☐ Burgundy Bluet (*Enallagma dubium*)
- ☐ Big Bluet (*Enallagma durum*)
- ☐ Stream Bluet (*Enallagma exsulans*)
- ☐ Skimming Bluet (*Enallagma geminatum*)
- ☐ Neotropical Bluet (*Enallagma novaehispaniae*)
- ☐ Arroyo Bluet (*Enallagma praevarum*)
- ☐ Claw-tipped Bluet (*Enallagma semicirculare*)
- ☐ Orange Bluet (*Enallagma signatum*)
- ☐ Slender Bluet (*Enallagma traviatum*)
- ☐ Vesper Bluet (*Enallagma vesperum*)

- ☐ Painted Damsel (*Hesperagrion heterodoxum*)

- ☐ Desert Forktail (*Ischnura barberi*)
- ☐ Pacific Forktail (*Ischnura cervula*)
- ☐ Plains Forktail (*Ischnura damula*)
- ☐ Mexican Forktail (*Ischnura demorsa*)

☐ Black-fronted Forktail (*Ischnura denticollis*)
☐ Citrine Forktail (*Ischnura hastata*)
☐ Lilypad Forktail (*Ischnura kellicotti*)
☐ Western Forktail (*Ischnura perparva*)
☐ Fragile Forktail (*Ischnura posita*)
☐ Furtive Forktail (*Ischnura prognata*)
☐ Rambur's Forktail (*Ischnura ramburii*)
☐ Eastern Forktail (*Ischnura verticalis*)

☐ Southern Sprite (*Nehalennia integricollis*)
☐ Everglades Sprite (*Nehalennia pallidula*)

☐ Caribbean Yellowface (*Neoerythromma cultellatum*)

☐ Duckweed Firetail (*Telebasis byersi*)
☐ Desert Firetail (*Telebasis salva*)

PETALTAILS (PETALURIDAE)
☐ Gray Petaltail (*Tachopteryx thoreyi*)

DARNERS (AESHNIDAE)
☐ Lance-tipped Darner (*Aeshna constricta*)
☐ Arroyo Darner (*Aeshna dugesi*)
☐ Variable Darner (*Aeshna interrupta*)
☐ Sedge Darner (*Aeshna juncea*)
☐ Blue-eyed Darner (*Aeshna multicolor*)
☐ Paddle-tailed Darner (*Aeshna palmata*)
☐ Persephone's Darner (*Aeshna persephone*)
☐ Turquoise-tipped Darner (*Aeshna psilus*)
☐ Shadow Darner (*Aeshna umbrosa*)

☐ Amazon Darner (*Anax amazili*)
☐ Common Green Darner (*Anax junius*)
☐ Comet Darner (*Anax longipes*)
☐ Giant Darner (*Anax walsinghami*)

☐ Springtime Darner (*Basiaeschna janata*)

☐ Fawn Darner (*Boyeria vinosa*)

☐ Blue-faced Darner (*Coryphaeschna adnexa*)
☐ Regal Darner (*Coryphaeschna ingens*)

☐ Swamp Darner (*Epiaeschna heros*)

☐ Taper-tailed Darner (*Gomphaeschna antilope*)
☐ Harlequin Darner (*Gomphaeschna furcillata*)

☐ Bar-sided Darner (*Gynacantha mexicana*)

☐ Twilight Darner (*Gynacantha nervosa*)

☐ Cyrano Darner (*Nasiaeschna pentacantha*)

☐ Riffle Darner (*Oplonaeschna armata*)

CLUBTAILS (GOMPHIDAE)
☐ Broad-striped Forceptail (*Aphylla angustifolia*)
☐ Narrow-striped Forceptail (*Aphylla protracta*)
☐ Two-striped Forceptail (*Aphylla williamsoni*)

☐ Stillwater Clubtail (*Arigomphus lentulus*)
☐ Bayou Clubtail (*Arigomphus maxwelli*)
☐ Jade Clubtail (*Arigomphus submedianus*)
☐ Unicorn Clubtail (*Arigomphus villosipes*)

☐ Southeastern Spinyleg (*Dromogomphus armatus*)
☐ Black-shouldered Spinyleg (*Dromogomphus spinosus*)
☐ Flag-tailed Spinyleg (*Dromogomphus spoliatus*)

☐ White-belted Ringtail (*Erpetogomphus compositus*)
☐ Yellow-legged Ringtail (*Erpetogomphus crotalinus*)
☐ Eastern Ringtail (*Erpetogomphus designatus*)
☐ Blue-faced Ringtail (*Erpetogomphus eutainia*)
☐ Dashed Ringtail (*Erpetogomphus heterodon*)
☐ Serpent Ringtail (*Erpetogomphus lampropeltis*)

☐ Banner Clubtail (*Gomphus apomyius*)
☐ Plains Clubtail (*Gomphus externus*)
☐ Tamaulipan Clubtail (*Gomphus gonzalezi*)
☐ Pronghorn Clubtail (*Gomphus graslinellus*)
☐ Cocoa Clubtail (*Gomphus hybridus*)
☐ Ashy Clubtail (*Gomphus lividus*)
☐ Sulphur-tipped Clubtail (*Gomphus militaris*)
☐ Gulf Coast Clubtail (*Gomphus modestus*)
☐ Oklahoma Clubtail (*Gomphus oklahomensis*)
☐ Ozark Clubtail (*Gomphus ozarkensis*)
☐ Rapids Clubtail (*Gomphus quadricolor*)
☐ Cobra Clubtail (*Gomphus vastus*)

☐ Dragonhunter (*Hagenius brevistylus*)

☐ Arizona Snaketail (*Ophiogomphus arizonicus*)

☐ Pale Snaketail (*Ophiogomphus severus*)
☐ Westfall's Snaketail (*Ophiogomphus westfalli*)

☐ Five-striped Leaftail (*Phyllogomphoides albrighti*)
☐ Four-striped Leaftail (*Phyllogomphoides stigmatus*)

☐ Gray Sanddragon (*Progomphus borealis*)
☐ Common Sanddragon (*Progomphus obscurus*)

☐ Least Clubtail (*Stylogomphus albistylus*)

☐ Brimstone Clubtail (*Stylurus intricatus*)
☐ Laura's Clubtail (*Stylurus laurae*)
☐ Russet-tipped Clubtail (*Stylurus plagiatus*)
☐ Arrow Clubtail (*Stylurus spiniceps*)

SPIKETAILS (CORDULEGASTRIDAE)
☐ Apache Spiketail (*Cordulegaster diadema*)
☐ Pacific Spiketail (*Cordulegaster dorsalis*)
☐ Twin-spotted Spiketail (*Cordulegaster maculata*)
☐ Arrowhead Spiketail (*Cordulegaster obliqua*)

CRUISERS (CORDULIIDAE: MACROMIINAE)
☐ Stream Cruiser (*Didymops transversa*)

☐ Allegheny River Cruiser (*Macromia alleghaniensis*)
☐ Bronzed River Cruiser (*Macromia annulata*)
☐ Illinois River Cruiser (*Macromia illinoiensis*)
☐ Gilded River Cruiser (*Macromia pacifica*)
☐ Royal River Cruiser (*Macromia taeniolata*)

EMERALDS (CORDULIIDAE: CORDULIINAE)
☐ Stripe-winged Baskettail (*Epitheca costalis*)
☐ Common Baskettail (*Epitheca cynosura*)
☐ Dot-winged Baskettail (*Epitheca petechialis*)
☐ Prince Baskettail (*Epitheca princeps*)
☐ Mantled Baskettail (*Epitheca semiaquea*)
☐ Robust Baskettail (*Epitheca spinosa*)
☐ Florida Baskettail (*Epitheca stella*)

☐ Selys' Sundragon (*Helocordulia selysii*)
☐ Uhler's Sundragon (*Helocordulia uhleri*)

☐ Alabama Shadowdragon (*Neurocordulia alabamensis*)
☐ Smoky Shadowdragon (*Neurocordulia molesta*)

☐ Umber Shadowdragon (*Neurocordulia obsoleta*)
☐ Cinnamon Shadowdragon (*Neurocordulia virginiensis*)
☐ Orange Shadowdragon (*Neurocordulia xanthosoma*)

☐ Fine-lined Emerald (*Somatochlora filosa*)
☐ Coppery Emerald (*Somatochlora georgiana*)
☐ Mocha Emerald (*Somatochlora linearis*)
☐ Texas Emerald (*Somatochlora margarita*)
☐ Ozark Emerald (*Somatochlora ozarkensis*)
☐ Mountain Emerald (*Somatochlora semicircularis*)
☐ Clamp-tipped Emerald (*Somatochlora tenebrosa*)

SKIMMERS (LIBELLULIDAE)
☐ Red-tailed Pennant (*Brachymesia furcata*)
☐ Four-spotted Pennant (*Brachymesia gravida*)
☐ Tawny Pennant (*Brachymesia herbida*)

☐ Pale-faced Clubskimmer (*Brechmorhoga mendax*)

☐ Gray-waisted Skimmer (*Cannaphila insularis*)

☐ Amanda's Pennant (*Celithemis amanda*)
☐ Red-veined Pennant (*Celithemis bertha*)
☐ Calico Pennant (*Celithemis elisa*)
☐ Halloween Pennant (*Celithemis eponina*)
☐ Banded Pennant (*Celithemis fasciata*)
☐ Faded Pennant (*Celithemis ornata*)
☐ Double-ringed Pennant (*Celithemis verna*)

☐ Checkered Setwing (*Dythemis fugax*)
☐ Mayan Setwing (*Dythemis maya*)
☐ Black Setwing (*Dythemis nigrescens*)
☐ Swift Setwing (*Dythemis velox*)

☐ Western Pondhawk (*Erythemis collocata*)
☐ Flame-tailed Pondhawk (*Erythemis peruviana*)
☐ Pin-tailed Pondhawk (*Erythemis plebeja*)
☐ Eastern Pondhawk (*Erythemis simplicicollis*)
☐ Great Pondhawk (*Erythemis vesiculosa*)

☐ Plateau Dragonlet (*Erythrodiplax basifusca*)
☐ Seaside Dragonlet (*Erythrodiplax berenice*)
☐ Black-winged Dragonlet (*Erythrodiplax funerea*)

☐ Red-faced Dragonlet (*Erythrodiplax fusca*)
☐ Little Blue Dragonlet (*Erythrodiplax minuscula*)
☐ Band-winged Dragonlet (*Erythrodiplax umbrata*)

☐ Blue Corporal (*Ladona deplanata*)

☐ Golden-winged Skimmer (*Libellula auripennis*)
☐ Bar-winged Skimmer (*Libellula axilena*)
☐ Comanche Skimmer (*Libellula comanche*)
☐ Bleached Skimmer (*Libellula composita*)
☐ Neon Skimmer (*Libellula croceipennis*)
☐ Spangled Skimmer (*Libellula cyanea*)
☐ Yellow-sided Skimmer (*Libellula flavida*)
☐ Eight-spotted Skimmer (*Libellula forensis*)
☐ Slaty Skimmer (*Libellula incesta*)
☐ Widow Skimmer (*Libellula luctuosa*)
☐ Needham's Skimmer (*Libellula needhami*)
☐ Hoary Skimmer (*Libellula nodisticta*)
☐ Twelve-spotted Skimmer (*Libellula pulchella*)
☐ Four-spotted Skimmer (*Libellula quadrimaculata*)
☐ Flame Skimmer (*Libellula saturata*)
☐ Painted Skimmer (*Libellula semifasciata*)
☐ Great Blue Skimmer (*Libellula vibrans*)

☐ Marl Pennant (*Macrodiplax balteata*)

☐ Ivory-striped Sylph (*Macrothemis imitans*)
☐ Straw-colored Sylph (*Macrothemis inacuta*)
☐ Jade-striped Sylph (*Macrothemis inequiunguis*)

☐ Hyacinth Glider (*Miathyria marcella*)

☐ Spot-tailed Dasher (*Micrathyria aequalis*)
☐ Three-striped Dasher (*Micrathyria didyma*)
☐ Thornbush Dasher (*Micrathyria hagenii*)

☐ Orange-bellied Skimmer (*Orthemis discolor*)

☐ Roseate Skimmer (*Orthemis ferruginea*)

☐ Blue Dasher (*Pachydiplax longipennis*)

☐ Red Rock Skimmer (*Paltothemis lineatipes*)

☐ Wandering Glider (*Pantala flavescens*)
☐ Spot-winged Glider (*Pantala hymenaea*)

☐ Slough Amberwing (*Perithemis domitia*)
☐ Mexican Amberwing (*Perithemis intensa*)
☐ Eastern Amberwing (*Perithemis tenera*)

☐ Common Whitetail (*Plathemis lydia*)
☐ Desert Whitetail (*Plathemis subornata*)

☐ Filigree Skimmer (*Pseudoleon superbus*)

☐ Blue-faced Meadowhawk (*Sympetrum ambiguum*)
☐ Variegated Meadowhawk (*Sympetrum corruptum*)
☐ Saffron-winged Meadowhawk (*Sympetrum costiferum*)
☐ Black Meadowhawk (*Sympetrum danae*)
☐ Cardinal Meadowhawk (*Sympetrum illotum*)
☐ Cherry-faced Meadowhawk (*Sympetrum internum*)
☐ Striped Meadowhawk (*Sympetrum pallipes*)
☐ Band-winged Meadowhawk (*Sympetrum semicinctum*)
☐ Yellow-legged Meadowhawk (*Sympetrum vicinum*)

☐ Aztec Glider (*Tauriphila azteca*)

☐ Evening Skimmer (*Tholymis citrina*)

☐ Striped Saddlebags (*Tramea calverti*)
☐ Carolina Saddlebags (*Tramea carolina*)
☐ Antillean Saddlebags (*Tramea insularis*)
☐ Black Saddlebags (*Tramea lacerata*)
☐ Red Saddlebags (*Tramea onusta*)

Glossary

abdomen — The hindrearmost major section of the insect body; in the Odonata, long, slender, and consisting of ten segments (Fig. 4).

accessory — genitalia The structures beneath the second abdominal segment of male odonates, where females attach their abdomens for mating (Fig. 4) (cf. **secondary genitalia**).

acuminate — Pointed; said of a structure that tapers to a point.

aestivation — The state in which an organism is metabolically inactive and/or physically dormant during the summer (cf. **diapause, obligate diapause**).

alar — Referring or pertaining to wings.

anal crossing — An apparent basal crossvein within the cubito-anal interspace.

analloop — A group of cells in the basal area of the dragonfly hindwing that is distinctively shaped in the form of a circle or boot (Fig. 6).

andromorphic — In females, having a color form that is similar to that of males of the same species (cf. **gynomorphic**).

Anisoptera — The suborder of Odonata to which the dragonflies belong.

antealar carina (pl. **carinae**) — A ridge anterior to the wing bases on the pterothorax of an odonate.

antehumeral — Situated on the mesepisternum between the middorsal carina and the mesopleural suture (Fig. 5), as for example a longitudinal stripe.

antenodal crossvein — Any of several crossveins, proximal to the nodus, that connect the costa, subcosta, and radius (Fig. 6).

apex (pl. **apices**) — The part of an insect structure (e.g. appendage, joint, or segment) farthest from the base of that structure.

apical — At or pertaining to the apex (cf. **distal, medial, proximal**).

apterostigmatous — Lacking a pterostigma.

arculus — A transverse vein located at or just anterior to the base of the triangle in the odonate wing (Fig. 6).

auricle — An ear-shaped projection on of the second abdominal segment (one on each side) of male dragonflies, except in the Libellulidae.

basal plate — A sclerite, medially divided or bilobed in dragonflies, at the base of the ovipositor in darners (Aeshnidae) and spreadwings (Lestidae) (Fig. 5).

bifid — Divided, as a structure, into two parts or lobes.

brace vein — A slanted crossvein proximal to the posterobasal corner of the pterostigma, in an odonate wing.

bridge — A nearly longitudinal vein extending from vein Rs beyond the nodus proximally to vein M just distal to the median fork.

bridge crossvein — A crossvein between the bridge and veins M and M_2, basal to the fusion of the bridge with Rs.

bursa (**bursa copulatrix**) — A saclike expansion of the dorsal side of the vagina, where spermatozoa are deposited by the odonate female prior to fertilization.

carina (pl. **carinae**) — An elevation or ridge in the cuticle of an insect exoskeleton.

caudal appendage — One of several structures at the end of the odonate abdomen—the cerci, epiproct, and paraprocts—that are found in both sexes (Fig. 5).

cell — A closed area in an insect wing that is bound by veins (Fig. 6).

cercus (pl. **cerci**) — One of a pair of appendages at the tip of the odonate abdomen: in males, it forms one of the superior (dorsal) caudal appendages that are used to clasp the female; in females, it is leaflike or conical and easily broken (Figs. 4, 5; cf. **claspers**).

chitin — A major polysaccharide constituent of arthropod cuticle.

circumtropical — Distributed throughout the world's tropics.

claspers — The appendages at the distal end of a male odonate that are used to grasp the female's head (Anisoptera) or prothorax (Zygoptera) during mating; claspers consist of cerci and epiproct in Anisoptera and cerci and paraprocts in Zygoptera.

club — An enlarged area of abdominal segments 7–9, particularly in clubtails (Gomphidae).

clypeus — The anterior sclerite above the labrum and below the frons on the odonate head (Fig. 5).

copulation — The act of mating, wherein the male holds the female as in tandem, but the tip of the female's abdomen swings around and up to contact the second segment of the male's abdomen, where the accessory genitalia are located (cf. **tandem position, wheel position**).

costa — The longitudinal vein running along the leading edge of the odonate wing (Fig. 6).

coxa (pl. **coxae**) — The basalmost segment of an insect leg (Fig. 5).

crepuscular — Active at dusk, or in some cases at daybreak.

crossvein — Any of the short veins running between the longitudinal veins and their branches in an odonate wing (Fig. 6).

cubito-anal interspace — The space formed in the basal area of the wing between the cubitus and the anal vein.

cuticle — The outer layer of the insect exoskeleton.

damselfly — Any member of the suborder Zygoptera, characterized by having relatively narrow wings; fore and hind-wings of same shape; small, widely separated compound eyes; and the body generally smaller and more slender than a dragonfly's.

degree-day — The number of degrees above a minimum temperature acceptable for growth, multiplied by time in days, representing a measure of physiological process through a combination of time and temperature.

diapause — A genetically programmed delay in development that is not the direct result of prevailing conditions (cf. **aestivation, obligate diapause**).

dichromatic — Having two distinct color forms, as among individuals within a species.

dimorphic — Having two forms; exhibiting, for example, differences between the sexes of a given species.

distal — Situated farthest from the base of a body structure (cf. **apical, medial, proximal**).

dorsal — Situated on, or pertaining to, the dorsum.

dorsum — The top or back side of the body or other structure.

dragonfly — Any member of the suborder Anisoptera, characterized by having relatively broad wings (the hindwings broader basally than the forewings); large compound eyes, touching in most groups; and the body generally larger and more robust than a damselfly's.

ecotone — A transitional area between two adjacent, and in some cases very different, ecological communities.

emarginate — Having its edge interrupted at some point, as the margin of a sclerite that has an incision or notch.

emergence — The departure of a larva from the water to undergo metamorphosis into an adult, its exuviae left behind in the process.

endemic — Having a distribution restricted to a particular region.

endophytic oviposition — The laying of eggs into plant tissue (cf. **exophytic oviposition**).

epiproct — A dorsal projection from abdominal segment 10: in dragonflies, an inferior caudal appendage (Fig. 4); in damselflies, usually reduced to a small buttonlike structure.

exophytic oviposition — The laying of eggs by some dragonflies, either by dropping them freely into the water or attaching them superficially on aquatic plants (cf. **endophytic oviposition**).

exoskeleton — The outer, hard part of an insect, including the legs and wings.

exuviae (sing. and pl.) — The cast-off skin from any of the several larval molts of an insect, including that produced by transformation into an adult.

femur (pl. **femora**) — The first (basal) long leg segment of an insect (Fig. 4).

flavescent — Appearing yellow or yellowish.

flight season — The period of the year during which adults of a given odonate species (or other flying insects) have been found to occur.

forage — To search actively for food.

forcipate — Appearing forceps-like, said for example of two

structures that curve inward to meet, or nearly meet, at their tips.

fossa — A pit or depression in a surface.

frons — The front of the head, essentially the face (Figs. 4, 5).

gaff — The short region beyond the posterior angle of the triangle in the hindwing of a member of the Anisoptera, where the distal branch of vein A_1 is fused with the posterior branch of vein Cu_2.

gena (pl. **genae**) — The area of the head analogous to the cheeks, above the mandibles and lateral to the clypeus.

guarding — The defense of a female, by a male, against attack by other males while she lays eggs.

gynomorphic — In females, having a color form that is distinctly different from that of males of the same species (cf. **andromorphic**).

hamule — One of two ventrally projecting, paired structures housed in a capsule under the second abdominal segment of a male odonate, serving to hold the female's abdomen in place during copulation.

hastate — Spear-shaped; more or less triangular with sharp basal lobes extending away from the axis.

hemimetabolous — In insects, having an incomplete metamorphosis, one incorporating gradual development, external wing development, and aquatic larvae but nonaquatic adults (cf. **naiad, nymph**).

humeral — Of or pertaining to the shoulder area; in insects, the dorsolateral area of the thorax.

humeral stripe — A dark stripe along the mesopleural suture (Figs. 4, 5).

humeral suture — The suture along the mesopleuron (Fig. 5).

immature — An adult odonate past the teneral stage but not yet exhibiting the full brilliance of mature coloration; often seen some distance from water and thus not seeking mating opportunities (cf. **teneral**).

in copula — Joined in copulation.

instar — An individual larval stage, one of several successive stages marked by molts.

intercalary — Additional, or inserted between, as a wing vein or other structure.

interpleural suture — The suture separating the mesopleuron and the metapleuron.

interspecific — Occurring between different species.

intraspecific — Occurring between individuals of the same species.

labium — The lower lip of an odonate, employed in the capture of prey.

labrum — The upper lip of an odonate (Fig. 5).

larva (pl. **larvae**) — The initial, aquatic stage of an odonate, following its emergence from the egg and terminating in its emergence as an adult (cf. **naiad, nymph**).

lentic — Of, relating to, or living in bodies of standing water: ponds, lakes, pools.

lotic — Of, relating to, or living in bodies of running water: rivers, streams, creeks.

maculation — The pattern of spots or other markings on an insect.

mandible — One of the large toothlike structures used by an insect for biting and chewing (Fig. 5).

mature — Of reproductive age, with full adult coloration.

medial — Situated on, or pertaining to, the middle of a body structure (cf. **apical, distal, proximal**).

mesepimeron — The sclerite between the humeral suture and the mesepisternum (Fig. 5).

mesepisternal tubercle — A small tubercle on the mesepisternum, just behind the mesostigmal plates, in some damselflies (*Argia* and *Enallagma*).

mesepisternum — The area anterior and dorsal to the wings on the mesothorax (Fig. 5).

mesopleural stripe — A dark stripe along the length of the mesopleural suture (cf. **humeral stripe**; Fig. 5).

mesopleural suture — Suture dividing the mesepisternum from the mesepimeron.

mesopleuron — The side of the mesothorax.

mesostigmal plate — A small, often triangular sclerite on the dorsal anterior edge of the mesepisternum (Fig. 5).

mesothorax — The second or middle thoracic segment (cf. **metathorax, prothorax**).

metamorphosis — The process of transformation through successive life stages, basically from larva to adult in insects.

metapleural stripe — A stripe along the suture dividing the metepisternum and the metepimeron (Fig. 5).

metapleuron — The side of the metathorax.

metathorax — The third, hindmost segment of the thorax (cf. **mesothorax, prothorax**).

metepimeron — The area of the metapleuron posterior to the metapleural suture and extending to the metathoracic venter (Fig. 5).

metepisternum — The anterior area of the metapleuron, lying between the interpleural and metapleural sutures (Fig. 5).

middorsal — Situated on, or pertaining to, the middle of the dorsum.

molt — One of the successive times in an insect's lifespan when the exuviae is shed, permitting additional growth.

naiad — An individual of the larval stage of an insect, properly referring only to members the Ephemeroptera (mayflies), Odonata, and Plecoptera (stoneflies).

nodus — An indentation or notch along the front margin of the wing of an odonate, usually medially located (Figs. 4, 6).

nominate — A subordinate taxon (subspecies, subgenus) that bears the same name as its subsuming higher taxon (species, genus), for example *Lestes disjunctus disjunctus*.

nymph — The larval stage of a paurometabolous insect; commonly used when referring to an odonate though odonates are technically hemimetabolous (cf. **larva, naiad**).

obelisk — A posture adopted for thermoregulation by some skimmers (Libellulidae) and clubtails (Gomphidae) to lessen exposure of the body to the sun, thereby helping to keep cooler on hot summer days. The abdomen is pointed directly toward the sun.

obligate diapause — A period of arrested development occurring during an unfavorable season or time of the year (cf. **aestivation, diapause**).

occiput — The area of the head lying between the vertex and the membranous neck region that articulates the head with the prothorax (Figs. 4, 5).

ocellus (pl. **ocelli**) — One of the smaller, simple eyes of an insect, situated in odonates between the large compound eyes, and used for light detection (Fig. 5).

Odonata — The order of insects consisting of the dragonflies and damselflies.

odonate — Being, or pertaining to, a dragonfly or damselfly.

ommatidium (pl. **ommatidia**) — A single facet, among the many, of a compound eye of an insect.

oviposit — Lay eggs.

oviposition — The act or process of laying eggs.

ovipositor — The complex structure at the posterior end of some female Odonata—damselflies, darners (Aeshnidae), and petaltails (Petaluridae)—that functions in endophytic oviposition (Figs. 4, 5).

paranal cells — The single row of cells at the base of an odonate wing, immediately behind the anal vein (Fig. 6).

paraproct — One of the pair of lobes or processes projecting ventrolaterally from abdominal segment 10 of odonates; in male damselflies they form the inferior (ventral) caudal appendages used to clasp the female during mating (Fig. 5); in dragonflies they are usually rounded and little modified.

peduncle — The basal segment of the penis, where spermatozoa are stored prior to copulation.

petiolate — Having a narrow stemlike base.

planate — The intercalated wing vein, formed by the longitudinal alignment of cell margins, that lies behind a major longitudinal vein but does not extend to the base of the vein or the margin of the wing.

polymorphic — Having more than two distinct forms, as among individuals within a given species.

postclypeus — In Odonata, the horizontal surface of the clypeus (Fig. 5).

postnodal crossveins — The crossveins, distal to the nodus, that connect the costa, vein R_1, and vein M_1 of an odonate wing.

pronotum — The dorsal portion of the prothorax, the first thoracic segment.

prothorax — The small first segment of the thorax, situated just after the head and before the larger thoracic area bearing the legs and wings (cf. **mesothorax, metathorax**).

proximal — Situated nearest the base of a body structure (cf. **medial, distal**).

pruinose — Having a waxy or powdery covering, the covering exuding from cuticle in odonates, chiefly in older individuals, and turning areas of the body light blue, gray, or white.

pterostigma — The thickened blood-filled cell at the front of the wingtip in most Odonata (Fig. 4).

pterothorax — The fused meso- and metathoracic segments

of an odonate, those thoracic segments posterior to the prothorax that bear the wings.

quadrangle — A cell in Zygoptera wings bounded by vein M, Cu, and arculus and a crossvein between M and Cu (Fig. 6).

quadrate — Four-sided; roughly square.

radial planate — The planate lying just behind the vein Rs of an odonate wing, beginning a little distal to the nodus.

radial sector — The posterior branch of the radius.

radius — The third longitudinal vein of an odonate wing, lying posterior to the subcosta.

riparian — Of or pertaining to the bank of a stream or river, usually taken to include the trees and other vegetation bordering the water.

sclerite — A hard, chitinous plate on an insect's exoskeleton bounded by membrane or sutures.

secondary genitalia — The structures beneath the second abdominal segment of male odonates, where females attach their abdomens for mating (Fig. 4; cf. **accessory genitalia**).

seta (pl. **setae**) — A hair or bristle on an insect.

sigmoid — S-shaped; often used to characterize the caudal appendages of male damselflies.

spine — A pointed, immovable projection on an insect's leg.

spiracle — An opening on the lateral body wall, of both thorax and abdomen, whereby air passes in and out of an insect's tracheal system (Fig. 5).

spur — A pointed, movable projection on an insect's leg.

sternite — A sclerite located ventrally on the thorax or abdomen of an insect.

sternum (pl. **sterna**) — The ventral area of any abdominal segment of an insect.

stylus (pl. **styli**) — A small process at the end of the ovipositor valve (Fig. 5).

subcosta — The second, usually unbranched, longitudinal wing vein of an odonate, lying posterior to the costa and anterior to the radius.

subgenital plate — A flat or scoop-shaped sclerite covering the genital opening of female clubtails (Gomphidae) and skimmers (Libellulidae) on the sternite of abdominal segment 9.

subtriangle — A cell or cells composing a triangular area in an odonate wing just proximal to the triangle (Fig. 6).

sulcus — A groove.

supertriangle — An elongate area, anterior to the triangle, composed of one or more cells (Fig. 6).

suture — A line or groove separating two sclerites.

symphoretic — Being in a commensal relationship such that certain larvae, as those of Chironomidae (Diptera), live on the exoskeleteton of larger hosts and are transported apparently without significant benefit or harm to their hosts, but with several possible benefits to the chironomid.

tandem position — A configuration in which a male and a female are linked together, either in flight or at rest (cf. **wheel position**).

tarsus (pl. **tarsi**) — The third major segment of the insect leg, made up of several subsegments (Fig. 4).

teneral — An odonate immediately after emergence from the larval state, a condition in which it is soft and relatively pale, and often has a shimmer to the wings (cf. **immature**).

tergite — A dorsal sclerite, or part of a segment, on the insect abdomen.

tergum — The upper or dorsal surface of any abdominal segment on an insect.

territoriality — The active defense of a small area.

thorax — The second major body section of an odonate, bearing the wings and legs (Fig 4).

tibia (pl. **tibiae**) — The second major segment of the insect leg, usually longer and thinner than the femur (Fig. 4).

torifer — The dorsal basal area on abdominal segment 10 in male dancers (*Argia*).

torus (pl. **tori**) — A dorsoapical, median protuberance on abdominal segment 10 in male dancers (*Argia*).

triangle — A small cell or group of cells in the basal area of the forewing or hindwing of an odonate (Fig. 6).

trochanter — A short segment of the insect leg between the coxa and the femur (Fig. 5).

valves — The external sheath of the ovipositor.

vein — One of the hollow tubes in the wings of an insect, providing strength and framework, through which blood is pumped at the time of emergence.

ventral — Situated on the bottom or underside of the body or other structure.

vertex — The top of the head, between the compound eyes (Figs. 4, 5).

vulvar lamina — A plate under abdominal segment 9 of female odonates that serves to hold eggs in place during exophytic oviposition, the form of which is often diagnostic of species.

wheel position — A configuration, in odonates during copulation, in which the male grasps the female with the tip of his abdomen behind her head (damselflies) or on the head (dragonflies) and she brings the tip of her abdomen up to make contact with his secondary genitalia (cf. **tandem position**).

Zygoptera — The suborder of Odonata to which the damselflies belong.

Bibliography

Field Guides and Other Reference Works

Beckemeyer, R.J., and D.G. Huggins. 1997. Checklist of Kansas dragonflies. Kansas School Naturalist 43:1–15.

———. 1998. Checklist of Kansas damselflies. Kansas School Naturalist. 44:1–15.

Berger, C. 2004. Wild Guide: Dragonflies. Mechanicsburg, Pa.: Stackpole Books, 124 pp.

Biggs, K. 2000. Common dragonflies of California: A beginner's pocket guide. Sebastopol, Calif.: Azalea Creek Publishing, 96 pp.

———. 2004. Common dragonflies of the Southwest: A beginner's pocket guide. Sebastopol, Calif.: Azalea Creek Publishing, 160 pp.

Brooks, S. 2003. Dragonflies. Washington, D.C.: Smithsonian Books, 96 pp.

Corbet, P.S. 1963. A biology of dragonflies. Chicago: Quadrangle Books, 247 pp.

———. 1980. Biology of Odonata. Annual Review of Entomology 25:189–217.

———. 1999. Dragonflies: Behavior and ecology of Odonata. Ithaca: Cornell University Press, 829 pp.

Curry, J.R. 2001. Dragonflies of Indiana. Indianapolis: Indiana Academy of Science, 304 pp.

Dunkle, S.W. 1989. Dragonflies of the Florida Peninsula, Bermuda, and the Bahamas. Gainesville, Fla.: Scientific Publishers, 154 pp.

———. 1990. Damselflies of Florida, Bermuda, and the Bahamas. Gainesville, Fla.: Scientific Publishers, 148 pp.

———. 2000. Dragonflies through binoculars: A field guide to dragonflies of North America. New York: Oxford University Press, 266 pp.

Glotzhober, R.C., and D. McShaffrey. 2002. The dragonflies and damselflies of Ohio. Columbus: Ohio Biological Survey, 364 pp.

Legler, K., D. Legler, and D. Westover. 1998. Color guide to common dragonflies of Wisconsin. 429 Franklin St., Sauk City, WI 53583.

Manolis, T. 2003. Dragonflies and damselflies of California. Berkeley: California Natural History Guides, 201 pp.

Mead, K. 2003. Dragonflies of the North woods. Duluth, Minn.: Kollath-Stensaas Publishing, 203 pp.

Needham, J.G., M.J. Westfall, Jr., and M.L. May. 2000. Dragonflies of North America, Revised Edition. Gainesville, Fla.: Scientific Publishers, 939 pp.

Nikula, B., J.L. Loose, and M.R. Burne. 2003. A field guide to the dragonflies and damselflies of Massachusetts. Westborough, Mass.: Massachusetts Division of Fisheries and Wildlife, 197 pp.

Nikula, B., J. Sones, D. Stokes, and L. Stokes. 2002. Beginner's guide to dragonflies. Boston: Little, Brown, 159 pp.

Paulson, D.R. 1999. Dragonflies of Washington. Seattle Audubon Society, 32 pp.

Paulson, D.R., and S.W. Dunkle. 1999. A checklist of North American Odonata. University of Puget Sound Occasional Papers 56:1–86.

Rosche, L. 2002. Dragonflies and damselflies of northeast Ohio. Cleveland Museum of Natural History, 94 pp.

Silsby, J. 2001. Dragonflies of the world. Washington, D.C.: Smithsonian Institution Press, 216 pp.

Walton, R.K., and R.A. Forster. 1997. Common dragonflies of the northeast. (Video, NHS, 7 Concord Greene #8, Concord, MA 01742).

Westfall, M.J., Jr., and M.L. May. 1996. Damselflies of North America. Gainesville, Fla: Scientific Publishers, 649 pp.

Westfall, M.J., Jr., and K.J. Tennessen. 1996. Odonata, in R.W. Merritt and K.W. Cummins, eds., An introduction to the aquatic insects of North America, 3rd ed. Dubuque, Iowa: Kendall/Hunt, 862 pp.

Odonata Societies

Dragonfly Society of the Americas
Dr. and Mrs. T.W. Donnelly
2091 Partridge Lane
Binghamton, NY 13903

Societas Internationalis Odonatologica (S.I.O.) Foundation
Editor of Odonatologica, Bastiaan Kiauta
P.O. Box 256
7520 AG Bilthoven
The Netherlands

Worldwide Dragonfly Association
Jill Silsby
1 Haydn Avenue
Purley
Surrey, CR8 4AG
United Kingdom

Sources Cited in Text

Abbott, J.C. 1996. New and interesting records from Texas and Oklahoma. Argia 8:14–15.

———. 2001. Distribution of dragonflies and damselflies (Odonata) in Texas. Transactions of the American Entomological Society 127:189–228.

Abbott, J.C., R.A. Behrstock, and R.R. Larsen. 2003. Notes on the distribution of Odonata in the Texas panhandle, with a summary of new state and county records. Southwestern Naturalist 48:444–448.

Abbott, J.C., and K.W. Stewart. 1998. Odonata of the south-central Nearctic region, including northeastern Mexico. Entomological News 109:201–212.

Aguilar, A.C. 1992. Comportamiento reproductivo y policromatismo en *Ischnura denticollis* Burmeister (Zygoptera: Coenagrionidae). Bulletin of American Odonatology 1:57–64.

———. 1993. Population structure in *Ischnura denticollis* (Burmeister) (Zygoptera: Coenagrionidae). Odonatologica 22: 455–464.

Ahrens, C. 1935. A new record for *Archilestes grandis* (Odonata: Agrionidae sensu Selysii). Entomological News 46:183.

———. 1938. A list of dragonflies taken during the summer of 1936 in western United States (Odonata). Entomological News 49:9–16.

Albright, P.N. 1952. Contributions to the knowledge of the Odonata of Texas with particular attention to the area around San Antonio. M.S. thesis, Trinity Univ., San Antonio, 48 pp.

Alcock, J. 1979. Multiple mating in *Calopteryx maculata* (Odonata: Calopterygidae) and the advantage of non-contact guarding by males. Journal of Natural History 13:439–446.

———. 1982. Post-copulatory mate guarding by males of the damselfly *Hetaerina vulnerata* Selys (Odonata: Calopterygidae). Animal Behavior 30:99–107.

———. 1983. Mate guarding and the acquisition of mates in *Calopteryx maculata* (P. de Beauvois) (Zygoptera: Calopterygidae). Odonatologica 12:153–160.

———. 1987a. The effects of experimental manipulation of resources on the behavior of two calopterygid damselflies that exhibit resource-defense polygyny. Canadian Journal of Zoology 65:2475–2482.

———. 1987b. Male reproductive tactics in the libellulid dragonfly *Paltothemis lineatipes*: Temporal partitioning of territories. Behavior 103:157–173.

———. 1989a. The mating system of *Libellula saturata* Uhler (Anisoptera: Libellulidae). Odonatologica 18:89–93.

———. 1989b. Annual variation in the mating system of the dragonfly *Paltothemis lineatipes* (Anisoptera: Libellulidae). Journal of the Zoological Society of London 218:597–602.

———. 1990. Oviposition resources, territoriality and male reproductive tactics in the dragonfly *Paltothemis lineatipes* (Odonata: Libellulidae). Behavior 113:251–263.

Anadu, D.I., H.U. Anaso, and O.N.D. Onyeka. 1996. Acute toxicity of the insect larvicide abate (Temephos) on the fish *Tilapia melanopleura* and the dragonfly larvae *Neurocordulia virginiensis*. Journal of Environmental Science and Health, B 31:1363–1375.

Anholt, B.R. 1997. Sexual size dimorphism and sex-specific survival in adults of the damselfly *Lestes disjunctus*. Ecological Entomology 22:127–132.

Arnaud, P.H. 1972. Mass movement of *Sympetrum corruptum* [sic] (Hagen) (Odonata: Libellulidae) in central California. Pan-Pacific Entomologist 48:75–76.

Artiss, T., T.R. Schultz, D.A. Polhemus, and C. Simon. 2001. Molecular phylogenetic analysis of the dragonfly genera *Libellula*, *Ladona* and *Plathemis* (Odonata: Libellulidae) based on mitochondrial cytochrome oxidase I and 16s rRNA sequence data. Molecular Phylogenetics and Evolution 18: 348–361.

Baird, J.M., and M.L. May. 1997. Foraging behavior of *Pachydiplax longipennis* (Odonata: Libellulidae). Journal of Insect Behavior 10:655–678.

Baker, R.L., and H.F. Clifford. 1982. Life cycle of an *Enallagma boreale* Selys population from the boreal forest in Alberta, Canada (Zygoptera: Coenagrionidae). Odonatologica 11: 317–322.

Ballou, J. 1984. Visual recognition of females by male *Calopteryx maculata* (Odonata: Calopterygidae). Great Lakes Entomologist 17:201–204.

Barber, B., and V. Elia. 1994. *Tholymis citrina*; a recent record from Florida and an historical record from Texas. Argia 5:10–11.

Barlow, A.E. 1991. New observations on the distribution and behavior of *Tachopteryx thoreyi* (Hag.) (Anisoptera: Petaluridae). Notulae Odonatologica 3:131–132.

Barr, J.E. 1981. Two new state records for Odonata in Louisiana. Proceedings of the Louisiana Academy of Science 44:164.

Beatty, G.H. 1945. Odonata collected and observed in 1945 at two artificial ponds at Upton, New Jersey. Bulletin of the Brooklyn Entomological Society 40:178–187.

———. 1946. Dragonflies (Odonata) collected in Pennsylvania and New Jersey in 1945. Entomological News 57:1–10, 50–56, 76–81, 104–111.

Beatty, A.F., and G.H. Beatty. 1970. Gregarious (?) oviposition of *Calopteryx amata* Hagen (Odonata). Pennsylvania Academy of Science 44:156–158.

Bell, R., and W.H. Whitcomb. 1961. *Erythemis simplicicollis* (Say), a dragonfly predator of the bollworm moth. Florida Entomologist 44:95–97.

Belle, J. 1972. Een opmerking over het gedrag van *Enallagma cyathigerum* (Charp.) Bij het eierleggen (Odonata). Entomologische Berichten Amsterdam 32:39.

Bennefield, B.L. 1965. A taxonomic study of the subgenus *Ladona* (Odonata: Libellulidae). Kansas University Science Bulletin 45:361–396.

Bick, G.H. 1950. The dragonflies of Mississippi (Odonata: Anisoptera). American Midland Naturalist 43:66–78.

———. 1951. The nymph of *Libellula semifasciata* Burmeister. Proceedings of the Entomological Society of Washington 53:247–250.

———. 1957. The Odonata of Louisiana. Tulane Studies in Zoology 5:71–135.

———. 1963. Reproductive behavior in *Enallagma civile* and *Ar-*

gia apicalis. Proceedings of the North Central Branch of the Entomological Society of America 18:110–111.

———. 1966. Threat display in unaccompanied females of the damselfly *Ischnura verticalis* (Say). Proceedings of the Entomological Society of Washington 68:271.

———. 1972. A review of territorial and reproductive behavior in Zygoptera. Contactbrief Nederlandse Libellenonderzoekers (Suppl). 10:1–14.

———. 1978. New state records of United States Odonata. Notulae Odonatologica 1:17–19.

———. 1983. Odonata at risk in conterminous United States and Canada. Odonatologica 12:209–226.

Bick, G.H., and J.F. Aycock. 1950. The life history of *Aphylla williamsoni* Gloyd (Odonata, Aeschnidae). Proceedings of the Entomological Society of Washington 52:26–32.

Bick, G.H., and J.C. Bick. 1957. The Odonata of Oklahoma. Southwestern Naturalist 2:1–18.

———. 1958. The ecology of the Odonata at a small creek in southern Oklahoma. Journal of the Tennessee Academy of Science 33:240–251.

———. 1963. Behavior and population structure of the damselfly *Enallagma civile* (Hagen) (Odonata: Coenagrionidae). Southwestern Naturalist 8:57–84.

———. 1965a. Demography and behavior of the damselfly *Argia apicalis* (Say) (Odonata: Coenagriidae). Ecology 46:461–472.

———. 1965b. Color variation and significance of color in reproduction in the damselfly *Argia apicalis* (Say) (Zygoptera: Coenagriidae). Canadian Entomologist 97:32–41.

———. 1965c. Sperm transfer in damselflies (Odonata: Zygoptera). Annals of the Entomological Society of America 58:592.

———. 1970. Oviposition in *Archilestes grandis* (Rambur) (Odonata: Lestidae). Entomological News 81:157–163.

———. 1972. Substrate utilization during reproduction by *Argia plana* Calvert and *Argia moesta* (Hagen) (Odonata: Coenagrionidae). Odonatologica 1:3–9.

———. 1980. A bibliography of reproductive behavior of Zygoptera of Canada and conterminous United States. Odonatologica 9:5–19.

———. 1982. Behavior of adults of dark-winged and clear-winged subspecies of *Argia fumipennis* (Burmeister) (Zygoptera: Coenagrionidae). Odonatologica 11:99–107.

———. 1995. A review of the genus *Telebasis*, with descriptions of eight new species (Zygoptera: Coenagrionidae). Odonatologica 24:11–44.

Bick, G.H., and L.E. Hornuff. 1965. Behavior of the damselfly *Lestes unguiculatus* Hagen (Odonata: Lestidae). Proceedings of the Indiana Academy of Science 75:110–115.

———. 1966. Reproductive behavior in the damselflies *Enallagma aspersum* (Hagen) and *Enallagma exsulans* (Hagen) (Odonata: Coenagrionidae). Proceedings of the Entomological Society of Washington 68:78–85.

Bick, G.H., and D. Sulzbach. 1966. Reproductive behavior of the damselfly. *Hetaerina americana* (Fabricius) (Odonata: Calopterygidae). Animal Behavior 14:156–158.

Bick, G.H., J.F. Aycock, and A. Orestano. 1950. *Tauriphilia australis* (Hagen) and *Miathyria marcella* (Selys) from Florida and Louisiana (Odonata, Libellulidae). Proceedings of the Entomological Society of Washington 52:81–84.

Bick, G.H., J.C. Bick, and L.E. Hornuff. 1976. Behavior of *Chromagrion conditum* (Hagen) adults (Zygoptera: Coenagrionidae). Odonatologica 5:129–141.

Bird, R.D. 1932a. Dragonflies of Oklahoma. Publication of the University of Oklahoma Biological Survey 4:51–57.

———. 1932b. *Platycordulia xanthosoma* Williamson (Odonata: Corduliidae). Entomological News 43:234–235.

———. 1933. Dragonfly hunting in Oklahoma. Scientific Monthly 36:371–377.

Blair, W.F. 1950. The biotic provinces of Texas. Texas Journal of Science 2:93–117.

Blair, W.F., and T.H. Hubbell. 1938. The biotic districts of Oklahoma. American Midland Naturalist 20:425–454.

Blust, M.H. 1980. Life history and production ecology of *Stylogomphus alibistylus* (Hagen): (Odonata: Gomphidae). M.S. Thesis, Univ. Delaware, 70 pp.

Borror, D.J. 1930. Notes on the Odonata occurring in the vicinity of Silver Lake, Logan County, Ohio, from June 25 to September 1, 1930. Ohio Journal of Science 30:411–415.

———. 1934. Ecological studies of *Ariga moesta* Hagen (Odonata: Coenagrionidae) by means of marking. Ohio Journal of Science 34:97–108.

———. 1942. A revision of the Libellulinae genus *Erythrodiplax* (Odonata). Contrib. Zool. Entomol., No. 4, Biol. Series. Ohio State University, 286 pp.

———. 1953. A migratory flight of dragonflies. Entomological News 64:204.

Bray, W.L. 1901. The ecological relations of the vegetation of western Texas. Botany Gazette 32:195–217, 262–291.

———. 1905. Vegetation of the sotol country in Texas. Bulletin of the University of Texas 60:1–24.

Brimley, C.S. 1903. A list of dragonflies from North Carolina. Especially from the vicinity of Raleigh. Entomological News 14:150–157.

Bruner, W.E. 1931. The vegetation of Oklahoma. Ecological Monographs 1:99–188.

Buchecker, H. 1876. Systema entomologiae sistens insectorum classes, genera, species. Pars 1. Odonata (Fabric) europ. München, 16 pp.

Buchholtz, C. 1955. Eine vergleichende Ethologie der orientalischen Calopterygiden (Odonata) als Beitrag zu ihrer systematischen Deutung. Zeitschrift für Tierzuchtung und Zuchtungsbiologie 12:364-386.

Bulankova, E. 1997. Dragonflies (Odonata) as bioindicators of environmental quality. Biologia, Bratislava 52:177–180.

Burton, J.F. 1996. Movements of the dragonfly *Libellula quadrimaculata* Linnaeus, 1758 in North-west Europe in 1963 (Odonata, Libellulidae). Atalanta 27:175–187.

Butler, T., J.E. Peterson, and P.S. Corbet. 1975. An exceptionally early and informative arrival of adult *Anax junius* in Ontario (Odonata: Aeschnidae). Canadian Entomologist 107: 1253–1254.

Byers, C.F. 1925. Odonata collected in Cheboygan and Emmet counties, Michigan. Papers of the Michigan Academy of Science, Arts and Letters 5:389–398.

———. 1930. A contribution to the knowledge of Florida Odonata. University of Florida Publications, Biological Science Series 1, 327 pp.

———. 1936. The immature form of *Brachymesia gravida*, with notes on the taxonomy of the group (Odonata: Libellulidae). Entomological News 47:35–37, 60–64.

———. 1937. A review of the dragonflies of the genera *Neurocordulia* and *Platycordulia*. Miscellaneous Publications of the Museum of Zoology, University of Michigan 36:1–36.

———. 1939. A study of the dragonflies of the genus *Progomphus* (Gomphidae) with a description of a new species. Proceedings of the Florida Academy of Science 4:19–85.

Calvert, P.P. 1893. Catalogue of the Odonata (dragonflies) of the vicinity of Philadelphia, with an introduction to the study of this group of insects. Transactions of the American Entomological Society 20:152–272.

———. 1898a. Further notes on the new dragonfly *Ischnura kellicotti* (Odonata). Entomological News 9:211–213.

———. 1898b. The odonate genus *Macrothemis* and its allies. Proceedings of the Boston Society of Natural History 28:301–332.

———. 1899. Odonata from Tepic, Mexico, with supplementary notes on those of Baja California. Proceedings of the California Academy of Science 3:371–418.

———. 1901–08. Odonata. In: Biologia Centrali Americana: Insects Neuroptera. London, R.H. Porter & Dulau & Co., 420 pp.

———. 1902. Neuroptera-Odonata *In:* A list of the insects of Beulah, New Mexico, ed. H. Skinner. Transactions of the American Entomological Society 29:42–43.

———. 1907. The differentials of three North American species of *Libellula*. Entomological News 18:201–204.

———. 1915. The supposed dimorphic female of *Ischnura verticalis* Say *In:* Miscellaneous notes on Odonata, M.B. Lyon. Entomological News 26:56–62.

———. 1926. Relations of a late autumnal dragonfly (Odonata) to temperature. Ecology 7:185–190.

———. 1929. Different rates of growth among animals with special reference to the Odonata. Proceedings of the American Philosophical Society 68:227–274.

Campanella, P.J. 1975. The evolution of mating systems in temperate zone dragonflies (Odonata: Anisoptera) II: *Libellula luctuosa* (Burmeister). Behavior 54:278–310.

Campanella, P.J., and L.L. Wolf. 1974. Temporal leks as a mating system in a temperate zone dragonfly (Odonata: Anisoptera) I. *Plathemis lydia* (Drury). Behavior 51:49–85.

Cannings, R.A. 1983. *Libellula subornata* (Odonata: Libellulidae) in Canada. Journal of the Entomological Society of British Columbia 80:54–55.

———. 1988. *Pantala hymenaea* (Say) new to British Columbia, Canada, with notes on its status in the northwestern United States (Anisoptera: Libellulidae). Notulae Odonatologica 3:31–32.

———. 1989. *Enallagma basidens* Calvert, a dragonfly new to Canada, with notes on the expansion of its range in North America (Zygoptera: Coenagrionidae). Notulae Odonatologica 3:49–64.

Cannings, R.A., and K.M. Stuart. 1977. The dragonflies of British Columbia. Handbook of the British Columbia Provencial Museum, No. 35, 254 pp.

Cannings, R.A., S.G. Cannings, and R.J. Cannings. 1980. The distribution of the genus *Lestes* in a saline lake series in central British Columbia, Canada (Zygoptera: Lestidae). Odonatologica 9:19–28.

Carle, F.L. 1980. A new *Lanthus* (Odonata: Gomphidae) from eastern North America with adult and nymphal keys to American Octogomphines. Annals of the Entomological Society of America 73:172–179.

———. 1981. A new species of *Ophiogomphus* from eastern North America, with a key to the regional species (Anisoptera: Gomphidae). Odonatologica 10:271–278.

———. 1982. The wing vein homologies and phylogeny of the Odonata: A continuing debate. Societas Internationalis Odonatologica Rapid Communications 4:1–66.

———. 1983. A new *Zoraena* (Odonata: Cordulegastridae) from eastern North America, with a key to the adult Cordulegastridae of America. Annals of the Entomological Society of America 76:61–68.

Carle, F.L. 1986. The classification, phylogeny and biogeography of the Gomphidae (Anisoptera). I. Classification. Odonatologica 15:275–326.

———. 1992. *Ophiogomphus (Ophionurus) australis* spec. nov. from the Gulf Coast of Louisiana, with larval and adult keys to American *Ophiogomphus* (Anisoptera: Gomphidae). Odonatologica 21:141–152.

———. 1993. *Sympetrum janeae* spec. nov. from eastern North America, with a key to Nearctic *Sympetrum* (Anisoptera: Libellulidae). Odonatologica 22:1–16.

Carle, F.L., and K.M. Kjer. 2002. Phylogeny of *Libellula* Linnaeus (Odonata: Insecta). Zootaxa 87:1–18.

Carpenter, F.M. 1992. Treatise on invertebrate paleontology. Part R. Arthropoda 3. Volumes 3&4: Superclass Hexapoda. Boulder, Colorado: 655 pp.

Carpenter, V. 1991. Dragonflies and damselflies of Cape Cod. Cape Cod Museum of Natural History, Natural History Series No. 4, 79 pp.

Carter, W.T. 1931. The soils of Texas. Texas Agricultural Experiment Station Bulletin 431, 192 pp.

Castella, E. 1987. Larval Odonata distribution as a describer of fluvial ecosystems. Advances in Odonatology 3:23–40.

Catling, P.M. 1996. Evidence for the recent northward spread of *Enallagma civile* (Zygoptera: Coenagrionidae) in southern Ontario. Proceedings of the Entomololgical Society of Ontario 127:131–133.

———. 1998. Evidence for a recent northward spread of *Enallagma civile* in New York state. Argia 10:16.

Catling, P.M., and P.D. Pratt. 1997. An expanding "race" of the azure bluet, *Enallagma aspersum* in Ontario? Argia 9:16–17.

Chao, H. 1954. Classification of the Chinese dragonflies of the family Gomphidae (Odonata). Part II. Acta Entomologica Sinica 4:23–84.

Chippindale, P.T., K.D. Varshal, D.H. Whitmore, and J.V. Robinson. 1999. Phylogenetic relationships of North American dragonflies of the genus *Ischnura* (Odonata: Zygoptera: Coenagrionidae) based on sequences of three mitochondrial genes. Molecular Phylogenetics and Evolution 11: 110–121.

Chivers, D.P., B.D. Wisenden, and R.J.F. Smith. 1996. Damselfly larvae learn to recognize predators from chemical cues in the predator's diet. Animal Behavior 52:315–320.

Chovanec, A., and R. Raab. 1997. Dragonflies (Insecta, Odonata) and the ecological status of newly created wetlands-examples for long-term bioindication programmes. Limnology 27:381–392.

Clark, W.H. 1979. *Neurocordulia xanthosoma* (Williamson) from New Mexico, A significant range extension (Anisoptera: Corduliidae). Notulae Odonatologica 1:72.

Claus-Walker, D.B., P.H. Crowley, and F. Johansson. 1997. Fish predation, cannibalism, and larval development in the dragonfly *Epitheca cynosura*. Canadian Journal of Zoology 75:687–696.

Convey, P. 1990. Influences of the choice between territorial and satellite behaviour in male *Libellula quadrimaculata* Linn. (Odonata: Libellulidae). Behavior 106:125–141.

Cook, C., and J.J. Daigle. 1985. *Ophiogomphus westfalli* spec. nov. from the Ozark region of Arkansas and Missouri, with a key to the *Ophiogomphus* species of eastern North America (Anisoptera: Gomphidae). Odonatologica 14:89–99.

Cope, E.D. 1880. The zoological position of Texas. Bulletin of the U.S. National Museum 17:1–51.

Corbet, P.S. 1963. A biology of dragonflies. Quadrangle Books, Chicago, 247pp.

Cordero, A. 1990. The inheritance of female polymorphism in the damselfly *Ischnura graellsii* (Rambur) (Odonata: Coenagrionidae). Heredity 64:341–346.

Cordero, A., and J.A. Andres. 1996. Colour polymorphism in odonates: females that mimic males? Journal of the British Dragonfly Society 12:50–60.

Currie, N. 1963. Mating behavior and local dispersal in *Erythemis simplicicollis*. Proceedings of the North Central Branch of the Entomological Society of America 18:112–115.

Daigle, J.J. 1994. The larva and adult male of *Somatochlora georgiana* Walker (Odonata: Corduliidae). Bulletin of American Odonatology 2:21–26.

Davis, W.T. 1898. Preliminary list of the dragonflies of Staten Island, with notes and dates of capture. Journal of the New York Entomological Society 6:195–198.

———. 1929. Notes on the dragonflies of the genus *Neurocordulia*. Bulletin of the Brooklyn Entomological Society 37: 449–450.

———. 1933. Dragonflies of the genus *Tetragoneuria*. Bulletin of the Brooklyn Entomological Society 28:87–104.

DeBano, S.J. 1993. Territoriality in the dragonfly *Libellula saturata*: mutual avoidance or resource defense? Odonatologica 22:431–441.

———. 1996. Male mate searching and female availability in the dragonfly *Libellula saturata*: Relationships in time and space. Southwestern Naturalist 41:293–298.

DeMarmels, J. 1984. The genus *Nehalennia* Selys, its species and their phylogenetic relationships (Zygoptera: Coenagrionidae). Odonatologica 13:501–527.

———. 1988. Odonata del Estado Tachira. Revista Cientifica Unet 2:91–111.

———. 2002. A study of *Chromagrion* Needham, 1903, *Hesperagrion* Calvert 1902, and *Zoniagrion* Kennedy, 1917: three monotypic North American damselfly genera with uncertain generic relationships (Zygoptera: Coenagrionidae). Odonatolgica 31(2):139–150.

DeMarmels, J., and H. Schiess. 1977. Zum Vorkommen der Zwerglibelle *Nehalennia speciosa* (Charp. 1840) in der Schweiz (Odonata: Coenagriondae). Vierteljahresschrift der Naturforschenden Gesellschaft in Zürich 122:339–348.

Dice, L.R. 1943. The biotic provinces of North America. University of Michigan Press. Ann Arbor, 78 pp.

Dickerson, J.E., Jr., J.V. Robinson, J.T. Gilley, and J.D. Wagner. 1982. Intermale aggression distance of *Plathemis lydia* (Drury) (Odonata: Libellulidae). Southwestern Naturalist 27: 457–458.

Doerksen, G. 1979. Notes on mating and oviposition in *Enallagma cyathigerum* (Charpentier) (Odonata: Zygoptera: Coenagrionidae). Abst. Pap. 5th Int. Symp. Odonatologica Montreal, p. 12.

Dolny, A., and J. Asmera. 1989. Prispevek k ekologickemu hodnoceni vazek. Studia. Oecologica 2:9–15.

Donnelly, T.W. 1961. The Odonata of Washington, D.C. and vicinity. Proceedings of the Entomological Society of Washington 63:1–13.

———. 1963. Possible phylogenetic relationships among North and Central American *Enallagma*. Proceedings of the North Central Branch of the Entomological Society of America 18:116–119.

———. 1964. *Enallagma westfalli*, a new damselfly from eastern Texas, with remarks on the genus *Teleallagma* Kennedy. Proceedings of the Entomological Society of Washington 66: 103–109.

———. 1968. A new species of *Enallagma* from Central America (Odonata: Coenagrionidae). Florida Entomologist 51: 101–105.

———. 1973. The status of *Enallagma traviatum* and *westfalli* (Odonata: Coenagrionidae). Proceedings of the Entomological Society of Washington 75:297–302.

———. 1978. Odonata of the Sam Houston National Forest and vicinity, east Texas, United States, 1960–1966. Notulae Odonatologica 1:6–7.

———. 1984. A new species of *Macrothemis* from Central America with notes on the distinction between *Brechmorhoga* and *Macrothemis*. Florida Entomologist 67:169–174.

———. 1989. *Protoneura sulfurata*, a new species of damselfly from Costa Rica, with notes on the circum-Caribbean species of the genus (Odonata: Protoneuridae). Florida Entomologist 72:436–441.

———. 1992. Taxonomic problems (?) with *Tetragoneuria*. Argia 4:11–12.

———. 1993. Cannibalism in *Anax junius!* Argia 5:15.

———. 1995. *Orthemis ferruginea*—An adventure in Caribbean biogeography. Argia 7:9–12.

Donnelly, T.W., and K.J. Tennessen. 1994. *Macromia illinoiensis* and *georgina*: A study of their variation and apparent subspecific relationship (Odonata: Corduliidae). Bulletin of American Odonatology 2:27–61.

Dosdall, L.M., and D.W. Parker. 1998. First report of a symphoretic association between *Nanocladius branchicolus* Saether (Diptera: Chironomidae) and *Argia moesta* (Hagen) (Odonata: Coenagrionidae). American Midland Naturalist 139:181-185.

Dunkle, S.W. 1978. Notes on adult behavior and emergence of *Paltothemis lineatipes* Karsch, 1890 (Anisoptera: Libellulidae). Odonatologica 7:277–279.

———. 1979. Ocular mating marks in female nearctic Aeshnidae (Anisoptera). Odonatologica 8:123–127.

———. 1981. The ecology and behavior of *Tachopteryx thoreyi* (Hagen) (Anisoptera: Petaluridae). Odonatologica 10:189–199.

———. 1983. Polychromatism in female Aeshnidae. Selysia 12:3–4.

———. 1984a. Head damage due to mating in *Ophiogomphus* dragonflies (Anisoptera: Gomphidae). Notulae Odonatologica 2:63–64.

———. 1984b. Novel features of reproduction in the dragonfly genus *Progomphus* (Anisoptera: Gomphidae). Odonatologica 13:477–480.

———. 1985a. Larval growth in *Nasiaeschna pentacantha* (Rambur) (Anisoptera: Aeshnidae). Odonatologica 14:29–35.

———. The taxonomy of the *Libellula vibrans* group (Odonata: Libellulidae). Transactions of the American Entomological Society 111:399–405.

———. 1988. Fusion between cerci and abdomen in male gomphini dragonflies (Anisoptera: Gomphidae). Odonatologica 17:55–56.

———. 1989. Dragonflies of the Florida peninsula, Bermuda and the Bahamas. Sci. Publ. Natur. Guide No. 1. Gainesville, Florida, 154 pp.

———. 1990. Damselflies of Florida, Bermuda and the Bahamas. Scientific Publishers Nature Guide No. 3. Gainesville, Florida, 148 pp.

———. 1992a. Distribution of dragonflies and damselflies (Odonata) in Florida. Bulletin of American Odonatology 2:29–50.

———. 1992b. *Gomphus (Gomphurus) gonzalezi* spec. nov., a new dragonfly from Texas and Mexico (Anisoptera: Gomphidae). Odonatologica 21:79–84.

———. 1995. Conservation of dragonflies (Odonata) and their habitats in North America. Proceedings of the International Symposium on the Conservation of Dragonflies and their Habitats. P.S. Corbet, S.W. Dunkle, and H. Ubukata (eds.), pp. 23–27. Japanese Society for Preservation of Birds, Kushiro, 70 pp.

———. 1998. Another *Orthemis discolor* record from Texas. Argia 10:7.

———. 2000. Dragonflies through binoculars: A field guide to dragonflies of North America. Oxford University Press: New York, 266 pp.

Dunson, W.A. 1980. Adaptations of nymphs of a marine dragonfly, *Erythrodiplax berenice*, to wide variations in salinity. Physiological Zoology 53:445–452.

Erickson, C.J. 1989. Interactions between the dragonfly *Hagenius brevistylus* Selys and the damselfly *Calopteryx maculata* (P. de Beauv.) (Anisoptera: Gomphidae; Zygoptera: Calopterygidae). Notulae Odonatologica 3:59–60.

Erickson, C.J., and M.E. Reid. 1989. Wingclapping behavior in *Calopteryx maculata* (P. de Beauvois) (Zygoptera: Calopterygidae). Odonatologica 18:379–383.

Eriksen, C.H. 1960. The oviposition of *Enallagma exsulans* (Odonata: Agrionidae). Annals of the Entomological Society of America 53:439.

Evans, M.A. 1995. Checklist of the Odonata of New Mexico with additions to the Colorado checklist. Proceedings of the Denver Museum of Natural History 3:1–6.

Ferguson, A. 1940. A preliminary list of the Odonata of Dallas County, Texas. Field and Laboratory 8:1–10.

Fincke, O.M. 1987. Female monogamy in the damselfly *Ischnura verticalis* Say (Zygoptera: Coenagrionidae). Odonatologica 16:129–143.

———. 1994. Female colour polymorphism in damselflies: failure to reject the null hypothesis. Animal Behavior 47:1249–1266.

Fisher, E.G. 1940. A list of Maryland Odonata. Entomological News 51:37–42, 67–72.

Fitzhugh, G.H., and J.H. Marden. 1997. Maturational changes in troponin T expression, Ca^{2+}-sensitivity and twitch contraction kinetics in dragonfly flight muscle. Journal of Experimental Biology 200:1473–1482.

Flint, O.S. 2000. *Nehalennia pallidula* in Texas! Argia 12:3–4.

Forsyth, A., and R.D. Montgomerie. 1987. Alternative reproductive tactics in the territorial damselfly *Calopteryx maculata*: sneaking by older males. Behavioral Ecology and Sociobiology 21:73–81.

Foster, E. 1914. Odonata observed at Mound Louisiana. Unpublished manuscript, Tulane University.

Fried, C.S., and M.L. May. 1983. Energy expenditure and food intake of territorial male *Pachydiplax longipennis* (Odonata: Libellulidae). Ecological Entomology 8:283–292.

Frost, S.W. 1971. *Pachydiplax longipennis* (Odonata: Anisoptera): Records of night activity. Florida Entomologist 54:205.

Furtado, J.I. 1973. Annotated records of some dragonflies (Odonata) from Ontario. Canadian Field Naturalist 87:463–466.

Garcia-Diaz, J. 1938. An ecological survey of the freshwater insects of Puerto Rico. I. The Odonata: with new life histories. Journal of Agriculture, University of Puerto Rico 22:43–97.

Garman, P. 1917. The Zygoptera, or damselflies, of Illinois. Bulletin of the Illinois State Laboratory Natural History Survey Bulletin 39:411–587.

——. 1932. The genus *Archilestes* in Kentucky. Entomological News 43:85–92.

Garrison, R.W. 1976. Multivariate analysis of geographic variation in *Libellula luctuosa* Burmeister (Odonata: Libellulidae). Pan Pacific Entomologist 52:181–203.

——. 1978. A mark-recapture study of imaginal *Enallagma cyathigerum* (Charpentier) and *Argia vivida* Hagen (Zygoptera: Coenagrionidae). Odonatologica 7:223–236.

——. 1982. *Paltothemis cyanosoma*, a new species of dragonfly from Mexico (Odonata: Libellulidae). Pan Pacific Entomologist 58:135–138.

——. 1984. Revision of the genus *Enallagma* in the United States west of the Rocky Mountains, and identification of certain larvae by discriminant analysis. University of California Publications in Entomology 105:1–129.

——. 1986. The genus *Aphylla* in Mexico and Central America, with a description of a new species, *Aphylla angustifolia* (Odonata: Gomphidae). Annals of the Entomological Society of America 79:938–944.

——. 1990. A synopsis of the genus *Hetaerina* with descriptions of four new species (Odonata: Calopterygidae). Transactions of the American Entomological Society 116:175–259.

——. 1994a. A synopsis of the Genus *Argia* of the United States with keys and descriptions of new species, *Argia sabino, A. leonorae,* and *A. pima* (Odonata: Coenagrionidae). Transactions of the American Entomological Society 120:287–368.

——. 1994b. A revision of the new world genus *Erpetogomphus* Hagen *in* Selys (Odonata: Gomphidae). Tijdschrift voor Entomologie 137:173–269.

——. 1997. Odonata. *In* Nomina Insecta Nearctica. Vol. 4. Non-holometabolous Orders. Entomological Information Service, Rockville, Md., pp. 551–579.

——. 1999. The genus *Neoneura*, with keys and description of a new species, *Neoneura jurzitzai* spec. nov. (Zygoptera: Protoneuridae). Odonatologica 28:343–375.

Glas, G., and M. Verdonk. 1972. Zomerkamp drenthe. Libellen. Jaarboek. C.J.N. 1971, pp. 151–158.

Gloyd, L.K. 1932. Four new dragonfly records for the United States. Entomological News 43:189–190.

——. 1940. On the status of *Gomphaeschna antilope* (Hagen) (Odonata). Occasional Papers of the Museum of Zoology, University of Michigan 415:1–14.

——. 1951. Records of some Virginia Odonata. Entomological News 62:109–114.

——. 1958. The dragonfly fauna of the Big Bend region of Trans-Pecos, Texas. Occasional Papers of the Museum of Zoology, University of Michigan 593:1–23.

——. 1959. Elevation of the *Macromia* group to family status. Entomological News 70:197–205.

——. 1968. Union of *Argia fumipennis* and *A. violacea* and the recognition of 3 subspecies. Occasional Papers of the Museum of Zoology, University of Michigan 658:1–6.

——. 1980. The taxonomic status of the genera *Superlestes* and *Cyptolestes* Williamson 1921 (Odonata: Lestidae). Occasional Papers of the Museum of Zoology, University of Michigan 694:1–3.

Gloyd, L.K., and M. Wright. 1959. Odonata. Chapters 34 (pp. 917–940) *In:* H.B. Ward and G.C. Whipple, Freshwater Biology (W.T. Edmondson ed.) second edition. Wiley, New York.

Gower, J.L., and E.J. Kormondy. 1963. Life history of the damselfly *Lestes rectangularis* with special reference to seasonal regulation. Ecology 44:398–402.

Grether, G.F. 1995. Natural and sexual selection on wing coloration in the rubyspot damselfly *Hetaerina americana*. Ph.D. dissertation, University of California, Davis.

——. 1996a. Sexual selection and survival selection on wing coloration and body size in the rubyspot damselfly *Hetaerina americana*. Evolution 50:1939–1948.

——. 1996b. Intrasexual competition alone favors a sexually dimorphic ornament in the rubyspot damselfly *Hetaerina americana*. Evolution 50:1949–1957.

Grether, G.F., and R.M. Grey. 1996. Novel cost of a sexually selected trait in the rubyspot damselfly *Hetaerina americana*: conspicuous to prey. Behavioral Ecology 7:465–473.

Grieve, E.G. 1937. Studies on the biology of the damselfly *Ischnura verticalis* Say, with notes on certain parasites. Entomologica Americana 17:121–153.

Hagen, H.G. 1875. Synopsis of the Odonata of America. Proceedings of the Boston Society of Natural History 18:20–96.

Hardy, H.T., Jr., 1966. The effect of sunlight and temperature on the posture of *Perithemis tenera* (Odonata). Proceedings of the Oklahoma Academy of Science 46:41–45.

Harp, G.L. 1983. New and unusual records of Arkansas Anisoptera. United States. Notulaca Odonatologica 2:26–27.

——. 1985. Further distributional records for Arkansas Anisoptera. Proceedings of the Arkansas Academy of Science 39:131–135.

——. 1986. Protracted oviposition by *Hetaerina titia* (Drury) (Zygoptera: Calopterygidae). Notulae Odonatologica 2:132–133.

Harp, G.L., and P.A. Harp. 1996. Previously unpublished Odonata records for Arkansas, Kentucky and Texas. Notulae Odonatologica 4:127–130.

Harp, G.L., and J.D. Rickett. 1977. The dragonflies (Anisoptera) of Arkansas. Proceedings of the Arkansas Academy of Science 31:50–54.

Harrison, J.F., and J.R.B. Lighton. 1998. Oxygen-sensitive flight metabolism in the dragonfly *Erythemis simplicicollis*. Journal of Experimental Biology 201:1739–1744.

Harvey, I.F., and S.F. Hubbard. 1987. Observations on the reproductive behavior of *Orthemis ferruginea* (Fabricius) (Anisoptera: Libellulidae). Odonatologica 16:1–8.

Harwell, J.E. 1951. Notes on the Odonata of northeastern Texas. Texas Journal of Science 3:204–207.

Hilton, D.F.J. 1989. Incidence of androchromotypic female *Ischnura ramburi* (Odonata: Coenagrionidae) in the Hawaiian Islands. Entomological News 100:147–149.

Hinnekint, B.O.N. 1987. Population dynamics of *Ischnura elegans* (Vander Linden) (Insecta: Odonata) with special reference to morphological colour changes, female polymorphism, multiannual cycles and their influence on behaviour. Hydrobiologica 146:3–31.

Holland, C.W. 1944. Physiographic divisions of the Quarternary lowlands of Lousiana. Proceedings of the Louisiana Academy of Science 8:11–24.

Hornuff, L.E. 1951. A further study of *Aphylla williamsoni* Gloyd (Odonata, Aeschnidae). Proceedings of the Lousiana Academy of Science 14:39–44.

——. Functional morphology of the external genitalia of Nearctic damselflies (Odonata: Zygoptera). Thesis, University of Oklahoma, 34 pp.

Howe, R.H., Jr. 1917. Manual of the Odonata of New England. Memoirs of the Thoreau Museum of Natural History, parts I & II 2:1–138

Ingram, B.R. 1976. Life histories of three species of Lestidae in North Carolina, United States (Zygoptera). Odonatologica 5:231–244.

Ingram, B.R., and C.E. Jenner. 1976. Life histories of *Enallagma hageni* (Walsh) and *Enallagma aspersum* (Hagen) (Zygoptera: Coenagrionidae). Odonatologica 5:331–245.

Jacobs, M.E. 1955. Studies on territorialism and sexual selection in dragonflies. Ecology 36:566–586.

Johnson, C. 1961. Breeding behavior and oviposition in *Hetaerina americana* (Fabricius) and *H. titia* (Drury) (Odonata: Agriidae). Canadian Entomologist 93:260–266.

——. 1962a. Breeding behavior and oviposition in *Calopteryx maculatum* (Beauvois) (Odonata: Calopterygidae). American Midland Naturalist 68:242–247.

——. 1962b. A description of territorial behavior and a quantitative study of its function in males of *Hetaerina americana* (Fabricius) (Odonata: Agriidae). Canadian Entomologist 94:178–190.

——. 1962c. A study of territoriality and breeding behavior in *Pachydiplax longipennis* Burmeister (Odonata: Libellulidae). Southwestern Naturalist 7:191–199.

——. 1963. Interspecific territoriality in *Hetaerina americana* (Fabricius) and *H. titia* (Drury) (Odonata: Calopterygidae) with a preliminary analysis of the wing color pattern variation. Canadian Entomologist 95:575–582.

——. 1964a. Polymorphism in the damselflies *Enallagma civile* (Hagen) and *E. praevarum* (Hagen). American Midland Naturalist 72:408–416.

——. 1964b. Mating expectancies and sex ratio in the damselfly *Enallagma praevarum* (Odonata: Coenagrionidae). Southwestern Naturalist 9:297–304.

——.1964c. The inheritance of female dimorphism in the damselfly *Ischnura damula*. Genetics 49:513–519.

——. 1964d. Seasonal ecology of *Ischnura damula* Calvert (Odonata: Coenagrionidae). Texas Journal of Science 16:50–61.

——. 1965. Mating and oviposition of damselflies in the laboratory. Canadian Entomologist 97:321–326.

——. 1966a. Environmental modification of the behavior for habitat selection in adult damselflies. Ecology 47:674–676.

——. 1966b. Genetics of female dimorphism in *Ischnura demorsa*. Heredity 21:453–459.

——. 1966c. Improvements for colonizing damselflies in the laboratory. Texas Journal of Science 18:179–183.

——. 1968. Seasonal ecology of the dragonfly *Oplonaeschna armata* Hagen (Odonata: Aeshnidae). American Midland Naturalist 80:449–457.

——. 1972a. The damselflies (Zygoptera) of Texas. Bulletin of the Florida State Museum, Biological Sciences 16:55–128.

——. 1972b. An analysis of geographic variation in the damselfly *Argia apicalis* (Zygoptera: Coenagrionidae). Canadian Entomologist 104:1515–1527.

——. 1973a. Variability, distribution and taxonomy of *Calopteryx dimidiata* (Zygoptera: Calopterygidae). Florida Entomologist 56:207–222.

——. 1973b. Distributional patterns and their interpretation in *Hetaerina* (Odonata: Caloptyerygidae). Florida Entomologist 56:24–42.

——. 1973c. Ovarian development and age recognition in the damselfly *Argia moesta* (Hagen 1861) (Zygoptera: Coenagrionidae). Southwestern Naturalist 9:297–304.

——. 1974. Taxonomic keys and distributional patterns for nearctic species of *Calopteryx* damselflies. Florida Entomologist 57:231–248.

——. 1975a. Variability in the damselfly *Lestes sigma* Calvert (Zygoptera: Lestidae). Texas Journal of Science 26:165–169.

——. Johnson, C. 1975b. Polymorphism and natural selection in ischnuran damselflies. Evolutionary Theory 1:81–90.

Johnson, C., and M.J. Westfall, Jr. 1970. Diagnostic keys and notes on the damselflies (Zygoptera) of Florida. Bulletin of the Florida State Museum, Biological Sciences 15:45–89.

Johnson, D.M. 1986. The life history of *Tetragoneura cynosura* (Say) in Bays Mountain Lake, Tennessee, United States (Anisoptera: Corduliidae). Odonatologica 15:81–90.

Johnson, D.M., R.E. Bohanan, C.N. Watson, and T.H. Martin. 1984. Coexistence of *Enallagma divagans* and *Enallagma traviatum* (Zygoptera: Coenagrionidae) in Bays Mountain Lake: an in situ enclosure experiment. Advances in Odonatology 2:57–70.

Johnson, D.M., and P.H. Crowley. 1980. Habitat and seasonal segregation among coexisting odonate larvae. Odonatologica 9:297–308.

Johnson, D.M., P.H. Crowley, R.E. Bhohanan, C.N. Watson, and T.H. Martin. 1985. Competition among larval dragonflies: A field exclosure experiment. Ecology 66:119–128.

Johnson, J., and D.R. Paulson. 1998. *Enallagma civile* recorded in Oregon. Argia 10:22–23.

Johnson, J.H. 1982. Diet composition and prey selection of *Cordulegaster maculata* Sel. larvae (Anisoptera: Cordulegastridae). Notulae Odonatogica 1:151–153.

Jordan, F., and A.C. McCreary. 1996. Effects of an odonate predator and habitat complexity on survival of the flagfish *Jordanella floridae*. Wetlands 16:583–586.

Juritza, G. 1978. Unsere Libellen: die Libellen Europas in 120 farbfotos. Bunte Kosmos-Taschenfuhrer. Franchkh, Stuttgart.

Kellicott, D.S. 1890. Dragonflies congregating at night. Entomological News 1:146.

——. 1899. The Odonata of Ohio. Bulletin of Ohio State University (IV) (5):1–116.

Kennedy, C.H. 1915a. Interesting western Odonata. Annals of the Entomological Society of America 8:297–303.

——. 1915b. Notes on the life history and ecology of the dragonflies (Odonata) of Washington and Oregon. Proceedings of the U.S. National Museum 49:259–345.

——. 1917. Notes on the life history and ecology of the dragonflies (Odonata) of central California and Nevada. Proceedings of the U.S. National Museum 52:483–635.

——. 1919. A new species of *Argia* (Odonata). Canadian Entomologist 51:17.

——. 1921. Some interesting dragon-fly naiads from Texas. Proceedings of the U.S. National Museum 59:595–598.

——. 1922a. The morphology of the penis in the genus *Libellula* (Odonata). Entomological News 33:33–40.

——. 1922b. The phylogeny and the geographical distribution of the genus *Libellula* (Odonata). Entomological News 33:65–71, 105–111.

——. 1923a. The phylogeny and the distribution of the genus *Erythemis* (Odonata). Miscellaneous Publications of the Museum of Zoology, University of Michigan 11:19–22.

——. 1936. The habits and early stages of the dragonfly *Gomphaeschna furcillata* (Say). Proceedings of the Indiana Academy of Science 45:315–322.

Kielb, M.A., and M.F. O'Brien. 1996. Discovery of an isolated population of *Anax longipes* in Michigan (Odonata: Aeshnidae). Great Lakes Entomologist 29:161–164.

Kielb, M.A., E. Bright, and M.F. O'Brien. 1996. Range extension of *Stylogomphus albistylus* (Odonata: Gomphidae) for the upper peninsula of Michigan. Great Lakes Entomologist 29: 87–88.

Koen, J. 1937. The dragonflies of Reelfoot Lake area. Journal of the Tennessee Academy of Science 12:129–153.

Koenig, W.D. 1990. Territory size and duration in the white-tailed skimmer *Plathemis lydia* (Odonata: Libellulidae). Journal of Animal Ecology 59:317–333.

——. 1991. Levels of female choice in the white-tailed skimmer *Plathemis lydia* (Odonata: Libellulidae). Behavior 119: 193–224.

Koenig, W.D., and S.S. Albano. 1985. Patterns of territoriality and mating success in the white-tailed skimmer *Plathemis lydia* (Odonata: Anisoptera). American Midland Naturalist 114:1–12.

——. 1987a. Breeding site fidelity in *Plathemis lydia* Drury (Anisoptera: Libellulidae). Odonatologica 16:249–259.

——. 1987b. Lifetime reproductive success, selection, and the opportunity for selection in the white-tailed skimmer *Plathemis lydia* (Odonata: Libellulidae). Evolution 41: 22–36.

Kormondy, E.J. 1958. Catalogue of the Odonata of Michigan. Miscellaneous Publications of the Museum of Zoology, University of Michigan 104:1–43.

——. 1959. The systematics of *Tetragoneuria*, based on ecological, life history, and morphological evidence. Miscellaneous Publications of the Museum of Zoology, University of Michigan 107:1–79.

——. 1960. New North American records of anisopterous Odonata. Entomological News 71:121–130.

Kriegsman, C.O., and P.E. Lutz. 1965. Life-history of *Anax junius* Drury (Odonata). ASB Bulletin 12:48.

LaRivers, I. 1938. An annotated list of the Libelluloidea (Odonata) of southern Nevada. Pomona Coll. Journal of Entomology and Zoology 30:73–85.

——. 1946. Some dragonfly observations in alkaline areas in Nevada. Entomological News 57:209–217.

Lebeuf, L., And J.G. Pilon. 1977. Cycle biologique de *Enallagma boreale* Selys (Odonata: Coenagrionidae) en milieu conditionne. Annales de Société Entomol. du Québec 22:78–118.

Lebuis, M.A., and J.G. Pilon. 1976. Analyse comparative de la faune odonatologique de quatre milieux de la région de Saint-Hippolyte, Comte de Prevost. Annales de la Société Entomologique du Québec 21:3–25.

Leonard, J.W. 1940. *Lanthus albistylus* (Hagen), a new record for Michigan, with ecological notes on the species (Odonata: Gomphinae). Occasional Papers of the Museum of Zoology, University of Michigan 414:1–6.

Leong, J.M., and J.E. Hafernik. 1992a. Hybridization between two damselfly species (Odonata: Coenagrionidae): Morphometric and genitalic differentiation. Annals of the Entomological Society of America 85:662–670.

——. 1992b. Seasonal variation in allopatric populations of *Ischnura denticollis* (Burmeister) and *Ischnura gemina* (Kennedy) (Odonata: Coenagrionidae). Pan Pacific Entomologist 68:268–278.

Logan, E.R. 1971. A comparative ecological and behavioral study of two species of damselflies, *Enallagma boreale* (Selys) and *Enallagma carunculatum* Morse (Odonata: Coenagrionidae). Washington State University, Pullman, Ph.D. dissertation, 82 pp.

Lohmann, H. 1992. Revision der Cordulegastridae. 1. Entwurf einer neuen klassifizierung der familie (Odonata: Anisoptera). Opuscula Zoologica Fluminensia 96:1–18.

Lombardo, P. 1997. Predation by *Enallagma* nymphs (Odonata, Zygoptera) under different conditions of spatial heterogeneity. Hydrobiologia 356:1–9.

Lounibos, L.P., R.L. Escher, L.B. Dewald, N. Nishimura, and V.L. Larson. 1990. Odonata associated with water lettuce (*Pistia stratioides* L.) in south Florida. Odonatologica 19:359–366.

Louton, J.A. 1982. Lotic dragonfly (Anisoptera: Odonata) nymphs of the southeatern United States: Identification, distribution and historical biogeography. University of Tennessee, Ph.D. dissertation, 357 pp.

Lutz, P.E., and A.R. Pittman. 1970. Some ecological factors influencing a community of adult Odonata. Ecology 51: 279–284.

Macan, T.T. 1974. Twenty generations of *Pyrrhosoma nymphula* (Sulzer) and *Enallagma cyathigerum* (Charpentier) (Zygoptera: Coenagrionidae). Odonatologica 3:107–119.

Mackinnon, B.I., and M.L. May. 1994. Mating habitat choice

and reproductive success of *Pachydiplax longipennis* (Burmeister) (Anisoptera: Libellulidae). Advances in Odonatology 6:59–77.

Mahato, M., and D.M. Johnson. 1991. Invasion of the Bays Mountain Lake dragonfly assemblage by *Dromogomphus spinosus* (Odonata: Gomphidae). Journal of the North American Benthological Society 10:165–176.

Marden, J.H. 1995. Large-scale changes in thermal sensitivity of flight performance during adult maturation in a dragonfly. Journal of Experimental Biology 198:2095–2102.

Martin, R.D.C. 1939. Life histories of *Agrion aequabile* und *Agrion maculatum* (Agriidae). Annals of the Entomological Society of America 32(3): 601–619.

Mauffray, B. 1997. The dragonflies and damselflies (Odonata) of Louisiana. Bulletin of the American Odonatology 5:1–26.

May, M.L. 1977. Thermoregulation and reproductive activity in tropical dragonflies of the genus *Micrathyria*. Ecology 58: 787–798.

———. 1980. Temporal activity patterns of *Micrathyria* in Central America (Anisoptera: Libellulidae). Odonatologica 9: 57–74.

———. 1984. Energetics of adult Anisoptera, with special reference to feeding and reproductive behavior. Advances in Odonatology 2:95–116.

———. 1992. Morphological and ecological differences among species of *Ladona* (Anisoptera: Libellulidae). Bulletin of American Odonatology 1:51–56.

———. 1995a. Dependence of flight behavior and heat production on air temperature in the green darner dragonfly *Anax junius* (Odonata: Aeshnidae). Journal of Experimental Biology 198:2385–2392.

———. 1995b. Simultaneous control of head and thoracic temperature by the green darner dragonfly *Anax junius* (Odonata: Aeshnidae). Journal of Experimental Biology 198: 2373–2384.

———. 1995c. A preliminary phylogenetic analysis of the "Corduliidae." Abstr. Pap. 13th Int. Symp. Odonatologica, Essen, p. 36.

———. 1995d. The subgenus *Tetragoneuria* (Anisoptera: Corduliidae: *Epitheca*) in New Jersey. Bulletin of American Odonatology 2:63–74.

———. 1997. Reconsideration of the status of the genera *Phyllomacromia* and *Macromia* (Anisoptera: Corduliidae). Odonatologica 26:405–414.

———. 1998a. *Macrothemis fallax*, a new species of dragonfly from Central America (Anisoptera: Libellulidae), with a key to male *Macrothemis*. International Journal of Odonatology 1:137–153.

———. 1998b. Body temperature regulation in a late-season dragonfly, *Sympetrum vicinum* (Odonata: Libellulidae). International Journal of Odonatology 1:1–13.

———. 2002. Phylogeny and taxonomy of the damselfly genus *Enallagma* and related taxa (Odonata: Zygoptera: Coenagrionidae). Systematic Entomology 27:387–408.

May, M.L., and F.L. Carle. 1996. An annotated list of the Odonata of New Jersey with an appendix on nomenclature in

the genus *Gomphus*. Bulletin of American Odonatology 4: 1–35.

McCafferty, W.P. 1979. Swarm feeding by the damselfly *Hetaerina americana* (Odonata: Calopterygidae) on mayfly hatches. Aquatic Insects 1:149–151.

McLachlan, R. 1896. Oceanic migration of a nearly cosmopolitan dragon-fly (*Pantala flavescens*, F). Entomologists Monthly Magazine 32:254.

McMillan, V. 1984. Dragonfly monopoly. Natural History 93: 32–39.

McPeek, M.A. 1989. Differential dispersal tendencies among *Enallagma* damselflies (Odonata) inhabiting different habitats. Oikos 56:187–195.

———. 1990a. Determination of species composition in the *Enallagma* damselfly assemblages of permanent lakes. Ecology 71:83–98.

———. 1990b. Behavioral differences between *Enallagma* species (Odonata) influencing differential vulnerability to predators. Ecology 71:1714–1726.

———. 1995. Morphological evolution mediated by behavior in the damselflies of two communities. Evolution 49:749–769.

———. 1997. Measuring phenotypic selection on an adaptation: Lamellae of damselflies experiencing dragonfly predation. Evolution 51:459–466.

———. The consequences of changing the top predator in a food web: a comparative experimental approach. Ecological Monographs 68:1–23.

McPeek, M.A., and J.M. Brown. 2000. Building a regional species pool: diversification of the *Enallagma* damselflies in eastern North America. Ecology 81:904–920.

McPeek, M.A., K. Schrot, and J.M. Brown. 1996. Adaptation to predators in a new community: swimming performance and predator avoidance in damselflies. Ecology 77:617–629.

McVey, M.E. 1981. Lifetime reproductive tactics in a territorial dragonfly *Erythemis simplicicollis*. Rockefeller University, New York, Ph.d. Dissertation, 410 pp.

———. 1985. Rates of color maturation in relation to age, diet, and temperature in male *Erythemis simplicicollis* (Say) (Anisoptera: Libellulidae). Odonatologica 14:101–114.

———. 1988. The opportunity for sexual selection in a territorial dragonfly, *Erythemis simplicicollis* In T.H. Clutton-Brock, ed., Reproductive Success. University of Chicago Press, Chicago.

McVey, M.E., and B.J. Smittle. 1984. Sperm precedence in the dragonfly *Erythemis simplicicollis*. Journal of Insect Physiology 20:619–628.

Mesterton-Gibbons, M., J.H. Marden, and L.A. Dugatkinss. 1996. On wars of attrition without assessment. Entomol. Journal of Theoretical Biology 181:65–83.

Meulenbrock, J.L. 1972. Nogmaals het gedrag van *Enallagma cyathigerum* (Charp.) (Odonata). Entomol. Ber. Amsterdam 32:69.

Miller, P.L. 1982. Observations on the reproductive behavior of *Celithemis eponina* Drury (Libellulidae, Odonata) in Florida. Entomologists Monthly Magazine 117:209–212.

Mitchell, R. 1962. Storm-induced dispersal in the damselfly *Is-*

chnura verticalis (Say). American Midland Naturalist 68: 199–202.

Molner, D.R., and R.J. Lavigne. 1979. Odonata of Wyoming. University of Wyoming Agricultural Experiment Station SM-37, 142 pp.

Montgomery, B.E. 1925. Records of Indiana dragonflies, I. Proceedings of the Indiana Academy of Science 34:383–389.

———. 1933. Notes on New Jersey Dragonflies. Entomological News 44:40–44.

———. 1937. Records of Indiana dragonflies, IX. 1935–1936. Proceedings of the Indiana Academy of Science 46:203–210.

———. 1942. The distribution and relative seasonal abundance of the Indiana species of *Enallagma*. Proceedings of the Indiana Academy of Science 51:273–278.

———. 1943. *Sympetrum internum*, new name for *Sympetrum decisum* auct., nec Hagen (Odonata: Libellulidae). Canadian Entomologist 75:57–58.

———. 1966. Distribution of Odonata. *In:* Chandler, L., The origin and composition of the insect fauna, Chapt. 20, Natural Features of Indiana, Indiana Academy of Science, pp. 348–349.

Moore, A.J. 1987. Behavioral ecology of *Libellula luctuosa* Burmeister (Anisoptera: Libellulidae). 2. Proposed functions for territorial behaviors. Odonatologica 16:385–391.

———. 1989. The behavioral ecology of *Libellula luctuosa* (Burmeister) (Odonata: Libellulidae): III. male density, OSR, and male and female mating behavior. Ethology 80:120–136.

———. 1990. The evolution of sexual dimorphism by sexual selection: the separate effects of intrasexual selection and intersexual selection. Evolution 44:315–331.

Morgan, A.H. 1930. Field book of ponds and streams. Putnam, New York, 448pp.

Morse, A.P. 1895. New North American Odonata. Psyche 7: 207–211.

Moskowitz, D.P., and D.M. Bell. 1998. *Archilestes grandis* (Great Spreadwing) in central New Jersey, with notes on water quality. Bulletin of American Odonatology 5:49–54.

Moss, S.P. 1992. Oviposition site selection in *Enallagma civile* (Hagen) and the consequences of aggregating behavior (Zygoptera: Coenagrionidae). Odonatologica 21:153–164.

Mulhern, R.C. 1971. Dragonflies of northeast Louisiana. Northeast Louisiana University, Monroe, Masters thesis, 69 pp.

Muttkowski, R.A. 1910. Catalogue of the Odonata of North America. Bulletin of the Public Museum, City of Milwaukee. 1:1-207.

———. 1911. Studies in *Tetragoneuria*. Bulletin of Wisconsin Natural History Society 9:91–134.

———. 1915. Studies in *Tetragoneuria*, II. Bulletin of the Wisconsin Natural History Society 13:49–61.

Needham, J.G. 1901. Aquatic insects in the Adirondacks. New York State Mus. Bull. 47, Odonata, pp. 429–540.

———. 1903. Life histories of Odonata sub-order Zygoptera. Part 3 in Aquatic Insects of New York State. New York State Mus. Bull. 68:218–279.

———. 1905. Two elusive dragonflies. Entomological News 16:3–6.

———. 1943. Life history notes on *Micrathyria* (Odonata). Annals of the Entomological Society of America 36:185–189.

———. 1946. Some dragonflies of early spring in south Florida. Florida Entomologist 28:42–47.

———. 1950. Three new species of North American dragonflies with notes on related species. Transactions of the American Entomological Society 76:1–12.

Needham, J.G., and C.A. Hart. 1901. The dragon-flies (Odonata) of Illinois. Part I. Petaluridae, Aeschnidae and Gomphidae. Bulletin of the Illinois State Laboratory, Natural History 6, 94 pp.

Needham, J.G., and H.B. Heywood. 1929. A handbook of the dragonflies of North America. Springfield, Illinois, Charles C. Thomas, 378 pp.

Needham, J.G., and M.J. Westfall, Jr. 1955. A manual of the dragonflies of North America (Anisoptera), including the Greater Antilles and the provinces of the Mexican border. Berkeley, University of California Press, 615 pp.

Needham, J.G., M.J. Wesfall, Jr., and M.L. May. 2000. Dragonflies of North America. Revised Edition. Gainesville, Florida: Scientific Publishers, 939pp.

Nikula, B. 1998. *Neoneura amelia*, new to the United States. Argia 9:11–12.

Novelo-G., R. 1981. Comportamiento sexual y territorial en *Orthemis ferruginea* (Fab.) (Odonata: Libellulidae). Thesis, Univ. Nac. Aut. Mex.

———. 1992. Biosystematics of the larvae of the genus *Argia* in Mexico (Zygoptera: Coenagrionidae). Odonatologica 13: 429–449.

Novelo-G., R., and E. Gonzales-S. 1984. Reproductive behavior in *Orthemis ferruginea* (Fab.) (Odonata: Libellulidae). Folia Entomologica Mexicana 59:11–24.

———. 1991. Odonata de la Reserva del la Biosfera de la Michila, Durango, Mexico, Parte II. Náyades. Folia Entomologica Mexicana 81:105–164.

O'Briant, P. 1972. A study of behavioral and reproductive patterns of adult *Lestes vigilax* Hagen (Odonata: Lestidae). Journal of the Elisha Mitchell Scientific Society 88: 196.

O'Donnell, S. 1996. Dragonflies (*Gynacantha nervosa* Rambur) avoid wasps (*Polybia aequatorialis* Zavattari and *Mischocyttarus* sp.) as prey. Journal of Insect Behavior 9:159–162.

Opler, P.A. 1971. Mass movement of *Tarnetrum corruptum* (Odonata: Libellulidae). Pan-Pacific Entomologist 47:223.

Orr, R.L. 1998. A bit of 1997 migratory *Anax junius* data from Maryland. Argia 10:13–14.

Ortenburger, A.I. 1928a. Plant collections representative of some typical communities of western Oklahoma. Proceedings of the Oklahoma Academy of Science 410:49–52.

———. 1928b. Plant collections representative of some typical communities of eastern Oklahoma. Proceedings of the Oklahoma Academy of Science 410:53–57.

Pajunen, V.I. 1966. Aggressive behavior and territoriality in a population of *Calopteryx virgo* L. (Odonata: Calopterygidae). Annales Zoologici Fennici 3:201–214.

Parr, M.J. 1976. Some aspects of the population of the damselfly *Enallagma cyathigerum* (Charpentier) (Zygoptera: Coenagrionidae). Odonatologica 5:45–57.

Parr, M.J., and M. Palmer. 1971. The sex ratios, mating frequencies and mating expectancies of three coenagrionids (Odonata: Zygoptera) in northern England. Entomologia Scandinavia 2:191–204.

Patrick, O.R., and P.E. Lutz. 1969. A life history study of the damselfly *Ischnura posita* (Hagen). Journal of the Elisha Mitchell Scientific Society 85:131.

Paulson, D.R. 1966. The dragonflies (Odonata: Anisoptera) of southern Florida. University of Miami, Ph.D. dissertation, 603 pp.

———. 1969. Oviposition in the tropical dragonfly genus *Micrathyria* (Odonata, Libellulidae). Tombo 12:12–16.

———. 1973. Temporal isolation in two species of dragonflies, *Epitheca sepia* (Gloyd, 1933) and *E. stella* (Williamson, 1911) (Anisoptera: Corduliidae). Odonatologica 2:115–119.

———. 1974. Reproductive isolation in damselflies. Systematic Zoology 23:40–49.

———. 1983. A new species of dragonfly, *Gomphus (Gomphurus) lynnae* spec. nov., from the Yakima River, Washington, with notes on the pruinosity in Gomphidae (Anisoptera). Odonatologica 12:59–70.

———. 1984. Odonata from the Yucatan penninsula, Mexico. Notulae Odonatologica 2:33–38.

———. 1997. The dragonflies of Washington. Bulletin of American Odonatology 4:75–90.

———. 1998a. Variation in head spines in female *Ophiogomphus*, with a possible example of reproductive character displacement (Anisoptera: Gomphidae). Bulletin of American Odonatology 5:55–58.

———. 1998b. The distribution and relative abundance of the sibling species *Orthemis ferruginea* (Fabricius, 1775) and *O. discolor* (Burmeister, 1839) in North and Middle America (Anisoptera: Libellulidae). International Journal of Odonatology 1:89–93.

———. 2002. Comments on the *Erythrodiplax connata* (Burmeister, 1839) group, with the elevation of *E. fusca* (Rambur, 1842), *E. minuscula* (Rambur, 1842), and *E. basifusca* (Calvert, 1895) to full species (Anisoptera: Libellulidae). Bulletin of American Odonatology 6:101–110.

Paulson, D.R., and R.A. Cannings. 1980. Distribution, natural history and relationships of *Ischnura erratica* Calvert (Zygoptera: Coenagrionidae). Odonatologica 9:147–153.

Paulson, D.R., and S.W. Dunkle. 1996. Common names of North American dragonflies and damselflies, adopted by the Dragonfly Society of the Americas. Argia 8(1).

Paulson, D.R., and R.W. Garrison. 1977. A list and new distributional records of Pacific Coast Odonata. Pan Pacific Entomologist 53:147–160.

Penn, G.H. 1951. Seasonal variation in the adult size of *Pachydiplax longipennis* (Burmeister). Annals of the Entomological Society of America 44:193–197.

Pezalla, V.M. 1979. Behavioral ecology of the dragonfly *Libellula pulchella* Drury (Odonata: Anisoptera). American Midland Naturalist 102:1–22.

Phillips, E.C. 1996. Habitat preference of large predatory aquatic insects (Megaloptera and Odonata) in Ozark streams of Arkansas. Texas Journal of Science 48:255–260.

———. 2001. Life history, food habits and production of *Progomphus obscurus* Rambur (Odonata: Gomphidae) in Harmon creek of east Texas. Texas Journal of Science 53:19–28.

Pither, J., and P.D. Taylor. 1998. An experimental assessment of landscape connectivity. Oikos 83:166–174.

Polhemus, D., and A. Asquith. 1996. Hawaiian damselflies: A field identification guide. Hawaii Biol Surv. Handbook, Bishop Museum Special Publication 90, 122 pp.

Price, A.H., R.L. Orr, R. Honig, M. Vidrine, and S.L. Orzell. 1989. Status survey for the big thicket emerald dragonfly (*Somatochlora margarita*). Draft Report. Texas Parks Wild. Dept. Coop. Agreement No. 14-16-0002-86-925, Amndt. No. 7.

Pritchard, A.E. 1935. Two new dragonflies from Oklahoma. Occasional Papers of the Museum of Zoology, University of Michigan 319:1–10.

———. 1936. Notes on *Somatochlora ozarkensis* Bird. (Odonata, Libellulidae, Corduliinae). Entomological News 47:99–101.

———. 1964. The prey of adult dragonflies in northern Alberta. Canadian Entomologist 96:821–825.

Pritchard, G., and A. Kortello. 1997. Roosting, perching, and habitat selection in *Argia vivida* Hagen and *Amphiagrion abbreviatum* (Selys) (Odonata: Coenagrionidae), two damselflies inhabiting geothermal springs. Canadian Entomologist 129:733–743.

Provonsha, A.V. 1975. The Zygoptera (Odonata) of Utah with notes on their biology. Great Basin Naturalist 35:379–390.

Provonsha, A.V., and W.P. McCafferty. 1973. Previously unknown nymphs of western Odonata (Zygoptera: Calopterygidae, Coenagrionidae). Proceedings of the Entomological Society of Washington 75(4):449–454.

Reichholf, J. 1973. A migration of *Pantala flavescens* (Fabricius, 1798) along the shore of Santa Catarina, Brazil (Anisoptera: Libellulidae). Odonatologica 2:121–124.

Richardson, J.M.L., and R.L. Baker. 1997. Effect of body size and feeding on fecundity in the damselfly *Ischnura verticalis* (Odonata: Coenagrionidae). Oikos 79:477–483.

Ris, F. 1910. Libellulinen monographisch bearbeitet. Volume 1. Collections zoologiques du Baron Edm. de Sélys Longchamps, Catalogue systematique et descriptif. Coll. Sélys Longchamps 11:245–384.

———. 1930. A revision of the Libellulinae genus *Perithemis* (Odonata). Miscellaneous Publications of the University of Michigan Museum of Zoology 21:1–50.

Rivard, D., J.G. Pilon, and S. Thiphrakesone. 1975. Effect of constant temperature environments on egg development of *Enallagma boreale* Selys (Zygoptera: Coenagrionidae). Odonatologica 4:271–276.

Robert, A. 1939. Notes sur les odonates de Nominingue. *Lestes eurinus* Say et *Enallagma vesperum* Calvert dans le Quebec. Naturaliste Canada 66:47–64.

———. 1963. Les libellules du Quebec. Minist. Tourisme, Chasse et Peche, Prov. Quebec, Serv. Faune, Bull. 1, 223 pp.

Robertson, H.M. 1985. Female dimorphism and mating behavior in a damselfly, *Ischnura ramburii*: female mimicking males. Animal Behavior 33:805–809.

Robey, C.W. 1975. Observations on breeding behavior of *Pachydiplax longipennis* (Odonata: Libellulidae). Psyche 82:89–96.

Robinson, J.V. 1983. Effects of water mite parasitism on the demographics of an adult population of *Ischnura posita* (Hagen) (Odonata: Coenagrionidae). American Midland Naturalist 109:169–174.

Robinson, J.V., and R. Allgeyer. 1996. Covariation in life-history traits, demographics and behavior in ischnuran damselflies: the evolution of manandry. Biology Journal of the Linnean Society 58:85–98.

Robinson, J.V., and B.L. Frye. 1986. Survivorship, mating and activity pattern of adult *Telebasis salva* (Hagen) (Zygoptera: Coenagrionidae). Odonatologica 15:211–217.

Robinson, J.V., and W.H. Jordan. 1996. Ontogenetic color change in *Ischnura kellicotti* Williamson females (Zygoptera: Coenagrionidae). Odonatologica 25:83–85.

Robinson, J.V., and K.L. Novak. 1997. The relationship between mating system and penis morphology in ischnuran damselflies (Odonata: Coenagrionidae). Biology Journal of the Linnean Society 60:187–200.

Robinson, J.V., C.C. Bronstad, and C.H. Bronstad. 1985. Contrasting diurnal and nocturnal perching sites of *Ischnura posita* (Hag.) (Zygoptera: Coenagrionidae). Notulae Odonatologica 2:85–87.

Robinson, J.V., J.E. Dickerson, and D.R. Bible. 1983. The demographics and habitat utilization of adult *Argia sedula* (Hagen) as determined by mark-recapture analysis (Zygoptera: Coenagrionidae). Odonatologica 12:167–172.

Robinson, J.V., D.A. Hayworth, and M.B. Harvey. 1991. The effect of caudal lamellae loss on swimming speed of the damselfly *Argia moesta* (Hagen) (Odonata: Coenagrionidae). American Midland Naturalist 125:240–244.

Robinson, J.V., L.R. Shaffer, D.D. Hagermeier, and N.J. Smatresk. 1991. The ecological role of caudal lamellae loss in the larval damselfly *Ischnura posita* (Hagen) (Odonata: Zygoptera). Oecologia 87:1–7.

Roemhild, G. 1975. The damselflies (Zygoptera) of Montana. Montana State University Agricultural Station Research Report 87:1–53.

Rohdendorf, B.B. 1991. Fundamentals of Paleontology. Vol. 9. Arthropoda, Tracheata, Chelicerata, Order Odonata, pp. 81–100. Smithsonian Inst. Lib. & National Science Foundation, Washinton, D.C., 894 pp.

Root, F.M. 1924. Notes on dragonflies (Odonata) from Lee County, Georgia, with a description of *Enallagma dubium*, new species. Entomological News 35:317–324.

Sanborn, A.F. 1996. The cicada *Diceroprocta delicata* (Homoptera: Cicadidae) as prey for the dragonfly *Erythemis simplicicollis* (Anisoptera: Libellulidae). Florida Entomologist 79:69–70.

Savard, M., and C. Girard. 1996. Première mention de *Libellula (Plathemis) lydia* Drury (Odonata: Libellulidae) dans le bas-saint-laurent et note sur sa repartition au Quebec. Fabreries 21:88–90.

Sawchyn, W.W., and C. Gillott. 1974. The life histories of three species of *Lestes* (Odonata: Zygoptera) in Saskatchewan. Canadian Entomologist 106:1283–1293.

Schaefer, P.W., S.E. Barth, and H.B. White, III. 1996a. Predation by *Enallagma civile* (Odonata: Coenagrionidae) on adult sweetpotato whitefly, *Bemisia tabaci* (Homoptera: Aleyrodidae). Entomological News 107:275–276.

———. 1996b. Incidental capture of male *Epiaeschna heros* (Odonata: Aeshnidae) in traps designed for arboreal *Calosoma sycophanta* (Coleoptera: Carabidae). Entomological News 107:261–266.

Schiemenz, H. 1953. Die Libellen unserer Heimat. Urania, Jena.

Schmidt, E. 1985. Habitat inventorization, characterization and bioindication by a "Representative spectrum of Odonata species (RSO)." Odonatologica 14:127–133.

Scudder, G.G.E., R.A. Cannings, and K.M. Stuart. 1976. An annotated checklist of the Odonata (Insecta) of British Columbia. Syesis 9:143–161.

Shaffer, L.R., and J.V. Robinson. 1993. Ontogenetic differences in intraspecific aggression of damselfly larvae: *Ischnura posita* (Hagen) (Zygoptera: Coenagrionidae). Odonatologica 22:311–317.

Sherman, K. 1983. The adaptive significance of postcopulatory mate guarding in a dragonfly, *Pachydiplax longipennis*. Animal Behavior 31:1107–1115.

Shiffer, C.N. 1968. Homeochromatic females in the dragonfly *Perithemis tenera*. Proceedings of the Pennsylvania Academy of Science 42:138–141.

Smith, R.F., and A.E. Pritchard. 1956. Odonata. *In*: Aquatic insects of California. R.L. Usinger, ed. Berkeley, University of California Press, pp. 106–153.

Sternberg, K. 1996. Colours, colour change, colour patterns and 'Cuticular windows' as light traps—their thermoregulatoric and ecological significance in some *Aeshna* species (Odonata: Aeshnidae). Zoologischer Anzeiger 235:77–88.

Stroud, H.B., and G.T. Hanson. 1981. Arkansas geography: The physical landscape and the historical-culture setting. Library of Congress, N.Y., pp. 11–38.

Sublette, J.E., and M.S. Sublette. 1967. The limnology of playa lakes on the Llano Estacado, New Mexico and Texas. Southwestern Naturalist 12:369–406.

Susanke, G.R., and G.L. Harp. 1991. Selected biological aspects of *Gomphurus ozarkensis* (Westfall) (Anisoptera: Gomphidae). Advances in Odonatology 5:143–151.

Svihla, A. 1961. An unusual ovipositing activity of *Pantala flavescens* Fabricius. Tombo 4:18.

Switzer, P.V. 1997a. Factors affecting site fidelity in a territorial animal, *Perithemis tenera*. Animal Behavior 53:865–877.

———. 1997b. Past reproductive success affects future habitat selection. Behavior, Ecology and Sociobiology 40:307–312.

Tennessen, K.J. 1973. A preliminary report on the systematics of *Tetragoneuria* (Odonata: Corduliidae) in the southeastern United States. University of Florida, masters thesis, 198 pp.

———. 1975. Reproductive behavior and isolation of two sympatric coenagrionid damselflies in Florida. University of Florida, Ph.D. dissertation, 78 pp.

———. 1979. Distance traveled by transforming nymphs of *Tetragoneuria* at Marion County Lake, Alabama, United States (Anisoptera: Corduliidae). Notulae Odonatologica 1:63–65.

———. 1998. When is an ovipositor not an ovipositor? Argia 10:14.

Tennessen, K.J., and S.A. Murray. 1978. Diel periodicity in hatching of *Epitheca cynosura* (Say) eggs (Anisoptera: Corduliidae). Odonatologica 7:59–65.

Tharp, B.C. 1926. Structure of Texas vegetation east of the 98th meridian. University of Texas Bulletin 2606:1–100.

———. 1939. The vegetation of Texas. Texas Academy of Science Publication, Natural History Non-technical Series 1:1–74.

Thornthwaite, C.W. 1948. An approach toward a rational classification of climate. Geography Review 38:55–94.

Tinkham, E.R. 1934. The dragonfly fauna of Presidio and Jeff Davis counties of the Big Bend Region of Trans-Pecos, Texas. Canadian Entomologist 66:213–218.

Trottier, R. 1966. The emergence and sex ratio of *Anax junius* Drury (Odonata: Aeshnidae) in Canada. Canadian Entomologist 98:794–798.

———. 1967. Observations on *Pantala flavescens* (Fabricius) (Odonata: Libellulidae) in Canada. Canadian Field-Naturalist 81:231.

———. 1971. Effect of temperature on the life cycle of *Anax junius* (Odonata: Aeshnidae) in Canada. Canadian Entomologist 103:1671–1683.

Turner, L.M. 1935. Notes on forest types of northwestern Arkansas. American Midland Natural History 16:417–421.

Turner, P. E. 1965. Migration of the dragonfly *Tarnetrum corruptum* (Hagen). Pan-Pacific Entomologist 41:66–67.

Valley, S. 1998. Notes from Oregon. Argia 10:9.

Van Damme, K., and H.J. Dumont. 1999. A drought-resistant larva of *Pantala flavescens* (Fabricius, 1798) (Odonata: Libellulidae) in the Lençois Maranhenses, NE-Brazil. International Journal of Odonatology 2:69–76.

Vidrine, M.F., C.M. Allen, and H.D. Guillory. 1992a. List of parish distribution records of Odonata in Louisiana. Louisiana Environmental Profesional 9:19–39.

———. 1992b. Flight records of Odonata in southwestern Louisiana. Louisiana Environmental Profesional 9:40–53.

Viosca, P., Jr. 1933. Louisiana out-of-doors, A handbook and guide. New Orleans, published by author, 187 pp.

Voshell, J.R., and G.M. Simmons. 1978. The Odonata of a new reservoir in the southeastern United States. Odonatologica 7:67–76.

Waage, J.K. 1972. Longevity and mobility of adult *Calopteryx maculata* (Beauvois, 1805) (Zygoptera: Calopterygidae). Odonatologica 1:155–162.

———. 1974. Reproductive behaviour and its relation to territoriality in *Calopteryx maculata* (Beauvois). Behaviour 47:240–256.

———. 1975. Reproductive isolation and the potential for character displacement in the damselflies *Calopteryx maculata* and *C. aequabilis* (Odonata: Calopterygidae). Systematic Zoology 24:24–36.

———. 1978. Oviposition duration and egg deposition rates in *Calopteryx maculata* (Beauvois) (Zygoptera; Calopterygidae). Odonatologica 7:77–88.

———. 1979a. Adaptive significance of postcopulatory guarding of mates and non-mates by male *Calopteryx maculata* (Odonata). Behavior, Ecology, Sociobiology 6:147–154.

———. 1979b. Dual function of the damselfly penis: sperm removal and transfer. Science 203:916–918.

———. 1980. Adult sex ratios and female reproductive potential in *Calopteryx* (Zygoptera: Calopterygidae). Odonatologica 9:217–230.

———. 1983. Sexual selection, ESS Theory and insect behavior: some examples from damselflies (Odonata). Florida Entomologist 66:19–31.

———. 1984. Female and male interactions during courtship in *Calopteryx maculata* and *C. dimidiata* (Odonata: Calopterygidae): influence of oviposition behaviour. Animal Behaviour 32:400–404.

———. 1986. Sperm displacement by two libellulid dragonflies with disparate copulation durations (Anisoptera). Odonatologica 15:429–444.

———. 1988. Reproductive behavior of the damselfly *Calopteryx dimidiata* Burmeister (Zygoptera: Calopterygidae). Odonatologica 17:365–378.

Wakana, I. 1959. On the swarm and the migratory flight of *Pantala flavescens*, an observation in Kawagoe area. Tombo 1:26–30.

Walker, E.M. 1912. The North American dragonflies of the genus *Aeshna*. University of Toronto Studies, Biological Series 11, 213 pp.

———. 1913. Mutual adaptation of the sexes in *Argia moesta putrida*. Canadian Entomologist 45(9): 277–279.

———. 1917. The known nymphs of the North American species of *Sympetrum* (Odonata). Canadian Entomologist 9:409–418.

———. 1925. The North American dragonflies of the genus *Somatochlora*. University of Toronto Studies, Biological Series 26, 202 pp.

———. 1941. List of the Odonata of Ontario with distributional and seasonal data. Transactional of the Royal Canadian Institute 23:201–265.

———. 1951. *Sympetrum semicinctum* (Say) and its nearest allies (Odonata). Entomological News 62:153–163.

———. 1953. The Odonata of Canada and Alaska. Vol. 1. Part I: General, Part II: The Zygoptera-damselflies. University of Toronto Press, Toronto, 292 pp.

———. 1958. The Odonata of Canada and Alaska. Vol. 2. Part III: The Anisoptera-four families. University of Toronto Press, Toronto, 318 pp.

———. 1966. On the generic status of *Tetragoneuria* and *Epicordulia* (Odonata: Corduliidae). Canadian Entomologist 98:897–902.

Walker, E.M., and P.S. Corbet. 1975. The Odonata of Canada and Alaska. Vol. 3. Part III: The Anisoptera-three families. University of Toronto Press, Toronto, 307 pp.

Waltz, E.C. 1982. Alternative mating tactics and the law of diminishing returns: The satellite threshold model. Behavior, Ecology, Sociobiology 10:75–83.

Warren, A. 1915. A study of the food habits of the Hawaiian dragonflies on Pinau. College of Hawaii Publication, Bulletin 3:1–45.

Weichsel, J.I. 1987. The life history and behavior of *Hetaerina americana* (Fabricius) (Odonata: Calopterygidae). Ph.D. dissertation, University of Michigian, 208 pp.

Wellborn, G.A., and J.V. Robinson. 1987. Microhabitat selection as an antipredator strategy in the aquatic insect *Pachydiplax longipennis* Burmeister (Odonata: Libellulidae). Oecologia 71:185–189.

Westfall, M.J., Jr. 1941. Notes on Florida Odonata. Entomological News 52:31–34.

———. 1943. The synonymy of *Libellula auripennis* Burmeister and *Libellula jesseana* Williamson, and a description of a new species, *Libellula needhami* (Odonata). Transactions of the American Entomological Society 69:17–31.

———. 1957. A new species of *Telebasis* from Florida (Odonata: Zygoptera). Florida Entomologist 40:19–27.

———. 1974. A critical study of *Gomphus modestus* Needham, 1942, with notes on related species (Anisoptera: Gomphidae). Odonatologica 3:63–73.

Westfall, M.J., Jr., and M.L. May. 1996. Damselflies of North America. Scientific Publishers, Gainesville, Fla., 649 pp.

Westfall, M.J., Jr., and K.J. Tennessen. 1979. Taxonomic clarification within the genus *Dromogomphus* Selys (Odonata: Gomphidae). Florida Entomologist 62:266–273.

Whedon, A.D. 1914. Preliminary notes on the Odonata of southern Minnesota. Fifteenth Report of the State Entomologist of Minnesota, pp.77–103.

Whitehouse, F.C. 1941. British Columbia dragonflies (Odonata), with notes on distribution and habits. American Midland Naturalist 26:488–557.

Williams, C.E. 1976. *Neurocordulia (Platycordulia) xanthosoma* (Williamson) in Texas (Odonata: Libellulidae: Corduliinae). Great Lakes Entomologist 9:63–73.

———. 1977. Courtship display in *Belonia croceipennis* (Selys) with notes on copulation and oviposition (Anisoptera: Libellulidae). Odonatologica 6:283–287.

———. 1979a. *Gomphaeschna furcillata* (Say), a new state record for Texas (Anisoptera: Aeshnidae). Notulae Odonatologica 1:50.

———. 1979b. An apparent size difference between northern United States and Texas specimens of *Macromia pacifica* Hag. (Anisoptera: Macromiidae). Notulae Odonatologica 1:49–50.

———. 1979c. Observations on the behavior of the nymph of *Neurocordulia xanthosoma* (Williamson) under laboratory conditions (Anisoptera: Corduliidae). Notulae Odonatologica 1:44–45.

Williams, F.X. 1937. Notes on the biology of *Gynacantha nervosa* (Aeschninae), a crepuscular dragonfly in Guatemala. Pan Pacific Entomologist 13:1–8.

Williamson, E.B. 1898. A new species of *Ischnura*. Entomological News 9:209–211.

———. 1899a. Habits of *Ischnura kellicotti* (Order Odonata). Entomological News 10:68–69.

———. 1899b. The dragonflies of Indiana. Indiana Department of Geology & Natural Resources, 24th Ann. Rept., pp. 233–333.

———. 1900a. On the habits of *Tachopteryx thoreyi* (Order Odonata). Entomological News 11:398–399.

———. 1900b. The dragon-flies of Indiana. Annual Report of the Department of Geology, Indiana. 24:231–333, 1003 1011.

———. 1906a. Copulation of Odonata. I. Entomological News 17:143–150.

———. 1906b. *Plathemis subornata* (Odonata). Entomological News 17:351.

———. 1908. A new dragonfly (Odonata) belonging to the Cordulinae, and a revision of the classification of the subfamily. Entomological News 19:428–434.

———. 1909. The North American dragonflies (Odonata) of the genus *Macromia*. Proceedings of the U.S. National Museum 37:369–398.

———. 1915. Notes on neotropical dragonflies, or Odonata. Proceedings of the U.S. National Museum 48:610–638.

———. 1916. On certain *Acanthagrion*, including three new species (Odonata). Entomological News 27:313–325, 349–358.

———. 1917. The genus *Neoneura* (Odonata). Transactions of the American Entomological Society 43:211–246.

———. 1920. Notes on Indiana Dragonflies. Proceedings of the Indiana Academy of Science 30:99–104.

———. 1922a. Indiana *Somatochlora* again. Entomological News 33:200–207.

———. 1922b. Notes on *Celithemis* with description of two new species (Odonata). Occasional Papers of the Museum of Zoology, University of Michigan 108:1–20.

———. 1923a. Notes on American species of *Triacanthagyna* and *Gynacantha*. Miscellaneous Publications of the Museum of Zoology, University of Michigan 9, 80 pp.

———. 1923b. Notes on the genus *Erythemis* with a description of a new species (Odonata). Miscellaneous Publications of the Museum of Zoology, University of Michigan 11:1–18.

———. 1932. Dragonflies collected in Missouri. Occasional Papers of the Museum of Zoology, University of Michigan 240:1–40.

Wilson, C.B. 1909. Dragonflies of the Mississippi River collected during the pearl mussel investigations on the Mississippi River, July and August, 1907. Proceedings of the U.S. National Museum 36:653–671.

———. 1911. Dragonflies of Jamaica. John Hopkins University Circular 2:47–51.

———. 1912. Dragonflies of the Cumberland Valley in Kentucky and Tennessee. Proceedings of the U.S. National Museum 43:189–200.

———. 1920. Dragonflies and damselflies in relation to pond-fish culture, with a list of those found near Fairport, Iowa. Bulletin of the U.S. Bureau of Fisheries 36:181–264.

Wisenden, B.D., D.P. Chivers, and R.J.F. Smith. 1997. Learned recognition of predation risk by *Enallagma* damselfly larvae (Odonata, Zygoptera) on the basis of chemical cues. Journal of Chemical Ecology 23:137–151.

Wright, M. 1943a. A comparison of the dragonfly of the lower delta of the Mississippi River with that of the marshes of the central Gulf Coast. Ecological Monographs 13:481–497.

———. 1943b. Dragonflies collected in the vicinity of Florala, Alabama. Florida Entomologist 26:30–31, 49–51.

Young, A.M. 1967. The flying season and emergence period of *Anax junius* in Illinois (Odonata: Aeshnidae). Canadian Entomologist 99:886–889.

———. 1980. Observations on feeding aggregations of *Orthemis ferruginea* (Fabricius) in Costa Rica (Anisoptera: Libellulidae). Odonatologica 9:325–328.

Zimmerman, E.C. 1948. Insects of Hawaii. Vol. 2. Apterygota-Thysanoptera. University of Hawaii Press, Honolulu.

Photo Credits

Photographs are by the author except as follows:

Robert D. Barber—35d, 44a.

Roy J. Beckemeyer—38e.

Robert A. Behrstock—1d (top), 1e (top), 1f, 2b, 2c, 2d, 2e, 3a, 3b, 3d, 3e, 3f, 4d, 4f, 6b, 6c, 6f, 7a (bottom), 7c, 7f, 9d, 9e (bottom), 9f (top), 10a, 10b (top), 10c (top), 10e (top), 10f (bottom), 11a, 11b, 11d, 11e, 12b, 12d, 13d, 14c (top), 14f, 15a, 15b, 15d (top), 15f, 16d, 17a, 17c (top), 17d, 17e, 18a, 18c, 18f (bottom), 19c, 19d, 19f (bottom), 20b, 20c (top), 20d, 21a, 21b, 21c, 21d, 21f, 22a, 22d, 22f, 23a, 23b, 23d, 23e, 23f, 24c, 29e, 30c, 31d, 31e, 31f, 32a, 32c, 32d, 32e, 33a, 34b, 34c, 34d, 36a, 37f, 39a (top), 39c, 39d, 39e, 40c, 40e, 42b, 42c, 43b, 44c, 46f, 47e (top), 47f, 48a (left), 48c, 48d, 49a, 49b, 49c, 49e, 50d (top), 51a, 51d, 52a, 52b, 52d, 53a, 53b, 53c, 53e, 54a, 54b (top), 54c, 54f, 55b, 55c, 55d, 55f, 56b, 56f, 57a, 57f, 57f (inset), 58b, 60b, 60e (left), 60f, 61a (top), 61d, 61e, 62a, 63c, 64b, 64e.

Omar R. Bocanegra—15c, 29b, 37e.

Robert A. Cannings, Royal British Columbia Museum—26c, 26d.

George P. Doerksen, Royal British Columbia Museum—14c.

Thomas W. Donnelly—63b.

Sidney W. Dunkle—6a, 9f (bottom), 10c (bottom), 10e (bottom), 12c, 13c, 14c (bottom), 15e, 25b, 25c, 25e, 25f, 27a, 28a, 29d, 30a, 33b, 34a, 34c, 36c, 38b, 40d, 40f, 41b, 42a, 42f, 44b, 44e, 44f, 45b, 45d, 46a, 48a (right), 52e, 63f, 64a, 64d.

Randy L. Emmitt—35c, 57d, 63e.

Ted L. Eubanks—51c.

Rosser W. Garrison—58c.

R. Stephen Krotzer—17c (bottom), 30b, 30e, 30f, 32b, 35a, 35b, 38c, 40a, 40b, 41e, 43c, 45a, 46c, 46d, 48f.

Blair J. Nikula—2f, 4c, 4e, 11f, 12e, 13b, 14b, 14e (top), 15d (bottom), 16c, 18b, 25d, 26a, 26b, 27c, 29c, 30d, 34f, 37b, 37c, 38d, 44d, 45e, 45f, 59a (top).

Dennis R. Paulson—14e (bottom), 20c (bottom), 27e, 28b, 38a, 46b, 49d, 49f, 58e.

Clark N. Shiffer—16a, 18d, 22b, 38f.

Netta C. Smith—47c, 52f (right), 59b, 62d, 62e, 62f.

Curtis E. Williams—1a, 1b, 27b, 42e, 43d, 55f.

Index

English names are printed in roman type; scientific names are in *italics*. **Bold** numbers refer to main entries; *italicized* numbers refer to an illustration on that page.